ADHESIVES IN MANUFACTURING

MANUFACTURING ENGINEERING
AND MATERIALS PROCESSING

A Series of Reference Books and Textbooks

SERIES EDITORS

Geoffrey Boothroyd

Department of Mechanical Engineering
University of Massachusetts
Amherst, Massachusetts

George E. Dieter

Dean, College of Engineering
University of Maryland
College Park, Maryland

1. Computers in Manufacturing, *U. Rembold, M. Seth, and J. S. Weinstein*
2. Cold Rolling of Steel, *William L. Roberts*
3. Strengthening of Ceramics: Treatments, Tests, and Design Applications, *Henry P. Kirchner*
4. Metal Forming: The Application of Limit Analysis, *Betzalel Avitzur*
5. Improving Productivity by Classification, Coding, and Data Base Standardization: The Key to Maximizing CAD/CAM and Group Technology, *William F. Hyde*
6. Automatic Assembly, *Geoffrey Boothroyd, Corrado Poli, and Laurence E. Murch*
7. Manufacturing Engineering Processes, *Leo Alting*
8. Modern Ceramic Engineering: Properties, Processing, and Use in Design, *David W. Richerson*
9. Interface Technology for Computer-Controlled Manufacturing Processes, *Ulrich Rembold, Karl Armbruster, and Wolfgang Ülzmann*
10. Hot Rolling of Steel, *William L. Roberts*
11. Adhesives in Manufacturing, *Gerald L. Schneberger*

OTHER VOLUMES IN PREPARATION

ADHESIVES IN MANUFACTURING

edited by

Gerald L. Schneberger

GMI Engineering and Management Institute
Flint, Michigan

MARCEL DEKKER, INC.   New York and Basel

Library of Congress Cataloging in Publication Data

Main entry under title:

Adhesives in manufacturing.

   (Manufacturing engineering and materials
processing ; 11)
   Includes index.
   1. Adhesives. I. Schneberger, Gerald L.
II. Title. III. Series.
TP968.A29   1983   668'.3   83-7867
ISBN 0-8247-1894-1

MARCEL DEKKER, INC.
270 Madison Avenue, New York, New York 10016

Current printing (last digit):
10 9 8 7 6 5 4 3 2 1

PRINTED IN THE UNITED STATES OF AMERICA

Preface

Adhesives in Manufacturing has been written primarily to serve production and development people. While basic researchers may find its
contents interesting and of value, its major thrust is to provide information and understanding for those who choose to use adhesives.

The chapters have been grouped to first introduce Fundamentals
and then consider Specific Types of Adhesives. These sections are
followed by chapters which consider the Techniques of Bonding Practice and the Performance, Durability, and Testing of bonds. It is
hoped this arrangement will enable readers to progress in a logical
fashion from the general to the specific.

The authors chosen to prepare chapters for *Adhesives in Manufacturing* are a select group of extremely knowledgeable persons who
represent adhesive formulators, distributors, and consumers. Collectively they possess an immense amount of knowledge and practical
experience. They have attempted to prepare material geared to manufacturing and process selection personnel—people whose knowledge of
chemistry is limited but whose need to work with chemicals is great.

It is a pleasure to thank the authors for their excellent overall
attitude of cooperation and enthusiasm during the months of writing
and editing which preceded publication. The support and assistance
of the staff at Marcel Dekker, Inc. was also effective and is appreciated.

Finally, it is with the greatest gratitude that the efforts of
Mrs. Mary Rensberger, Editorial Assistant for the project, are recognized. Without her organizational ability and conscientious attention
to detail, the endeavor would never have been brought to fruition.

The editor would be pleased to receive comments and construc-
tive suggestions concerning *Adhesives in Manufacturing* at any time.

Gerald L. Schneberger

Contributors

John R. Adams, B.S., M.E. Director of Education, Training Division, Binks Manufacturing Company, Franklin Park, Illinois

Werner Bohm Director of Marketing, Packaging and Product Assembly Division, Nordson Corporation, Atlanta, Georgia

Justin C. Bolger, Sc.D., M.S., B.S. Vice President and General Manager, Polymer Products Division, Amicon Corporation, Lexington, Massachusetts

Donald H. Buckley, D. Eng., D. Juris., B.S. Chief, Tribology Branch, National Aeronautics and Space Administration Lewis Research Center, Cleveland, Ohio

Richard Chait, P.E., B.S., M.S., Ph.D. Chief, Materials Integrity and Testing Technology Division, Army Materials and Mechanics Research Center, Watertown, Massachusetts

Edward T. Clegg, A.B., M.B.A. Chemist, Engineering Standardization Branch, Army Materials and Mechanics Research Center, Watertown, Massachusetts

Garry O. DeFrayne, B.S., Ch.E., M.A. Senior Materials Engineer, Adhesives Laboratory, Chrysler Corporation, Highland Park, Michigan

Robert D. Dexheimer, B.A.* Market Manager, Henkel Adhesives Company, Minneapolis, Minnesota

James A. Graham, B.A. Business Manager, Adhesives Division, Chemical Product Group, Lord Corporation, Erie, Pennsylvania

---

*Retired

J. W. Hagan, B.S. Senior Development Scientist, Specialty Chemicals Division, Union Carbide Corporation, Bound Brook, New Jersey

Martin Hauser, B.S., Ph.D. Vice President, Environmental Health and Safety, Industrial Group, Loctite Corporation, Newington, Connecticut

Girard S. Haviland, CMfgE, B.ME. Manager, Product Engineering, Loctite Corporation, Newington, Connecticut

Barry R. Killick, B.S.A.e. Chief Engineer, Pyles Division, Kent-Moore Corporation, Wixom, Michigan

Harold Koski, A.B., M.A.* Associate Professor of Chemistry, Department of Science and Mathematics, GMI Engineering and Management Institute, Flint, Michigan

Jan V. Lindyberg, B.S., M.S. Technical Marketing Specialist, RTV Products Department, General Electric Company, Silicone Products Division, Waterford, New York

Gregory M. MacIver, B.S., Ch.E. Product Sales Manager, Adhesives Operations, Chemical Division, Goodyear Tire and Rubber Company, Ashland, Ohio

Dale Meinhold, B.S., M.S. Associate Professor of Mathematics, Department of Science and Mathematics, GMI Engineering and Management Institute, Flint, Michigan

J. Dean Minford, B. S., M. Litt., Ph.D. Scientific Associate, Product and Process Engineering Division, Alcoa Laboratories, Alcoa Center, Pennsylvania

J. Merle Nielsen, B.S., M. S., Ph.D. Materials Engineer, Materials Information Services, Corporate Research and Development, General Electric Company, Schenectady, New York

Paul O. Nielsen, B.S. President, Paul O. Nielsen, Inc., Warren, Pennsylvania

W. A. Pletcher, B.S., M.S., Ph.D. Laboratory Manager, Industrial Products Laboratory, Adhesives, Coatings and Sealers Division, 3M Company, St. Paul, Minnesota

---

*Retired

Gerald L. Schneberger, P.E., A.B., M.S., Ph.D. Director, Continuing Education and Professional Services, GMI Engineering and Management Institute, Flint, Michigan

Kenneth C. Stueben, Ph.D. Senior Research Scientist, Specialty Chemicals Division, Union Carbide Corporation, Bound Brook, New Jersey

J. M. Tancrede, Ph.D.* Project Scientist, Coatings Materials Division, Union Carbide Corporation, Bound Brook, New Jersey

Douglas P. Thompson, II, B.S. Sales Engineer, Adhesives Operations, Chemical Division, Goodyear Tire and Rubber Company, Ashland, Ohio

William F. Thomsen, B.S., M.A. Technical Representative, Technical Service and Development Division, Eastman Chemical Products, Inc., Kingsport, Tennessee

Leonard R. Vertnik , B.S. Manager of Technical Services, Henkel Adhesives Company, Minneapolis, Minnesota

William K. Westray, P.E., B.S., M.S.† Marketing Manager, Magnus Metal Processing Division, Economics Laboratory, Inc., St. Paul, Minnesota

Edmund J. Yaroch, B.ChE. Laboratory Manager, Automotive, Construction and Convenience Packaging Products Laboratory, Adhesives, Coatings and Sealers Division, 3M Company, St. Paul, Minnesota

Robert H. Young, Ph.D., M.Sc., B.Sc.‡ Group Leader, Research and Development, Union Carbide Corporation, Bound Brook, New Jersey

Present affiliations
*Senior Research Chemist, Specialties Technology Division, Exxon Chemical Company, Baton Rouge, Louisiana
†Marketing Director, Barnes Drill Company, Rockford, Illinois
‡Section Leader, Research and Development, Weyerhaeuser Company, Tacoma, Washington

Contents

ix

BONDING PRACTICE

PERFORMANCE, DURABILITY, AND TESTING

FUNDAMENTALS

1
Adhesive Concepts and Terminology

Gerald L. Schneberger *GMI Engineering and Management Institute,
Flint, Michigan*

I. INTRODUCTION

A knowledge of basic concepts and terminology is helpful in deriving
the maximum benefit from the use of adhesives and facilitating communication between designers, users, and producers. This chapter,
therefore, contains a brief description of the more important terms and
ideas involved in bonding. They are arranged in no particular order
of significance and many of them are described or discussed in further detail in other chapters of the book.

II. SURFACES

Surfaces are complex and dynamic. They are never flat despite the
best attempts at polishing. Industrially bonded surfaces often have a
surface roughness of 50–100 microinches (12,700–25,400 μm) or more.
    Rough surfaces are neither automatically good nor bad for bond-
ing. The smoother the surface, the fewer crevices and microvoids are
present to retain trapped air and prevent good contact of the adhesive with the surface. On the other hand, a rough surface provides
more actual surface area for contact with the adhesive and may even
allow some mechanical interlocking as the adhesive solidifies in cracks
and crevices.
    The most important property of a surface, as far as bonding is
concerned, is cleanliness. Unless surfaces are clean, that is, free of
loosely held foreign material, the bond may fail unexpectedly and/or
at low loads for no apparent reason. Typical surface contaminants
include dirt, oil, grease, mold release, oxides, moisture, gases, and
lubricants.
    Surfaces are dynamic in the sense that their chemical composition
is frequently subject to change with time. Some metals may oxidize
(corrode) by reacting with atmospheric moisture and oxygen which
permeate the bondline. Occasionally, plastic surfaces may become
coated with chemicals that migrate from the interior of the part or
which reach the surface from the external atmosphere by diffusion
through the bondline. In either case, a competition for attachment to
the surface develops between the adhesive and the newly arrived
chemical. Strong, long-lasting adhesives are generally those whose
chemical structure is such that they can compete favorably for surface
attachment sites with other chemicals which may find their way to the
surface.
    Surfaces also have a different composition than might ordinarily
be expected. Most commonly, bonded metal surfaces exist, not as free
metals, but as hydrated metal oxides. Freshly milled aluminum surfaces, for example, begin forming a transparent oxide layer immediately when exposed to the atmosphere. Iron and ferrous alloys do
likewise. Applying adhesives to a metal-oxide surface is not necessarily bad. If the metal oxide is tightly bound to the metal and is co-

hesively strong and chemically stable, then the resulting adhesive bond may be both strong and durable. Aluminum, for example, is frequently oxidized under strictly controlled conditions (anodized) prior to bonding in order to provide the type of surface just described.

III. ADHESION

Adhesion is the phenomenon which allows two objects to retain surface contact nonmagnetically. Most theories of adhesion can be expressed in terms of electrical attractions. Essentially, certain minute locations on one surface possess or acquire positive or negative electrical charges which then align themselves adjacent to an oppositely charged location on the other surface. The ensuing unlike charge attractions result in adhesion of the surfaces. An adhesive is simply a material applied as a liquid, which allows the establishment of unlike charge attractions and then solidifies to provide the bond with significant internal strength.

IV. WETTING

Wetting refers to the ability of a liquid to intimately contact the surface over which it flows. Poor wetting is typified by water on a waxed surface. The water has a tendency to "bead up" because the water molecules are more vigorously attracted to themselves than they are to the waxed surface. Good wetting is essential for strong adhesive bonds. Adhesives may contain wetting agents, that is, chemicals which increase the ability of the adhesive to contact and be attracted to the surfaces being bonded. Wax, grease, oil, and fluorocarbons are notoriously difficult to wet. It is for this reason that most surfaces are subjected to some sort of degreasing treatment prior to bonding.

Surfaces which are not easily wet by water will be difficult to bond. The "water-break free" test is often used to determine the bondability of surfaces. Water is poured on the surface and observed to see whether it breaks up into droplets or remains as a single puddle. If individual droplets are formed, the surface is presumed to be contaminated with grease or oil. Some fluorocarbons, hydrocarbons, and silicones can not be wet by water even when they are clean. The test works well for nearly everything else, however.

V. BOND STRENGTH

Bond strength refers to the amount of force required to break the bond. It represents the force required to overcome either the adhe-

sive strength of the bond or the cohesive strength of the adhesive
material or of one of the bonded objects.

VI. FATIGUE LIFE

Fatigue life, which is usually expressed as the number of cycles required for failure, is the number of times a stress can be applied and
released before the bond fails. The test is usually performed with an
electrically driven instrument capable of subjecting the bond to tension
or bending at a very high frequency. The underlying assumption is
that the harmful effect of such stresses will be cumulative, and although they are performed at a very high frequency, it will still be
possible to correlate the results with service conditions.

VII. HEAT LIMITATIONS

Most structural adhesives begin to lose strength significantly at temperatures above 300°F (149°C). A few are available which can tolerate
continuous exposure at 400°F (204°C) or higher. The basic difficulty
experienced by most adhesives at high temperatures is simply that the
chemical bonds which are an inherent part of their structure begin to
break down as the temperature increases. Invariably, those products
which have extremely high heat stability also contain chemical bonds
of unusual thermal stability. Sometimes heat-stable fillers can be added to the formulation in order to raise the service temperature.

VIII. DEBONDING

Debonding is the process by which an adhesive looses its attachment
to a surface. Debonding usually occurs when water or some other
chemical displaces the adhesive from surface attachment sites. The
process is usually a gradual one although it may not be noticed until
the adhesive and the adherend separate completely, and a catastrophic failure results.

IX. UNSTRESSED AGING DATA

Unstressed aging data describe how bond properties of interest, possibly impact or shear strength, change with time using specimens that
have not been stressed prior to testing. Engineers frequently request
that strength-retention data be collected for a particular adhesive
bonding system. As a result, bonds are made and tested to failure at
various times after joining. The results obtained frequently depend
heavily upon whether the samples were simply stored under ambient

conditions or whether they were cyclically stressed before testing.
Since most adhesively bonded products will experience stresses in
service, it is often misleading to base any decision concerning them
on test results obtained from unstressed samples.

X. NONDESTRUCTIVE TESTING

Nondestructive testing refers to a number of tests for bond strength
or durability which are performed on assemblies without destroying
the bond. Nondestructive tests are often carried out by subjecting
the bond to ultrasonic vibration or other physical stimuli and measuring the physical response of the system. Techniques range in complexity from a hammer-tap sound test to electronic frequency analysis.

XI. EXPANSION COEFFICIENTS

The expansion coefficient of a material is a number which indicates
how the material changes dimensionally with changes in temperature.
An expansion coefficient of 6×10^{-5} cm/cm/°C means that a 1-cm segment will increase 6×10^{-5} cm in length for each degree Celsius of increased temperature. Similarly the part will shrink by the same
amount for each degree of temperature decrease. This concept is important in adhesive bonding because adhesives are frequently used to
join materials with dissimilar expansion behaviors. Furthermore, the
adhesive itself undergoes expansion and contraction as the temperature rises and falls. When the mismatch of expansion behavior is too
great, severe stresses result at the bondline, and a separation or
fracture of the adhesive may occur. Increasing the flexibility of the
adhesive is a common technique for resisting expansion-contraction
failure. In general, flexible adhesives are better able to resist such
stresses because they can shrink and stretch with the moving adherends. On the other hand, flexible formulations often have a greater tendency to creep. The expansion behavior of adhesives can often
be modified by the addition of fillers or additives.

XII. CREEP

The tendency of any material to undergo a permanent change in dimension when subjected to mechanical stress is known as creep.
Creep is particularly a problem with thermoplastic polymers. The tendency of a material to creep results primarily from the flexibility of its
molecular segments. Creep may often be retarded or reduced to an
acceptable level with fillers or small amounts of crosslinking.

XIII. THERMOPLASTIC

Polymers which soften when heated are called thermoplastic. With
very few exceptions, they are also soluble in at least a few solvents.
Thermoplastic materials usually have a tendency to creep under load,
and they are generally not used for critical structural application.

XIV. THERMOSETTING

Plastics which do not deform readily under load and which resist solution in nearly all solvents are called thermosetting. The molecular
chain segments of thermosetting materials are chemically bound to one
another (crosslinked) in such a fashion that it is difficult for segments
to move independently. In effect, the chain segments lock each other
in place, and permanent deformation is difficult or impossible at ordinary temperatures. Solvents are also unable to dissolve thermosetting
resins.

XV. SURFACE CONVERSION

Surface conversion refers to any process in which the surface to be
bonded is changed from its condition as manufactured or assembled to
one which is more suitable for bonding. More than simple cleaning is
implied by surface conversion. Typically, a ferrous surface might be
converted to zinc phosphate, or an aluminum surface to aluminum
oxide. Phosphatizing, anodizing, chromating, and polymer oxidation
are typical of conversion processes widely used industrially. The advantage of surface conversion is simply that the surface is changed
from a more or less unknown condition to one of fairly well understood
composition and behavior. Thus, some assurance is gained that the
adhesive is always being applied to a uniform surface.

XVI. SCANNING ELECTRON MICROSCOPY

This is a surface observation technique using a microscope which de-
pends upon electron beams rather than light rays for magnification.
It gives extremely high magnifications with excellent depth of field.
Magnification up to 100,000 or more are achieved with relative ease.
The equipment is expensive and somewhat sophisticated, but its ac-
quisition may be justified in certain critical situations.

XVII. POT LIFE

The period of time after mixing a two- or three-component adhesive
during which the mixture is still workable is the pot life. Once the

pot life has been exceeded, the viscosity of the formulation becomes
so great that spreading and wetting is difficult.

XVIII. VISCOSITY

Viscosity is the ability of a liquid to resist flow. In practical terms,
solutions of high viscosity are difficult to stir, pump, and spray.
Low-viscosity formulations, on the other hand, may be excessively
runny and difficult to apply uniformly to vertical surfaces. This
problem is often solved through the addition of thixotropic materials.

XIX. THIXOTROPY

A temporary decrease in viscosity upon stirring is called thixotropy.
In more technical terms, it is accurate to say that the viscosity of
thixotropic materials is shear-rate dependent. Thixotropic additives
are often used to give an adhesive enough body to resist sagging on
vertical surfaces and yet become runny enough to flow easily under
the stress of brushing, rolling, or spraying during application.

XX. GLASS TRANSITION TEMPERATURE

The glass transition temperature T_g is the temperature above which an
adhesive (or other polymeric material) becomes somewhat rubbery. At
temperatures above the T_g, polymers are somewhat ductile while at
temperatures far below the transition point, they are often brittle.
Adhesive formulators generally select resins having transition temperatures which are a compromise between flexible-rubbery and brittle-
glassy structures. Adhesives designed for use on flexible materials
almost always have transition temperatures below the anticipated use
temperature to insure that the bondline will remain flexible in service.

XXI. SHELF LIFE

Shelf life is the period of time for which an adhesive may be stored
without deteriorating to the point that performance will be unsatisfactory. Shelf life for reactive adhesives can usually be extended by
refrigeration.

XXII. CURING

Curing is the process during which an adhesive changes from a liquid
to a solid. Classically, curing means the formation of a crosslinked
polymer film. More loosely, the term is sometimes used to describe

any method of forming a solid adhesive film, including solvent loss and linear polymerization.

XXIII. SAG RESISTANCE

Sag resistance refers to the ability of the adhesive to stay in place on a vertical surface without running or "sagging" on the surface. Fillers are sometimes added to adhesives to give them more body and improve their sag resistance.

XXIV. RESIN

The major polymer or polymer-forming material used in an adhesive is the resin. Some resins are natural products such as starch or rubber. Other resins, such as epoxies and urethanes, are produced by chemical synthesis.

XXV. POLYMER

The plastic-like material which forms the bulk of most adhesive films is a polymer. Polymers are made by the repetitive joining of many small chemical units (molecules). The resulting high molecular weight material is known as a polymer. Most of the bulk properties of an adhesive material, such as heat resistance and tensile strength, are primarily due to the nature of the polymer used in the adhesive.

XXVI. MOLD RELEASE AGENT

Mold release agents are materials used to reduce the adhesion of a polymer to the mold in which parts are formed. Release agents are the opposite of adhesives. Their task is to reduce or prevent adhesion. Release agents may either be applied to a mold surface before it is filled (external release agents) or mixed into the molding material before it is used (internal release agents). In either case, it is essential that mold release agents be removed from surfaces before bonding is attempted.

XXVII. OUTGASSING

Outgassing is the formation of gas bubbles in an adhesive film while it is curing. This generally undesirable situation may arise because gaseous products are formed while the adhesive cures or because curing conditions, such as temperature, may cause bubbles of dissolved air to be released. In general, severe outgassing should be avoided.

XXVIII. CURING AGENT

Curing agents are chemicals which react with an adhesive to form a solid adhesive film. Curing agents are generally used in much larger amounts than catalysts. They may be considered coreactants with the adhesive resin since they are consumed during the curing reaction.

XXIX. HARDENER

Hardener is a term which is more or less synonymous with catalyst. Hardeners are mixed with adhesive resins in very small quantities to initiate the curing reaction.

XXX. FILLER

A material which is added to an adhesive formulation to influence its properties in some specific fashion is called a filler. Fillers are generally particulate materials, such as finely ground glass, metal, clay, or other inorganic chemicals. They are used to control such properties as impact resistance, heat stability, conductivity, sag resistance, etc.

XXXI. ADHESIVE FAILURE

Adhesive failure is the failure of a bond at the interface between the adhesive and the adherend.

XXXII. COHESIVE FAILURE

The failure of a bond within either the adhesive or the adherend, rather than at the interface between them, is known as cohesive failure.

2
Surfaces

Donald H. Buckley *National Aeronautics and Space Administration
Lewis Research Center, Cleveland, Ohio*

I. INTRODUCTION

When a solid surface is examined either microscopically with a scanning electron microscope or mechanically with a surface profilometer,
it is found to contain irregularities; that is, the surface is not flat and
smooth. A depiction of a surface displaying these irregularities, or
asperities as they are commonly called, is presented in Fig. 1a.

Nearly all real surfaces contain the asperities except brittle,
single-crystal materials that have been cleaved along natural cleavage
planes and metallic pin tips that have been field evaporated in the
field ion microscope. Even with brittle materials, the cleavage process results in the generation of surfaces that contain cleavage steps,
and it is only the terraces between these steps that are atomically
smooth.

The actual shape and distribution of surface asperities has been
the object of considerable research. An excellent review of the subject can be found in Williamson [1].

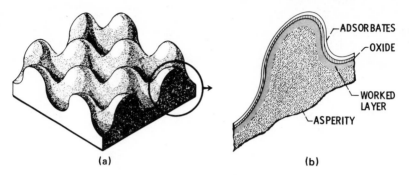

Fig. 1. (a) Surface topography; (b) detail of surface topography.

The surfaces of the asperities are not atomically clean but contain
surface films (Fig. 1b). For metals and alloys these films generally
consist of oxides and adsorbed gases—usually water vapor, carbon
monoxide, and carbon dioxide. With many nonmetals the surface films
may simply consist of other adsorbates. All of the reacted and adsorbed film materials can exert a strong effect on the mechanical and
metallurgical behavior of the solids to which they adhere, as indicated
by the collection of papers appearing in Westwood and Stoloff [2].

In addition to the films present on the surface of a solid, the
surficial (near surface) layers of the solid itself may vary considerably in structure from the bulk of the solid. With crystalline solids
these layers may consist of recrystallized material, strain hardened
regions, and/or textured regions. These surficial layers develop
when any type of finishing or polishing of the surface is done, particularly when that surface is a metal. These layers can also be a region rich in bulk impurities [3]. In amorphous solids these layers may
contain voids and microcracks.

II. SURFACE PROFILE

A careful examination of industrially prepared metal and alloy surfaces
with surface profilometers (devices capable of revealing the surface
topography) indicates the true microroughness of these surfaces.
They are, on a microscale, rough—as indicated by the surface profile traces of Fig. 2. The surface represented by Fig. 2a, steel finished by mechanical wire brushing, is extremely rough and contains
many large irregularities. The surface of Fig. 2b, steel prepared by
conventional industrial grinding, is much smoother, but many micro-
irregularities still exist. When a steel surface is very carefully lapped, smaller and fewer asperities are produced. This is demonstrated
in the surface profile trace of Fig. 2c, the smoothest of the three

(a) Brushed

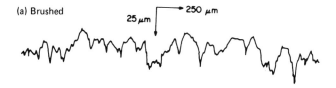

(b) Ground

(c) Lapped

Fig. 2. Profiles of steel surfaces finished by three different methods. (a) Brushed; (b) ground; (c) lapped.

traces presented. Even the lapping, however, leaves some surface roughness.

Extremely smooth, asperity-free surfaces are shown in the profile traces of Fig. 3a and b for mica and quartz surfaces. Both surfaces were generated by cleavage. A crack which developed during the cleavage process is visible in the profile trace of the mica surface.

Normally, metal surfaces cannot be generated with the smoothness reflected in the profiles of Fig. 3a and b. Such surfaces can be generated in the field ion microscope, but this only provides an atomically smooth surface over an area reflected by the pin tip radius of 500–1000 Å. Deposition of metal films on quartz surfaces as that in Fig. 3c can be used to obtain an asperity-free metal surface. The results of such an approach are shown in the surface profile trace of Fig. 3d for iron vapor deposited in vacuum onto quartz. The surface is smooth compared to those in Fig. 2.

Very frequently, for example, when a greater surface area would promote such things as adhesion, it is desirable not to have extremely smooth surfaces. A host of different methods—mechanical, chemical, and physical—can be used to increase surface area.

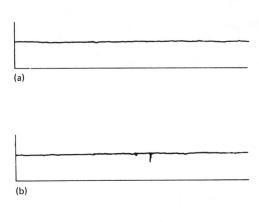

(a)

(b)

(c)

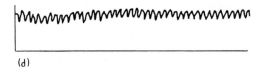

(d)

Fig. 3. Surface profiles of various materials. (a) Mica; (b) quartz;
(c) iron on quartz; (d) 1.0-μm standard surface roughness.

      One of the most commonly used mechanical methods is to sand or
bead blast the surface, thereby removing material through erosive
wear. An example of such a surface is shown in the photomicrograph
of Fig. 4. The surface looks like waves on water. This kind of surface topography may be very useful in certain bonding applications,
catalysis, and chemical processes.
      Among the chemical surface-roughening methods, chemical etch-
ing is used. The particular reagents must be selected on the basis of
the surface to be etched.

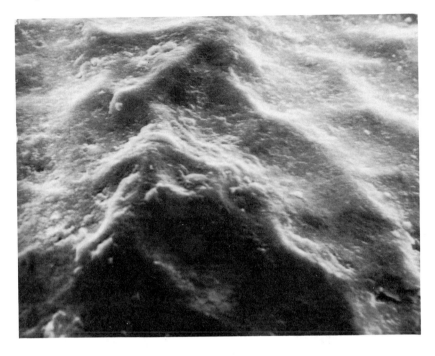

Fig. 4. Photomicrograph of a glass-bead blasted aluminum surface.
1 cm = 0.01 mm.

A very good technique for the roughening of surfaces involves
the use of inert gas ions which are made to bombard a surface and
generate a surface texture. Ion etching, as it is called, is particularly versatile in that both the bombarding species and the ion energies
can be selected to vary the surface topography. In addition, the incident angle of the ion beam can be varied. The different types of
surface textures that can be developed by this technique are indicated
in the photomicrographs of silver surfaces shown in Fig. 5. In the
upper portion of the figure the various surface orientations are depicted on the unit triangle. In the three micrographs in the lower
portion of the figures, the ion etching effect on three principal planes
of silver—namely, the (100), (110), and (111) surfaces—are shown.
The photomicrographs of Fig. 5 indicate that considerable variation in
surface texture can be developed with ion beam etching.

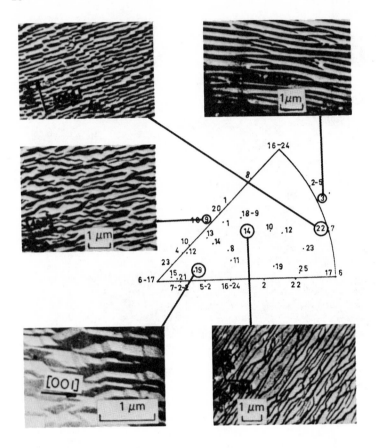

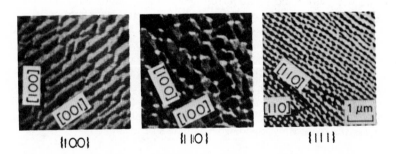

Fig. 5. Ion beam etched silver crystal surfaces with normally incident Ar$^+$ at 8 keV; dose 7×10^{18} ions.

III. CRYSTALLINE STRUCTURE

The crystal structure of ideal surfaces for most practically used materials is generally one of three major types: body-centered cubic, face-centered cubic, or close-packed hexagonal. All engineering surfaces vary from these ideal structures. Most real surfaces have grain boundaries which develop during the solidification of crystalline solids. These grain boundaries are, in a strict sense, defects which exist in the bulk solid and extend to the surface. They are atomic bridges linking the crystal structure of the two adjacent grains. Because of their role they do not possess a regular structure; they are highly active and are very energetic. Grain boundaries are large defects, readily observable on real surfaces. In addition to these there are, however, many lesser defects that may exist. These include subboundaries, twins, dislocations, interstitials, and vacancies.

Subboundaries are low-angle grain boundaries where only a slight mismatch in the orientation of adjacent grains occurs. When the crystal lattices of the adjacent grains are not parallel but are slightly tilted one toward the other, the defect is referred to as a tilt boundary. When the lattices are parallel, but one is rotated about a simple crystallographic axis relative to the other with the boundary being normal to this axis, the defect is a twist boundary.

The twin boundary occurs where there is only a degree or two of mismatch between the twins with the twins being mirror images of each other. They are frequently seen on the basal planes of hexagonal metals with deformation.

Dislocations are atomic line defects existing in crystalline solids. They may actually be in the subsurface and terminate with one end at the surface or they may be in the surface. There are those dislocations that are entirely along a line where an extra half plane of atoms exists; these are called edge dislocations. In addition, there are screw dislocations that form along a spiral dislocation line. A screw dislocation can be seen on the surface as a wedge of atoms which is the terminus of the spiral. The small-angle boundaries or subboundaries referred to earlier are generally composed of edge dislocations. It is the presence of these defects which cause crystalline solids to deviate so markedly from theoretical strength.

Some of the crystalline defects that may be found on a solid surface are presented schematically in Fig. 6. The vacant lattice site (Fig. 6a) is simply the absence of one atom from a crystal lattice site. The interstitial (also shown in Fig. 6a) is an extra atom crowded into the crystal lattice. Edge and screw dislocations are shown in Fig. 6b, and the small-angle boundary in Fig. 6c.

Mechanically finished surfaces generally have undergone a high degree of strain and thus may contain a large amount of lattice distortion which generates high concentrations of dislocations. While the initial presence of dislocations causes a reduction in strength, their

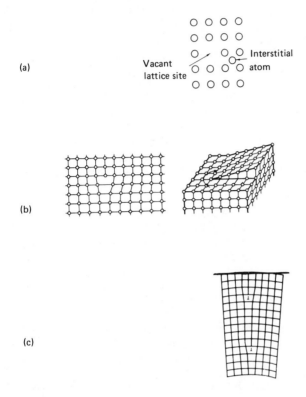

Fig. 6. Crystalline defects in solids. (a) Vacancy and interstitial crystal defects. (b) Edge and screw dislocations. (c) Small-angle boundary composed of edge dislocations indicated by T symbols.

multiplication and interaction during deformation produces an increase in surficial strength.

With plastic deformation of real surfaces, the strain produces a reduction, generally, of the recrystallization temperature of the material at the surface. In many materials the combination of strain and temperature can bring about surface recrystallization, which has an annealing effect. Annealing relieves the lattice strain and the stored energy; a sharp reduction in the concentration of surface defects results. In a dynamic, nonequilibrium system, such as that encountered in mechanical activity on surfaces, the surface layers may be strained many times, recrystallized, and then strained again.

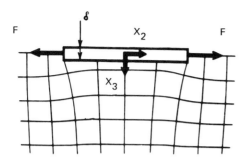

Fig. 7. Lattice distortion under compressed surface stress.

The application of surface forces that may be associated with the techniques used to develop a desired surface topography can, in the absence of sufficient energy to produce recrystallization, result in a lattice distortion, such as that indicated in Fig. 7. The application of a force normal to the surface, that is, a compressive force, causes forces X_2 and X_3 to develop. The resultant force normal to the surface, X_3, will cause distortion in the crystal lattice.

IV. CHEMISTRY OF SURFACES

Clean surfaces of solids are extremely chemically active. The surface atoms of an elemental metal, for example, are highly energetic. They are bound by like atoms everywhere but at the free surface. This boundary is depicted schematically in Fig. 8. A copper atom, for example, which lies in a (111) plane in the bulk of the solid will have a coordination number of 12, that is, it is bonded to 12 nearest neighbors. That same copper atom at the surface, however, will have a coordination number of only 9. It has only nine nearest neighbors, three less than when it is in the bulk solid. Thus, the energy that normally would be associated with bonding to three additional like atoms is now available at the surface. This energy, expressed over an area consisting of many atoms in the surface lattice, is referred to as the surface energy.

Another way of looking at surface energy is to understand it as being the energy necessary to generate a new solid surface. This can be accomplished by the separation of adjacent planes in the solid. The energy required for separation is a function of atomic packing. For example, in copper the atomic packing density is greatest in (111) planes (greatest number of nearest neighbors within the plane). As a

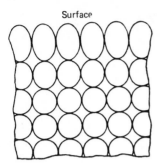

Fig. 8. Surface atoms (schematic). Since these atoms are not entirely surrounded by others, they possess more energy than internal atoms.

result the bonding force between adjacent (111) planes is least and, therefore, the surface energy of new (111) surfaces generated, say by cleavage, is less than it is for other planes, such as the (110) and (100) planes. This lesser binding strength is also reflected in the distance between adjacent planes—it being greater between adjacent (111) planes than between other planes, such as the (110) and (100) planes.

    Because the atoms at the surface have this unused energy, they can interact with each other, with other atoms from the bulk, and with species from the environment. One of the things these surface atoms can do, because they are not bound as rigidly as atoms in the bulk, is alter their lattice spacing at the surface (Fig. 9). This process is

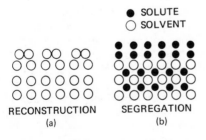

Fig. 9. Possible chemical surface events. (a) Reconstruction; (b) segregation.

commonly called reconstruction. Low-energy electron diffraction (LEED) studies have revealed reconstructed surfaces in some crystalline solids, but not in others.

Another event that can occur in solids containing more than a single element (e.g., alloys) is the diffusion to and segregation at the surface of atoms from the bulk. In a simple binary alloy, for example, the solute atoms can diffuse from the near-surface regions to the surface and completely cover the surface of the solvent. This has been observed for many binary alloy systems including aluminum in copper, tin in copper, indium in copper, aluminum in iron, and silicon in iron. The process of surface segregation is depicted schematically in Fig. 9.

The segregation mechanism is not really understood. One hypothesis is that the solute segregates on the surface because segregation reduces the surface energy. A second theory is that the solute produces a strain in the crystal lattice of the solvent, and that, because of this unnatural lattice state, there is a necessary driving force to eject the solute atoms from the bulk. The result of this is segregation at the surface.

With ionic crystalline solids, surface charging can occur due to unequal vacancy energies. Near-surface vacancies in these solids may be either positively or negatively charged. The energy associated with the formation of these vacancies varies, and an excess of one or the other type of vacancy occurs. This excess causes a voltage between the surface and the interior of the solid. The overall energy of the system is decreased when there is an excess of positively charged ion vacancies in the solid, and these migrate toward the surface. When in the bulk (Fig. 10a), they will bring about an internal arrangement of vacancy charges (Fig. 10b) such that the positive charge exists near the surface, and the Debye charged layer will form. This layer will have no effect on the behavior of the solid surface, as in, for example, its interaction with an adhesive.

A great deal of concern must be given to the surface of polymers just as to the surfaces of metals and ionic solids. There are a host of variables in the preparation process that can alter the chemistry of a polymer surface, just as the mode of preparation can, for example, alter the topography of a metal surface. With polymers, not only is the physical mode of surface preparation important, but also the environment in which the surface is prepared can pronouncedly affect polymer surface chemistry.

The effect of the environment in which polymers are prepared has been carefully studied with electron spectroscopy for chemical analysis (ESCA). Electron spectroscopy for chemical analysis, or x-ray photoemission spectroscopy (XPS) as it is currently called, permits analysis of surfaces by measuring both the energies associated with electron levels in atoms and molecules and shifts in electron ener-

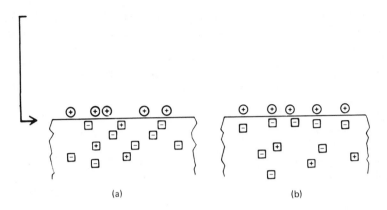

Fig. 10. Formation of Debye layers in ionic crystals due to unequal vacancy-formation energies where E^-E^+. Positive-ion vacancies are represented by - because of their effective negative charge, and negative ion vacancies by +. (a) Uniform distribution of positive-ion and negative-ion vacancies, which are present in different amounts because of different energies of vacancy formation. A voltage is formed between the surface and the interior of the crystal. (b) The overall energy of the crystal is decreased when the excess positive-ion vacancies migrate toward the surface, forming a Debye space-charge layer. The interior of the crystal below the Debye layer now possesses equal numbers of positive-ion vacancies and negative-ion vacancies.

gies accompanying various chemical reactions. It is a surface-sensitive tool.

Films of high-density polyethylene were prepared from powders by hand-pressing the powder between sheets of clean aluminum foil at 200°C, the minimum temperature needed to ensure plastic flow. Samples from identical powders were prepared in three different environments: in air, in nitrogen, and, after being pumped down to 10^{-14} Torr, in pure nitrogen or argon. The surfaces were examined with XPS for both oxygen (O_{1s}) and carbon (C_{1s}).

The O_{1s} and C_{1s} spectra corresponding to samples from the three modes of preparation are markedly different (Fig. 11a-c). (ATR) and (TIR) experiments did not reveal the presence of any oxygen functions (-OH, $>C=O$, C - O - C, etc.), and, in fact, the spectra were virtually identical. This demonstrates the great power of XPS to distinguish minute differences in samples when such differences are localized at or near the surface. Comparison with the data for low-density polyethylene and with model monomer systems shows that the O_{1s} signal arises from $>C=O$ environments. The three methods of prepar-

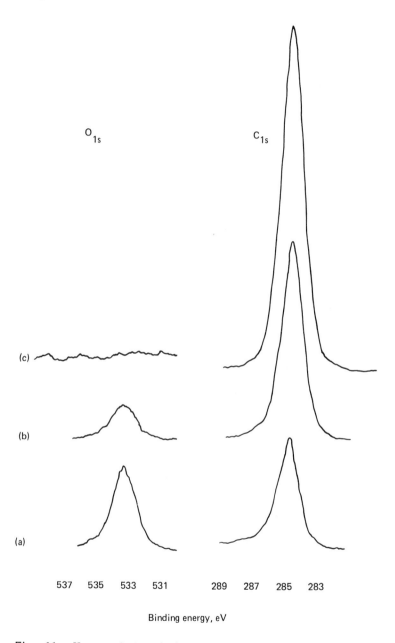

Fig. 11. X-ray photoemission spectroscopy analysis of C_{1s} and O_{1s} levels of high-density polyethylene films pressed (a) in air; (b) in nitrogen; (c) after gasification in a 10^{-14} Torr vacuum, in pure nitrogen or argon.

ation clearly indicate that "unoxidized" surfaces may most readily be prepared by excluding all traces of oxygen during the pressing stage. On leaving samples exposed to the atmosphere for some time, hydrogen from extraneous water in the atmosphere bonds to the surface $\diagup C \!=\! O$ groups, and the O_{1s} peak then acquires the characteristic doublet nature. Again, ATR and TIR do not reveal any changes since the hydrogen bonding is localized at the surface [4].

A. Chemisorption

In addition to the characteristics the solid surface itself displays, the surface can interact with the environment. This interaction is extremely important because it alters the surface chemistry, physics, metallurgy, and mechanical behavior. If a metal surface is very carefully cleaned in a vacuum system, and then a gas such as oxygen is admitted to the system, the gas will adsorb on the metal surface. This interaction results in strong bonds being formed between the metal and the adsorbing species. With the exception of inert gases, this adsorption results in bonding which is chemical in nature. The process is referred to as chemisorption. It is indicated schematically in Fig. 12. Once adsorbed, these films are generally difficult to remove.

Where the species adsorbing on a clean surface is elemental, the adsorption is direct. The atoms in the surface of the solid retain their individual identity, as do the atoms of the adsorbate; yet each is chemically bonded to the other. When the adsorbing species is molecular, chemisorption may be a two-step process—dissociation of the molecule on contact with the energetic, clean surface followed by adsorption of the dissociated constituents.

Chemisorption is a monolayer process. Furthermore, the bond strengths that exist between the adsorbing species and the solid surface are a function of chemical activity of the solid surface (surface energy), the degree of surface coverage of that adsorbate or another adsorbate, the reactivity of the adsorbing species, and its structure.

The surface energy of the solid surface is important because the more energetic the surface, the stronger is the tendency to chemisorb. The effect of surface energy can be demonstrated by examining various crystallographic planes of a single metal. In general, the high-energy, low−atomic-density planes will chemisorb environmental species much more rapidly than will the high-density, low-surface-energy planes. This has been demonstrated experimentally to be the case. Hydrogen sulfide will adsorb more readily on (110) and (100) surfaces of copper than on (111) surfaces.

When different solid surface materials are considered, adsorption differences are also observed. For example, copper, silver, and gold are the noble metals, and many of their properties are considered very similar. Yet with respect to chemisorption of a gas, such as

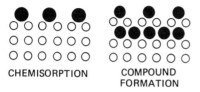

Fig. 12. Chemisorption and compound formation on solid surfaces.

oxygen, considerable differences are observed. Oxygen will chemisorb relatively strongly on copper, weakly on silver, and not at all on
gold.

The reactivity of the adsorbing species is also very important.
Examination of the halogen family indicates that fluorine will adsorb
more strongly than chlorine, chlorine more strongly than bromine, and
bromine more strongly than iodine.

The structure of the adsorbing species is also very significant in
surface bonding and chemisorption. This can be demonstrated with
the adsorption of simple hydrocarbons. Something as simple as the
degree of bond saturation in the molecule will make a difference. If
ethane, ethylene, and acetylene are adsorbed on an iron surface, the
tenacity of the resulting chemisorbed films is in direct relation to the
degree of bond unsaturation. Acetylene is much more strongly bound
to the surface than ethylene is, which, in turn, is more strongly
bound than ethane is. The simple explanation for this is that the unsaturated carbon-to-carbon bonds break on adsorption and bond to
the iron. The greater the number of carbon-to-carbon bonds, the
greater the number of potential bonds from the hydrocarbon molecule
to iron.

B. Compound Formation

Compound formation on the surface of solids is extremely important.
The naturally occurring oxides present on metal surfaces prevent
their destruction when in contact with other solids. Furthermore,
their presence can alter deformation behavior. Solids and liquids can
readily react with clean solid surfaces to form compounds whose presence can alter surface properties.

Chemisorbed films can often interact with a surface to form chemical compounds. When this occurs, the surface material and the adsorbate lose their individual identities and form an entirely new sub-

stance with its own properties. Unlike chemisorption, which is simply a monolayer process, constituents from the environment can react with the solid surface by diffusion of the solid surface material into the compound and diffusion of the environmental species into the film. The compound can continue to thicken on the surface if the film is porous and allows for the two-way diffusion to occur (Fig. 12).

An example of the formation of a porous compound on a surface is the rust, or iron oxide, produced in the oxidation of iron in a moist air environment. The oxidation process will continue to consume the iron. In contrast, the oxidation of aluminum to form aluminum oxide results in the formation of a thin, dense oxide which, because of the film's density, retards diffusion and further growth.

C. Metallurgical Effects

Surface behavior is altered by the presence of grain boundaries on the surface of crystalline solids as well as by other surface defects. For example, grain boundaries, in addition to having a chemical effect because of their high energy, also influence mechanical properties. The microhardness is generally higher in grain boundaries than it is in grains [5]. This hardness effect is seen in the data for iron in Fig. 13 which shows that, as the grain boundary is approached, the hardness increases. The hardness is at a maximum directly in the boundary. The increase is marked and not marginal.

The increase in hardness seen in the grain boundary of Fig. 13 can be explained on the basis of what has already been discussed relative to grain boundaries. They are regions of high dislocation concentration and lattice strain. Strained metal has a higher dislocation concentration than the annealed or strain-free material, and the concentration generally will increase with increased deformation until recrystallization occurs.

Dislocations, like grain boundaries, are higher energy sites on a surface than nondislocation areas. It is for this reason that strained metal surfaces are chemically more active than annealed or strain-free surfaces are. This chemical activity of dislocation sites can be demonstrated by a technique called etch pitting. Certain chemical agents will react on the surface of a material, and, because of the more energetic nature (i.e., greater reaction rates) of the dislocations, these sites will be preferentially etched. The result is that pits are left on the surface at each dislocation site. From these pits it is possible to identify the location and concentration of the dislocations. The literature lists the necessary reagents required for etch pitting various materials.

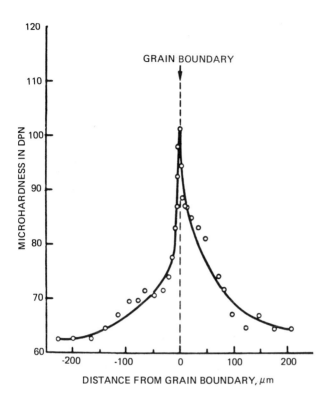

Fig. 13. Microhardness traverse across a grain boundary in iron containing 0.02% oxygen. (From Ref. 5.)

V. SURFACE FILM EFFECTS

The properties of solid surfaces are markedly altered by the presence of foreign substances. An atomically clean metal surface has certain characteristic chemical, physical, and metallurgical properties. Just as soon as something interacts chemically with that surface, those properties are changed. This fact is extremely important to understand because most real surfaces are found, not in the atomically clean state, but rather with film(s) present on their surface, as has already been discussed in reference to Fig. 1.

It can safely be said that the wide variations found in the literature for the surface properties of materials can be directly attributed to the effect of these films. Some surfaces are more strongly influenced by these films than others, and the specific film composition will produce varying effects.

The presence of oxides on metal surfaces has been observed to produce surface hardening. Roscoe, in some very fine experiments conducted in the 1930s, demonstrated this effect with cadmium oxide present on a cadmium surface [6]. Since that time it has been demonstrated by other investigators who observed dislocations emerging at the surface with the oxide impeding their mobility.

While oxides and some other films produce surface hardening, other surface films increase ductility. For example, water on alkali halide crystals allows an otherwise brittle solid to deform plastically. Water has the same effect on ceramics.

Magnesium oxide (MgO) is normally a very brittle material with a Knoop surface hardness in the clean state of about 750 kg/mm^2. When MgO is cleaved under a hydrocarbon such as toluene so as to exclude moisture, a hardness value of this magnitude is measured for the MgO surface. If, however, the MgO is cleaved in moist air where the freshly generated clean surface can interact with the moisture in the air, a different result is obtained (Fig. 14).

Figure 14 presents the hardness of MgO as a function of indentation time in two environments, dry toluene and moist air. The increased surface ductility in the presence of water is striking. Not only is there an appreciable difference in hardness, but also that difference increases with increasing indentation time. The hardness in moist air decreases with increasing indentation time; in dry toluene, the hardness is unchanged. It is this change with time that makes the film effect a true surface property and not simply a lubricating effect produced by the water.

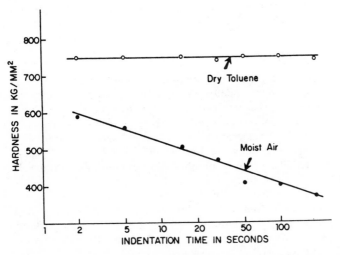

Fig. 14. Illustration of the time dependence of microhardness of MgO in moist air. (From Ref. 7.)

In the late 1920s, the Russian researchers V. I. Likhtman et al. found that the presence of certain organic molecules on the surface of solids produced a softening effect [8]. Mechanical behavior was altered by these films. Such substances as oleic acid in Vaseline oil are examples of materials examined. This effect is important because many of the materials studied are commonly found substances. The surface softening can be very beneficial in certain instances, such as in the arrest of the formation of fatigue cracks.

VI. CHARACTERIZATION OF SURFACES

A. Microscopy

The nature and character of the surface is extremely important to the understanding of the performance of that surface in adhesive systems. Microscopy has been, and probably still is, the most common technique employed for the characterization of surfaces. The magnification of surfaces to identify structure dates back at least to the sixteenth century. With simple lenses the features of surfaces could be magnified 100 times.

The optical microscope was developed in the 1700s, and its availability really initiated the effective characterization of surfaces. It became the most effective tool for the study of surfaces, and it remains that to this day. With the aid of oil immersion, the ordinary optical microscope can yield detailed surface features at magnifications ranging to about 500–1000. Thus, the character and structure of grain boundaries in metals and alloys are readily identifiable.

The development of the electron microscope was a notable advance because it permitted magnifications from 10 to 100,000. For the first time, atomistic features of surfaces were identifiable, and rows of atoms on the surface could be readily seen. Today, electron microscopy is used routinely to identify and characterize dislocation structures in materials.

In the 1950s, with the development of the field ion microscope by Müeller and Tsong [9], it became possible to characterize surface structure at the level of the individual atom. At this point in the development of research instrumentation it must be said that, for structural analysis, the field ion microscope is the ultimate tool because it identifies individual atom sites on a solid surface. Thus, each individual white spot in the field ion micrograph of Fig. 15 represents an individual atom site while the rings represent atomic planes.

The field ion microscope has been adapted for use in adhesive studies and in the characterization of surfaces in such studies. Figure 15 is a photomicrograph of a tungsten surface. The micrograph reveals the atoms and planes. The ring just to the upper right of center is the (110) plane of tungsten. Each row out from the center

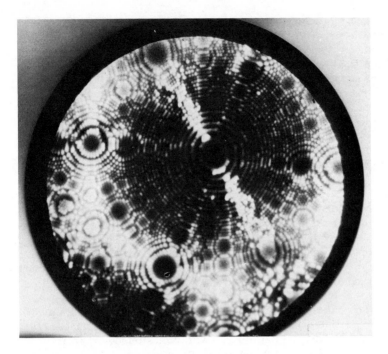

Fig. 15. Field ion micrograph of a tungsten surface (16.0 kV) prior to contact.

ring represents the layer of atoms next nearest to the surface. The field ion microscope is so sensitive that it can be used to detect the absence of a single atom from the surface or, conversely, the presence of an extra or a foreign atom.

In recent years the atom probe has been developed for the determination of single-atom chemistry. When used in conjunction with the field ion microscope, the atom probe can not only characterize the structural arrangement of individual atoms in a solid surface, but also determine the chemistry of an individual atom present in the surface.

B. Etching

Etching is the interaction of surfaces with chemical agents such as acids or bases. In simple metals in the polycrystalline form, the crystallographic orientation of each adjacent grain at the surface will vary.

The energy of these surfaces will vary, and they will, therefore, react at differing rates with a particular chemical agent. Thus, in a crude way, one can distinguish the more atomically dense surface planes from the less dense. The more dense planes have lower surface energies and, therefore, will not be as readily attacked as the less dense planes.

Etching can be, and has been, very effectively used as a technique for revealing grain boundaries in metals. Grain boundaries are sites of higher energy than the surfaces of the individual grains and, therefore, they will be preferentially attacked. Furthermore, different phases in alloys will etch differently at the surface. The chemical reagents for such etching are available in standard metallurgical handbooks.

In addition to identifying orientation, grain boundaries, and phases, etchants can be used to identify atomistic defects such as dislocations, the atomic line defects in crystalline materials referred to earlier. The proper etchant can not only indicate the concentration of dislocations at the surface, but also reveal their locations. Furthermore, etchants reveal the atomic plane upon which a dislocation lies.

A wealth of useful information can be acquired about a surface by the use of the optical microscope and etching techniques. The use of simple, chemical spot tests in conjunction with the foregoing can provide insight into the chemistry of the solid surface. These tests require the use of readily available chemical reagents which, when applied to the surface, will reveal the metallic element present in the surface. Many standard college chemistry textbooks on inorganic qualitative analysis list the required materials.

The chemical spot test can be used to detect adhesive transfer to surfaces where dissimilar materials are in contact. A permanent pattern of an element's distribution on a surface can be obtained if a porous paper, such as filter paper, is impregnated with the chemical reagents. The paper is pressed against the surface to be analyzed; then the paper is moistened. A map of the location of the elements thus appears on the paper.

C. Analytical Surface Tools

The field ion microscope and the atom probe have already been discussed. These tools have been used in adhesive studies to characterize surfaces before and after adhesive contacts.

A host of analytical tools has been developed in recent years to characterize the real nature of solid surfaces. Some operate on the principle of atomic arrangement in the surface layers of crystalline solids. One such device is LEED. This surface tool can be very useful to those interested in surfaces. It analyzes, by an electron dif-

fraction technique, the general atomic arrangement of the outermost
surface layers of a solid. A rudimentary understanding of atomic
arrangement will assist in understanding the nature of the surface
structure.

With the exception of amorphous carbon, glasses, and some
polymers, nearly all materials—including metals, alloys, ceramics,
solid lubricants, and graphitic carbon are crystalline. This means
that the atoms or molecules are arranged in accordance with particular structures that can readily be identified. These structures are
mostly cubic and hexagonal. The cubic structure can be further subdivided into the face-centered cubic structure and the body-centered
cubic structures.

Metals such as copper, nickel, silver, gold, platinum, and aluminum have a face-centered cubic structure, while metals such as
iron, tantalum, niobium, vanadium, and tungsten have a body-centered cubic structure. A number of metals—including zinc, cadmium, co-
balt, rhenium, zirconium, and titanium—have a hexagonal crystal
structure.

The atoms making up the faces of the cube for the face-centered
cubic and the body-centered cubic are referred to as the (100) surfaces. They constitute planes of atoms that can move relative to each
other when the crystal is deformed plastically and are, therefore, also
referred to as slip planes within the crystal. Under applied stresses
these [the (100)] planes are one of the sets of planes most commonly
observed to slip over one another in the body-centered cubic system.

A third set of planes in the cubic system are the (111) planes.
They are the planes on which slip and cleavage in face-centered cubic
materials are most frequently observed to occur. There are other
planes that may be observed as well. For example, if one were to use
x-ray diffraction, electron channeling, or some other technique for
the determination of crystallographic orientations, a host of different
crystallographic planes would appear in each grain of a polycrystalline
sample analyzed.

The (111) planes in the face-centered cubic crystal are the
planes of closest atomic packing, as already mentioned. The atoms
occupy the least area for the number of planes involved. When a
(111) plane is present on a surface, it has the lowest surface energy
in the face-centered cubic system and, therefore, is the least likely
to interact chemically with environmental constituents.

The (110) planes in the face-centered cubic system are the least
densely packed. The two outer rows of atoms as well as the center
row are outermost with the two in-between rows below these. The
(110) planes have higher surface energies than the (111) planes and
are, therefore, much more reactive. Furthermore, because they are
less densely packed, their mechanical behavior, such as elastic modulus and microhardness, is also less than is observed for the (111)
planes.

On any particular crystallographic plane present on the surface, the atomic packing can vary with direction of movement. For example, on a (111) surface, two basic directional packing variations exist. Surface energies also vary with these two directions.

1. Low-energy electron diffraction (LEED)

Low-energy electron diffraction is a very widely used surface tool for characterization of the surface atomic structure seen on crystalline solids. Because the device detects the surface crystal structure, single crystals are generally studied although large-grained poly-crystals can also be examined. Low-energy electrons in the range of

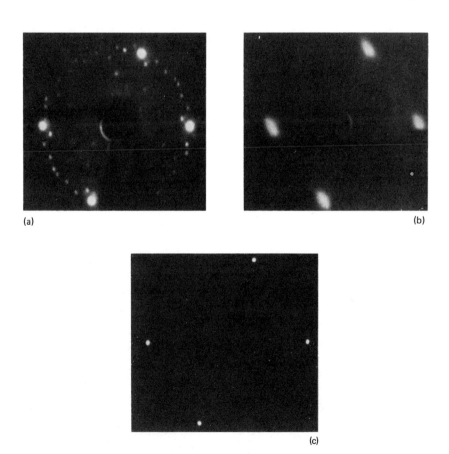

(a)   (b)

(c)

Fig. 16. Low-energy electron diffraction patterns of iron (011) sur-face (110 V). (a) Carbon contaminants; (b) argon bombarded; (c) clean surface.

20–400 eV are diffracted from the surface crystal lattice, producing a reciprocal image of the lattice on a phosphorus screen.

Figure 16 contains three LEED patterns from an iron (011) surface. The photograph and pattern in the upper left corner is for the iron surface with oxide removed. Upon oxide removal and heating of the iron in a vacuum, the surface becomes covered with a film which produces a ring structure of diffraction spots on the surface. Auger electron spectroscopy analysis (which will be discussed in Sec. VI.C.2) identified the surface film as graphitic carbon. Carbon segregates from the iron bulk to the surface.

When the iron is argon-ion bombarded, the carbon disappears. Four diffraction spots, in a rectangular array representative of the clean iron (011) surface, remain. The carbon has been removed by the argon-ion bombardment. The iron diffraction spots are not sharp but rather fuzzy and elongated because the argon bombardment produces strains which can be removed by very mild heating for a short period followed by cooling to room temperature. This produces the clean iron surface with the diffraction pattern of the lower photograph in Fig. 16.

Low-energy electron diffraction can, as indicated, be an effective tool in identifying a clean metal surface and describing its structure and its condition or state. It can also be used to identify the

 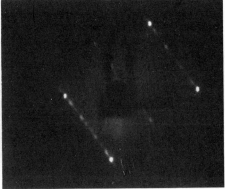

Fig. 17. Low-energy electron diffraction patterns obtained with two polymer-forming hydrocarbons on iron (011) surface, 1000 L exposure. (a) CH_2-O-CH_2; (b) CH_2-CHCl.

structure of films formed on a clean surface. Two different polymer-forming hydrocarbon molecules produce entirely different structures when adsorbed on a clean iron surface. These films and their structures are presented in Fig. 17. The two molecules, ethylene oxide and vinyl chloride, both contain two carbon atoms, but one molecule contains oxygen, and the other chlorine. A closely packed structure of ethylene oxide completely covers the iron surface.

2. Auger electron spectroscopy

Although LEED is very useful for structural surface analysis, it does not give any indication of the chemistry of the surface. Other surface tools must be used to obtain this information. A very effective tool for this purpose is Auger emission spectroscopy (AES) analysis. It has the ability to analyze for all the elements present on a surface except hydrogen and helium. It is sensitive to surface coverages of an element such as oxygen of as little as one-hundredth of a monolayer. It analyzes to a depth of four or five atomic layers.

The basic mechanism in both AES and LEED analysis involves the use of a beam of electrons, but the energy of the electrons used in AES is higher—usually 1500–3000 eV. The incident electrons strike the sample surface, penetrate the electron shells of the outermost surface atoms, and cause the ejection of a second electron called an Auger electron. The ejected electron carries with it an energy characteristic of the atom from which it came. Thus, by measuring the energy of the ejected electron, it is possible to identify its source element.

The electron energies detected can be recorded on a strip-chart recorder or on an oscilloscope. The details of AES analysis and the type of data generated can be found in Kane and Larrabee [10].

Figure 18 presents an Auger spectrum for an iron (011) surface. An ordinary iron surface with normal surface contaminants present will yield a spectrum such as that displayed in Fig. 18. The surface contains the elements sulfur, carbon, oxygen, and iron. The carbon and sulfur have two possible sources of origin: impurities in the bulk iron which have segregated to the surface and adsorbates from the environment. For example, the carbon can arrive on the surface as carbon monoxide or carbon dioxide from the environment, or it can diffuse to the surface from the bulk.

The oxygen peak of Fig. 18 could result from the iron oxides present on the surface or, again, from adsorbates, such as carbon compounds or water vapor. The three iron peaks originate from the iron oxides and the iron metal.

If the surface represented in Fig. 18 were bombarded with argon ions, the surface contaminants—sulfur, carbon, and oxygen—would be knocked off, leaving only iron to be detected by the Auger spectrometer. A low-energy iron peak would appear at the left end of the spectrum after the sulfur, carbon, and oxygen had been removed.

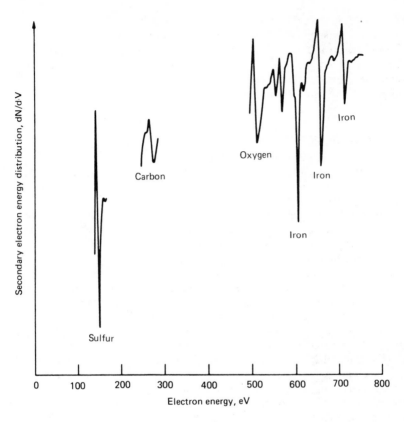

Fig. 18. Auger electron spectroscopy analysis of iron (011) surface.

This low-energy peak is easily lost when the surface is contaminated and is, therefore, usually seen only when the surface is clean.

In addition to supplying information on the identity of elements present on a surface, AES can give insight into the form in which an element exists on a surface. For example, carbon can arrive at a surface from many sources. It can diffuse from a bulk metal or alloy and segregate at the surface. It can be present as adsorbed carbon monoxide, or it can exist in the crystalline form of graphite. The form of the carbon can be extremely important to adhesive bonding characteristics.

Analysis of the shapes of Auger peaks can provide considerable information about the source of an element, such as, for example, carbon. Carbon-peak shape analysis can indicate whether the carbon comes from the bulk solid carbide, adsorbed carbon containing gases (e.g., CO or CO_2), or graphite. The Auger peaks are all for carbon,

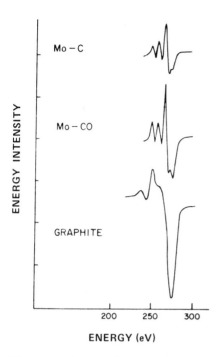

Fig. 19. Auger electron spectroscopy of carbon segregated at a Mo (110) surface during initial cleaning, in CO on a clean Mo (110) surface (Mo-CO), and in graphite.

but their shape differences tend to identify the carbon source. These differences can be seen in Fig. 19 for the various forms of carbon on a molybdenum surface.

3. X-ray photoemission spectroscopy

While AES can give some indication of the surface structure from which an element came, it is rather limited in this area. Its principal function is elemental surface analysis. There are other surface tools that can determine the molecular structure from which an element came. One such tool is XPS.

With XPS an x ray, rather than an electron beam, is used as the energy source. The x-ray beam is monochromatic, and it causes electrons with kinetic energies characteristic of the surface atoms to be ejected from the specimen. A spectrum containing the elements present is obtained by plotting the total number of electrons ejected from the surface as a function of kinetic energy. X-ray photoemission

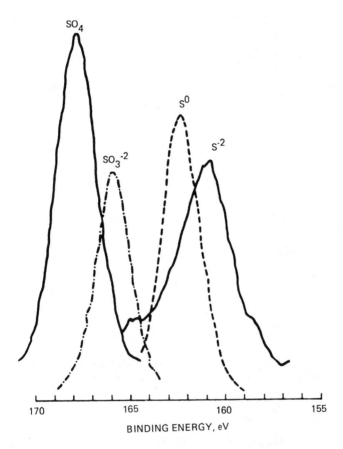

Fig. 20. Sulfur (2p) XPS of representative sulfur types.

spectroscopy gives binding energies of the elements, and from these binding energies it is possible to identify the nature of the compounds in which these elements exist. The binding energy of the electrons ejected from the surface is determined by the chemical environment and is, roughly, a function of the atomic charge.

The binding energy measured with XPS will be altered by changing the particular elements bound to the element being examined. This change is demonstrated for sulfur in the data of Fig. 20. Elemental sulfur (S^0) has a characteristic binding energy of 162.5 eV. Negatively charged sulfur (S^2) has a readily measurable lower binding energy. When oxygen is bound to the sulfur, there is an increase in the sulfur binding energy. Furthermore, the amount of oxygen bound to the sulfur will affect the observed binding energy. The SO_4^{2-} structure

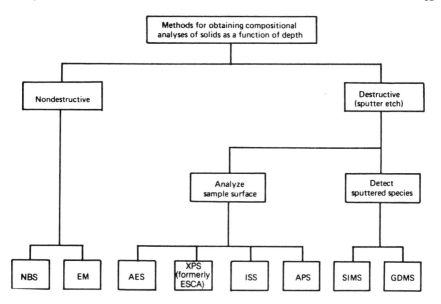

Fig. 21. Some techniques for chemical characterization of surfaces.

has a greater binding energy than the SO_3^{2-} structure. This differ-ence in binding energy can be used to distinguish between sulfur bound in these two states.

4. Other techniques

A host of surface tools have been developed for the analysis and chem-ical characterization of surfaces. At the time of this writing the auth-or knows of over 70 such tools. The number will undoubtedly have increased by the time of publication of this chapter, and there is no doubt that the number will continue to grow. A few of the more com-monly used techniques are indicated in Fig. 21. The techniques indi-cated are separated into destructive and nondestructive classes.

The nondestructive techniques are nuclear back-scattering spec-troscopy (NBS) and electron microprobe (EM). Auger electron spec-troscopy (AES), x-ray photoemission spectroscopy (XPS), ion-scat-tering spectroscopy (ISS), and appearance potential spectroscopy (APS) are destructive only if sputter etching or depth profiling is used.

Two techniques, which are definitely destructive to the surface, are secondary ion-mass spectroscopy (SIMS) and glow-discharge mass spectroscopy (GDMS). These techniques detect the species sputtered from the surface and, thus, analyze material that has been removed

Table 1 Comparative Table for the Various Techniques Used for the Chemical Characterization of Surfaces

	NBS	EM	AES	XPS	ISS	SIMS	GDMS	APS
Destructive to sample (in general)	No	No	No	No	No	Yes	Yes	No
Elements that can be detected	Heavy	$Z \geq 3$	$Z \geq 3$	$Z \geq 3$	$Z \geq 3$	All	All except He, Ne	$Z \geq 3$
Elemental identification[a]	F	G	E	E	E	G	G	E
Sensitivity (typical, in monolayers)	50	5	~ 0.01	<0.01	~ 0.01	<1	~ 1	≤ 0.1
Detectability (i.e., ppm)[b]	NA[b]	100	<1	NA	NA	1	100	NA
Results are (in principle)	Abs[c]	Abs	Abs	Abs	Abs	Abs	Abs	Abs
Depth probed (in Å)	10^4	$10^4 - 10^5$	15-20	15-75	3	$\sim 5 \times 10^4$	$10-10^4$	~ 10
Depth distribution of elements	Yes	Yes	Y/d[d]	Y/d	Yes	Yes	Yes	Y/d
Chemical (i.e., binding) information	No	Yes	Yes	Yes	No	No	No	Yes

[a]E, excellent; G, good; F, Fair.
[b]NA, not applicable.
[c]Absolute.
[d]Y/d, Yes, if destructive.

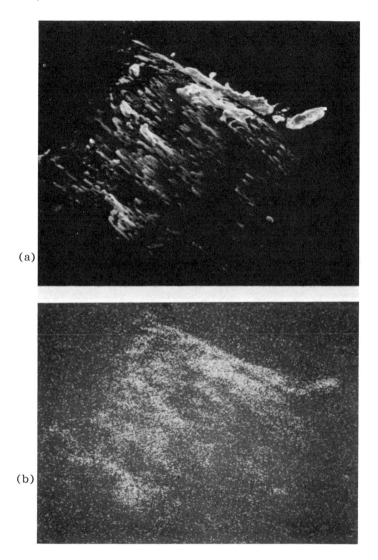

(a)

(b)

Fig. 22. Silicon (111) surface after contact with gold. (a) Scanning electron micrograph (175X); (b) x-ray map.

from the surface. The details of the use of these techniques can be found in Kane and Larrabee [10].

The capabilities of the instruments presented in Fig. 21 are set forth generally in Table 1. All of these devices have certain limitations. Note that among them they can detect all elements except hydrogen and helium, can provide excellent chemical identification, have sensitivities of surface elements to as little as 0.01 monolayer, and give chemical information. Their disadvantage, not indicated in Table 1, is that they must be operated in a vacuum.

Probably the most versatile tool available for use on surfaces, and the one requiring the least technical interpretation skill is the scanning electron microscope (SEM). It is extremely useful to the surface analyst because it provides a view of surface depth features, such as the asperities or surface irregularities discussed in the first part of this chapter. It can also identify the topography of surfaces where adhesion has taken place with adhesives.

When the SEM is incorporated with x-ray energy dispersive analysis, both topography and chemistry can be determined. The x-ray analysis is not a surface analytical tool, but it can provide considerable information when material transfer has taken place, as in adhesion. The effectiveness of this combination can be seen in Fig. 22 where gold contacted and transferred to a single-crystal silicon surface. The upper figure is an SEM photomicrograph of the silicon surface after the transfer of gold, while the lower is an x-ray map of gold which reveals that the white areas in the upper photomicrograph are gold.

D. Adhesion

When two atomically clean surfaces are brought into solid-state contact, adhesion across the interface always occurs. The interfacial bond strength is, with a few exceptions, stronger than the cohesive bonds of the cohesively weaker of the two solid surfaces, and when surface separation is attempted, fracture will occur not at the interface, but in the cohesively weaker material. Thus, strong adhesion can be achieved by simply atomically cleaning the solid surfaces. There is, however, the multitude of practical joint systems where the generation of atomically clean surfaces in a vacuum is not feasible. Nevertheless, much can be done to improve adhesion by careful surface preparation both when performing two-body adhesion and when employing adhesives.

The adhesion of polymers to metal surfaces is of interest with respect to both two-body and single-body adhesion, that is, where the polmyer is the adhesive. Polytetrafluoroethylene (PTFE) has an extremely low surface energy, is difficult to get to adhere to metal surfaces, and is, therefore, an ideal polymer for measuring adhesion to metals.

The field ion microscope (FIM) is a powerful tool for studying the adhesion process, particularly of polymers to metals. A combination of high magnification and a resolution of $2-3$ Å permits the adhesion process to be studied in atomic detail.

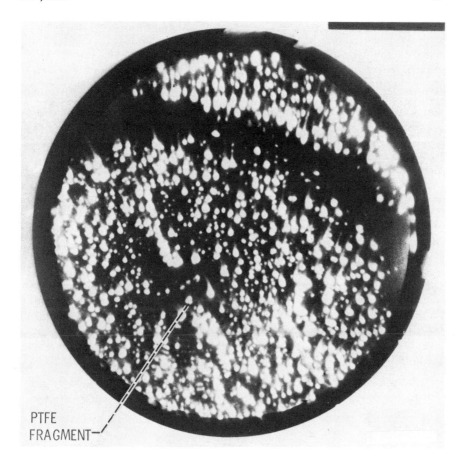

PTFE
FRAGMENT

Fig. 23. Tungsten after PTFE contact (16.5 kV, helium image gas).

A series of PTFE-tungsten contacts were made with atomically clean tungsten contacting PTFE at loads between 20 and 30 g, and the force of adhesion was measured. Figure 15 is an FIM picture of a clean tungsten tip. Figure 23 is an FIM picture taken after contact with PTFE for a few seconds. Many extra image points are apparent on the postcontact micrograph, particularly on the (110) plane shown in this figure. Adsorbed or adhered atoms can be observed because the geometry of the extra atoms on the surface of the plane creates points of localized field enhancement resulting in increased probability of ionization and, hence, greater brightness. Thus, the clusters visible in the figure are fragments of PTFE which adhered to the tung-

sten surface after separation had occurred. The other bright image
spots also represent PTFE on the metal surface, but their clusterlike
nature cannot be resolved. The fragments of PTFE have the appearance of the end of PTFE chain that is normal to the (110) plane. The
fact that the fragments are stable at the very high electric field required for helium-ion imaging implies that the bond between the PTFE
and the tungsten is very strong. Otherwise field desorption of the
adhered PTFE would occur.

To obtain a measure of the bonding between the PTFE and the
tungsten, the forces of adhesion were measured in terms of an adhesion coefficient for a number of contacts over varying periods of contact time. The results are summarized in Fig. 24. For short contact
times the forces of adhesion were immeasurably small. After 2 min,
however, the force of adhesion increased markedly. At contact times
of 4–6 min, adhesion coefficients approaching those for clean metals
in contact with each other were obtained.

It was observed the negligible adhesive force was obtained when
a second contact was made with a previously contacted tip, indicating
that the adhesive polymer-to-metal bond is stronger than the cohesive
polymer bond. Polymer radicals can be expected to occur as a result
of the breaking of chains by the chemical interaction of polymer and
metal. Thus, for PTFE contacting a clean tungsten surface, the possibility of reactive valence states of carbon atoms in PTFE bonding to
tungsten exists.

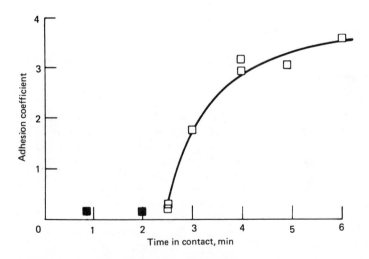

Fig. 24. Adhesion coefficients for PTFE-tungsten contacts. Solid
symbols denote immeasurably small force.

A heavily loaded tungsten-PTFE (approximately three times more load ∿1 mg) contact gave the rather surprising result that extensive deformation of the tungsten occurred. The deformation extended far into the bulk of the material.

Mechanical contacts with a polyimide polymer contacting tungsten tips were made in vacuum of 10^{-9} Torr with both light and heavy loads. At light loads the results obtained were analogous to those obtained with PTFE. Random distribution of bright spots was visible, indicative of polymer fragments adhering to the tungsten. The spots (polymer fragments) were particularly heavily clustered on the (110) surface as was observed with PTFE.

From the data of Figs. 23 and 24 it is obvious that a low—surface-energy polymer such as PTFE can form strong adhesive bonds to a metal surface when the metal surface is clean and the contact pressure is very high. Surface analytical tools such as those described previously are very useful in identifying the degree of surface cleanliness.

Strong adhesive bond forces can develop between polymer and metal surfaces even when the metal surface is not atomically clean. The application of compressive surface forces can bring about strong adhesive bonding. The pressing of polymeric materials between metal foils can bring about strong adhesive bonding of polymers to metals with normal oxides present on the metal surfaces.

When high-density polyethylene is pressed against aluminum foil, XPS analysis of the surface reveals transfer of the polyethylene to the aluminum surface. This transfer is demonstrated by the data of Fig. 25. Figure 25 consists of two XPS spectra. The upper one is the spectrum of the aluminum surface before being pressed by the high-density polyethylene. The lower spectrum is the spectrum of that same surface after it has been peeled away from the high-density polyethylene.

The main points that emerge from these studies are that the oxide layer in commercially produced foil is typically ∿20 Å thick and that a tenaciously held hydrocarbon-type layer is present at the surface which is not readily removed by either degreasing treatment or by heating under very high vacuum conditions. X-ray photoemission spectroscopy, therefore, provides a convenient tool for investigating the nature of the peeled surfaces. Figure 25 shows the O_{1s}, C_{1s}, and Al_{2p} levels for the surface of the aluminum foil used for pressing the polyethylene sample of Fig. 11b. (It should be stated at this stage that no trace of Al_{2p} core levels could be detected on this sample.) The most significant feature is that both the aluminum and oxygen core levels are of appreciable intensity in the peeled foil and this can only be interpreted on the basis of failure occurring very close to the aluminum surface. From the relative increase in the intensity of the peak due to the C_{1s} level (taken in conjunction with an escape depth of 10 Å for electrons of kinetic energy ∿968 eV), a reason-

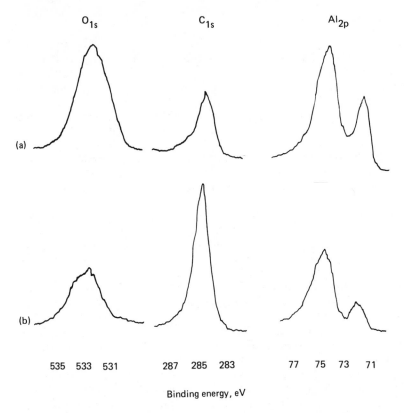

O$_{1s}$ C$_{1s}$ Al$_{2p}$

(a)

(b)

535 533 531   287 285 283   77 75 73 71

Binding energy, eV

Fig. 25. X-ray photoemission spectroscopy for C$_{1s}$, O$_{1s}$, and Al$_{2p}$ levels of aluminum surface. (a) Before; (b) after being peeled from pressed, high-density polyethylene.

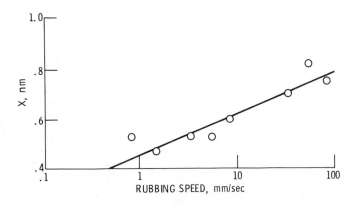

Fig. 26. Average thickness of PTFE transfer film on Ni sliding in vacuum: load, 2N; temperature, 24°C. (From Ref. 11.)

48

able estimate of ~ 10 Å can be made for the thickness of the polymer adhering to the peeled foil.

The adhesion of PTFE to tungsten in the atomically clean state has already been discussed. It has been found that a rubbing action between a metal and a PTFE surface will result in adhesive transfer of polymer (PTFE) to the metal. With polyethylene in contact with aluminum foils, adhesion of polymer to metal was achieved in the presence of surface oxides because of compressive loading, a mechanical activation of the adhesion process. A similar effect can be brought about by tangential motion of PTFE on the metal surface [11].

In Fig. 26 the PTFE film thickness is observed to increase with increased rubbing speed. X-ray photoemission spectroscopy analysis of the PTFE film transferred to the nickel surface revealed that the PTFE adhered to the nickel as a film and that the film was of the same composition as the bulk PTFE polymer. A small amount ($< 1\%$) of nickel fluoride (NiF_2) was present on the surface of the nickel, reflecting a chemical reaction to a limited extent of the PTFE with the nickel surface.

REFERENCES

1. J. B. P. Williamson, in *Interdisciplinary Approach to Friction and Wear*, NASA SP-181, 1968, p. 85.
2. *Environment-Sensitive Mechanical Behavior*, (A. R. C. Westwood and N. S. Stoloff, eds.), Gordon, New York, 1966.
3. E. K. Rideal, *Soc. Chem. Inc. (London)* 28:3 (1968).
4. D. T. Clark and W. J. Feast, *J. Macromol. Sci. Rev. Chem.* 12:191 (1975).
5. D. McLean, in *Grain Boundaries in Metals*, Clarendon, Oxford, 1957.
6. R. Roscoe, *Philos. Mag.* 21:399 (1936).
7. *Mechanical Properties of Intermetallic Compounds*, (J. H. Westbrook, ed.), New York, 1960.
8. V. I. Likhtman, P. A. Rebinder, and G. V. Karpenko, in *Effect of Surface-Active Media on the Deformation of Metals*, Chem. Pub., New York, 1960.
9. E. W. Müller and T. T. Tsong, in *Field Ion Microscopy; Principles and Applications*, Elsevier, New York, 1969.
10. *Characterization of Solid Surfaces*, (P. F. Kane and G. B. Larrabee, eds.), Plenum, New York, 1974.
11. D. R. Wheeler, *Wear*, 1980.

3

Polymer Structure and Adhesive Behavior

Gerald L. Schneberger *GMI Engineering and Management Institute, Flint, Michigan*

I. INTRODUCTION

Practicing engineers and technicians who design and produce adhesive bonds are sometimes overlooked in the current information explosion. There are a number of publications concerning the chemistry of polymers and the physics of adhesion [1,2]. These publications are most readily understood by practicing chemists and those who think easily in thermodynamic terms. There are also many how-to-do-it articles on bonding metals, plastics, and ceramics to themselves and to each other [3]. Such articles, however, often emphasize technique at the expense of understanding.

This chapter will attempt to reveal something of the world within the bondline to those whose knowledge of chemistry and physics has become more practical than theoretical. The emphasis will be on explaining why and how adhesives behave as they do.

II. THE NATURE OF POLYMERS

Polymers are very large molecules containing repetitive sequences of
atoms. Carbon, hydrogen, oxygen, and nitrogen are the elements
most commonly found in adhesive polymers. The physical properties
of an adhesive are very much influenced by the way atoms are arranged within the molecule.

A. Chemical Structure

Useful polymers have molecular weights which range from about ten
thousand to many millions. The atoms of these large molecules are

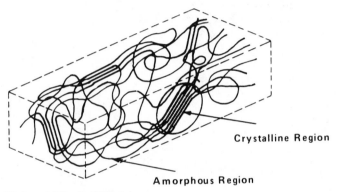

Crystalline Region

Amorphous Region

The "Fringed Micelle" Model

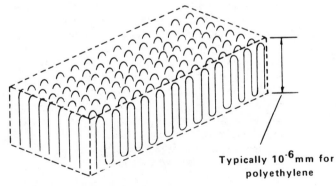

**Typically 10^{-6} mm for
polyethylene**

The "Folded Chain" Model

Fig. 1. Polymer crystallinity. The solid polymer has regions in which
the molecules are regularly arranged. Two possible arrangements are
shown. The concept of crystallinity applies only to linear polymers.

usually arranged in a linear or a crosslinked fashion. Typically, linear molecules are threadlike structures with diameters of about $1-2 \times 10^{-8}$ in. $(2.5 \times 10^{-7}$ mm) and lengths in the range of $6-60 \times 10^{-6}$ in. $(1.5-15 \times 10^{-4}$ mm). These molecules are intertwined and held together by the attractions of positive and negative electrical sites existing on their surfaces.

Linear molecules in the solid state may be crystalline, amorphous, or a combination of these phases. Crystallinity in polymers refers to regions where the chain segments have become ordered, usually parallel to one another (Fig. 1). Amorphous regions are those where the chain segments are intertwined at random. The widely used term, percent crystallinity, refers to the volume percent of a sample which is highly ordered, that is, crystalline. A 50% crystalline polymer has half its volume in the amorphous or nonordered form and half in the highly ordered or crystalline form. A percent crystallinity figure tells us nothing about the number, size, shape, or distribution of the crystalline regions.

Crosslinked molecules result when the linear structures, in effect, branch out and combine in three dimensions. The crosslinked structures are the result of strong chemical bonds between the linear segments of the molecule. The size and shape of crosslinked polymers reflect the fact that separate molecules really do not exist. Rather, the entire crosslinked material, be it adhesive film or molded part, is one giant molecule. In general, the concept of polymer crystallinity does not apply to crosslinked structures. Since the crosslink network grew at random as the resin cures, there is little chance that the molecular segments will achieve any sort of order. The term crosslink density describes the number or frequency of crosslinks per unit volume in a solid polymer and is a concept which is extremely useful in understanding the physical properties of adhesive bonds. Table 1 presents the structural features of some common adhesive polymers.

B. Physical Properties

The linear and/or crosslinked segments of a polymer molecule are in constant vibrational motion. This is true both for amorphous and crystalline linear polymers and for crosslinked polymers. The vibrations of various segments of the molecules are independent of one another. In crosslinked structures, vibrations are greatest between, rather than near, crosslink points.

Because of its molecular motion and imperfect packing, a bulk polymer sample has a certain amount of free volume or empty space at any temperature. The free volume which is often about $2-3\%$ of the total volume, increases with temperature because the molecular vibrations increase. This increase explains the thermal expansion behavior

Table 1   Structural Features of Some Common Adhesive Polymers

Polymer	Structure	Crystallinity
Polyethylene	Linear	High
Polyamide	Linear	Intermediate
Polyacrylate	Linear	Low
Epoxy	Crosslinked	Low
Polyester	Linear or Crosslinked	Variable

of most polymers (Table 2). Linear amorphous and lightly crosslinked polymers tend to expand more than crystalline or highly crosslinked structures because their molecular segments are further apart and, thus, can move more easily. Crystalline and highly crosslinked structures often have less free volume to begin with and, in addition, require more energy for vibration of their greater stiffness.

The actual amount of thermal expansion depends largely on how closely the molecules are packed and how strongly they attract each other. Plasticized polyvinyl chloride (PVC), in which the polymer molecules have been purposely separated by small weakly attracting molecules, has a linear coefficient of expansion which is three times greater than that of unplasticized PVC.

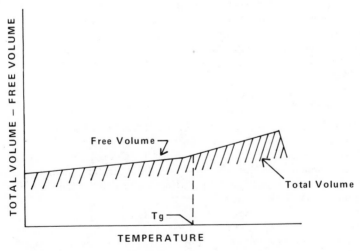

Fig. 2. The effect of temperature on total volume and free volume of a polymer. At the glass transition temperature T_g, the molecules become more flexible.

Table 2 Approximate Coefficients of Thermal Expansion
of Various Polymers

Polymer	Expansion coefficient
High-density polyethylene (above T_g) 95% crystalline	1.2×10^{-4} /°C
Natural rubber (above T_g) not crystalline	2.2×10^{-4} /°C
Epoxy resins (-50—50°C) crosslinked	6.8×10^{-5} /°C
Ebonite (25°C) crosslinked	6.8×10^{-5} /°C

Source: Data from Ref. 4.

Moisture, atmospheric gases, and many solvents are able to pene-trate the free space in a bulk polymer at rates which depend on tem-perature, size, and affinity of the penetrant for the polymer itself. Such diffusion rates usually increase with temperature, especially if the polymer undergoes a sudden increase in free volume.

Thermal motion, as well as free volume, influences the behavior of polymers. At low temperatures polymers exist as solids in which the molecular segments vibrate rather gently and independently. This is the case for both amorphous and crystalline regions. As the tem-perature of a polymer is increased, a point is reached at which the molecule suddenly becomes much more flexible. This increased flexi-bility occurs when the molecular vibrations become great enough to shake the adjacent chain segments apart.

The temperature required to cause this increase in molecular freedom is known as the glass transition temperature T_g, and it is accompanied by a sudden increase in total volume (Fig. 2). In more technical terms the T_g is the temperature of transition from a glassy to a rubbery state. As the temperature of a polymer is raised further and further above its T_g, the effective distance between molecular segments is increased. Flexibility, toughness, and solvent penetra-tion increase, while tensile strength and elastic modulus decrease, above the T_g.

For many linear polymers a temperature range is reached where the solid begins to liquify. This is the beginning of the melting range for commercial polymers. For crosslinked or strongly self-attractive linear polymers, the backbone bond energy may be exceeded before the melting range is reached. In such a case the polymer decomposes without melting. A common example of this behavior is the burning of wood or phenolic resins.

Internal stresses exist within the bulk polymer wherever chain segments are under greater-than-normal tension. The stresses arise during the formation of the solid polymer. They are often located at regions where part of a molecule is unequally attracted by its neighbors. Such unequal attractions can center around microvoids or cracks, which may act as stress risers. Sometimes residual stresses occur when the crosslink density changes rapidly from one location to another within the bulk polymer. These stressed locations are in many ways analogous to crystal defects in a metal. They represent a nonuniform atomic distribution which is often the center, or even the initiator, of catastrophic failure.

III. BOND FORMATION

Before discussing the behavior of adhesive bonds it will be useful to review the principles of bond formation. In virtually every case the adhesive is applied to the surface, allowed to flow out and wet, and then immobilized until solidification has taken place.

Effective wetting, that is, intimate contact of the adhesive with the adherend, requires surface gases, oils, moisture, etc., to be displaced by the larger, and generally more viscous, adhesive. A number of papers have reviewed this process [5,6], and it is necessary here to point out that the adhesive, if it is to wet easily, must be flowable. In practice this is accomplished in one of two ways. The adhesive may be applied as a nonpolymerized liquid which reacts, after wetting the surface, to form a polymer. This approach is typical of epoxies, phenolics, polyesters, and alpha-cyanoacrylates. The other technique is to apply the fully polymerized adhesive in a liquid or rubbery state. This procedure requires the polymer to be applied above its melting range, above its T_g, or in solution. Wetting will then be possible because the extensive molecular motion will permit the adhesive to compete with contaminants for attachment to surface sites. Hot melts and pressure sensitive adhesives exemplify this technique.

Bond strength develops as the adhesive solidifies. Hot melts solidify quickly and achieve their final strength as soon as they have cooled to their use temperature. Pressure-sensitive adhesives do not achieve high strengths because they are used above their T_gs. In this state their molecules are mobile enough to flow under load until the bond fails. Those adhesives which are polymerized within the bondline itself increase in strength as their molecule size increases up to a limit as shown in Fig. 3. This figure indicates that the ultimate tensile strength of a polymer reaches a maximum at about 500−1000 main-chain atoms. Larger molecules do not give higher strengths simply because their various segments act independently of each other.

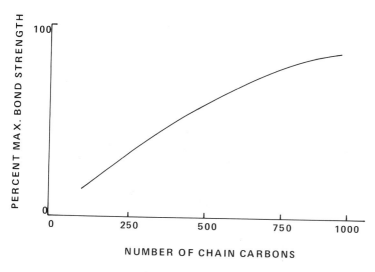

Fig. 3. Polymer chain length. This behavior is typical of many adhesive polymers.

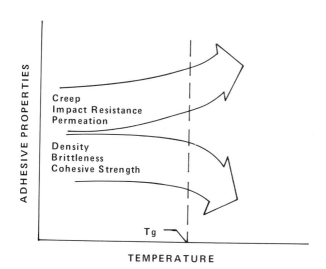

Fig. 4. General trends of some adhesive properties related to temperature or molecular mobility. The sharp change at the T_g reflects the abrupt increase in molecular motion.

IV. BOND BEHAVIOR

It is convenient to consider the bond behavior which results from molecular movement, internal stresses, degradation processes, and bond failure mechanisms.

A. Molecular Movement

Many properties of adhesive bonds are influenced by the mobility of the molecular chain structure (Fig. 4). When chain segments can move easily, they can deform under impact or assume new alignments under mechanical or thermal-expansion stress. This movement spreads the applied energy over a greater number of atoms and thus gives the bond a better chance to resist the stress; brittleness is reduced and flexibility increased.

Increased peel strengths result when a rigid resin is made more flexible. Permeation by moisture and solvents is also favored by molecular mobility because the adhesive polymer can "get out of the way" of the diffusing molecules.

Molecular mobility, however, is a little like exercise—too much can hurt you. When adhesive polymers move too much or too easily, cohesive strength and temperature resistance decrease while creep increases. Figure 4 shows that many bond properties change sharply at the T_g. This is to be expected since the amount of molecular motion in-

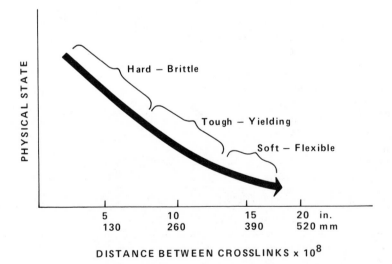

Fig. 5. Effect of crosslink density on the physical state of epoxy resins. (From Ref. 7.)

creases abruptly at this temperature. Practical adhesive formulations must, therefore, control molecular motion at some compromise value. Enough movement must be built in to prevent brittleness but not so much that creep or low cohesion becomes a problem.

Molecular flexibility can be controlled by

1. Crosslink density
2. T_g
3. Fillers
4. Plasticizers
5. Flexibilizers

Lower crosslink density means fewer branch points or motion-restricted atoms within the bulk adhesives and, thus, more chain flexibility. Figure 5 shows the relationship between crosslink density and the physical state of epoxy resins. In practice lower crosslink density may be achieved by using bulkier epoxy monomers or long-chain curing agents. In either case the increased molecular size reduces the number of available crosslink sites per unit volume and, thus, the resin becomes less brittle.

Using a resin with a lower T_g also increases the molecular flexibility, other things being equal (Table 3). Since the use temperature is farther above the T_g of low-T_g resins, the molecules are well into the rubbery region. Thus, they are more flexible and impact resistant than high-T_g-resins would be. Also, low-T_g resins retain their flexibility at low temperatures simply because the resin remains in its rubbery state until the temperature drops to the T_g. Silicone adhesives, which have good low-temperature impact properties, illustrate this behavior. Their T_g's are far below -100°F (-73°C). Natural rubber, on the other hand, may reach its T_g and become brittle around -100°F(-73°C).

The amount and particle size of fillers used in adhesives can also influence the ability of the polymer molecules to move when stressed. Increased levels of filler tend to make it more difficult for the molecules to move toward or away from each other. This may be due to void-space filling or physical entrapment of the polymer. As a result, coefficients of thermal expansion are usually reduced by the addition of fillers. Impact resistance is often improved by the use of filler since the filler phase may become load bearing or at least energy absorbing. Because the exact effect of a filler on a particular property is difficult to predict, changes in formulation should be undertaken only after extensive laboratory investigation.

Plasticizers and flexibilizers exert their softening effect on a polymer by inserting a cohesively weak region of material between the polymer molecules. The term plasticizer generally implies that the softening molecules are not actually bonded to the polymer itself. Such molecules are usually oils, esters, or small polymers which shear easily. They act like shortening in cookies. Flexibilizers are com-

Table 3   Approximate Glass Transition Temperatures and Low-
Temperature Flexibility of Some Adhesive Polymers

Polymer	T_g	Low-temperature flexibility
Polybutylacrylate	10°F (-12°C)	good
Natural rubber	-100°F (-73°C)	better
Silicone rubber	-150°F (-101°C)	best

pounds which react with the polymer. Because of their bulk they are
able to keep the chain segments apart. Thus, crosslinks per unit volume are reduced, and since the flexibilizer is cohesively weak, applied
stress will be able to cause movement within the bulk material.

B. Internal Stresses

Stresses which arise during cure or use can lead to failure of an adhesive because they may act as crack initiators. Cure-shrinkage stresses
and thermal-expansion stresses are two very common problems. Cure-
shrinkage stresses arise while the crosslink network is forming in a
resin or when solvent loss causes a reduction in resin volume. Expansion and contraction stresses occur on heating or cooling whenever
adhesives and adherends have different coefficients of thermal expansion. Table 4 shows that differences in thermal-expansion coefficients
are particularly great for metal-to-plastic and glass-to-fiber reinforced
polyester bonds.

Three common solutions to the internal stress problems are the
use of

1. Stress-resistant polymers
2. Fillers
3. Primers or adhesion promoters

Stress-resistant polymers are those which develop a controlled crosslink density or which are internally soft. An advantage of epoxies
is that the crosslink density can be fairly easily controlled. In addition, epoxies contain very little solvent, so that there is not much
shrinkage during cure. The use of flexibilizers in epoxy resins also
reduces stresses because the stress energy can be dissipated in the
easily deformed flexibilizer region of the resin.

Fillers are commonly used to adjust the thermal-expansion coefficient of an adhesive to a value close to that of the adherends. Fillers generally lower the expansion coefficient of a resin because they
themselves have low thermal-expansion coefficients. Therefore, the

Table 4 Coefficients of Thermal Expansion for Various Adherends

Material	Approximate coefficient of thermal expansion	Relative coefficient of thermal expansion
Steel	$1.4 \times 10^{-5}/°C$	1.0
Glass fibers	$1.4 \times 10^{-5}/°C$	1.0
Aluminum	$2.5 \times 10^{-5}/°C$	1.8
Phenolics, unfilled	$4.5 \times 10^{-5}/°C$	3.1
Polyesters, unfilled	$9.9 \times 10^{-5}/°C$	6.9
Epoxies, unfilled	$1.08 \times 10^{-4}/°C$	7.5

heat they soak up does not result in much increase in the volume of the resin. The filler particles, which may be as small as 1.5×10^{-5} mm (about 30 polymer-chain diameters), may also interfere with the movement of polymer molecules as discussed earlier. When the adherends have different expansion coefficients, the expansion coefficient of the adhesive should be adjusted to an intermediate value. Fillers can reduce the thermal-expansion coefficient of epoxy resins by as much as 75%. Thus, a filled epoxy would be a better adhesive for a steel-to-phenolic bond than the unfilled resin alone (Fig. 6).

The amount of filler actually used must be selected with care because many of the strength properties of an adhesive fall off after a

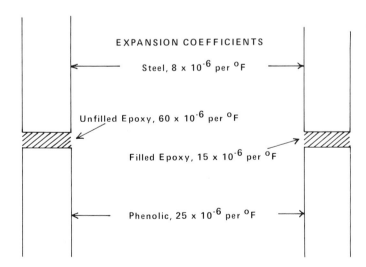

Fig. 6. Approximate thermal-expansion coefficients for a filled and unfilled epoxy bond between steel and a phenolic material.

certain filler level has been reached. This may be because stresses existing between the filler and the resin become critically important at high filler loadings.

Primers, or adhesion promoters, offer another approach to lowering stresses across a bond interface. Such compounds usually form thin films between the substrate and the adhesive (Fig. 7). Such materials may be somewhat flexible and generally have high adhesion for the materials which they separate. Thus, they can absorb expansion or impact stresses without adhesive failure. The use of acrylate automotive sealers over primers and of siloxane coupling agents for glass fibers in polyesters are examples of the application of this technique.

C. Degradation Process

Any chemical change within the adhesive leading to a lower bond strength can be considered a degradation process. Heat and moisture are probably the two most common causes of adhesive degradation. Water generally acts to cleave the polymer molecules—thus making them shorter—while heat may cause the polymer to become either larger or smaller.

The splitting of a polymer by reaction with water is called hydrolysis (Fig. 8). As the size of a polymer molecule is decreased, its cohesive strength will decline (Fig. 3). It has long been known that chemical bonds differ in their resistance to hydrolysis (Table 5).

The breakdown of an adhesive by reaction with moisture is often accelerated by traces of bases or acids. Thus, polyester adhesives do not function well in contact with acid or alkaline environments. Epoxies, by themselves, are resistant to most chemicals. The moisture resistance of an epoxy depends greatly on the curing agents and the

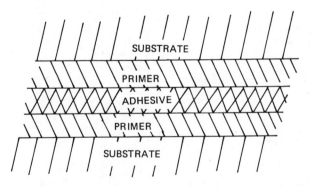

Fig. 7. The use of a primer or adhesion promoter to reduce expansion stresses across an interface.

Table 5   Relative Resistance of Various Chemical Bonds to Hydrolysis

Bond	Formula	Water-resistant	Found in
Carbon-carbon	C — C	yes	Most polymer chains except silicones
Hydrogen-carbon	C —H	yes	All organic polymers, especially elastomers
Ether	C—O—C	yes	Epoxies, phenolics
Ester	$-\overset{\displaystyle O}{\underset{}{C}}-O-\overset{H}{\underset{H}{C}}-$	no	Polyesters, acrylates
Amide	$-\overset{\displaystyle O}{\underset{}{C}}-\overset{}{\underset{H}{N}}-$	no	Polyamides

conditions used. Some crosslinking agents used with epoxies may in-troduce chemical structures which are slowly hydrolyzed. The same is true of polyurethanes. Phenolics, because they contain mostly ether and hydrocarbon structures, are quite moisture resistant.

The chemical products of polymer hydrolysis may be acidic. If so, the problem of substrate corrosion may occur. If hydrolysis pro-ducts are volatile, they may develop internal pressures, especially if the bond is heated. Polymer-decomposition products may also act to plasticize the bond and, thus, affect its physical properties.

Excessive heat may

1. Split polymer molecules
2. Continue crosslinking
3. Drive off plasticizer
4. Lower cohesive strength

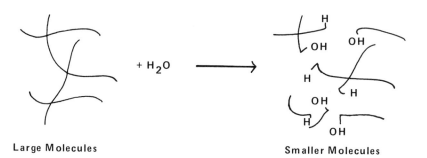

Large Molecules   Smaller Molecules

Fig. 8.   The degradation of polymer chains by reaction with water (hydrolysis).

Polymers become reactive at high temperatures. The same effects are often achieved by somewhat lower temperatures for longer times. Whether they react by chain decomposition or by additional crosslinking depends on the nature of the polymer, the temperature, and the availability of other reactive chemicals.

If the adhesive is a thermoset, crosslinking will probably continue if the temperature is raised above 300−400°F (149−204°C). Once crosslinking has been completed, the chain may react by burning, charring, or cracking. Sometimes fillers or additives in the resin become reactive when heated, and a whole host of possible reactions may then occur. Thermoplastic adhesives melt when heated and then char if oxygen is not present. They may burn or hydrolyze if exposed to atmospheric oxygen or water.

Continued exposure to temperatures a little below the charring or melting point often leads to embrittlement of the bond. This embrittlement may result from slow crosslinking or plasticizer loss. In either case, the bond may fail.

If heating brings a linear adhesive far above its T_g, the molecules will become so flexible that their cohesive strength will decline. When an adhesive is in this flexible condition, atmospheric moisture may be able to penetrate the adhesive film and act as a plasticizer. This penetration may occur even as the original plasticizer is being driven out. After cooling, the entrapped moisture which penetrated the film may be a less effective or a more effective plasticizer than the compound it replaced. Thus, the adhesive may become more brittle or softer than desired.

D. Bond Separation

It is not the author's intention to resolve the existing theories [8,9] of bond separation. Indeed, no one has, as yet, been able to do that. The intention here is to consider what happens to a polymer when an adhesive bond separates by any of several mechanisms.

When a bond separates cohesively within the adhesive, it is because adjacent molecular segments have been physically moved away from each other. This movement of polymer chain segments may result from the breaking of strong bonds *within* the chains or from the rupture of weak bonds *between* the chains. The bond separation may be slow (creep) or rapid (cracking).

Creep can occur when enough force is applied to a mass of linear molecules to cause them to disentangle or overcome their crystalline order. It is a slow process, much like stretching a marshmallow, and is typical of many thermoplastics. Creep behavior is more pronounced at temperatures above the T_g and for that reason, polymers with low T_g's cannot be used where large loads must be carried. Crosslinking reduces creep because the polymer segments are immobilized by the network structure and cannot slide over one another.

Cracking in polymers results when a localized stress becomes great enough to physically separate adjacent molecular segments in an otherwise hard, unyielding, adhesive mass. There is very little creep associated with the failure. The molecules bear the increasing stress until their breaking point is reached. Then they suddenly split apart, much as an ice cube shatters when crushed. Highly crystalline or highly crosslinked polymers are likely to crack rather than creep. This is especially true at low temperatures where very little molecular mobility is available to undergo deformation.

When bond separation is adhesive, that is, when the adhesive separates from the substrate, it is generally because a crack has followed the bond surface or because some chemical has displaced the adhesive from the adherend. Cracks may result from interal or external stresses. Displacing chemicals may come from the adhesive, the adherend, or the external environment.

Plasticizer migration, that is, the concentration at the surface of small molecules from within the adhesive, can result in weak boundary layers at the joint interface which lead to bond failure [10].

Competition between the adhesive and other chemicals for surface sites is a phenomenon which is not well understood in practice but

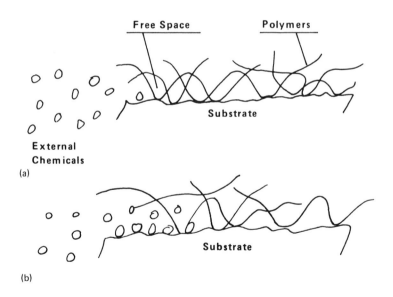

Fig. 9. Competition between an adhesive and other chemicals for surface sites leading to displacement of the adhesive from the surface. (a) Adhesive absorbed at surface sites. (b) Adhesive displaced from surface sites.

which has a firm theoretical basis. The concept of small molecules approaching a bond site and displacing an adhesive is more easily visualized when the real roughness of so-called "smooth" surfaces is considered.

Bonded surfaces really contain many peaks and valleys. It is difficult for an adhesive to displace the last traces of adsorbed surface contaminants from the valleys. As a result, adhesives will bond mostly at the peaks, leaving a certain amount of free space between the adhesive layer and the substrate. Since the adhesive molecules are in constant motion, the available free space is always changing in volume and shape. External moisture and solvent vapors can slowly work their way into such a bond interface. If these chemicals have a greater inherent affinity for the surface than does the adhesive, they may slowly displace the adhesive from the surface (Fig. 9).

Moisture and oxygen are the classic displacing agents since they often react with the adhesive or adherend as discussed earlier. If the penetrating agents cause swelling of the adhesive, the substrate, or both, then the resulting stresses may lead to bond failure.

REFERENCES

1. F. W. Billmeyer, Jr., in *Textbook of Polymer Science*, 2nd ed., Wiley, New York, 1971.
2. R. J. Good, *J. Adhesion* 4:133 (1972).
3. *Handbook of Adhesive Bonding* (C. V. Cagle, ed.), McGraw-Hill, New York, 1973.
4. W. J. Roff and J. R. Scott, in *Handbook of Common Polymers*, Chem. Rubber, Cleveland, 1971.
5. N. J. de Lollis, Adhesion theory and review, in *Handbook of Common Polymers*, Chem. Rubber, Cleveland, 1971.
6. W. A. Zisman, *J. Paint Tech.* 44:41 (1972).
7. H. Lee and K. Neville, in *Handbook of Epoxy Resins*, McGraw-Hill, New York, 1967.
8. J. J. Bikerman, in *The Science of Adhesive Joints*, 2nd ed., Academic, New York, 1968.
9. N. J. de Lollis, *Adhesives Age* (1969).
10. L. H. Sharpe, SAE Paper 700,067, SAE, Detroit, 1970.

4
Designing Adhesive Joints

Gerald L. Schneberger *GMI Engineering and Management Institute,*
Flint, Michigan

I. INTRODUCTION

This chapter is intended to provide general information about adhesives and joint behavior as well as some specific guidelines for designing joints. Designing an adhesive bond involves specifying its geometry and surface finish. It requires some familiarity with the adhesive to be used and the cure conditions to be employed. Bond width
and length, part dimensions, glueline thickness, and the types of
stresses anticipated are all important to the performance of the joint

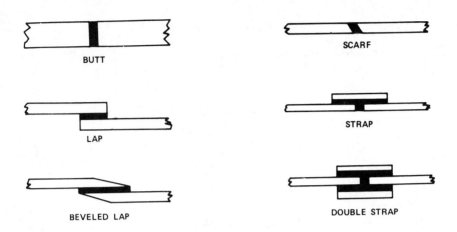

Fig. 1. Common adhesive joint geometries.

and need to be considered when designing bonds. To specify these
things appropriately, the designer needs to know as much as possible
about adhesive behavior and the nature of polymeric materials. Chapter 3 should be consulted for additional insight.

II. TERMS AND CONCEPTS

Several terms and concepts are important to a clear understanding of
this chapter.

A. Bond Geometries

Frequently encountered bond geometries are shown in Fig. 1. Butt
and lap joint geometries are the most common.

B. Load, Stress, and Strain

Load is the actual force brought to bear on a bond. It is the force
(usually mechanical) to which a bonded joint is subjected. It is usually expressed in terms of load per area or per linear dimension, i.e.,
kilograms per square cm. or kilograms per centimeter. Strain is the
reaction of the bonded assembly (or any material) to an applied stress.
It is commonly expressed in terms of centimeters of deformation per
centimeter of length, or as percentage elongation. For example, if
a force of 1200 kg sustained by a bonded area 2 cm wide and 3 cm long
causes a bond elongation of 0.025 cm, the situation can be described

as follows, The load is 1200 kg; the stress is 200 kg/cm^2 (1200 ÷ 6 cm^2); the strain is 0.0083 cm/cm (0.025 cm ÷ 3 cm) or 0.83% elongation.

C. Yield Point or Yield Strength

Yield point, or yield strength, refers to the extent to which a sample of material can be stretched without permanent deformation. If the stretching stress has been less than the yield strength, the part will return to its original dimension when the stress is removed. It has undergone elastic deformation and recovery. At the yield point, the part has moved beyond its elastic limit; when the stress is released, the part will not return to its original dimension but will retain an increase in length equal to the amount of deformation above the elastic limit. The stress required to exceed the elastic limit for any material is known as its yield strength, or yield point.

D. Adherend

The parts being bonded are the adherends.

E. Creep

The slow elongation of a bond due to continuous stretching of the adhesive or the adherend under load is known as creep.

III. JOINT STRESSES

Typically, bonds are simultaneously subjected to a combination of mechanical, thermal, and environmental (chemical) stresses.

A. Mechanical Stresses

Mechanical stresses result from tensile, compressive, shear, cleavage, and peel forces. Their typical distributions within bonds are illustrated in Figs. 2—6.

1. Tensile and compressive stresses

Tensile and compressive stresses (Fig. 2) may be considered a pulling apart or pushing together of the bonded assembly perpendicular to the glueline. Note that each square unit of the bonded surface contributes equally to the total joint strength. As a result, tensile and compressive strengths will be directly proportional to the bond area. If a 1-cm^2 bond will support 300 kg, then a 3-cm^2 bond will support 900 kg.

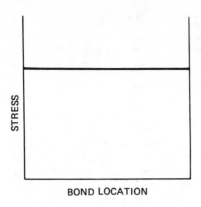

Fig. 2. Stress distribution in bonds undergoing tension and compression.

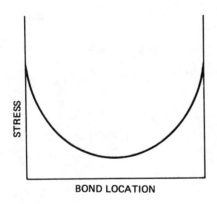

Fig. 3. Idealized tensile shear stress distribution in lap joints.

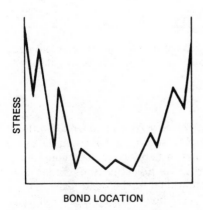

Fig. 4. Probable tensile shear stress distribution in lap joints.

Unfortunately, butt joints loaded purely in tension or compression are not common.

2. Lap shear stresses

Lap shear stresses are very commonly found in practical adhesive bonds. Tensile shear stresses are those which tend to slide the adherends away from each other along the bond axis. Figure 3 indicates that the resulting distribution of stress is definitely not uniform over the bonded area. As a result, the bond strength does not vary directly with the bond area. The actual localized stress experienced at the point of failure initiation is generally much greater than would be calculated using the load and the entire bond area. A 6-cm² bond which fails under a load of 72 kg will have some of its area subjected to lower, and some, to higher stress than the nominal 72 ÷ 6 = 12 kg/cm². Thus, an adhesive for a joint designed to withstand nominal lap shear loads of 12 kg/cm² must be formulated to withstand stresses which may be much higher. How much higher they may be depends on relationships such as those in Fig. 3. Because it assumes homogeneous surface conditions, the curve shown in this figure is highly idealized. If stresses were measured on extremely small adjacent areas of the bond, the curve would be somewhat like the one in Fig. 4. The sharp peaks which occur in this figure result from microdefects in the surface which function as stress-concentration points (stress risers). It is at these small areas of high stress that failure probably begins in most adhesive systems.

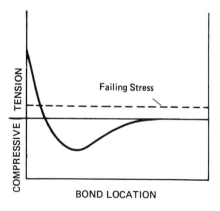

Fig. 5. Stress distribution in bonds undergoing cleavage.

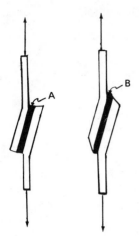

Fig. 6. (A) Cleavage stresses may be introduced in lap joints stressed in tensile shear. (B) The cleavage stress may be reduced by making the adherend ends thinner to allow them to deform.

3. Cleavage stresses

Cleavage stresses tend to perpendicularly separate bonded rigid parts
from one another at an end or edge of the bond. The situation is
much like splitting wood with a triangular wedge. The failure proceeds progressively from the end of the bond undergoing cleavage.
Thus resistance to failure is concentrated primarily at the edge or end
of the bond (Fig. 5).

Figure 6(A) indicates that cleavage stresses may be introduced at
the bond ends when a lap joint undergoes tensile shear loading. Such
loading acts to force the adherends into the same plane. Cleavage
forces result when the adherends deform to become coplanar. Such
forces can lead to failure of the bond at stress levels lower than those
expected with stiff, nondeformable adherends. Joint designs which
reduce the cleavage stresses by permitting both adherends to deform
with the same curvature should be used (Fig. 6B). Beveled lap joints
will often achieve this effect.

4. Peel stresses

Peel stresses are similar to cleavage stresses in that the stress is concentrated at the failing edge of the bond. They are more extreme than
cleavage stresses because one or both members of the bonded assembly
are flexible. The result is that bond failure will occur progressively

Table 1 Approximate Coefficients of Thermal Expansion for Various Materials

Material	Expansion coefficient $[(cm/cm/°C) \times 10^{-5}]$
Steel	1.2
Aluminum	2.5
Copper	1.7
Brass	1.9
Stainless steel	1.7
Glass	1.2
Polyethylene	10-25
Acrylonitrile butadiene styrene (ABS)	5-10
Polystyrene	6-8
Nylon, unfilled	8-9
Polyvinyl chloride, rigid	5-20
Polymethylmethacrylate	5-9
Phenolics	2-4
Epoxies	1-4
Urethanes	10-20

at relatively low loads from the end of the bond undergoing the cleavage stress. A rigid epoxy adhesive with a lap shear strength of 3000 psi (21,000 kPa) may have only 15 lb/in (42 kPa/cm) of peel strength.

Figures 3–6 illustrate that mechanical stresses applied to adhesive bonds are not distributed in a homogeneous fashion by all bond geometries. The exact shape of the stress distribution curves for shear, cleavage, and peel stresses will depend primarily on adherend stiffness and thickness, adhesive flexibility, and the yield points of both the adhesive and the adherend.

B. Thermal Stresses

Thermal stresses are primarily due to expansion and contraction forces which result from changing service temperatures. Thermal stresses can be severe, particularly when there is a mismatch of expansion coefficients between the adherend and the adhesive. Table 1 gives expansion coefficients for a number of routinely bonded materials. Table 2 presents similar data for a number of adhesives. A fivefold or greater difference in coefficients between an adherend and an epoxy adhesive is possible. Thus, when the adherend wants to move 0.01 mm, the adjacent adhesive will want to move 0.05 mm. The resulting

Table 2   Approximate Coefficients of Thermal Expansion for
Various Adhesives

Adhesive	Expansion coefficient $[(cm/cm/°C) \times 10^{-5}]$
Epoxy, unfilled	5-6
Epoxy, filled	2.5-4
Urethanes	10-20
Phenolics	2-4
Silicones	20-25
Anaerobics	12-14
Acrylics	13

stress may be enough to tear the adhesive or the adherend or cause
their separation. Sometimes the rapidity of a temperature change can
be at least as detrimental to bond strength as the extent of the
change.

C. Chemical Stresses

Chemical stresses result from the combined effects of all the chemicals
which influence the bond. Nearly all bonds are exposed to atmospheric oxygen and moisture, but many other chemicals may also be encountered. Solvents, perspiration, cleaning materials, oil, grease,
and oxides of sulfur, carbon, and nitrogen are typically encountered.
Adhesives may also be exposed to chemicals which migrate to the
bondline from within one or both of the adherends. Plasticizers are
typical examples.

IV. PERFORMANCE CRITERIA

An extremely important step in the process of designing an adhesive
joint is simply to state the performance expectations for the bond.
Communication with other departments within the company and with
vendors and customers will be greatly improved if the performance
requirements of the bond are specified as precisely as possible. Properties such as lap shear strength, impact strength, strength retention with aging at one or more temperatures, thickness tolerances,
humidity resistance, salt-spray resistance, solvent resistance, cure
time required to give handling strength, and the acceptable limits for
any of these tests are typical of what should be specified. Consult
Chap. 25 for additional information on this topic.

In addition to the bond criteria listed above, it will usually be appropriate to specify viscosity, percent solids, color, flash tempera-ture, and open time for the adhesive material itself. In some cases it is also necessary to specify the nature of the cleaning and surface preparation which is to be used. Chemical concentrations, tempera-tures, pressures, and times are frequently the most important param-eters of the surface-preparation process. In addition, it may be impor-tant to indicate the limits of surface roughness which can be tolerated.

Ideally, the test methods used to determine the foregoing adhe-sive and bond properties should also be described or identified in some well-understood manner, such as by American Society for Testing and Materials (ASTM) number.

When specifiying design criteria, it is important to avoid calling for unrealistic properties. It is not uncommon, for example, for design criteria to require a bond stronger than either of the materials joined. Thus, failure of one of the adherends is demanded rather than se-paration at the bond. If performance such as this is unrealistic, you will generally pay a cost penalty to achieve it. If one is bonding fiberglass sound-deadener pads to the inside of bulldozer cabs, it makes no sense to expect the bond to be stronger than the quarter-inch-thick steel in the roof panel. The designer should constantly watch for incongruities such as this.

V. DESIGN CONSIDERATIONS

A. Safety Factors

The appropriate safety factor for a given bond design varies primarily with the end use of the product. The safety factor refers to the ulti-mate strength of the bond compared to the maximum load expected in service. Some industries may traditionally use a safety factor of 150%, while others may demand 400—500%, depending on the application.

B. Geometry

Joint geometry refers primarily to the width of the bond, the length of the overlap, and the thickness of both the adhesive and the adherends. Figure 3 indicates that stresses are concentrated at the ends of a lap bond with the result that bond strength does not increase linearly with overlap. Figure 7 indicates that the affect of bond overlap on strength rises rapidly and then levels off.

The strength of a lap bond stressed in shear is directly propor-tional to the width of the overlap. Thus, doubling the bond width will double the strength. Given a choice, a designer would normally speci-fy a 2-in. (5.08-cm) width and a 1-in. (2.54-cm) overlap, rather than a 1-in. (2.54-cm) width and a 2-in. (5.08-cm) overlap, since the form-er geometry will lead to the stronger bond.

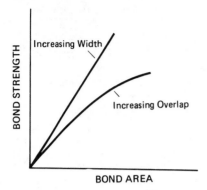

Fig. 7. The effect of overlap and width on bond strength.

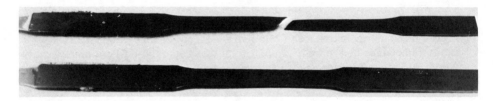

Fig. 8. Adhesive bond strength can exceed the ultimate strength of
the adherend as shown in the failed sample on the right.

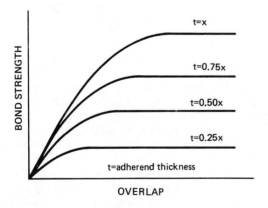

Fig. 9. The relationship between adherend thickness, overlap, and
bond strength for many metals. (From Ref. 1.)

Bond length and width are not the only geometric variables which influence bond strength. Two extremely important considerations are the thickness of the adherends and their yield strength. The load at which an adherend begins to yield depends on the stiffness of the material and its thickness. It is easily possible for the adhesive bond strength of thin members to exceed the yield and ultimate strength of the material (Fig. 8).

Figure 9 indicates a typical relationship between thickness, overlap, and bond strength for metals. The curves should be considered representative because the actual values will depend on the type of adhesive used, the method of surface preparation, the test procedure, and other variables. Figure 9, however, suggests that if the bond must withstand a load of 2000 lb (908 kg), metal thickness of something over 0.050 in. (0.13 cm) will be required. De Bruyne and Houwink [2] have analyzed the relationship between bond thickness, overlap, and stress. They have referred to the ratio of the square root of the thickness to the length of the overlap as the "joint factor". Figure 10 shows joint factor plotted against failing stress. Interestingly, this relationship holds for a number of different metals. Figure 10 indicates that as the metal thickness is increased or the overlap decreased, the stress (on a per area basis) approaches a limiting value which is characteristic of the adhesive. The relationship is essentially

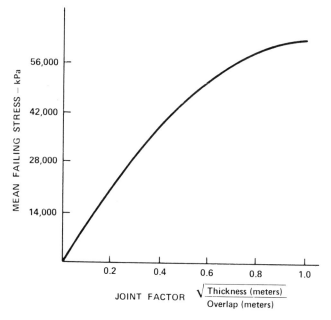

Fig. 10. Joint factor versus failing stress.

linear for joint factors up to 0.3. This means that the failing load is
independent of the overlap and depends (in a linear fashion) on the
square root of the thickness for bonds having a joint factor of 0.3 or
less.

C. Surface Smoothness

It is not unusual for surface smoothness to have an affect on bond
strength. The exact relationship between strength, durability, and
surface smoothness is difficult to predict and may vary from one adhesive to another. Rough surfaces may provide a mechanical-inter-
locking opportunity for the adhesive, but they may also trap small
volumes of air in their crevices and cause incomplete wetting by the
adhesive. It is possible that brittle adhesives are more compatible
with smooth surfaces than flexible adhesives are because the absence
of rough spots means that there are fewer stress risers to act as crack
initiators. More flexible adhesives are able to deform under stress
and resist cracking or tearing, so that roughness may be a less important factor in their use. Designers should ask for test results based
on a number of different roughnesses, including sandblasting if possible.

D. Cleaning Methods

The method chosen to clean the surface can have an affect on bond
strength. This is particularly true if rinsing is involved. Traditionally, metals have been cleaned prior to bonding with either an organic
solvent or an aqueous, alkaline cleaner. At the present time, there is
some movement away from organic cleaning formulations to low-temperature, aqueous systems. The result is that rinsing has become
extremely important in the removal of the last residues of alkaline
materials. It is important to specify the exact cleaning procedure to
be used. This is best done after a certain amount of experimentation.
A number of excellent references are available to guide the designer
in choosing cleaning processes [3,4].

E. Cure Considerations

The rate and extent of cure of an adhesive material can greatly influence both bond strength and durability. Generally speaking, designers should specify heat-cured adhesives if they wish to insure the
highest possible bond strengths because surface wetting is more thorough at elevated temperatures. Ideally, the rate of cure of an adhesive should be uniform with respect to time, and there should be no
gas production due to decomposition of the formulation or volatilization
of materials on the surface.

Most adhesives are formulated to cure within a rather broad "envelope" of time and temperature. The times and temperatures which will bring about the cure of a particular product are readily available from suppliers. Ultimate bond strength, however, may vary depending on which cure conditions are used. Thus, the designer should specify both the time and temperature, and if possible, the rate of temperature increase to which the parts will be subjected during the cure. Minimum times for holding the parts at a particular surface temperature are often specified.

Overcuring and undercuring are both undesirable. Overcuring leads to excessively brittle bonds which may be prone to crack, and undercuring results in an adhesive film which is cohesively weak and which may creep. It may be a mistake to assume that oven temperatures are uniform, especially if large installations are involved. They should regularly be checked for uniformity.

VI. SPECIFIC DESIGN GUIDELINES

A. General Considerations

In a recent publication [5], the American Iron and Steel Institute has identified some useful recommendations for joint design. They are as follows:

1. Always use the largest possible area, in keeping with reasonable cost.
2. Align the bondline so that it is stressed in the strongest direction.
3. Maximize shear and minimize peel and cleavage loads.
4. Design subassemblies so that only one bonding operation is required for the total assembly.
5. Remember to anticipate stresses (particularly peel or cleavage) which may be encountered during assembly or handling operations as well as those expected during service.
6. Avoid parts with complex curvatures.

B. Practical Joint Geometries

The joint cross sections shown in Fig. 11 have proved to be practical in a number of design applications. More complex geometries are rarely justified because of the preparation cost involved.

C. Formed, Fastened, and Welded Adhesive Bonds

Sometimes adhesive joints are combined with forming, fastening, or welding operations to result in a truly hybrid assembly design. The

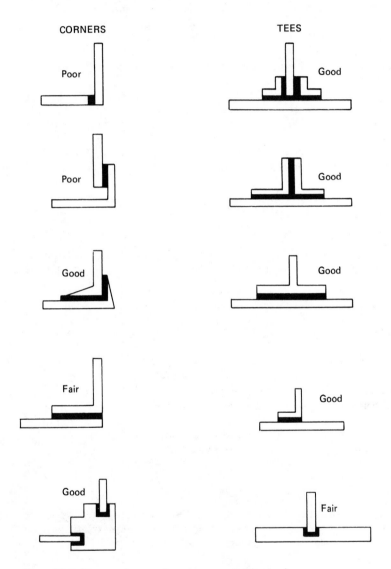

Fig. 11. Commonly used corner and tee designs.

Fig. 12. Adhesive bonding may be combined with formed, fastened, and spot-welded joints.

examples shown in Fig. 12 illustrate these principles. In general, the idea is to depend upon the adhesive for long-term fatigue resistance while using the forming, fastening, or welding operation to fixture the joint during cure and to reduce cleavage and peel problems in service.

REFERENCES

1. M. P. Petronio, in *Handbook of Adhesives* (I. Skeist, ed.), Reinhold, New York, 1962.
2. N. A. de Bruyne and H. Houwink, in *Adhesion and Cohesion*, Elsevier, London, 1951, p. 98.
3. R. C. Snogren, in *Handbook of Surface Preparation,* Palmerton, New York, 1974.
4. Metal surface treatments—cleaning, pickling and related processes, in *Encyclopedia of Chemical Technology*, Wiley, New York, 1967.
5. *Production Design Guide for Adhesive Bonding of Sheet Metal,* Am. Iron Steel Inst., Washington, D.C., n. d., p. 35.

5

Selecting An Adhesive: Why and How

Paul O. Nielsen *Paul O. Nielsen, Inc., Warren, Pennsylvania*

I. INTRODUCTION

The selection of a specific adhesive for a specific application can be likened to the process of selecting a marriage partner. Successful completion of questionnaires, suitability tests, interviews, trial evaluations, and tentative commitments theoretically lead to a final and lasting union. In matching an adhesive to prospective substrates, influ-

enced by design and performance requirements, the same procedures
of consideration in mating apparent compatibilities and evaluating prototypes or working models is, of course, the correct engineering
method to be employed. Unfortunately, as in some marriages, unfore-
seen stresses and unanticipated circumstances can nullify the best of
our intentions and expectations.

One of the major problems in selecting an adhesive for any given
application is the presence of so many attractive candidates. The
number of adhesive variations available to the engineer today is fan-
tastic. Adhesives offer such a broad range of mechanical and handling properties that it seems that there must be at least one which will
be best suited for any potential bonding application. On the other
hand, beneath the seemingly marvelous attributes of each adhesive
type, there are inherent, if subtle, dehabilitating characteristics of
which we must be aware.

Each category of adhesive has its individual appeal and appears
to possess some unique property—such as highest mechanical performance, greatest ease of use, lowest unit cost, greatest longevity, etc.
While any one of these claims may be valid for a tightly defined, totally controlled laboratory condition, in the real world of adhesive bonding, it is not unusual for the inadequacies of an adhesive to reduce its
promise to the point of nonusefulness. These inadequacies may not
necessarily be technical in nature. For example, the solution or sol-
vent-type adhesive which you have selected, after thorough and diligent examination, as the best and most economical for your application
may have to be rejected because of new local legislation barring volatile solvent-containing materials. Similarly, the cost of hoods, vents,
collecting towers, etc., necessary to comply with the law may suddenly
price the selected adhesive out of consideration.

One prime fact must be borne in mind when submitting oneself
to the rigor of selecting "the" adhesive for "the job": Maintaining
rigid criteria of nonnegotiable design and mechanical performance
greatly restricts the choice of adhesive and often imposes severe processing and/or cost penalties.

The practical engineer will recognize that logical compromises
between design, performance, handling characteristics, cost, etc., are
absolutely necessary to achieve profitable adhesive success.

II. ADHESIVES VERSUS OTHER FASTENING SYSTEMS

The first step in the adhesive selection process, which is often overlooked, should be the fundamental affirmation of the suitability of adhesive bonding, as opposed to other fastening methods, for your particular application. Threaded mechanical fastening, riveting, stapling,
welding, brazing, and soldering are just a few alternative joining
methods. These techniques are normally associated with metallic com-

ponents and structures. Heat staking, ultrasonic welding, and solvent bonding are joining methods used for plastic materials, but they cannot really be considered as adhesive systems since the classic definition of adhesive bonding is one in which two substrates are attached by a distinctly separate material entity.

Also, there are combinations of fastening systems, such as weldbonding, in which a continuous bead of adhesive is periodically reinforced or secured in the "fixturing" sense by a spot weld. The instantly stable spot weld holds the parts together while the adhesive cures. The resulting structure benefits from the brute strength of the weld and the fatigue-failure resistance of the adhesive. Another well-established combination is anaerobic adhesives and screws or bolts to reduce vibrational loosening.

All fastening and joining systems, including adhesives, fall into one of three general categories which describe the primary and typical ways in which they are most successfully used:

Periodic. The attachment of two members by occasional placement of through fasteners or site unions typified by rivets, screws, nuts, bolts, spot welds, etc. This is the most widely used joining technique for structures which require high mechanical strength and a minimum of sealing or other nonstrength performance properties.

Linear. A continuous or occasional edge bead attachment which is best represented by welding.

Area. An attachment achieved by full-face contact and union between the two mating surfaces. Soldering, brazing, and adhesive bonding are examples.

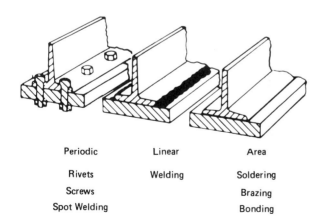

Periodic	Linear	Area
Rivets	Welding	Soldering
Screws		Brazing
Spot Welding		Bonding

Fig. 1. Periodic, linear, and area attachment systems.

Although adhesive bonding can be successfully employed in periodic or linear attachment applications, the main benefits and advantages of using adhesives are realized when they are used in "area" attachment designs. The three basic types of attachment configurations are shown in Fig. 1.

The diligent engineer will have gone through an extensive check list or trade study to determine the suitability and preference of adhesive bonding versus other important, traditional fastening systems prior to production commitment.

Each manufacturer must determine the type of trade study needed when selecting a joining method and materials for a given assembly. Every new product, combination of materials to be joined, rate of production, etc., will cause priorities to differ. A typical example of a trade-off analysis comparing riveting (periodic type attachment), welding (linear attachment), brazing, and adhesive bonding (both area attachments) is shown in Table 1. Joining-method features are weighed against manufacturing and performance considerations. Obvious points, such as the impracticality of brazing ceramics or of using organic adhesives for prolonged exposure at very high temperatures, would alter the study or preclude the consideration of certain materials and techniques. A numerical rating given to the elements of such an evaluation would indicate whether or not adhesive bonding could be used.

III. ADVANTAGES OF ADHESIVE BONDING

Many benefits accrue from the use of adhesive bonding. Some advantages, such as being able to attach components not otherwise joinable, are obvious, and their value is directly measurable. Other factors, such as eliminating corrosion between dissimilar materials, may or may not be of prime consideration in selecting a joining technique, but they definitely add a dimension to the worth of the assembly. A review of some of the more apparent beneficial aspects of adhesive use is provided below.

A. Fabrication of Smooth Exterior Surfaces

The fact that an adhesive is contained between joined surfaces eliminates the need for surface finishing, such as that required to grind down rivet heads, etc. Smooth exterior surfaces imply less drag and more efficient gaseous and fluid flow patterns over or around aerodynamic faces. This principle would apply to land, water, and aerospace vehicles. External aesthetics are achieved which are of considerable sales importance to many product lines, such as appliances, furniture, musical instruments, etc. The absence of protruding bolt or rivet heads also signifies that there are no concentrated stress

Table 1 How Joining Methods Compare[a]

	Riveting	Welding	Brazing	Adhesive bonding
Preliminary machining	P	E	P	E
With thin metals	P	P	F	E
Limits on metal combinations	F	P	P	E
Surface preparation	E	G	F	P
Tooling	E	F	F	F
Need for access to joint	P	P	E	E
Heat requirements	E	P	P	F−G
Stress distribution	P	F−G	E	E
Sealing function	P	F	E	G
Rate of strength development	E	E	E	P
Distortion of assembly	F	P	F	E
Final machining	G−E	F	E	E
Final heat treatment	E	F	F	E
Solvent resistance	E	E	E	F
Effect of temperature	E	E	E	P
Ease of repair	G	P	P	F
Level of skill required	E	G	E	E

[a]E, excellent; G, good; F, fair; P, poor.

sites in the attachment area. Forces are dispersed over the entire assembly face.

B. Protection Against Shock and Vibrational Failure

The elastomeric nature of most adhesives results in an energy-absorbing response to impact and fatigue forces. The ability of adhesively bonded structures to withstand significantly higher sonic-fatigue levels than mechanically fastened counterpart hardware in both subsonic and supersonic aircraft applications is a dramatic illustration of this property. Similarly, the transmission of sound itself is drastically reduced by passing through an adhesive interlayer. Since most polymeric adhesive systems are poor thermal conductors, they act to soak up stresses induced by thermal cycling, particularly in metallic structures. As mentioned earlier, the use of anaerobic adhesives as an interference fit material between threaded fasteners and the structures into which they are inserted or onto which they are attached reduces or prevents vibrational loosening. The automotive and farm implement industries use adhesives to great advantage in this type of application.

C. Use of Light-Gauge Materials

Stresses applied to a bonded structure are spread over a relatively
large area (total facial contact area between mating substrates) compared to stresses directed to and through local sites, such as the bolts
of a linear assembly. Thus, the total imposed force can be accepted
by components having lower bulk strength per unit area because their
cumulative load-bearing capability is probably greater than that pos-
sessed by a few through fasteners. Coincidentally, facing materials
have to be overdesigned in strength (made thicker and heavier) to
compensate for the reduction in strength caused by the need to drill
holes to allow for the passage of through fasteners.

The structurally efficient justification for using lightweight (thinner) surface materials in an adhesively bonded assembly is furthered
by the possibility of broadening the function of the facing. A thin
facing member may integrally incorporate a decorative exterior. Likewise, additional foil-thin top layers of specific-purpose material may
be adhesively incorporated into a composite, achieving certain required properties, such as surface reflectivity, conductivity, insulation, corrosion resistance, abrasion resistance, etc.

D. Joining of Dissimilar Materials

Galvanic corrosion, which occurs when metals of differing electromo-
tive force potential are in intimate contact with each other, is eliminated by an adhesive interlayer. Inherent properties of unmodified
organic adhesives include good electrical insulation and poor thermal
conductivity. Stresses caused by rapid thermal input or cycling are
reduced by slow heat transfer through the adhesive. Dimensional distortion of the assembly, caused by different coefficients of thermal
expansion, is reduced by the heat-sink action of the adhesive. This
ability, of course, is limited by the adhesive strength achieved with
the particular adhesive/substrate combination and by the cohesive
strength of the adhesive. It follows, then, that if extreme thermal
cycling is anticipated in a bonded assembly involving substrates of
widely differing expansion rates, an elastomeric adhesive will be preferred. The trade-off is between a "giving" adhesive with limited strength retention as temperatures rise, and a thermally stable
system depending on cohesive strength alone to resist the stresses as
the temperature rises.

Since bonded assemblies derive their prime strength from the
chemical interlocking of the adhesive with a substrate, the relative
modulus (stiffness) and chemical nature (metal, plastic, glass, ceramic, etc.) of the attached surfaces do not have to be as closely matched
as those of most mechanically attached components.

E. Simplified Joining of Contoured or Delicate Parts

Flat assemblies normally do not present mating problems even for mechanical fastening. Curved or multicontoured surfaces present an entirely different situation. Surface loading of bonded assemblies transfers stress through a continuous, gap-filling, substrate membrane,
which is connected to next-in-line components or faces. This fact allows the relaxation of fit tolerances. The reduced fit, or mating-gap
tolerance requirement can often result in lower component-preparation
costs and can simplify jig or fixturing setups.

Small sensing devices, such as strain gauges, are easily adhesively bonded to areas being monitored. Under these conditions no
induced surface stresses or flow-path disruptions, such as might happen if the devices were mechanically attached, occur. The elasticity,
or modulus, of the adhesive can be tailored to fit the particular test
requirements.

F. Reduced Cost of Manufacture and Assembly

Frequently, a mechanically fastened assembly requires secondary
operations, such as sealing or filling gaps between fastener sites.
Typical examples are storage bins, silos, water tanks, etc. Primary
mechanical strength or holding-power considerations do not totally fill
the functional requirements of the assembly. Adhesive bonding of
such assemblies satisfies the requirements for both mechanical
strength and end use. Similarly, the use of adhesives may eliminate the
need for the auxiliary attachment of thermal or acoustic insulation.
Savings of fabrication time and material cost by using adhesives in a
dual role are readily apparent. As mentioned earlier in this section,
eliminating the need to machine or countersink protruding rivet
heads or weld seams, etc., saves finishing time and costs.

IV. INTERDEPENDENCY OF USER AND SUPPLIER

For purposes of this text, the assumption is made that the user pro-
cures adhesives from other manufacturers. The technical expertise,
equipment, quality-control techniques, etc., required for the manufacture of adhesives of consistent quality are extremely specialized.
The critical factor in establishing a good support and working relationship with an adhesives manufacturer is open, concise, and frequent communications. By and large, adhesives are not purchased as
commodity items. Although you may procure adhesives to meet a company or federal specification, there are sufficient subtle differences
in handling or other nonstrength properties that distinguish adhesives
from sheet metal or bar stock, for example. The very fact that ad-
sives change—that is, flow, shrink, harden, etc.—during processing,

identifies them as specialty materials. As such, adhesive response characteristics and behavior during application and cure are as important as their final cured-state properties. The following paragraphs describe areas in which clear communication with the adhesive supplier is particularly important.

A. Acceptance Criteria for Adhesive Performance

Beyond the apparent strength requirement of an adhesive, there are many factors which must be defined. Processibility, durability, appearance, and environmental response are examples. The user must know as exactly as possible what behavior is desired from the adhesive and must share these expectations with the supplier. Some of the more important points which affect joint performance are discussed below.

B. What is the End Product?

The dialogue between user and supplier best starts with a statement of what the end product is and what its functional requirements are. Manufacturers often cloak their product-design projects in undue secrecy, which handcuffs the prospective adhesive supplier. Absolutely mandatory proprietary considerations aside, the start of the adhesive selection process should be as complete a description of the end product as is possible. Given this introduction, the adhesive supplier can review his experience with similar applications and more quickly make well-directed recommendations. Needless to say, an aura of confidentiality regarding the specific applications must be maintained. The point is that the better the definition of both end-product and expected part function, the more accurate the initial material identification will be.

C. Correctness of Design

In an ideal situation, the part designer is aware of the performance capability of adhesives in general. Adhesive technology allows the designer to consider options regarding the use of adhesives. The optimum adhesive joint is designed to handle loads in a mode which is consistent with the major strength feature of any given adhesive. Practicalities—such as geometry of mating members, cost of machining components, placement of attachments on the inside or outside, etc.—may influence the designer to use an adhesive in an unfairly demanding fashion. For example, an epoxy adhesive may be specified because of known high lap shear strength. In practice, however, attached members may be of such thinness or low modulus that the applied stresses cause deformation which translates the load into a peel format which, in turn, causes premature bond failure due to the inherently low peel strength of most epoxies. The knowledgeable designer can offer design options, such as beveled edges on skins, recess cut-fitting of mating members, use of load transfer stringers, etc., to compensate for such situations. Design considerations are more fully discussed in Chap. 4.

D. Strength Requirements

Requirement, rather than desire, is the law for the strength definition of an adhesive. Anything above the basic-design adhesive strength requirement will usually cost something in price, ease of application, or processing. One must certainly allow for surge loads, etc., but reasonable and accurate prediction of needed strength levels is necessary to allow the supplier to suggest possible materials. If a design calls for 800 psi (5600 kPa) of tensile shear strength, it would be irresponsible to insist upon 2500 psi (175,000 kPa) of adhesive performance. This action might preclude the use of a rapidly setting one-component cyanoacrylate and force the use of a higher strength epoxy involving two components, mixing and dispensing equipment, overnight cure times, etc. After exercising accurate engineering judgment, a design staff should not arbitrarily impose a safety factor of 100–200% on adhesive strengths without understanding the implications.

E. Environmental Survival Requirements

The adhesive user should be able to predict the environmental-exposure requirements of an assembly and relay that information to the supplier. Short-term exposure in a room at constant temperature and 25% relative humidity is a totally different situation from exposure to cycling temperature, fluctuating relative humidity, and ultraviolet (uv) radiation. The same mechanical strength may be required in both cases, and a designer may unwittingly select the same adhesive for both applications based only on that criterion (with one surviving and one failing). Disclosure of the entire expected environment-exposure profile of the bonded structure will permit the supplier to make certain that the adhesive formulation will provide the performance needed. For example, a uv absorber, or pigment, added to an adhesive could extend its exterior-service life 100% or more.

F. Service Conditions

One of the significant factors contributing to unanticipated early adhesive failure is the rate of load application. Various test procedures specify differing rates and increments of imposed loading. Different test procedures can result in conflicting and misunderstood strength data. Shock loading versus gradual loading, percent of ultimate load, frequency of load application, etc., are among the dynamic variables encountered. Other conditions besides strength which affect adhesive selection are the need for electrical conductivity, radiation resistance, vapor impermeability, etc. These and other service-performance conditions will certainly influence the user/supplier team in making the best adhesive selection.

G. Reliability

The reliability required of an adhesive bond is a somewhat nebulous property. Of course, a bonded assembly is expected to maintain structural integrity—but for how may use cycles, for how many exposures to intermittent proof loading? Is failure expected to be gradual or catastrophic? Permanence and reliability are not to be interchanged or equated. In adhesive terms, as with any organic or nonmetallic substance, there is no status quo for eternity. The measure of adhesive reliability can best be looked upon as a description of percent strength retention after defined loading, time, and/or conditional exposures.

H. Materials Being Bonded

Metals, plastics, rubber, glass, wood, ceramic, concrete, etc., can all be bonded to themselves or to each other. Material type, condition, porosity, finish, acidity, alkalinity, etc., influence adhesive selection. Certain adhesives have inherent affinities for certain surfaces and conditions. For example, even though epoxies may have higher lap shear strengths in metal-bonding applications, a lower strength structural acrylic might be used because it might be more reliably bonded to the oily surface condition of the metal. In order to use the epoxy to its higher strength advantage, the surfaces must have a higher level of cleanliness which requires more processing cost than the higher strength warrants.

Depending on the particular metal, a metallic surface may be relatively pure or have a top molecular layer of porous or plate-like oxides. Similarly the same plastic can have surfaces which differ in potential reactivity or bondability, depending on whether the item was machined, extruded, molded, etc. The type of surface material, the processing history, and the condition at the time of bonding are important pieces of information in the overall adhesive selection mosaic. (See Chap. 17 for a more complete discussion of surface treatment.)

I. Processing Limitations

As discussed earlier, adhesives are normally obtained as an intermediate product—one which must be further processed before attaining its stable use condition. Processing requires equipment, and controls must be available to satisfactorily use the adhesive. Different adhesives require different equipment, which might range from conventional spray guns to sophisticated metering, mixing, and dispensing apparatus or from simple scales and ovens to high-temperature and high-pressure autoclaves. Production rates and the total quantity of bonded assemblies to be produced will influence the choice of adhesives, particularly regarding handling characteristics. The designer must be aware of the impact of adhesive choice on production-facility equipment and processing capability.

J. Cost Considerations

Dollars per pound is the poorest way to compare the cost of adhesives. True material cost is best measured as dollars per application area, at a given thickness. Assuming a constant bondline thickness, the calculations below show that an adhesive with a specific gravity of 1.1 (9.17 lb/gal) costing $2.00/lb is less expensive per application than an adhesive with a specific gravity of 1.8 (14.83 lb/gal) costing $1.50/lb.

Adhesive 1 (0.01-in. bondline)

$$\frac{\$2.00}{lb} \times \frac{9.17 \text{ lb}}{gal} \times \frac{1 \text{ gal}}{231 \text{ in.}^3 \text{ of adhesive}} \times \frac{1 \text{ in.}^3 \text{ of adhesive}}{100 \text{ in.}^2 \text{ bond area}/0.01\text{-in. thickness}}$$

$$= \frac{\$0.00079/\text{in.}^2}{0.01\text{-in. thickness}}$$

Adhesive 2 (0.01-in. bondline)

$$\frac{\$1.50}{lb} \times \frac{14.83 \text{ lb}}{gal} \times \frac{1 \text{ gal}}{231 \text{ in.}^3 \text{ of adhesive}} \times \frac{1 \text{ in.}^3 \text{ of adhesive}}{100 \text{ in.}^2 \text{ bond area}/0.01\text{-in. thickness}}$$

$$= \frac{\$0.00096/\text{in.}^2}{0.01\text{-in. thickness}}$$

The actual material cost is usually the least important factor in selecting an adhesive. Among the costs to be included in the expense of using any given adhesive are the labor of measuring, mixing, applying; equipment amortization; quality-assurance procedures; original development and testing; specification control; etc. The fact that some adhesives require minimum handling—no weighing, no mixing, no application equipment—justifies their higher dollars-per-pound or dollars per job cost, assuming performance is satisfactory.

K. Review Prior to Production

The production arm of the user organization plays a vital role in the selection and application of adhesive. The selection of an adhesive should be coordinated with the production department. This pre-use liaison will ensure that the proper equipment is available for both manufacturing and surface preparation, if required. Many factors, such as working life of the adhesive once ready for application versus how much time is required to apply the adhesive and mate the parts, must be reconciled. Work life (pot life) and its relationship to cure times varies according to adhesive type. Anaerobic and cyanoacrylate adhesive are one-component systems which can be left on one surface for an almost indefinite time. They cure almost instantly when contained between the two surfaces to be bonded. Solvent- or solution-type adhesives begin to cure or harden if left exposed on one surface. Epoxies and polyurethanes have a short pot life compared to cure time in thin-film bonding applications. Some adhesives may have high "grab" and require little or no tooling, whereas slow-curing adhesives may require fixturing during the hardening process. Provision must be made for part storage commensurate with the cleanliness requirements of the adhesive to be used and for proper surface treatment of the parts to be bonded.

Dimension tolerances required for mating parts may vary considerably, depending on the adhesive used. For example, some cyanoacrylates may not bond properly if the gap between surfaces exceeds 0.001 in. (0.0025 cm), whereas a gap-filling epoxy may be used over almost any distance.

V. DESCRIPTION OF PRIMARY ADHESIVE TYPES

There are two basic categories of polymeric resin materials used in the manufacture of industrial structural adhesives. One group of materials is *thermoplastic* in nature. These materials are capable of being repeatedly softened by heat. They harden upon cooling. Examples of thermoplastic adhesives are hot melts, dispersions (synthetic rubber, latex), solutions (acetates, acrylics), and cyanoacrylates.

The second group of materials are *thermosetting* in nature. These materials undergo a chemical reaction through the action of a catalyst or other stimulus, becoming, once they have cured, a relatively infusible substance. Phenolics, epoxies, polyurethanes, structural acrylics, silicones, and anaerobics are examples of thermosetting adhesives.

Other chapters of this text are devoted to detailed descriptions of many of these adhesives. A brief synopsis of major attributes (both positive and negative) of primary adhesive types is presented below to provide a gross overview.

Hot-melt adhesives
    Thermoplastic resins-100% solids
    Fast application (hot)-fast set (cool)
    Rigid to flexible bonds
    Bonds permeable or impermeable surfaces
    Nonpolar-poor adhesion-poor wetting
    Low cost-materials and labor
    Requires special dispensing equipment
    Moisture insensitive-attacked by solvents
    Low heat resistance-degrade
    Poor creep resistance-deformation at 5 psi at 125°F (35 kPa at 50°C)
Dispersion or solution adhesives
    Thermoplastic resins-20 to 80% solids
    Very easy to apply-long shelf life
    Set by evaporation-slow or fast
    Low material cost
    Excellent wetting and penetration
    No special equipment required
    Moderate clamp pressure needed
    Low strength under load-creep
    Require permeable surfaces
    Heat, solvent, and moisture sensitive
Cyanoacrylate adhesives
    Thermoplastic-100% solids-liquid
    Easy to apply, cure
    Fast cure-from 30 sec to 5 min
    Low viscosity-needs well-mated surfaces
    Good tensile strength-poor impact
    Only simple equipment needed
    Will not normally bond permeable materials
    High material cost-poor shelf life
    Poor solvent and water resistance
    Handling hazards-bonds to skin
Phenolic or urea adhesives
    Thermoset resins-100% solids or solution
    Gas during cure-solvent evaporation
    High tensile strength-low impact
    Highly polar-excellent adhesion
    Long shelf life-easy application
    Require high-pressure, oven cure
    Low material cost
    Excellent wetting and penetration
    High shrinkage stresses-brittle
Polyurethane adhesives
    Thermoset-100% solids-liquid
    Flexible sealant/adhesive-low creep

Bond permeable or impermeable surfaces
One or two components-room-temperature or oven cure
Good flexibility at low temperature
Moisture-sensitive components
May undergo reversion with heat or moisture
Moderate material cost
Some toxic formulations
Special equipment to mix, de-air, dispense

Anaerobic adhesives
Thermoset-100% solids-liquid
One component-moderate cost
Easy application-simple cure
Machinery and structural grades
Good cohesive, low adhesive strength
Excellent for bolts, nuts, and static joints
Require primer for many materials
Not suitable for permeable surfaces
Will not cure where air contacts
Crazes some plastics

Silicone adhesives
Thermoset-100% solids-liquid
Rubberlike-high impact and peel
Retain properties from -100 to +400°F (from -70 to +200°C)
Excellent sealant-low stress
One component-short shelf life
Easy application-1- to 5-day cure
Resist most chemicals and solvents
Simple equipment needed
Some by-products—reversion possible
High material cost

Epoxy adhesives
Thermoset-100% solids-liquid
Variable pot life and cure time
Long shelf life-one or two components
Rigid or flexible bonds-highly polar
Least creep under load-least shrinkage
Resist most chemicals and solvents
Widest variety of formulations available
Some formulations dermatitic
Require careful controls-clean surfaces
Need equipment to weight, mix, dispense, and clamp

Structural acrylic adhesives
Thermoset-100% solids-liquid
Two components-no mixing required
Critical application-limited to 25-ml gap
Short work life after application
Fast set time

Tough flexible bonds
Minimum surface preparation
Excellent solvent and chemical resistance
Odorous—ventilation required

VI. ADHESIVE SELECTION PHILOSOPHY

There are no absolutely perfect solutions to any adhesive problem.
Fortunately, with the variety of industrial adhesives available today,
at least one "best compromise" adhesive can usually be identified for
most every legitimate application. A flexible and willing-to-compro-
mise attitude is necessary to avoid precluding the use of a system
which, if fairly evaluated, can solve your problem. An attitude which
insists on the use of a specific adhesive regardless of the impact on
design or performance or cost is foolhardy. It is understandable that
an engineer or design staff may prefer the use of certain adhesives
due to past favorable personal or inherited experience. This cannot
be totally faulted since familiarity with any technological subject fos-
ters the ability to work closer to performance limits. The optimum
strategy is to approach the adhesive selection process by establishing
a complete grid of acceptance criteria. This grid, or screening analysis, will rate the various elements which totally add up to a successful
adhesive application. The main divisions, or items, of this screening
grid would include, but not necessarily be limited to, the following:

Item	Comment
Mechanical strength	The foremost consideration. Must meet real minimum level. Higher performance superfluous unless available at no higher overall cost.
Detail dimension tolerance	What is the cost impact on toler-ances of ± 0.0005 in. (0.0012 cm) for details needed with adhesive A versus ± 0.005 in. (0.012 cm) for adhesive B?
Alternate materials to be bonded	Will switching from an acetal to a glass-filled nylon allow a more reliable bond?
Quality assurance	Procurement and process control specifications on one-component versus two-component adhesives.

Item	Comment
	Surface preparation and cleanliness level requirements.
Production quantity	Number of assemblies per unit time versus working time. Storage limitations, warm rooms to accelerate cures, bonding-fixture recycle time, etc.
Personnel training	Workman competency-level requirements vary with the type and form of adhesive. The placement of a stable film adhesive versus the often critical weighing, mixing, and dispensing of a two- or three-component paste adhesive.
Safety	Handling of solvent systems. Some adhesives more dermatitic than others, etc.
Overall costs	Material cost plus the sum of at least all of the preceding items.

The adhesive candidate which gets the highest rating in the above screening analysis and meets all other objectives of handling, etc., becomes the "best compromise" system. This selection is normally of a generic nature. The final adhesive characterization now becomes an optimization within the adhesive-type category itself. Communication between user and supplier cannot be overemphasized at this point. Minor formula variations—such as increasing viscosity by use of a thixotropic agent, color coding by pigment addition, etc.—can be accomplished. In the case of two-component systems, certain ingredients may be switched from the resin side to the hardener side or vice versa to change to a more favorable mixing ratio. Custom packaging in premeasured kits or disposable cartridges is possible. A given adhesive may be supplied as a two-component package, in premixed fashion as a paste, preimpregnated in a carrier reinforcement, or as an unsupported film. In the case of room-temperature curing or limited life systems, these materials would be stored under refrigeration.

It is impossible to detail the options a manufacturer has available in the production of adhesive materials. Only by open discussion and review of basic requirements and desired features will the user fully tap the resources of the supplier.

VII. ADHESIVE SELECTION EXAMPLES

The following examples of the adhesive selection process are based on actual industrial applications. These illustrations have been abbreviated to point out the primary factors involved in the adhesive choice.

Example 1

Initial description of requirements

Bonding molded plastic to elastomeric material
Assembly will see high temperature [200°F (90°C)] and humidity (100% rh).
Severe service conditions

Solution

Not possible with limited information presented.

Further information

End product is shower spray head
Acetal ring to be bonded to neoprene diaphragm.
1/2-in. square (3.12 cm^2) bond area.
5-year service life.

Final recommendation

Many adhesives will meet the strength requirement. Solution or solvent adhesives ruled out because of nonporous substrates. Silicone considered too slow-curing for production rate requirements. Urethane requires unavailable application equipment. Epoxies do not bond well to acetal without surface treatment. Design does not suit anaerobics. Cyanoacrylates would not survive long-term humidity exposure. "Best compromise" candidate is structural acrylic adhesive.

Example 2

Initial description of requirements (same as Example 1)

Bonding molded plastic to elastomeric material
Assembly will see high temperature [200 °F (90°C)] and humidity (100% rh).
Severe service conditions.

Solution

Not possible with limited information presented.

Further information

End product is disposable dish scrubber
Polystyrene to be bonded to cellulosic sponge.
Give-away item.

Final recommendation

Most adhesives too expensive relative to total product worth or too time consuming in application or curing requirements. Cyanoacrylate, which would otherwise be suitable crazes the plastic surface causing an appearance problem. Liquid pressure-sensitive adhesive can be used since one surface is porous.

Example 3
    Initial description of requirements
        Bond aluminum attachment clip to steel curtain panel
        Long-term exterior weatherability needed.
        Must have field repair capability
    Solution
        Strength, weatherability, and cost factors reduce con-
        sideration to polyurethane or epoxy. Polyurethane not
        convenient for field installation. Epoxy has marginal
        peel-failure resistance. Further analysis required.
    Further information
        Redesign of attachment clip increased bonding area by
        100% and transferred loading to tension, rather than
        peel, stresses.
    Final recommendation
        Redesign allows confident use of epoxy for production
        plus premeasured kit for field installation and repair.

VIII. SUMMARY

The obvious answer to what adhesive to use for any given application
is: the one that performs the required mechanical function and sur-
vives the operational and environmental exposure over the lifetime
of the assembly at the best overall economy of material, processing,
and quality-assurance cost. This answer may appear generalized but,
indeed, is quite specific since a logical progression through the
screening factors discussed in this chapter will lead to definite "best
compromise" choices which may be further optimized by close user/
supplier coordination.

    An excellent guideline reference to types of adhesives used for
joining various materials may be found in Chap. 21.

6

Hazardous Materials — An Overview

J. Merle Nielsen *General Electric Company, Schenectady, New York*

I. POTENTIAL AND ACTUAL HAZARDS

A hazardous material is one that can cause injury or harm. Since *potential hazard* is inherent in all materials, some degree of risk will exist in their use. You can drown in water. Although hydrocarbons are useful as solvents, chemicals, and fuels, they are always a fire and explosion hazard. Although zinc compounds are necessary in the enzymes of the human body, they are toxic when excessively inhaled or ingested. Although poly vinyl chloride (PVC) is a highly useful plastic, epidemiological evidence has shown that long exposure at high levels to the vinyl chloride monomer from which it is made can result in rare angiosarcoma of the liver in 20 years. Obviously, hazards vary in ease of detection and control, but any material which is used without suitable precautions can become an "unreasonable risk" to health, safety, property, or the environment.

The *actual hazard* of a material depends, not only on its potential hazard, but also on the conditions under which it is used: the surroundings, the amounts and concentrations, the degrees of subdivision and dispersion, the presence or absence of other materials, the temperature, the pressure, the user, the precautions that are taken in its use, etc. In addition to immediate effects, one must be concerned

with the future. For example, how would the unlimited use of halocarbon aerosol propellents affect the earth's ozone layer in the next 50
years? What hazards might occur from the unrestrained use of di-
chlorodiphenyltrichloroethane (DDT)? What restrictions should be
placed on consumer use of saccharin, given its very low level of carcinogenicity in the rat? Where do we store high-level radioactive
wastes? Unthinking carelessness in using materials imposes unnecessary risks.

To use a material is to invite along its "package" of potential
hazards, including those that may be poorly understood or entirely
unknown. To avoid "unreasonable risk" with a material it is necessary
to obtain sufficient information about its nature and properties (including identification of what is not yet known), so that it can be handled in a safe and healthful manner, allowing a margin of safety for
the inevitable unknown. Some materials do pose greater hazards than
others, but a potential hazard, no matter how severe or threatening,
need not be harmful. When "the nature of the beast" is understood,
a proper cage can be prepared. Risk can be determined and controlled for any material usage. The economic-political-social decision of
"unreasonable risk" can then define the precautions required if a
hazardous material is used. Science and technology can define the
risk, but society must choose, *and be willing to pay for*, the level of
risk limitation it desires.

II. HAZARDOUS MATERIALS REGULATION

Some regulation of hazardous materials has been carried out by the
federal government for many decades at a comparatively low level. In
the early 1970s Congress established the basic law and regulatory
powers for the government to enter all stages of materials use in the
production process—from new materials and raw materials to the dis-
posing of wastes and controlling emissions and effluents—for the
worthy purposes of ensuring the health and safety of workers and the
public and of preserving and restoring the environment, as related
to hazardous materials, for example:

> (An employer must) furnish to each of his employees employment
> and a place of employment which are free from recognized
> hazards that are causing or likely to cause death or serious
> physical harm to employees.
> Occupational Safety and Health Act of 1970, Sec. 5(a)(1)

> The Administrator shall establish national programs for the pre-
> vention, reduction, and elimination of pollution...in cooperation
> with other Federal, State, and local agencies...
> The Federal Water Pollution Control Act, 1972, Sec. 104(a)

> It is declared to be the policy of Congress...to improve the
> regulatory and enforcement authority of the Secretary of Trans-
> portation to protect the Nation adequately against the risks to
> life and property which are inherent in the transportation of
> hazardous materials in commerce.
> Transportation Safety Act of 1974, Sect. 102

> The term "hazardous air pollutant" means an air pollutant...
> which in the judgment of the Administrator may cause, or con-
> tribute to, an increase in mortality or an increase in serious
> irreversible, or incapacitating reversible illness... The Admin-
> istrator shall prescribe an emission standard for such pollu-
> tant... [and] establish any such standard at the level which in
> his judgment provides an ample margin of safety to protect the
> public health from such hazardous air pollutants.
> Clean Air Act, As Amended 1972, Sec. 112

> Among the many chemical substances and mixtures which are
> constantly being developed and produced, there are some whose
> manufacture, processing, and distribution in commerce, use, or
> disposal may present an unreasonable risk of injury to health or
> the environment...
> Toxic Substances Control Act, 1976, Sec. 2(a)(1)

> Subtitle C of the Solid Waste Disposal Act, as amended by the
> Resource Conservation and Recovery Act of 1976, as amended
> (RCRA), directs the Environmental Protection Agency (EPA) to
> promulgate regulations to protect human health and the environ-
> ment from the improper management of hazardous waste.
> Issuance of Initial RCRA Regulations (Federal Register, 1980)

The scope of "hazardous materials" and the degree of control and detail of regulation were usually left to the agencies with recourse to the
judicial system as the check and balance. The states were usually
given the option of setting up and running their own programs and
methods of enforcement, so long as they met minimum federal standards. If a state were to fail to set up its own federally approved
program or to maintain federal approval, the federal agency would
then assume full control. As a result we now have "cradle to grave"
regulation of the "hazardous materials" used in our industrial society
to ensure that workers, the public, and the environment are not en-
dangered. This initiation of hazardous-materials regulation was probably necessary both because of public demand and because of the in-
ertia and economic disincentives of industrial self-regulation. Establishment of hazardous-materials regulation has been of broad interest
internationally in the past decade.

Promulgation of regulations was just getting underway in the
1970s. By the beginning of the 1980s it had become a deluge. On the
average, over 150 regulations pertaining to chemicals appeared in the

Federal Register each working day in 1980, an increase of 70% over the
1978 rate. Compliance costs for regulatory requirements have become
very large and are continuing to grow, diverting both capital and
manpower from production. It is obvious that where regulation is
needed, the cost must be borne; but an unnecessary proliferation of
regulations saps the economy.

The agencies have almost invariably followed the adversary approach (as opposed to consensus and cooperation between differing interests) in establishing new regulations. The philosophy and zeal of
the appointed administrators has been imposed on regulations. In
some cases agencies, lacking firm scientific evidence, have tried to
regulate on a speculative basis and to divorce economic considerations
from the regulation of hazardous materials. The Occupational Safety
and Health Agency's (OSHA's) court-rejected benzene standard and
its carcinogen policy are examples of such an attempt. Recent
changes in the presidency, the Congress, and the economy may soon
foster a new approach in this highly political area.

Speaking before the Chemical Manufacturers Association meeting
in Houston on October 27, 1980, Edwin H. Clark II—an environmental-
ist, educator, and Associate Assistant Administrator for Pesticides and
Toxic Substances for the Environmental Protection Agency (EPA) (also
EPA's representative on the government's Interagency Regulatory
Liaison Group)—stated that we may be "speeding toward idiocy in
dealing with toxics... It is difficult to act reasonably with the present government regulatory framework." The bureaucrats, he said,
dwell on "... [legal] process rather than substance. Only lawyers
benefit from this approach."

As an example of "process rather than substance" the inclusion
of monochlorobiphenyl among the severely regulated polychlorobi-
phenyls (PCBs) might be cited. Despite the protests of the former
manufacturer, this material has been legally defined as a PCB and
can, therefore, no longer be manufactured in the United States even
though it does not have the problems involving either the environmental persistence or the low rate of metabolism that led to the tight regulation of PCBs.

It is important for the heritage of hazardous-materials information which we pass on to the next generation to be soundly based and
evaluated, especially that which is codified into law, so that the needed protections can be provided. But we also *must* be careful not to
handicap ourselves in the use of materials by unwarranted restrictions.
Wisdom, judgment, common sense, and restraint are needed to prevent
emotion-driven, ill-conceived limitations. Wisdom, judgment, common
sense, and action are needed to provide necessary limitations. It is
to be hoped that we will be able to distinguish the two.

The Occupational Safety and Health Agency, organized under
the Department of Labor, received authority from Congress to regulate

hazards in the workplace. It began to do so with the adoption, almost
en masse, of the industrial "good practice" recommendations—which
had been established by various professional, industrial, and insurance organizations—as its initial standards. These recommendations
have been modified and added to during the years (Code of Federal
Regulations, Title 29). The EPA was organized as an independent
agency and was given regulatory jurisdiction outside the workplace.
Over the years its responsibilities have been increased, for example,
under the Clean Air Act (CAA), the Clean Water Act (CWA), the
Toxic Substances Control Act (TSCA), the Resource Conservation and
Recovery Act (RCRA), and the "Superfund." It has broad powers to
regulate any material that may be hazardous to health or to the environment. The Department of Transportation (DOT), under its Materials Transportation Bureau (MTB) and Office of Hazardous Materials Operations (OHMO), has broad powers to regulate the conditions
and limitations of shipping hazardous materials (Code of Federal Regulations, Title 49). The Department of Transportation and EPA (under
RCRA) share the regulatory authority for hazardous-waste shipment.

III. HAZARDOUS MATERIALS INFORMATION

A. Introduction

> ...(Y)e shall not bend too much over this vapor but turn away
> therefrom and bind up the mouth.

The physician Ulrich Ellenbog recorded this statement in his 1473 book
which was addressed to the goldsmiths of the city of Augsburg concerning their use of amalgams (mercury alloys) in their work. He also
indicated that this work should be done, as much as possible, in the
open air and not in closed rooms (quoted in Goldwater, 1972). In order that lessons provided in the past do not have to be learned again,
it is important to properly record, preserve, index, provide access to,
and transmit (to those who need it) hazardous-materials information.

Through the centuries a heritage of information on the confrontation of chemical materials and biology has been developed. The data
in this multidiscipline area have been obtained from many years of observations, including tragic overexposures and accidents. An extensive international literature on health, safety, and environmental hazards exists from medicine, science, and technology. In the past few
decades--with the help of safety specialists, industrial hygienists,
public-health professionals, epidemiologists, toxicologists, pharmaco-
logists, ecologists, environmentalists, etc., and with a push from regulatory agencies—the expansion in scope and depth of this literature
has greatly accelerated. Even so, the knowledge we now have points
out many significant areas of ignorance.

B. Establishing Toxic Hazard Levels

Determining which materials are hazardous and the quantitative limits of hazard are fundamental problems. There has been no difficulty in finding agreement on any hazard for which cause and effect can be readily recognized; however the subtle hazards, the borderline and threshold effects, and the variations in response of those exposed to the same hazard offer difficulties and uncertainties. Our increased analytical skills make us aware of minute exposures which were previously ignored because they could not be detected or measured. In most cases we do not know what hazards are posed by such minute exposures. Our statistical analysis techniques indicate possible low-level hazards that would not otherwise be identifiable, but cause and effect in these cases can remain elusive.

Data obtained from human exposures best relates to human health and safety questions, but animal models are used as best approximations when human data are unavailable.

1. Epidemiology

Much human toxicological experience with materials has been derived from clinical evidence and from before the fact (prospective) and after the fact (restrospective) studies, which are carried out to find out what effects the exposures experienced have had or might have on a selected population. These studies are of two types: morbidity studies, which relate to the illnesses of persons who have experienced toxic exposures, and mortality studies, which relate to the deaths of persons who have experienced toxic exposure. Such studies require the comparison of an exposed group with a similar, but unexposed, group. The statistical information derived is used to indicate possible toxic hazards to man.

On January 23, 1974 B. F. Goodrich reported the deaths of three former workers at their Louisville, Kentucky PVC plant from rare angiosarcoma of the liver. This cancer is so rare that it, alone, pinpointed a problem related to exposures in the plant. Through epidemiological study of the years 1961–1976, 18 angiosarcoma deaths in the United States and 48 worldwide were discovered. All the deceased has been exposed to vinyl chloride monomer at high levels; most had worked as PVC-reactor cleaners. The information developed on vinyl chloride led to a new OSHA standard which dropped exposure limits in the workplace from hundreds of ppm in air for systemic toxicity protection to 1 ppm for carcinogenicity protection.

Laboratory evidence from test animals in 1961 had indicated possible testicular atrophy from exposure to 1,2-dibromo-3-chloropropane (DBCP), a pesticide, but the published information was not adequately communicated. In July, 1977, sperm-sample test results from Occidental Chemical's Lathrop, California plant were sent to Donald Wharton at the University of California. Wharton was the director of the Labor Occupational Health Project. His aid was enlisted in tracing down the cause of the observed worker infertility. Later epidemiological studies indicated that 95 workers nationwide, including Shell and Dow production workers, had become infertile from DBCP exposure. (Israeli workers were also reported sterile from DBCP exposure.) The epidemiological information led to a new OSHA standard which established a 1 ppb exposure limit. Dow reported in 1980 that of 29 workers at its Magnolia, Arkansas plant who had been found sterile in 1977 after DBCP exposure, 15 had significantly recovered fertility, 10 still had zero sperm counts, and 4 had not been retested.

Many epidemiological studies do not produce the clear-cut results that these two established. Often no statistically significant outcome is found, and no cause and effect relationship can be established.

2. Animal testing for indications of human toxic hazards

In the absence of human data, laboratory studies with animals (following established procedure) have been used to indicate possible toxic hazards to man. In the older literature, studies were mostly for acute lethal effects: A group of test animals was given a single dose of (or a short exposure to) the material to be tested and then observed for up to 2 weeks to see what happened. The result obtained was either the mean lethal dose LD_{50} or the mean lethal concentration LC_{50} (for a defined time) which was the dose or concentration that would kill 50% of the animals in the test group.

Larger numbers of animals are used in subacute testing with dosages administered over 30−90 days in a manner that does not usually kill the test animals. Careful observation of the animals during the test period coupled with careful dissection and pathological examination of each animal, following its sacrifice during or after the exposure period, gives exposure physiological effect time data which may indicate possible hazards for man.

Chronic toxicity studies are carried out like subacute studies except that testing must be carried out for a major fraction of the normal lifespan of the test animal, and large colonies of carefully controlled test animals (hundreds of animals) must be used to obtain statistically significant results. Two years or more are usually required for chronic testing in the rat; four years in the dog. Tests are also carried out with successive generations of test animals to determine genetic and intrauterine effects on offspring. Dose-response data on the production of disease in the heart, liver, kidney, lungs, etc., as well as information on carcinogenicity (cancer production) and teratogenicity (fetal damage), are sought. Some strains of test animals are more sensitive to toxic effects than others. The age of the animal and the route of entry of the chemical into the body also affect toxicity results.

Much difficulty can be encountered in ensuring that uniform water, air, diet, temperature, sanitary conditions, etc., are maintained during chronic testing of an animal colony. Control groups must be maintained as identically as possible with respect to the exposed animals to help make known the existence of uncontrolled environmental conditions that could affect the test results. Such testing can be very expensive. Running a "complete profile," including metabolic studies, on the effect of a single chemical compound on at least two species of mammal (one not a rodent) has been estimated to cost well over one million dollars. Minimal chronic screening may cost over one hundred thousand dollars.

It must be remembered that testing with animals is still only an approximation for man. Significant differences can exist between the biochemistry of man and that of animal models. The problems of analogy become especially great when an attempt is made to compare very high doses in test animals to very low doses in man. On the other hand, a chemical which is able to enter a biological system and cause changes in the genetic material of living cells must be seriously considered as a threat to man.

3. Evaluated data

At the end of World War II the American Conference of Governmental Industrial Hygienists (ACGIH) responded to the need for professionally evaluated data for guidelines for safe worker exposure to airborne materials in the workplace. Starting in 1947, ACGIH published annually its *Threshold Limit Values* (TLVs), also called 8-hour Time Weighted Averages (8-hr TWAs). Threshold Limit Values were defined as workplace exposures for a single airborne material that could be endured safely for 8 hr/day, 5 days/week, by a normal worker for a lifetime of work without significant ill effect. The established TLVs were reviewed annually for revision and additions as new data became available and as experience with materials increased. Ceiling levels (*C* prefix) were also established for those materials requiring a maximum permissible exposure. It should be remembered that TLVs can be invalidated if additional exposure has occurred by ingestion or skin absorption, if exposure to a second toxic material occurs simultaneously, or if a hypersensitive person is exposed. Exposures in the workplace should be kept as low as feasible (below the TLV) to further increase the safety margin in favor of the worker. It should also be remembered that some TLVs have been established only to prevent transitory discomfort or irritation and are not based on toxicological necessity.

In 1970 the 1968 TLVs for several hundred materials were adopted as legal safety and health requirements and became the Permissible Exposure Limits (PELs) to be enforced by OSHA. The ACGIH continues to review annually its TLVs, and its current list now differs appreciably from that of OSHA. (Its current revisions are probably better practice than old OSHA values.) In addition to the TLV Committee of ACGIH, the Z-37 Committee of the American National Standards Institute (ANSI), the National Institute of Occupational Safety and Health (NIOSH), and certain contractors to OSHA, NIOSH, and EPA have been criteria developers—that is, they have been recognized as competent in evaluating data and making recommendations for standards.

IV. POTENTIAL HAZARDS WITH MATERIALS

The following is an organized view of the kinds of potential hazards to be expected with materials. (Classifications can be difficult to make and sometimes overlap. A single material may exhibit several of the hazards.)

A. Physical-State Effects

1. Compressed gases and pressurized hot vapors

Pressurized gases (including hot vapors like steam) can be hazardous. Equipment failure of hoses, gauges, safety devices, pipes, or pressure vessels can occur. Gases can leak slowly, or they can be quickly released and can expand to fill enclosed spaces. They can then readily mix with air to become an explosive hazard, displace air to become an asphyxiation hazard, or exhibit the toxic, anesthetic, corrosive, thermal, or other properties of the gas or vapor. Rigid safety procedures must be used with pressurized systems. In a fire, pressurized cylinders can be ruptured violently or can release their hazardous contents as safety devices are activated.

When pressurized at over 2 atm (absolute), acetylene gas becomes shock sensitive and can violently decompose spontaneously. In order to ship it safely at 18 atm a special cylinder is required, filled with a porous solid and acetone to dissolve the acetylene under pressure. When acetylene is released from such a container, the pressure must be controlled at two atmospheres (absolute) or less.

2. Liquified gases, cryogenic liquids

At atmospheric pressure, materials which are normally gaseous can be stored as liquids by maintaining them at low temperatures. Many liquefied gases can also be stored as liquids at room temperature (i.e., NH_3, SO_2, C_3H_8) by pressurization. (The maximum pressure of this type of system is the vapor pressure of the liquid at the storage temperature.) However, cryogenic liquids (i.e., N_2, O_2, He) have critical temperatures so low ($-147.1°C$ for N_2) that they *cannot* be maintained as liquids by pressure at normal temperatures. A stored cryo-

genic liguid must be well insulated and vented during storage, or
violent vessel rupture can occur.

Upon release or spill, liquefied gases rapidly pick up heat from
their surroundings and quickly vaporize, expand, and diffuse to become gaseous hazards. Contact with the liquid can freeze body tissues. Cryogenic liquids can cool and embrittle structural materials,
so that they can shatter on impact.

3. Particulates and aerosols

Materials in a finely divided state can pose hazards that they do not
exhibit in bulk form. Sluggishly combustible materials, if suspended
in air as a fine mist or solid particulate, can become a fire or explosion
hazard. Fine aluminum powder dispersed in air can be explosive.

Air-dispersed particulates can be carried into the environment
or inhaled into the respiratory tract. Large particles are filtered out
in the nasal passages, but solid particles below 5 μm in diameter or
fine-diameter fibers much longer than 10 μm can penetrate deep into
the lungs. The diseases silicosis and asbestosis result from excessive
inhalation of crystalline silica dust or fine asbestos fibers, respectively.

Particulate materials can be injurious to the eyes. They can contact the eye surface as an abrasive irritant, or penetrate the eye as a
projectile (e.g., from a grinding wheel or other tool).

4. Radioactive materials

The ionizing radiation that is emitted from the nucleus of an unstable
isotope of an element, including alpha particles, beta particles, pro-
tons, neutrons, gamma radiation, and x-rays, can be damaging to tissue or lethal, depending on the intensity of the radiation and time of
exposure. Increasing the distance from the radiation source or providing a shield to absorb the radiation can reduce exposure to external radiation, depending on the type and level of energy. Internal
radiation exposure from inhaled or ingested radioactive substances
also depends on the time retained in the body, the amounts that are
retained, and the biological concentrations that can occur (like the
concentration of radioactive iodine in the thyroid gland).

Radiation energy initiates chemical reactions in the cells of the
body that would not otherwise occur. At low levels the body defenses
normally handle the disruptions, providing protection from the background radiation that we all experience. However, at excessively high
rates of radiation, cell damage can build up to produce injury and
even death.

B. Chemical Reactivity

1. Corrosive materials

Materials that cause "eating away" or deterioration or destruction of other materials are the corrosives. Strong acids and bases are corrosives. Hydrochloric acid corrosion of carbon steel, concentrated sodium hydroxide corrosion of aluminum, and phenol corrosion of body tissues are examples of the effect these materials can have on their environment.

2. Incompatible materials

When mixed or placed in contact, incompatible materials can undergo vigorous reactions and/or produce hazardous reaction products.

a. Highly exothermic reactions. Reactions which release a lot of energy, causing uncontrolled pressure and temperature increases when two or more materials are mixed, include the heat of solution of sodium hydroxide or of concentrated sulfuric acid in water, acid-base reactions, certain polymerization reactions, hypergolic reactions (including materials which ignite on contact, such as acetone and 85% nitric acid), pyrophoric reactions (involving materials which quickly self-ignite at or near room temperature when exposed to air), etc.

b. Oxidizing agents. Contact or intimate mixing of oxidizing agents, such as concentrated nitric and sulfuric acids, dichromates, permanganates, perchlorates, nitrates, hypochlorites, etc., with reducing agents or combustible materials can result in fire, explosion, or detonation hazards. Elevated oxygen content in air increases fire hazard. Ozone is a powerful oxidizing agent. Any material which will burn in air will also burn in nitrogen dioxide, but it might also explode.

   Materials which are strong oxidizing agents must be clearly identified and kept away from incompatibles.

c. Materials which generate hazardous reaction products. A few examples of the generation of hazardous reaction products follow. When carbon-containing materials are oxidized in a deficiency of air, toxic carbon monoxide gas can be released. An explosive mixture of hydrogen and air can result from the reaction of hydrochloric acid and carbon steel. Sodium hydrosulfide reacts with mineral acids to liberate toxic hydrogen sulfide, and calcium fluoride reacts with mineral acids to produce toxic, corrosive hydrogen fluoride. The reaction of mercury or silver salts with acetylene in ammonia solution produces acetylides, which are sensitive and violently reactive when dry. The autoxidation of diisopropyl ether by oxygen from air at room temperature can produce a slow buildup of peroxides in the ether. In a few weeks or months a dangerous concentration of reactive peroxides can exist.

3. Flammable and combustible materials

Materials which will vaporize at room temperature or when heated to produce an air-fuel mixture that can be ignited are fire and/or explosion hazards. Ignitable gas, mist, or dust dispersions in air can be similarly hazardous.

The lowest liquid temperature that will produce sufficient volatilization of a liquid to produce an ignitable fuel-air mixture above the liquid surface is called the flash point. Flash points are used to indicate the case of ignition of liquids—the lower flash point usually indicating the greater flammability hazard. The safety requirements and restrictions for packaging, transportation, labeling, storage, use, and waste disposal of flammable and combustible liquids are based on their flash points. Closed-cup flash points are used to define flammable [below 100°F (37.8°C)] and combustible [100°F (37.8°C) or above] liquids. Those materials that can "flash" at or below room temperature are the most hazardous. Materials with high flash points require heating to become ignitable, but when heated, a combustible material can be a very severe fire hazard.

4. Unstable materials

Materials which are inherently unstable need special attention when they are used. Instability can arise from a special self-reaction or self-decomposition pathway when it is activated, from an intimate mixture of oxidizing and reducing agents, or from their combination in a single, energetic molecule. Explosives like gunpowder or 2,4,6-trinitrotoluene (TNT) exemplify the latter type. A homogeneous mixture of either alcohols or glycerine with concentrated hydrogen peroxide becomes a shock-sensitive, powerful explosive.

When kept cool, clean, and pure, concentrated hydrogen peroxide (> 70% in water) can be handled safely; however, if contaminated or heated, 70% and 90% hydrogen peroxide can reach self-decomposition temperatures above 500°F (260°C) and 1360°F (740°C), respectively. Acetylene gas at pressures above 2 atm (absolute) is sensitive to violent self-decomposition. Styrene monomer can be prevented from violently polymerizing during room temperature storage in contact with air by the addition of a small amount (10—15 ppm) of 4-t-butylcatechol. If the inhibitor is removed or becomes exhausted, uncontrolled exothermic polymerization proceeds. Organic peroxides are unstable compounds themselves, as well as oxidizing agents. If acetone is mixed with peroxides, it can form shock-sensitive crystals which can decompose violently.

C. Toxic Effects of Materials

All materials are toxic if "too much" is imposed on body defenses. The dose or exposure (intake per unit time) determines the toxic effect that can result from a given material by a given route of entry into the body.

The normal human body is able to cope with a single exposure to a limited amount of a toxic material; but when the amount imposed on the body's defenses exceeds certain levels, and/or when exposure to more than one material occurs (additive or synergistic effects), the body cannot handle the problem, and injury results. This threshold level for injury can differ for each toxic substance or mixture and for each individual. It can also vary for a given individual at different times, depending on his metabolic situation. The body can build up a tolerance to certain toxic materials, such as arsenic or zinc, as exposures are increased; the resulting tolerance may be either long-lasting or short-term (quickly lost when reduced levels of the material are experienced). Illness and disease (past or current), fatigue, stress, nutritional status, etc., can all affect the detoxifying mechanisms of the body. Sex and genetic factors can influence the body's ability to resist damage from certain materials. Toxic material can contact the body as airborne material (gas, vapor, or particulate), as liquids, or as solids.

1. Acute, chronic, and cumulative toxicity

The assault on the body's defenses by a toxic material may result from a single or short-term, relatively massive confrontation (acute toxicity) or from repeated or continuous contact with lower, *apparently* tolerable levels for a much longer period of time (chronic toxicity). Some toxic materials, such as lead, can be retained in the body (cumulative toxicity) with no apparent problems until tolerance levels are exceeded or until a new or intensified stress reduces the effectiveness of the body's defenses. Since early poisoning symptoms are not specific in many cases, they can easily be mistaken for something else, like simple fatigue. The real problem may not be recognized until damage is more extensive.

2. Local contact effects and body entry

a. Skin. Over 60% of known industrial injuries due to chemicals are skin-related injuries. Usually these are minor, involving only mild irritation; but they can also involve skin penetration and more serious forms of dermatitis with skin eruptions, infection, sensitization, allergic reaction, etc. Some materials can be corrosive to the skin, and many can produce chemical burns which are comparable to heat destruction of tissue. Solvents can attack the skin as liquids or as vapors. Their action may be strictly local, with damage occurring only at the position of contact, or, for those chemicals or solvents which are able to penetrate the skin and enter the bloodstream, the effect can also be systemic.

The least damage to be expected from solvent contact is the removal of some of the natural fats and oils of the skin, perhaps leaving the skin dry or brittle and more subject to cracking and bacterial infection. This kind of damage is greater for repeated or prolonged contact and for those solvents that more readily dissolve the fats and oils of the skin. For example, aromatic hydrocarbons can be expected to be more damaging than aliphatic hydrocarbons. In addition, the solvent contacting the skin may be irritating or corrosive to tissue; or, by removing the protective fats and oils, it may make the skin sensitive to attack or penetration by materials that would normally be repulsed without difficulty.

Where skin is damaged, for example, by cuts, cracks, or abrasion, contact with chemicals offers an increased hazard since removal of the skin barrier provides a more direct pathway to the bloodstream. Solvents and other materials can also be forced through the skin by mechanical or hydraulic pressure. People with open cuts, abrasions, burns, lesions, etc., should not work where chemical or solvent contact with the wound is likely to occur. Solvents can carry with them dissolved impurities which have their own effects on the body. Certain solvents, such as dimethyl sulfoxide (DMSO), can very readily transport dissolved materials through the skin barrier.

Wearing of contaminated work clothing, carrying oily rags in the pockets, failure to properly wash or bathe after exposure to certain materials, etc. (all of which allow prolonged contact), can lead to skin and/or systemic effects.

b. Eyes. In addition to the possibility of solid particulate irritation and injury, the surface of the eye is subject to the effects of direct contact with gases, liquids, and soluble solids. Many materials cause temporary pain, discomfort, irritation, and lachrymation when they enter the eye. Others can produce long-lasting damage or permanent loss of sight.

The corrosive materials, acids and alkalis, cause burns and destruction of tissue; the alkalis are especially destructive, not only damaging the cornea, but also changing the pH of the interior fluids of the eye. Ammonium hydroxide, because of its penetrating ability, may have more serious effects than sodium hydroxide which is a stronger alkali.

Materials which can denature or damage the protein structures of the transparent cornea or the lens produce clouding of these, hence, either temporary or permanent vision loss, as reported for methyl silicate.

c. Lungs. An average worker will inhale about $10-14$ m^3 of air in an 8-hr day of light work activity. Whatever is in this volume of air comes in contact with the mucous membranes of the respiratory tract and the lungs where damage to, or absorption into, the body can occur. The air we breathe carries vapors, gases, and particulates, such as mists, dusts, fumes, and fibers. The vapors and gases can interact with the mucous and tissue in the respiratory tract and/or pass through membranes into the bloodstream. The respirable particulates (below about 5 μm in diameter) can be carried deep into the lungs and deposited. The larger particulate are trapped in the mucous of the upper respiratory passages, transported to the nose or esophagus, and either expelled or swallowed.

Particulates deposited in the lungs are dissolved by body fluids if they are reasonably soluble and are carried into the body. Some of the insoluble particulates, especially those of smaller size, are engulfed by the body scavengers, the phagocytes, and transported to the "mucous elevator," to the lymph system, or to the bloodstream. The insoluble particulates remaining can apparently be benign and inert or can produce *pneumoconiosis* (literally, the pathological response of the lung to dust). In the latter case, the lung begins to grow scar tissue (fibrosis) to enclose the particles, and this thickened tissue reduces lung flexibility and the effectiveness of gas exchange between the small air sacs (alveoli) and the blood. *Asbestosis* and *silicosis* are pneumoconioses produced by excessive exposure to fine asbestos fibers (> 5 μm long) or to crystalline silica particles (<5 μm in diameter), respectively, which will show up in x-rays. These progressive diseases produce such inflexible and thickened walls in the lung when fully developed that it becomes increasingly difficult for oxygen to get to the bloodstream and for carbon dioxide to get from the blood into the lungs for exhaling.

All of the materials entering the respiratory system, except those expelled from the body with mucous, are able to exert their chemical and toxological effects on the body. Airborne particulates are often absorbed via the gastrointestinal (GI) tract even though their original entry into the body was via the respiratory system.

Before they are cleared from the respiratory tract and lungs, inhaled materials can produce irritation and inflammation of the lung tissue, resulting in pulmonary edema, bronchitis, pneumonitis, etc., which are reversible as long as the exposure is not too great. The degree of irritation and inflammation from these pulmonary irritants depends on which materials are involved and the degree of exposure, but the effects of these materials on the lungs do not show up in x-rays. Nor do effects of inhaled toxic materials which are absorbed into the blood from the lungs and distributed throughout the body show up on x-rays.

Chronic exposures to cadmium by inhalation are known to produce long-term damage to the walls of the alveoli, reducing both their resiliency and elasticity and resulting in the condition known as *emphysema*, in which effective lung inhalation-exhalation capacity is irreversibly and excessively reduced. Cadmium also enters the bloodstream to produce systemic effects (particularly affecting the kidneys).

d. Gastrointestinal tract. As indicated, inhaled materials collected in mucous in the respiratory tract are often swallowed with the mucous and can exert their effects on the GI tract. Other materials are directly ingested through the mouth. The health hazard from direct ingestion is rather minor in the industrial use of materials, except in those cases in which poor hygienic practices allow the transfer of toxic materials, for example, lead compounds, to the mouth.

Although minor in industry, ingestion is a real hazard for which training and countermeasures should be established. Corrosive acids and alkalis can irritate, damage, and destroy tissue; toxic materials can produce systemic effects and poisoning, depending on dose, time, and efficiency of absorption from the GI tract into the bloodstream. Medical direction, before and after the fact, is needed to respond properly to cases of ingestion. First aid procedures must be established for handling possible ingestion of materials used.

Mention needs to be made of the aspiration hazard which can exist, especially with halocarbon and hydrocarbon solvents and light oils. If vomiting spontaneously results after ingestion (or is used to remove the ingested material from the stomach), aspiration of liquid into the lungs can occur. Depending on the molecular weight, surface tension, chemical reactivity, toxicity properties, and amount aspirated, certain liquids can suffocate by vaporizing to fill the lungs (displacing air), can cause rapid respiratory paralysis, cardiac arrest, and asphyxia, or can produce pulmonary edema and hemorrhage. For example, aspiration of one ounce of xylene into the lungs can be fatal.

3. Systemic effects

When absorption into the body via the skin, lungs, GI tract, or by multiple entry ways occurs, a material can be transported throughout the body by the bloodstream. The particular nature of the chemical and the amounts involved determine what can occur. The chemistry and processes of the body (metabolism) immediately proceeds to utilize, store, and/or get rid of the chemical. A given material can be handled by the body in various ways. Favored metabolic pathways may be used until over-loaded with too much material to handle; then other (or multiple) pathways may be used. Breakdown products, modified materials, and unchanged materials can be eliminated from the body in the urine, feces, exhaled air, sweat, or other bodily excretions; they can be stored, for example in the fatty tissues, or react with particular structures, enzymes, or organs of the body.

Some materials resist chemical changes in the body. For example, carbon monoxide (CO) is not metabolized but is eliminated unchanged from the lungs. When it enters the bloodstream, it complexes with the hemoglobin of the red cells of the blood to form carboxyhemoglobin, preventing the complexing of oxygen and reducing the oxygen-carrying capacity of the blood. (Carbon monoxide is about 250 times as active in forming a hemoglobin complex as is oxygen; that is, if air with a mixture of 250 parts of oxygen and 1 part of CO by volume were breathed, about half of the hemoglobin sites would be complexed with CO, causing oxygen deficiency in the cells of the body.) Breathing 200 ppm CO for a few hours produces headache and mental dullness; breathing 600 ppm produces headache in less than an hour, unconsciousness in 1—1.5 hr, and death in 4 hr. When CO-free air is breathed, CO can be gradually displaced from the hemoglobin and exhaled from the lungs. The process is apparently completely reversible so long as no brain damage or heart attack has occurred due to oxygen starvation. The solvent, methylene chloride is readily absorbed from the lungs and is stored in the fat of the body. It is metabolized to CO and produces a background level of carboxyhemoglobin as long as it is in the body. Heavy smokers can carry a burden of over 5% carboxyhemoglobin in their blood from smoking alone.

The normal person is readily able to eliminate in urine the lead compounds taken into the body through food, water, and inhalation of ambient dust. Those who have high lead exposures or who have suffered kidney disease (reduced lead elimination capability) may have a buildup of lead in the tissues of the body that may result in cumulative lead poisoning with many toxic effects that can be severe. Cadmium compounds have lower systemic hazards by ingestion than by inhalation because they tend to induce vomiting when ingested and, thus, are not absorbed so much. However, when cadmium compounds are absorbed into the body, in addition to other poisonous effects, they damage the kidneys. (Those who have excessive body burdens of lead, cadmium, or other trace elements are sometimes medically treated with "chelating agents" which are molecules that can form multiple chemical bonds to metal ions to pull them away from deposits in body tissues and solubilize the ions in the body fluids, so that their rate of excretion in the urine will increase.) Fluoride ion is readily eliminated in the urine, and body buildup occurs only with sustained excessive intake. An excess of fluoride in the body causes calcium fluoride deposits and calcium deficiency and can be manifest in bone damage.

Systemic effects of many solvents are manifest in the central nervous system (CNS), resulting in headache, dizziness, "drunkenness," and unconsciousness. Some solvents also have specific systemic hazards. For example, methyl alcohol is metabolized to formaldehyde and formic acid in the body; it is capable of causing permanent damage to the optic nerve and, possibly, the retina of the eye, producing blindness. Carbon tetrachloride severely damages the liver and kidneys; it is especially hazardous to those who already have liver or kidney problems. Its systemic effects are potentiated by simultaneous exposure to ethyl alcohol.

When an outbreak of peripheral polyneuropathy (a disease of the nerve cells which is manifest by weakness and reduced sensation and control at body extremities) occurred among workers in a Colum-

MATERIAL SAFETY DATA SHEET

CORPORATE RESEARCH & DEVELOPMENT

SCHENECTADY, N. Y. 12305

Phone: (518) 385-4085   DIAL COMM 8*235-4085

No. _____ 397

n-HEXANE
Revision A

Date July 1979

SECTION I. MATERIAL IDENTIFICATION

MATERIAL NAME: n-HEXANE
DESCRIPTION: n-Hexane or a mixed isomer solvent with substantial levels of n-hexane
OTHER DESIGNATIONS: Hexane, $CH_3(CH_2)_4CH_3$, C_6H_{14}, ASTM D1836, CAS# 000 110 543
MANUFACTURER: Available from many sources, including:
          Exxon Company, USA and Phillips Petroleum Company

SECTION II. INGREDIENTS AND HAZARDS	%	HAZARD DATA

Typical Composition:
  n-Hexane (major component)
  Other Hexanes (minor component or nil) }
  Other Saturated Hydrocarbons (C_5 to C_7)
  Olefinic Hydrocarbons (C_5 to C_7)
  Aromatic Hydrocarbons
	>98
	
	Trace
	Trace
	<0.1
		8-hr TWA 25 ppm* or
		  90 mg/m3
		
		Human, Inhalation

*ACGIH (1979 Intended Changes List) reduced the TLV because of possible nerve cell damage which has recently been established. Current OSHA 8-hr TWA is 500 ppm.		TCLo 1400 ppm
		  n-hexane
		(central nervous system)
		
		Mouse, inhalation
		  LCLo 120 g/m3

SECTION III. PHYSICAL DATA

Boiling point at 1 atm, deg F --- ca 152-156*   Specific gravity (20/4C) -- ca 0.66*
Vapor pressure at 60 F, mm Hg --- ca 100*   Volatiles, % -------------- 100
Vapor density (Air=1) ---------- 3   Molecular weight --------- 86.20
Water solubility --------------- Insoluble

Appearance & Odor: A clear, colorless, mobile fluid. Mild hydrocarbon odor.

*Precise values depend on the grade of the hexane.

SECTION IV. FIRE AND EXPLOSION DATA

			LOWER	UPPER
Flash Point and Method	Autoignition Temp.	Flammability Limits In Air		
<0 F (TCC)	500 F	Approx. % by volume	1.2	7.5

Extinguishing media: Use carbon dioxide, dry chemical or foam. Water may be ineffective
   in putting out fire and a water stream will spread flames; but a water spray should
   be used to cool fire-exposed containers to prevent pressure rupture.
This flammable liquid is a dangerous fire hazard, and it is a dangerous explosion hazard
   when heated.
Firefighters should use self-contained breathing equipment with proper eye and skin
   protection.

SECTION V. REACTIVITY DATA

This is a stable liquid in a closed container at room temperature. It does not polymerize.
This highly flammable liquid (OSHA Clas IB) must be kept away from heat, and sources of
   ignition. It is incompatible with oxidizing agents.
Thermal-oxidative decomposition products in air can include carbon monoxide.

GENERAL ⊛ ELECTRIC   Copyright© —1979 By General Electric Company

Fig. 1. *n*-hexane safety data sheet.

SECTION VI. HEALTH HAZARD INFORMATION

TLV 25 ppm n-Hexane (see Sect II

Excessive exposure to vapors of this material is irritating to the respiratory tract and can produce dizziness, numbness of extremities, intoxication and unconsciousness, depending on exposure level and time. In the body n-hexane can be metabolized (partially oxidized) to neurotoxins which cause nerve damage (peripheral polyneuropathy) in individuals repeatedly exposed above 1000 ppm over a period of months. (See N. Engl. J. Med. 285:82-85, 1971). Liquid contact with the eyes is irritating; skin contact can be irritating, defatting, and lead to dermatitis when repeated or prolonged. Ingestion is upsetting and irritating to the GI tract.

FIRST AID:
Eye contact: Flush eyes well with running water for 15 minutes. Get medical help if irritation persists.
Skin contact: Wash contact area with soap and water. Remove contaminated clothing promptly. Replace skin oils with lotions or creams.
Inhalation: Remove to fresh air. Restore breathing if required. Get medical help.
Ingestion: Get medical help immediately!

SECTION VII. SPILL, LEAK, AND DISPOSAL PROCEDURES

Establish plans and provide training prior to any emergency situation. When spills occur exclude workers from area except those assigned to clean-up who must have proper protection against inhalation of vapors or contact with liquid (see Sect. VIII). Provide maximum explosion-proof ventilation. Eliminate ignition sources. Flush hexane away from sensitive areas with a cold water spray. (Flush to ground not to the sewer! Small amounts of liquid (or absorbed liquid) can be allowed to evaporate with good ventilation or in a hood or open area; large spills should be picked up in a safe and appropriate manner for disposal.
DISPOSAL: Scrap material can be burned in an approved incinerator in accordance with Federal, State and local regulations.

SECTION VIII. SPECIAL PROTECTION INFORMATION

Provide general ventilation and local exhaust ventilation which is explosion proof and adequate to meet the action level or TLV requirements. For emergency or nonroutine exposures above the TLV use an approved organic vapor cartridge respirator for limited time below about 250 ppm, an approved gas mask (preferred) with organic vapor canister up to 2000 ppm and approved air-supplied or self-contained respirators for higher or unknown exposure levels. Use full-facepiece protection above 1000 ppm.
Prevent skin contact by use of impermeable gloves, aprons, boots, suits, etc as needed by the circumstances of use. Prevent eye contact by use of safety glasses, goggles, or face shield with goggles or glasses as the workplace circumstances may require.
Eyewash stations and safety showers should be readily available to areas of handling and use.

SECTION IX. SPECIAL PRECAUTIONS AND COMMENTS

Store in closed containers in a clean, cool, well-ventilated place, away from oxidizing agents and sources of heat and ignition. Protect containers from physical damage. Ground and bond containers for transfers to prevent static sparks. Use metal safety cans for handling small amounts. Storage and handling conditions must follow OSHA regulations for OSHA Class IB flammable liquid. No smoking in areas of storage or use.
Avoid breathing vapors! Prevent contact with skin or eyes! Do not ingest!
Use of preplacement exams and medical surveillance (especially for the nervous system and skin disorders) is recommended. Exposure monitoring and recordkeeping requirements which have been proposed by NIOSH for alkanes should be instituted.

DATA SOURCE(S) CODE: 1-9, 12, 19

APPROVALS: MIS, CRD
Industrial Hygiene and Safety
MEDICAL REVIEW: 12/79

GENERAL ELECTRIC

Fig. 1. (Continued)

MATERIAL SAFETY DATA SHEET

CORPORATE RESEARCH & DEVELOPMENT

SCHENECTADY, N. Y. 12305

Phone: (518) 385-4085   DIAL COMM 8*235-4085

SECTION I. MATERIAL IDENTIFICATION

MATERIAL NAME: HEXANE ISOMERS (Other Than n-Hexane)
DESCRIPTION: This material is a single hexane isomer or a mixture of hexane isomers
with a very low or negligible level of n-hexane.
MANUFACTURER: Available from many sources, including Phillips Petroleum Company

SECTION II. INGREDIENTS AND HAZARDS

	%	HAZARD DATA
2,2-Dimethylbutane or Neohexane (CAS# 000 075 832) ⎫ 2,3-Dimethylbutane or Diisopropyl (CAS# 000 079 298) ⎬ 2-Methylpentane or Isohexane (CAS# 000 107 835) ⎪ 3-Methylpentane (CAS# 000 096 140) ⎭	> 95	8-hr TWA 500 ppm*
n-Hexane (CAS# 000 110 543) (See MSDS #397)	< 1	8-hr TWA 25 ppm*
Other Saturated Hydrocarbons (C_5 to C_7)	Trace	
Unsaturated Hydrocarbons (C_5 to C_7)	Trace	
Aromatic Hydrocarbons	< 0.1	

*ACGIH (1979 Intended Changes List) 25 ppm for n-hexane
or 500 ppm for its isomers (excluding n-hexane).

SECTION III. PHYSICAL DATA

	2,2-Dimethyl Butane	2,3-Dimethyl Butane	2-Methyl Pentane	3-Methyl Pentane
Boiling point at 1 atm, deg C -----	49.7	57.9	60	63.3
Vapor pressure, mm Hg/Temp --------	400/31 C	400/39 C	400/41.6 C	100/10.5 C
Specific gravity at 20 C (H_2O=1) --	0.649	0.662	0.654	0.664
Flash point, deg F ---------------	-54	-20	-10	<20
Autoignition temperature, deg F ---	797	788	583	--
Flammability limits, LEL/UEL ------	1.2/7.0	1.2/7.0	1.2/7.0	--

These compounds and their mixtures are clear, colorless liquids which are insoluble
in water. Hydrocarbon odor.

SECTION IV. FIRE AND EXPLOSION DATA

Flash Point and Method	Autoignition Temp.	Flammability Limits In Air	LOWER	UPPER
Depends on components (See Sect. III)	Depends on components (See Sect. III)	Volume % (estimate)	1.2	--

Extinguishing media: Use carbon dioxide, dry chemical or foam. Water may be ineffective
in putting out fire and a water stream will spread flames; but a water spray should be
used to cool fire-exposed containers to prevent pressure rupture.
This flammable liquid is a dangerous fire hazard; it may be an explosion hazard when
heated.
Firefighters should use self-contained breathing equipment with proper eye and skin
protection.

SECTION V. REACTIVITY DATA

This is a stable liquid in a closed container at room temperature. It does not
polymerize.
This highly flammable liquid (OSHA Class IB) must be kept away from heat, and sources
of ignition. It is incompatible with oxidizing agents.
Thermal-oxidative decomposition products in air can include carbon monoxide.

GENERAL ⊕ ELECTRIC

Fig. 2. Hexane isomers safety data sheet.

SECTION VI. HEALTH HAZARD INFORMATION	TLV 500 ppm (See Sect II)

Excessive exposure to vapors of this material can be irritating to the respiratory tract and can produce dizziness, intoxication and unconsciousness, depending on exposure level and time. Liquid contact with the eyes is irritating; skin contact can be irritating, defatting, and lead to dermatitis when repeated or prolonged. Ingestion is upsetting and irritating to the GI tract.

FIRST AID:
Eye contact: Flush eyes well with running water for 15 minutes. Get medical help if irritation persists.
Skin contact: Wash contact area with soap and water. Remove contaminated clothing promptly. Replace skin oils with lotions or creams.
Inhalation: Remove to fresh air. Restore breathing if required. Get medical help.
Ingestion: Get medical help immediately.

SECTION VII. SPILL, LEAK, AND DISPOSAL PROCEDURES

Establish plans and provide training prior to any emergency situation. When spills occur exclude workers from area except those assigned to clean-up who must have proper protection against inhalation of vapors or contact with liquid (see Sect. VIII). Provide maximum explosion-proof ventilation. Eliminate ignition sources. Flush away from sensitive areas with a cold water spray. (Flush to ground not to the sewer!) Small amounts of liquid (or absorbed liquid) can be allowed to evaporate with good ventilation or in a hood or open area; large spills should be picked up in a safe and appropriate manner for disposal.
DISPOSAL: Scrap material can be burned in an approved incinerator in accordance with Federal, State and local regulations.

SECTION VIII. SPECIAL PROTECTION INFORMATION

Provide general ventilation and local exhaust ventilation which is explosion proof and adequate to meet the action level or TLV requirements. For emergency or nonroutine exposures above the TLV an approved organic vapor cartridge respirator can be used for a limited time below 1000 ppm, an approved gas mask (preferred) with organic vapor canister up to 4000 ppm, and approved air-supplied or self-contained respirators for higher or unknown exposure levels. Full-facepiece protection above 1000 ppm.
Prevent skin contact by use of impermeable gloves, aprons, boots, suits, etc as needed by the circumstances of use. Prevent eye contact by use of safety glasses, goggles, or face shield with goggles or glasses as the workplace circumstances may require. Eyewash stations and safety showers should be readily available to areas of handling and use.

SECTION IX. SPECIAL PRECAUTIONS AND COMMENTS

Store in closed containers in a clean, cool, well-ventilated place, away from oxidizing agents and sources of heat and ignition. Protect containers from physical damage. Ground and bond containers for transfers to prevent static sparks. Use metal safety cans for handling small amounts. Storage and handling conditions must follow OSHA regulations for OSHA Class IB flammable liquid. No smoking in areas of storage or use. Avoid breathing vapors! Prevent contact with skin or eyes! Do not ingest! Use of preplacement exams and medical surveillance (especially for the nervous system and skin disorders) is recommended. Exposure monitoring and recordkeeping requirements which have been proposed by NIOSH for alkanes should be instituted.

DATA SOURCE(S) CODE: 2-4, 6-8, 10-12, 19	APPROVALS: MIS, CRD
Judgments as to the suitability of information herein for purchaser's purposes are necessarily purchaser's responsibility. Therefore, although reasonable care has been taken in the preparation of such information, General Electric Company extends no warranties, makes no representations and assumes no responsibility as to the accuracy or suitability of such information for application to purchaser's intended purposes or for consequences of its use.	Industrial Hygiene and Safety
	MEDICAL REVIEW: 12/79

GENERAL ⊛ ELECTRIC

bus, Ohio, plant in the early 1970s, it was suspected that sustained, high exposures from careless, uncontrolled use of the solvent methyl butyl ketone (MBK) had been the cause. Detailed animal studies showed that the neuropathy did not result *directly* from MBK but from its metabolite in the body, 2,5-hexanedione. Studies with *n*-hexane had earlier shown that high, sustained exposures to this hydrocarbon also produced a peripheral polyneuropathy, and it was subsequently found that the same 2,5-hexanedione metabolite was involved:

$$CH_3\overset{\|}{\underset{O}{C}}CH_2CH_2CH_2CH_3 \longrightarrow CH_3\overset{\|}{\underset{O}{C}}CH_2CH_2\overset{\|}{\underset{O}{C}}CH_3 \longleftarrow CH_3CH_2CH_2CH_2CH_2CH_3$$

Methyl Butyl Ketone   2,5-Hexanedione   *n*-Hexane
                             Neurotoxin

The branched chain isomers of *n*-hexane and MBK and the 5-carbon related compounds do not produce this neurotoxic effect since they cannot be metabolized to the 2,5-diketone structure. Other 2,5-diketones have now been shown to have this neurotoxic effect, but 2,5-hexanedione appears to be the most potent. These studies have established a basis of understanding of the uniqueness of the effects of *n*-hexane and MBK as compared to the neurotoxic effects of related compounds (Figs. 1 and 2). In most cases metabolism detoxifies chemicals and facilitates their elimination from the body. In this case, metabolism produces a specific neurotoxin, which is damaging to the axons of peripheral nerve cells.

The body can respond to exposures to some chemical materials by development of *tolerance*, so that higher exposures are required to produce the same toxic effects that lower exposures had originally produced. For example, exposure to zinc oxide fumes can produce "fume fever", but the body quickly develops a tolerance, so that illness does not occur with a repeated exposure. (In the case of zinc "fume fever", the tolerance is quickly lost in a few days of nonexposure.) Other materials can produce a *sensitization* after a period of uneventful exposure, in which the body develops severe allergy-type responses to even minute traces of further exposure. Such sensitization can last for many years. Materials which are strong sensitizers include isocyanates [toluene 2,4-diisocyanate (TDI) and methylene bis-4-4'-phenyl diisocyanate (MDI)] which are polyurethane intermediates, certain amines (diethylene triamine and others which are epoxy curing agents), certain epoxies (butyl glycidyl ether), formaldehyde, and others.

4. Mutagens and carcinogens

Mutagens are materials which are capable of reacting with the genetic material of a living cell, deoxyribonucleic acid (DNA), to produce changes which are passed on in the heredity of the cell as it continues to divide. Such properties can be shown in laboratory tests that use bacteria or mammalian cells, such as the Ames Test, which uses a special mutant strain of *Salmonella typhimurium* that is not able to synthesize the essential amino acid histidine. When placed on a histidine-free culture plate, these bacteria do not grow. But if a mutagen is added which reverts the bacteria to histidine synthesizers, growth then occurs. This short-term test is useful because it correlates fairly well with chronic animal tests for carcinogens. Most materials which produce tumors or cancers in test animals also give a positive Ames Test; most materials which do not produce tumors or cancers in test animals, give a negative Ames Test.

Carcinogens are a special kind of mutagen which can change cells to give a new kind of growth (a neoplastic growth), different from the original cells. For governmental regulatory purposes a material is said to be carcinogenic if it produces either tumors or cancers in test animals and/or man. (Cancers are neoplasms that become highly invasive or malignant; they "send out colonies", so to speak. Tumors are neoplasms that are "benign" because they remain localized.) By 1978 more than 2300 materials had been reported to have produced tumors or cancers in man or in test animals. In the great majority of these cases it has been animal testing alone that has indicated a potential carcinogen for man. (Sec. III.B.2.)

Carcinogenic materials differ in potency. Some require very high dosage for very long periods of time (latency period) to produce low levels of benign tumors while other materials can produce malignant tumors in a shorter time at low dosage. Both saccharin and aflatoxins have been demonstrated to be carcinogenic in animal studies; saccharin is *very weak* in potency while aflatoxins are *very high* in potency. In both cases there is only limited information on direct human exposure. Carcinogens can vary in their potency in a given species or strain of test animal and in the young, as compared with the adult, animal. Long periods of time relative to the normal life span of the species (at least a 20-year latency period has been suggested for man) are usually needed for cancer to develop. Some materials must undergo metabolic activation by reaction with enzymes in the body to produce the "proximal" (functioning) carcinogen.

Cocarcinogens and *promoters* do not produce neoplasms by themselves. Cocarcinogens, acting in the presence of carcinogens, are able to increase the potency of the carcinogen and/or decrease the latency period before neoplasms develop in test animals exposed to a carcinogen. Ethanol acts as a cocarcinogen with vinyl chloride, causing a greatly increased rate of induced malignant tumors in animals. To be effective a carcinogenic promoter requires initial exposure of the cell to a carcinogen. This exposure is then followed by a lengthy, repeated exposure to the promoter before a neoplasm develops.

5. Teratogens and other reproductive effects

Materials which are systemically toxic to a pregnant woman can, of course, be toxic to the developing embryo or fetus if they are able to cross the placental barrier which separates the blood systems of mother and child. Babies have been born intoxicated or addicted to drugs because of the habits of their mothers. Transplacental effects can also occur because of exposures that are not injurious to the mother. A unique example is diethylstilbestrol. The daughters of women given this drug during pregnancy have a high risk of developing vaginal cancer at puberty.

In addition, a prospective mother may be exposed to materials at levels which are not at all injurious to her, but which can be injurious or devastating to a developing embryo or fetus. Such materials can be especially insidious during the early period of pregnancy (the first 40 days). An embryo or fetus can be severely affected by less than 10−20% of the dose level that would be toxic for the mother. Such materials are *teratogens*.

Before the 10th--14th day of pregnancy, embryotoxic effects are often embryocidal and result in spontaneous abortion. Less severe damage to cells may be bypassed and overcome as the embryo continues to develop. However, after the 14th day, *cell differentiation* and structural formation have begun, and cell damage after this critical point can be irreversible. Chemical attack on cells during the 14th− 40th day of pregnancy can slow specialized growth processes at a *critical* time in development or destroy essential structural cells (like the growing "bud" of an arm or leg), resulting in an obvious struc- tural defect without killing the fetus. The deformities of thalidomide babies are the classic example of *structural teratogenic effects*.

Behavioral teratogenic effects can also be caused by chemical at- tack on the growing fetus, both before and after the 40th day. Etha- nol has been established as a behavioral teratogen. Its effects can range from mental retardation in children of alcoholic mothers to mild- er effects in babies of social drinkers. Fetotoxic and teratogenic ma- terials can include lead, mercury, benzene, turpentine, carbon tetra- chloride, dimethylacetamide, aniline, and others. Much remains to be discovered in this area.

The delicate and intricate processes involved in reproduction can be interfered with by chemicals. "The Pill" is a prime example of this interference—in this case a planned, chemical interference. Male workers producing or using the pesticide DBCP experienced unexpect- ted sterility from exposure to this material. Epidemiological study identified at least 95 cases in which excessive DBCP exposure inter- fered in some manner with sperm production. The majority of the ex- posed men have recovered fertility, but some have not.

6. Insidious hazards

Toxic (and other) hazards can take us by surprise. The best defense is awareness that they exist, alertness to evidence, and maintained caution, keeping exposures to possible hazards at a minimum feasible level.

a. Unrecognized hazards. (1) These are hazardous properties or effects that no one has previously known or recognized. The carcinogenicity of vinyl chloride monomer and the special neurotoxic effects of n-hexane and MBK prior to the 1970s are examples.

(2) They also include known hazardous properties or effects for which information is inadequately communicated to and/or received by those potentially exposed. This hazard can have the same effects as (1) on those involved, but it is a different information problem, one that diligence and effort can certainly ameliorate. For example, although animal tests in 1961 had indicated a male reproductive hazard from DBCP, exposures in the 1970s produced a surprise of sterility for 95 workers using it.

(3) The exposure of hypersensitive individuals to conditions that experience has shown to be safe for "normal" persons can be an an unrecognized hazard. A very small percentage of people will have allergies and sensitivities that may make it hazardous for them to do what everyone else does. Individual differences must always be considered. Hypersensitivity can arise from past exposures, disease (both past and present), heredity, etc.

b. Known hazards without satisfactory warning. (1) The hazard of breathing high concentrations of toxic hydrogen sulfide gas is well known to oil-well drillers. Still, many fatalities have occurred when individuals have walked into ground depressions or sumps into which this heavier-than-air gas has seeped, displacing the air. At a high concentration the sense of smell is immediately fatigued, and, with a breath or two, paralysis of the breathing centers can occur. Unthinking rescuers collapse with the victims they were trying to help. Special care and caution is required for this kind of hazard. Training and availability of proper equipment are needed.

(2) Materials which require long periods of exposure or long periods after exposure for injury to be manifest can be hazardous. Cigarette smoking is a good example. Because the results are so long in coming, individuals choose present pleasures or convenience, even when hazards are well established. Such conditions can lead to carelessness or to a "so what" attitude.

c. Persistent materials. Stable materials that persist in the environment and which are highly fat-soluble and very slow to metabolize in biological systems can be hazardous, especially if they have been use-

ful materials produced in very large amounts. Such materials can collect in the food chain and appear in penguin fat and mother's milk.
Examples are DDT and PCBs. Their hazard is their persistence, if
nothing else. What hazards they may produce for man or the environment are not fully known.

V. INDUSTRIAL HAZARDS WITH MATERIALS

Materials hazards in a plant situation are the summation of all the previously discussed hazard potentials of materials for all the materials
assembled for the processing and fabrication operations of the plant.
The safety personnel, the industrial hygienist, the technologist, and
the management must unite their skills to carry out necessary handling
and processing in a manner that is safe, healthful, protective of the
environment, and protective of the plant.

The materials to be considered include component materials
(those that become part of a product), processing materials (fuel
gases, cutting fluids, solvents, pickling acids, etchants, etc.), and
peripheral materials (janitorial and maintenance supplies, boiler and
engine fuels, duplicating fluids and inks, etc.). Handling, storage,
work procedures, worker training, engineering controls, monitoring,
labeling, emergency and spill procedures, emissions and effluents,
waste disposal and transport, are all part of the industrial use of materials. Regulatory requirements are imposed on the whole.

In order to control materials hazards in the plant it is first
necessary to identify what materials are currently used or stored and
to obtain detailed information on each one. Such information can be
based on a safety data sheet (Figs. 1 and 2) obtained from the supplier or other source and supplemented as necessary to provide basic
hazard and handling information. Secondly, to assess hazard the
amounts of materials used and the processing conditions or operations
involving them must be identified and analyzed. Thirdly, the labeling, handling, transfer, and storage of materials for use must be considered, including container or bulk handling, packaging, and inventory control. How much is needed at one time? Where are materials
stored? In what amounts? Near what? How moved? By whom? Procedures for introducing new materials into the plant must be established.

With the knowledge of what is used, how and where it is used,
and the amounts involved, it is necessary to review engineering controls such as process and materials isolation, ventilation requirements,
electrical codes, fire-resistant construction, fire-extinguishing sys-
terms, drainage and diking, etc. This is done to ensure that the facility design is compatible with materials, material flow, and processing
requirements.

Possible exposure of workers to the materials must be assessed. Placarding of especially hazardous areas, operations, and materials is needed. Measurement and monitoring of airborne hazards during all phases of processing must be carried out to establish exposures and to ensure worker health and safety. Possibilities of skin contact, liquid splashing, particulate generation and dispersion, etc., must be considered with additional engineering controls and (as a last resort) special, personal protective equipment—such as chemical safety goggles, impermeable coveralls, and respirators—provided. Workers must be trained in the hazards of their work, especially in the hazardous properties of the materials they work with or near. They may need preemployment and regular medical examinations to ensure suitable and continued health in their work situation.

The impact of industrial materials use on the environment must be considered. Emissions into the air, effluents into the water, and hazardous waste disposal are all severely regulated. In addition to normal operating wastes from a plant, spills, accidents, and upsets must be considered and, in so far as possible, planned for, so that toxic or hazardous pollutants are not released to the environment. Secondary containment (to catch spills), reduced emission processing and/or materials [to meet EPA Guidelines and volatile organic compounds (VOCs) emission requirements], and water treating facilities [to obtain a permit under the National Pollutant Discharge Elimination System (NPDES)] may be required. In addition, to get rid of materials which have been legally defined as hazardous waste (HW), permits to generate, to store, to transport, and to operate treatment and disposal facilities must be obtained from EPA. The transporting of HW requires use of a manifest system and meeting of EPA and DOT regulations.

When all the requirements for the use of a material in an industrial application become clear, it may often be desirable to substitute, when possible, a less hazardous material. (In addition to substituting for the traditional reasons, that is, supply or supplier problems and material cost advantages, we now substitute to avoid health, safety, or waste disposal problems.) Some substitutions are easily carried out, but it may be difficult to substitute for a critical engineering material requiring a performance life of many years with which we have had long experience, such as asbestos. In some cases substitution is not feasible, so we work at improving processing, handling, and waste disposal to meet requirements for worker, customer, and environmental safety. We have to know the details of Food and Drug Administration (FDA), OSHA, DOT, EPA (CAA, CWA, TSCA, RCRA, "Superfund"), etc., requirements and the regulations for consumer product safety and product liability (and meet them) if we are to stay in business. When we do substitute materials or processing, we have

to be sure that we are not jumping from the frying pan into the fire.
We must obtain a broad scope of information, not only on material performance in our product, but also on health, safety, waste disposal,
etc., for the new material. A whole spectrum of materials properties,
not directly related to product performance, has now become a more
significant factor in the engineering equation. We have to become a
lot smarter about materials to carry out their industrial use in our
regulated environment.

BIBLIOGRAPHY

American Conference of Governmental Industrial Hygienists (ACGIH)
    (1980), Threshold Limit Values for Chemical Substances and
    Physical Agents in the Workroom Environment.
Bretherick, L. (1979), Handbook of Reactive Chemical Hazards, 2nd
    Edition, Butterworths, Boston.
Code of Federal Regulations, Titles 15,16,29,40,49, and others.
Federal Register (1980), Vol. 45, No. 98, May 19, 1980, p. 33066.
Fischbein, A., et al. (1979), Clinical Findings Among PCB-Exposed
    Capacitor Manufacturing Workers. In Health Effects of Halo-
    genated Aromatic Hydrocarbons W. J. Nicholson and J. A.
    Moore (Ed.), Annals of the New York Academy of Sciences,
    Volume 320, May 31, 1979, pp. 703ff.
Goldwater, L. G. (1972). Mercury-History of Quicksilver, York
    Press, Baltimore, p. 262.
Industrial Chemical News (1980), Vol. 1, No. 3, pp. 15-16.
Matthews, H. B. and Kato, S. (1979), The Metabolism and Disposi-
    tion of Halogenated Aromatics. In Health Effects of Halogena-
    ted Aromatic Hydrocarbons, pp. 134ff.
Maugh, T. H. II (1978). Chemical Carcinogens, The Scientific
    Basis for Regulation. Science 201: 1200-1205; Chemical Car-
    cinogens: How Dangerous Are Low Doses? Science 202:
    37-41.
National Institute of Occupational Health (NIOSH) (1974), An Identi-
    fication System for Occupationally Hazardous Materials, No.
    75-126, U. S. Department of Health, Education, and Welfare.
Parke, D. V. (1968). The Biochemistry of Foreign Compounds.
    Pergamon Press, Oxford.
Patty's Industrial Hygiene and Toxicology, 3rd Edition, Vol. 1
    (1978), Vol. 2A (1981), Vol. 2B (1981) (G. D. Clayton and
    F. E. Clayton Eds.), John Wiley & Sons, New York.
Sax, N. I. (1975), Dangerous Properties of Industrial Materials,
    4th Edition, Van Nostrand Reinhold Co., New York.
Spencer, P. S. and Schaumburg, H. H. (1980). Experimental and
    Clinical Neurotoxicology. Williams & Wilkins, Baltimore, MD.

Toxic Substances Control Sourcebook (1978), A. McRae and L. Welchel (Ed.), Aspen Systems Corporation, Germantown, MD

Wassermann, M. *et al.* (1979), World PCB's Map: Storage and Effects in Man and his Biologic Environment in the 1970's. In *Health Effects of Halogenated Aromatic Hydrocarbons.* pp. 69ff.

TYPES OF ADHESIVES

7

Structural Adhesives: Today's State of the Art

Justin C. Bolger *Amicon Corporation, Lexington, Massachusetts*

I. INTRODUCTION

The adjective "structural", as in structural adhesives, has come into
vogue during the twelve years since the writing of the original chapter on "Structural Adhesives for Metal Bonding" [1]. Adhesive manufacturers and their advertising departments now miss no opportunity
to use, or abuse, the word. Companies which formerly sold urethane,
acrylic, or anaerobic adhesives now call their products "structural

133

urethane", "structural acrylic", or "structural anaerobic" adhesives. Recently, this usage has further escalated, and these products are now called "second generation" or "third generation" structural adhesives.

Clearly, high-strength adhesives are growing rapidly in sales volume and in acceptance by progressively larger segments of industry. Adhesives which were once considered high-priced specialties used primarily by military contractors or aircraft manufacturers are now used to assemble passenger cars, trucks, buses, motors, appliance cabinets, computers, and other basic consumer products.

The purpose of this chapter is to provide an updated overview of today's state of the art in structural adhesives, as defined below. This chapter represents a complete rewrite of Ref. 1. Information on older, obsolete adhesive types has been eliminated, and new information has been added on the current generations of epoxy, acrylic, urethane, and other adhesives which are now growing in usage volume.

There is, however, still no universally accepted definition of the term "structural adhesive". Some writers try to differentiate between structural and nonstructural adhesives in terms of some arbitrary tensile shear strength procedure, such as ASTM D-1002. Here, we define structural adhesives as thermosetting resin compositions used to form permanent, load-bearing joints between two rigid, high-strength adherends.

Metals in general, and aluminum and its alloys in particular, comprise the most important class of adherends bonded by structural adhesives, but the same types of adhesives used for bonding metal are also used for bonding glass, oxides, ceramics, and glass-reinforced plastics. To make maximum use of the bond-strength capability of a structural adhesive, the cohesive strengths of the adherends must be high enough to cause failure in the adhesive rather than in one or the other adherend phase. Very few resin types provide the strength required for metal bonding, whereas a very large number of polymeric materials can be used to bond substances such as wood or paper, where relatively modest adhesive strengths are sufficient to cause the wood or paper, rather than the adhesive, to fail in service.

The most important feature of a structural adhesive is its ability to retain its load-bearing capability for very long periods under a wide variety of environmental conditions. Permanence, or durability, of an adhesive joint obviously depends on many factors other than the adhesive itself. These other factors include the design of the joint, the method of surface preparation, the extremes of temperature encountered in service, the loading history, and the exposure to water and other disruptive or corrosive media.

Despite early reluctance on the part of design engineers to use adhesives for "critical" bonding applications, it is now recognized that a small number of types of thermosetting resin adhesives, developed during the past 40 years, are able to provide useful bond strengths over an extremely wide use-temperature range, can bond virtually any high-strength material to another, and can survive many years of outdoor weathering, thermal cycling, and exposure to corrosive liquids and vapors. Although a few thermoplastic resin types might also be used for some of these applications (generally at the lower temperature extreme), the requirement that an adhesive joint be creep-resistant, that is, able to resist plastic deformation under long-term loading, generally means that a structural adhesive must become crosslinked during the process of forming and curing the joint.

II. STRUCTURE-PROPERTY RELATIONSHIPS IN THERMOSETTING ADHESIVES

The structure of any crosslinked, noncrystalline, polymeric adhesive can be represented by a "ladder model" in which the three-dimensional network is formed by tying together a relatively high—molecular-weight backbone polymer with random and relatively short crosslinking segments (Fig. 1). The properties of the network are then determined by the nature of the backbone-polymer chain segments, crosslink segments, and branch points. As any one, or any combination, of the structural changes shown in the left-hand column of Fig. 1 occurs, all of the properties shown in the right-hand column are changed in the direction shown. These basic property interrelationships, which are true regardless of the particular types of resins or

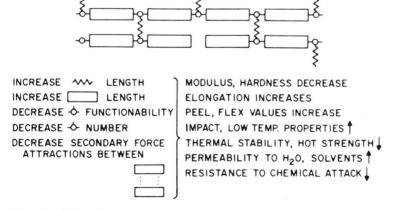

Fig. 1. "Ladder model" for thermosetting, noncrystalline adhesives: backbone-polymer chain segments, -☐- ; crosslink segments, ⋀⋀⋀; branch points, -○- . (From Ref. 1.)

reactants involved, are used subsequently to interpret the property combinations provided by commercially available structural adhesives.

A. Thermal and Chemical Stability

Consider first the ability of a polymer network to resist degradation at elevated temperatures or in the presence of oxygen, water, acids, bases, organic solvents, or other attacking media. For a predominantly aliphatic polymer, the basic $(CH_2)_n$ hydrocarbon building block is a relatively stable moiety. It resists attack in water over a very broad pH range and provides low permeability for water, oxygen, and polar penetrating solvents.

Useful structural adhesives can have, however, a very limited proportion of straight aliphatic carbon chains with the remainder of the polymer consisting of branch points and oxygen- or nitrogen-containing polar groups. Branch points are weak links from the standpoint of thermal stability. The hydrogen atom on a tertiary carbon is more labile than the hydrogen on a $-CH_2-$ unit. Hence, tertiary carbons can be dehydrogenated, particularly in the presence of oxygen, at temperatures lower than those required for the dehydrogenation of straight-chain hydrocarbons. When polyolefins degrade, chain scission begins at the branch points. Symmetrically substituted carbon atoms provide better thermal and oxidative stability than tertiary carbons do.

Other groups, which limit the use temperature of polymers and which a formulator normally tries to exclude from high-temperature adhesives, are aliphatic double bonds and strongly polarized carbon-oxygen or carbon-sulfur bonds. The isopropyl ether branch point formed in all cured epichlorohydrin-based epoxy resins, regardless of the curing agent, ultimately limits the use temperature of even the most highly crosslinked epoxy adhesives to long-term, load-bearing use below about 200°C.

1. Effect of substrate

Thermal and oxidative stability, as well as corrosion and water resistance, depend on the adherend surface as well as on the adhesive itself. Epoxy- and phenolic-based adhesives, for example, degrade less rapidly at elevated temperatures when in contact with glass or aluminum than when in contact with copper, nickel, magnesium, or zinc. As explained earlier [2], the divalent metals (Zn, Cu, Fe^{2+}, Ni, Mg, etc.) have a more basic (better proton acceptor) oxide surface than the higher valence metal oxides (Al, Fe^{3+}, Si) and hence serve to promote dehydrogenation reactions which lead to anion formation and chain scission. For any given metal, the method of surface preparation can determine oxide characteristics, and hence bond durability. The use of primers can also be decisive, as discussed in Sec. IV.

Hence, neither the durability of a joint nor the maximum use temperature of an adhesive can be predicted or taken directly from the manufacturer's literature. Almost invariably, strength values at elevated temperatures are reported based on tests with aluminum overlap joints—one of the most favorable cases. Only careful, long-term tests on the user's actual joint design, with the actual surface-treatment method, can give reliable design information for critical bonding applications.

2. Most stable linkages

The chemistry of high-temperature adhesives and coatings depends on several resonance-stabilized ring structures—most notably the benzene ring. The formulation of high-temperature adhesives reduces essentially to the problem of how to join benzene rings in a linear or three-dimensional network.

Direct benzene-benzene linkages are, of course, very stable—as evidenced by the temperatures at which coal tars are distilled and (ultimately) by the decomposition temperature of graphite. Aromatic ether and sulfone linkages are also relatively stable. But most high-temperature adhesives are made by extending benzene segments via heterocyclic rings to form polymers of the type shown in Eqs. (6) and (7). The imidazole ring, as in PBI polymers, has a resonance stabilization energy comparable in magnitude to benzene. Polyimides form interchain, secondary attractive forces which are strong enough to give linear polymers the rigidity (glass transition temperature T_g > 200°C) and creep-resistance of highly crosslinked thermosets.

A common feature of all of these aromatic structures is chain stiffness. None of the above linkages permits any significant degree of bond rotation or bond-angle deformation. As a result, high-temperature adhesives are invariably rigid—to the point of being brittle and readily fractured at room temperature. Attempts to toughen such resins via the addition of flexible, aliphatic chain segments invariably involve a serious sacrifice in thermal stability and hot strength.

Generally speaking, the only groups which offer *both* flexibility and high-temperature capability are the polysiloxanes. Silicone resins are prized for their ability to retain strength and flexibility over a temperature range which extends both well below and well above the range of epoxy- or phenolic-based thermosets. Although many laboratories have tried for many years to marry the best properties of silicones to epoxies or phenolics to obtain less brittle high-temperature adhesives, this effort has not yet yielded any structural adhesives able to progress beyond laboratory evaluation stages.

3. Resistance to hydrolysis

A special, but important, type of degradation, which can occur either in the bulk adhesive phase or near the adhesive interface, is hydroly-

Fig. 2. Elastomer samples after 30 days at 95% RH, 100°C, accelerated humidity aging test. The fluorinated polyacrylate (top left), the polyester-urethane (top right), and the anhydride-cured epoxy (center left) have all softened to a jelly-like liquid. The polyether-urethane (center right) has softened but retained its shape, while the nitrile-rubber—modified, one-component epoxy (bottom) has resisted hydrolytic attack.

sis—defined in chemistry tests as a double decomposition reaction that depends on the presence of ions formed from water. Hydrolysis requires the presence of water and, generally, also requires the presence of strong acid or (more frequently) strong base. The hydrolytic attack on an ester linkage in the presence of water at high pH is an important example of this mode of attack. It is initiated by attack on the electron-deficient atom in a polarized bond, as on the carbonyl carbon in Eq. (1).

$$R_1-\underset{\underset{\delta^+}{\overset{\overset{\displaystyle O\delta^-}{\|}}{C}}}{}-O-R_2 \xrightarrow[\text{H}_2\text{O}]{\text{OH}^-} \left[R_1-\underset{\underset{HO^-}{\overset{\overset{\displaystyle O}{\|}}{C}}}{}-OR_2 \right] \longrightarrow R_1\overset{\overset{\displaystyle O}{\|}}{C}-O^- + R_2OH \qquad (1)$$

Substitution of electron-withdrawing groups for the aliphatic R_1 and

R_2 groups can delocalize the charge at this point of attack, leading to reduced rates of hydrolysis. Thus, hydrolytic stability can vary from very good to completely unsatisfactory for adhesives flexible by the addition of polyesters, even for adhesives appearing to be similar in hardness and other physical properties.

Figure 2 shows how rapidly hydrolytic degradation can occur in a variety of flexible epoxy and polyurethane compounds. After exposure to 95% RH at 100°C for 30 days, as in the hydrolytic reversion test proposed by Gehimer and Nieske [3], the anhydride-cured epoxies, which are crosslinked almost entirely via aliphatic ester linkages, have both liquified to a jelly-like mass. The polyester-urethane has also liquified, whereas the polyether-urethane, having fewer ester linkages, has softened but has retained its shape. The nitrile-rubber-modified epoxy, however, having the same initial Shore A-2 hardness as the other four elastomers but containing fewer hydrolyzable linkages, has remained essentially unattacked.

B. Strength Properties of Structural Adhesives

Figures 3 and 4 illustrate the two most common strength tests for structural adhesives—the tensile, or overlap, shear test and the 90° or "T" peel test. Many previous authors [4–6] have discussed the significance of these tests. It is an unfortunate fact that many industrial-adhesive users, as well as some authors of adhesive data sheets, continue to regard peel and tensile shear values as inherent properties of an adhesive. In truth, tensile shear and peel strength depend on a virtually endless set of exterior variables—adherend composition, thickness, joint length and width, notches, flaws and end effects, bond- line thickness, cure conditions, cure pressure, and many others.

Adhesive joints can be loaded most usefully in pure shear. Strength in peel or cleavage is poor under the best of circumstances. This fact is perhaps best illustrated by the fact that most structural adhesives can support a 3000–5000-lb load under the tensile shear loading shown in Fig. 3, whereas very few adhesives can support more than a 5–10-lb load in the "T" peel mode of Fig. 4.

Rigid, glassy adhesives give low peel and tensile shear strengths because uneven stress concentrations permit initiation of fracture at low load levels. Very soft adhesives give low bond strengths due to a cohesive, tearing failure mechanism. Hence, the significance of "T" peel and tensile shear tests is that an adhesive must have a good balance of tensile strength *plus* elongation in order to give good test values.

Temperature causes predictable changes in bond strength, as shown by the parabolic shape of the strength values in Figs. 5 and 6. For every adhesive there will always be some low temperature at which brittle fracture occurs. There will always be some higher temperature,

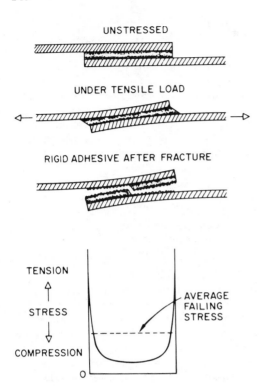

Fig. 3. Stress distribution and typical fracture pattern in overlap shear test.

above T_g, at which the adhesive becomes soft and fails in cohesion at low stress levels.

Figure 7 summarizes these hardness effects in relation to a number of important properties. In each case hardness is taken as an index of crosslink density as well as each of the other properties listed in the left-hand column of Fig. 1. As noted earlier, "T" peel and tensile shear strength go through a maximum on such plots. Silicones and most urethane elastomers, which are soft enough to register on the Shore A scale, are unable to give tensile shear strengths much in excess of about 600 psi on metals. Such adhesives are therefore rarely used for metal-to-metal bonding.

Structural-adhesive manufacturers normally class their products into one of the three categories shown in Fig. 7. The flexible grades—although offering the poorest chemical and thermal stability and, generally, providing the poorest resistance to bond degradation

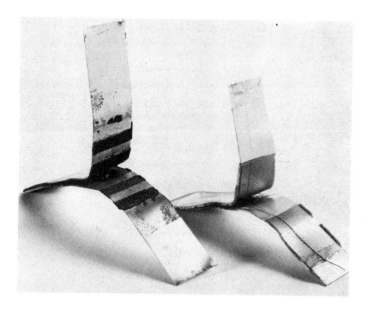

Fig. 4. Appearance of two liquid epoxy adhesives which both give about 35 lb/in. "T" peel strength. Left: Rigid (two-phase) nitrile-rubber—modified epoxy shows characteristic stop-start fracture bands. Right: Low modulus, single-phase, nitrile-rubber epoxy elastomer fails by appearing to tear smoothly away from one or the other metal surface. Wire was embedded to maintain constant 6-mil bond thickness.

in water—are normally recommended for bonding flexible structures, particularly where high peel strength is needed, where the system must cycle to low temperatures, or where the surfaces to be bonded differ widely in hardness or in expansion coefficient.

General purpose grades, intermediate in softness, normally are formulated to maximize overlap shear strength. The heat-resistant grades, which provide the best thermal, chemical, and moisture resistance, have the disadvantage of providing the lowest peel and impact strengths and are most likely to crack or cause cracking during thermal cycling.

C. Fatigue Strength and Endurance Limit

The durability of an adhesive joint depends not only on the adhesive and on the adherend surface, but also on a combination of time,

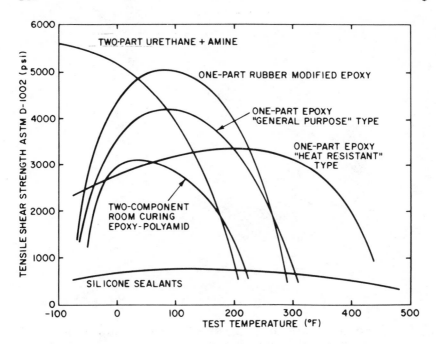

Fig. 5. Typical tensile strength for paste and liquid (100% solids) adhesives.

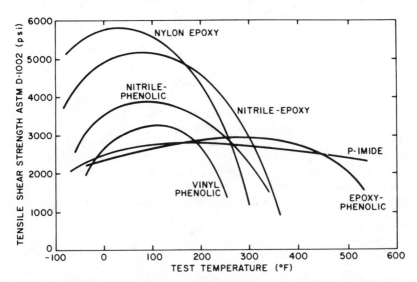

Fig. 6. Typical tensile shear strength data for tape, film, and solvent-based adhesives.

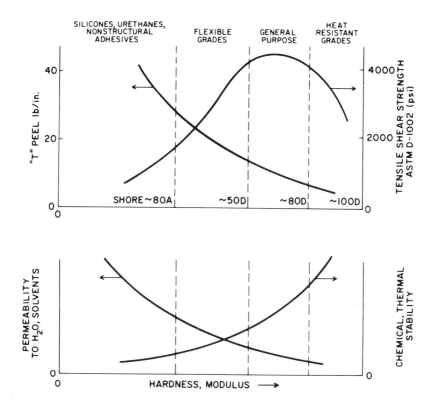

Fig. 7. Property trade-offs for single-phase, thermosetting adhesives.

stress, and environment. That is, a given joint which fails at a tensile shear stress of 4000 psi when pulled to failure by ASTM D-1002 cannot support the same stress level for a longer period of time without breaking. The same joint might fracture, for example, after several months at a constant stress of 2000 psi. It might fracture at even lower stress levels, in much shorter time, if maintained under constant load in a corrosive or high humidity atmosphere.

Because engineers frequently need to know the long-term "fatigue strength" or endurance limit of an adhesive joint, test procedures have been developed to maintain test specimens under constant stress for long periods in controlled environmental chambers. Lewis and Saxon [7] have advocated the concept of stress endurance limit (SEL), and their method of plotting breaking strength versus time to break, to obtain SEL, is shown in Fig. 8. The value of SEL always depends strongly on humidity. Many adhesive/adherend combinations give

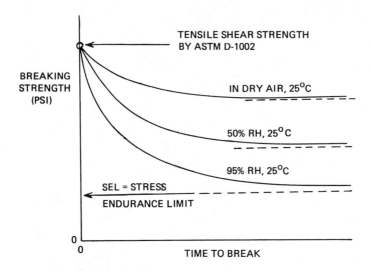

Fig. 8. Breaking strength of joints held at constant load in spring
loaded fixtures of Alcoa "napkin rings". The SEL predicts the long-
term load limit of a joint and is strongly dependent on temperature,
moisture, and surface preparation as well as on the adhesive itself.

very high initial strengths and may hold reasonable loads for long
periods in dry air but will fracture in minutes at low stress levels in a
high-humidity chamber.

D. Toughness and Impact Properties

The bond-strength predictions in Fig. 7 apply primarily to homogeneous, single-phase adhesives for which the strength trade-offs are predictable in terms of the ladder model of Fig. 1. Resourceful formulators have, however, known for some time that it is possible to make
polyphase adhesives whose combination of properties can give far better values than those predicted by a single-phase model of the type
shown in Fig. 7.

Almost all structural adhesives contain fibrous or particulate fillers, such as glass, other short fibers, or powdered aluminum, alu-
mina, or silica. Although adhesive manufacturers are suspected of
adding fillers to reduce costs, this suspicion is true only in a limited
number of cases. The proper use of a support mesh or of particulate
fillers is essential to provide most of the properties demanded of a
high-performance structural adhesive. Fillers control the rheology
and provide resistance to sag or flow during heat cure. Fillers reduce

shrinkage and moderate the peak exotherm during cure, reduce the thermal expansion coefficient (which is invariably much higher for all polymeric adhesives than for the metal or oxide adherends to be bonded), reduce permeability to and swelling by water, oxygen, or other penetrants. They also govern thermal and electrical conductivity, control dielectric properties, and provide pigmentation and decorative effects.

Particulate fillers are most widely used and are used at highest loadings, in rigid structural adhesives of the general-purpose and high-temperature types. Aluminum powder, in particular, is frequently used at high loadings since it can provide improvements in tensile strength as well as in heat resistance. It is not uncommon to be able to double the tensile shear strength of a rigid epoxy paste adhesive via the incorporation of aluminum powder. Aluminum-filled adhesives cannot be used for applications requiring a high dielectric breakdown strength or extended exposure to strong acids or base. Aluminum oxide and silica are normally used as fillers for such applications although they provide less tensile strength enhancement than aluminum does. These improvements in tensile shear strength are attributed, in part, to the reduction in shrinkage during cure and, in part, to the ability of these fillers to serve as crack stoppers, provided that the bond between the filler and the resin matrix is strong enough to prevent preferential crack propagation along the filler-resin interfaces.

Fillers are normally not used in high loadings in the flexible-adhesive grades where high peel strength and high deflection capabilities are the desired end properties and where the rubbery nature of the polymer matrix reduces the need to use fillers to moderate shrinkage stresses or crack propagation during fracture. In flexible adhesives, fillers are used only at levels sufficient to control flow, appearance, and electrical properties.

E. Reinforcement by Dispersed Rubber Phases

It has been known for over 20 years that one of the best ways to toughen rigid epoxy adhesives is via the incorporation of a discontinuous rubbery phase. Maximum toughening usually occurs when the rubber particles precipitate during cure as discrete particles on the order of 1000–2000 Å in diameter and when the rubber has functional groups which permit it to form covalent bonds at the rubber-epoxy interface. McGarry and Willner [8] and Lewis and Saxon [7] report that by proper addition of 5% by weight of a carboxy-functional nitrile rubber, the peel strengths and impact properties of several epoxy adhesives can be increased substantially. Willner [8] presents electron photomicrographs of cleavage of nitrile rubber—modified epoxy resins which clearly show the small, discrete spherical particles.

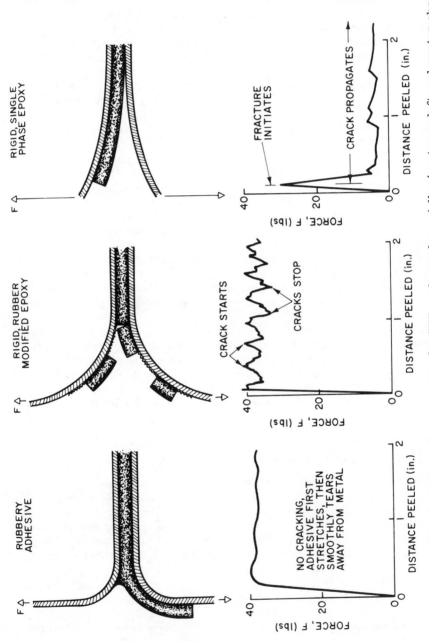

Fig. 9. Failure modes and force-distance plots for "T" peel samples. Adhesive types left and center above are also shown in Fig. 4 after fracture.

The crack-stopping effects of these dispersed rubber particles is illustrated by Figs. 4 and 9 which show "T" peel test results for two nitrile rubber—modified epoxy adhesives. Both contain 10 phr of a carboxy-functional nitrile rubber, and both give about 40 lb/in. "T" peel strength. But other physical properties, including the mechanisms whereby these high peel strengths are provided, are very different for these two adhesives.

The "single-phase", nitrile-epoxy adhesive was made by choosing an epoxy resin-curing agent combination in which the nitrile rubber is soluble. Hence, after cure, the adhesive is a clear, rubbery, single-phase resin having a Shore D hardness of 45 and an elongation of over 100%. In a typical "T" peel test this soft adhesive first stretches and then tears away from one or the other metal interface, giving the appearance shown in Fig. 4 (right) and the relatively smooth force-distance trace shown in Fig. 9. Since this flexible, single-phase, epoxy-rubber alloy achieves its high peel strength by bulk elongation, it shows all the other properties of the low-modulus, nonstructural adhesives of Fig. 7—low tensile shear strength (1500 psi) at room temperature, low hot strength, and poor bond-strength retention in water, solvents, acids, and bases.

The "two-phase", nitrile rubber—modified epoxy adhesive of Figs. 4 and 9 also contains 10 phr of the same carboxy-functional nitrile rubber. The rubber, however, is insoluble in this particular resin mixture (Epon 828 plus powdered dicyandiamide) and, therefore, precipitates as a dispersed phase. Since the rubber does not soften the continuous phase, this adhesive has a Shore D hardness of about 85, and gives tensile shear strength above 4000 psi. It has good hot strength plus the other properties characteristic of a rigid, general-purpose structural adhesive that are predicted in Fig. 7. Peel strength, however, is many times larger than predicted from the single-phase trade-offs (Fig. 7).

The mechanism for this peel-strength improvement can be deduced from Figs. 4 and 9. Without the rubber phase, a typical glassy adhesive first gives a momentary, high, peak-force value when the "T" peel test is started, corresponding to the force needed to deform the metal adherends and to initiate fracture in the adhesive. Once fracture has been initiated, it propagates rapidly ahead of the load front, normally traveling near one or the other interface, and the peel force recorded drops to, and remains at, a low value (typically in the range of 4—8 lb/in.).

With the rubber phase present, the peel force (Fig. 9) also rises abruptly to a sharp, peak-force value at the start of the test. Brittle fracture also appears to be initiated as before, but the fracture plane propagates only a short distance before being stopped by the rubber phase. Frequently, the plane of fracture shifts after each localized fracture from one interface to the other, give a distinctive "rail-

road track" fracture pattern shown in Fig. 4 (left). Figure 9 shows
the force trace which corresponds to the repeated crack stops and
starts in this "stick-slip" peel mode.

III. TAPE AND FILM ADHESIVES

Structural adhesives are most conveniently categorized by the physical
form in which they are used, with the major difference being between
adhesives which would be solids and those which would be liquids (or
spreadable pastes) in the absence of volatile, nonreactive solvents.

    1. Solid adhesives
        a. Film adhesives (unsupported)
        b. Tape adhesives (supported)
        c. Solid powders and preforms
        d. Solvent-based adhesives and primers
    2. "100% solids" paste and liquid adhesives
        a. One-component: long shelf life
            (1) Heat-cured
            (2) Cured by surface or anaerobic catalysis
            (3) Cured by exposure to ultraviolet (uv) light
        b. Two-component: short pot life
            (1) Room temperature-cured
            (2) Heat-cured

Of these, the most widely used types are the tape and film solid adhesives, discussed in this section, and the one-component (heat-
cured) and two-component (room temperature—cured) pastes discussed in Sec. IV.

Film adhesives are made by blending one or more high-molecular-
weight polymers—normally elastomers—with curing agents, fillers, and
other compounding ingredients and then extruding, calendering, or
casting the mixture in thin films, typically 5—10 mils thick. If the
mixture is cast or calendered into a mesh support, such as a woven or
nonwoven mesh of glass or other fibers, then the resulting supported
film is called a *tape* adhesive. Films and tapes may be either soft and
tacky or stiff and dry. They may be room temperature storable or
may require refrigeration between manufacture and time of use. Although early tape adhesives were made to cure at room temperature
after removal from dry ice storage, all are now cured at elevated temperature and pressure.

The distinguishing compositional feature of tape and film adhesives is that they contain a high proportion of a high-molecular-weight
polymer. Whereas a typical "100% solids" paste or liquid adhesive
must—to remain fluid and usable—contain only low-molecular-weight
resins, tape and film adhesives frequently contain polymers with

molecular weights of 20,000 or more. Based on the ladder model (Fig. 1), it follows that adhesives made from these high-molecular-weight linear polymers can be very much tougher and more resilient and can provide more recoverable elongation than the highly branched networks formed by curing the low-molecular-weight resins used in paste adhesives.

Figures 5 and 6 illustrate this point by comparing typical tensile shear data reported by the manufacturers for a variety of adhesive types. The best of the tape and film types have both higher peak values and broader service-temperature ranges than the best of the "100% solids" types.

Tape and film adhesives dominate the airframe market for structural adhesives. During the past 25 years, tape and film adhesives have been used for honeycomb bonding and other assembly procedures to an increasing extent on each succeeding generation of jet aircraft, from several hundred pounds of tape adhesives on early commercial jets, such as the Boeing 707, to several thousand pounds on large jets, such as the Boeing 747. The design of high-performance military aircraft such as the B-58 "Hustler" bomber or the swing-wing F-111 fighter was predicated on the extensive use of tape and film adhesives in the wing, tail, and fuselage structures. Frequently, designers will combine adhesive bonding with mechanical fastening to obtain advantages which cannot be provided by either technique alone. Combinations of bonding and riveting can provide better fatigue characteristics, more uniform load distribution, higher strength-to-weight ratios, and smoother skins than previous, all-riveted construction could [9].

The two primary reasons for the use of tape and film adhesives by the airframe industry are:

1. Handling and reliability advantages
2. Better toughness than pastes or other adhesive types.

The handling and reliability advantages stem from the fact that the tape or film adhesives come ready to use—no mixing, no degassing, no possibilities for errors in adding catalyst. Tapes can be tested both by the manufacturer and by the user. It is common practice for airframe manufacturers to run bond tests on small patches of material cut from *every* roll of tape shipped in by the manufacturer. Tape adhesives also make possible a variety of lay-up techniques in which large wing or tail structures can be assembled from skins, support spars, honeycombs, and edge members. These parts are interleaved with precut sheets of tape, assembled, inspected, clamped, cured, and reinspected to yield a virtually defect-free structure. The use of a support mesh also serves to control the bondline thickness of tape adhesives, thus avoiding thin, adhesive-starved areas where curvature or external pressure is largest.

The aircraft designer places an extremely high value on those properties of an adhesive associated with "toughness"—such as high peel strength, high impact, cleavage, and fatigue strength; ability to survive repeated cycling between high and low temperature extremes. A designer can compensate for low tensile shear strength by increasing the bonded overlap areas but cannot tolerate any progressive peel or cleavage fractures.

Although a bewilderingly large number of tape and film adhesives are now listed in manufacturers' catalogs, these adhesives are, in fact, composed of a relative small number of combinations of the three components shown in Table 1:

1. A high molecular-weight "backbone" polymer—which provides the elongation, toughness, and peel properties of the adhesive

2. A low molecular-weight crosslinking resin—invariably, either an epoxy or phenolic type

3. A curing agent (catalyst) for the crosslinking resin

Adhesives are named according to the identity of components 1 and 2 (e.g., nitrile-phenolic, nylon-epoxy) and are, therefore, also frequently referred to as "two-polymer" adhesives. Adhesives based on phenolic crosslinking resins liberate volatiles and, therefore, require high pressures to prevent void formation during cure, whereas adhesives based on epoxy curing agents require only sufficient pressure to maintain alignment and compensate for cure shrinkage.

A. Vinyl Phenolics

The first successful structural adhesive was developed in the United Kingdom during World War II for assembling de Havilland aircraft. Called the "Redux" adhesive system by its inventor (N. A. de Bruyne) at the Aero Corporation [10], it was made by spreading a resol phenolic solution on the part to be bonded, sprinkling a powdered polyvinyl formal resin on the surface, allowing the solvent to evaporate, closing the bond, and then curing under heat and pressure. Crude as this technique sounds by today's standards, the Redux adhesives provided very much higher bond strengths and bond durabilities than had ever before been obtained in any synthetic resin adhesive. They were the starting point from which modern structural adhesive systems have evolved. The Redux system first demonstrated the basic principle of toughening a phenolic resin with a high-molecular-weight linear polymer.

The "vinyl" in vinyl-phenolic adhesives refers either to a polyvinyl formal (PVF) or polyvinyl butyral (PVB) resin. Acetal resins of this type are widely used in heat-cured coatings, enamels, and primers, in air-drying wash primers, and in many other adhesive applications in addition to being used in vinyl-phenolic structural adhe-

Table 1 Most Important Tape and Film Adhesives

Designation	Backbone polymer	Crosslinking resin	Catalyst	High-pressure cure
Nylon-epoxy	Soluble nylon	Liquid epoxy	Dicy-type	No
Elastomer-epoxy	Nitrile rubber	Liquid epoxy	Dicy-type	No
Nitrile-phenolic	Nitrile rubber	Phenolic Novolac	Hexa, sulfur	Yes
Vinyl-phenolic	PVB or PVF	Resol phenolic	Acid	Yes
Epoxy-phenolic	Solid epoxy	Resol phenolic	Acid	Yes

Source: Ref. 1.

sives. Polyvinyl butyral resin, for example, is the major component of the adhesive used to laminate glass for automobile windshields and other "safety glass" uses. These acetals are prepared by partially hydrolyzing polyvinyl acetate to polyvinyl alcohol and then condensing with formaldehyde or butyraldehyde:

$$(2)$$

where for p-vinyl formal (PVF); $R = H$
for p-vinyl butyral (PVB); $R = -C_3H_7$

The molecular weights are high for the important adhesive grades, and the ratios of alcohol to acetate and acetal groups vary over wide limits. The importance of chain length can be appreciated by noting that Monsanto's Formvar 15/95B, although having only 6% alcohol units by weight, has an average of about 46 hydroxyl groups per chain because of its high degree of polymerization.

The alkylated resol-phenolic crosslinking resins are made by condensing phenol with a molar excess of formaldehyde in the presence of alcohol to produce a heat-reactive methylol functional resin of the formula:

$$(3)$$

where $-Z = -CH_2-$ or $-CH_2OCH_2-$
and $-R = -H$, $-CH_3$, or $-C_4H_9$

When heated above 300°F (150°C) this resin crosslinks by methylol-methylol condensation and by condensation with one of the hydroxyl groups on the PVB or PVF "vinyl" resin:

$$\underset{\text{(phenolic)}}{\overset{\overset{\text{H}}{\underset{|}{\text{O}}}}{\bigcirc}}\!\!-\!\text{CH}_2\text{OR} \;+\; \text{HO}-\underset{\underset{|}{\text{CH}_2}}{\overset{\overset{\text{CH}_2}{|}}{\text{CH}}} \;\xrightarrow[\Delta]{\text{H}^+}\; \underset{}{\overset{\overset{\text{H}}{\underset{|}{\text{O}}}}{\bigcirc}}\!\!-\!\text{CH}_2-\text{O}-\underset{\underset{|}{\text{CH}_2}}{\overset{\overset{\text{CH}_2}{|}}{\text{CH}}} \;+\; \text{ROH}\!\uparrow \qquad (4)$$

   In a typical recipe [11–13], 100 parts of PVB or PVF are combined with from 30 to 150 parts of the resol phenolic. Fillers and
catalysts, such as phosphoric or toluene sulfuric acid, may also be
added. The mixture is then calendered or solvent cast as a supported
or unsupported film. The higher the phenolic-to-acetal ratio, the
higher the use temperature the bond will withstand, but the lower the
flexibility and peel strength will be (as predicted by Fig. 7). Polyvinyl formal resin gives higher hot strength and tensile shear values,
but lower peel and flexibility, than PVB does at equal phenolic loadings. Solomon [14] and Minford [15] have shown the good long-term
bond-strength retention in water and other adverse environments of
vinyl-phenolic adhesives, such as Narmtape 105 and FM47. Despite
these generally good performance properties, however, the modern
trend is to move away from vinyl-phenolic adhesives toward one of the
newer types which provide cures at lower temperatures and pressures,
plus higher hot strength, higher peel strength, and other performance advantages which are not available in the vinyl-phenolic family.

B. Epoxy Phenolics

During the early 1950s Black, Blomquist, and others at the Forest
Products Laboratory (FPL) were engaged in a research program,
sponsored by the U.S. Air Force, aimed at developing higher temperature aircraft adhesives. These researchers were aware of the thermal limitations of the acetal resins and were the first to systematically
combine the (then) newly available high-molecular-weight epoxy resins
with resol phenolics to produce the tape adhesives now referred to as
epoxy phenolics [16,17].
   Solid epoxy resins derived from bisphenol-A (Table 2), have
about the same hydroxyl functionality as adhesive grades of PVB or
PVF resins do and undergo the same condensation reactions with phenolic resins. However, use of the epoxy polymer as the polyol, as in
the FPL 878 or 422J recipes shown in Table 3, results in a major improvement in thermal stability.
   The 442J formulation was designed for use on a glass mesh support, in which form it cured to give 2000–3000 psi tensile shear at
room temperature, good strength to 200°C, and excellent long-term
resistance to moisture. Adhesives of this type have limited storage

Table 2 Commercially Available Bisphenol-A—Type Epoxy Resins

	Epoxide equiv.[a]	g to esterify 1 mole[b] fatty acid	Av. mol wt.	n_{avg}	%OH[c]	Shell: Epon	CIBA: Araldite	DOW: Der	CELANESE Epi-Rez
Liquid resins	170-200[d]			<.1		815	506	334	504
	175-190[e]					826	6005	332	-
	175-210	85	380	0.1	20	828	6010	331	510
	225-290	105	470	0.5	17	834	6040	337	515
Solid resins	400-525	130	900	2.0	13	1001	6061	664	520
	870-1030	175	1400	3.7	10	1004	6084		530
	1500-2050	190	2900	8.8	9	1007	6097	667	540
	2400-4000	200	3750	12.0	8.5	1009	6099		550

[a]Epoxide equivalent = g resin per oxirane group.
[b]Assume a monobasic acid, equiv. wt $\geq$280.
[c]Assuming one oxirane yields two hydroxyls on esterification.
[d]Low viscosity grade, contains 7 to 10% butyl glycidyl ether.
[e]"High purity" grade with very low "n" value. Crystallizes at room temperature.
Source: Data from Ref. 1.

Table 3 Epoxy-Phenolic Adhesives: Compositions of Commercially Available Types

FPL 878	Epon 422J	Epon 422
20 Epon 1007 160 Resol phenolic 20 Solvent 3 Hydroxy-napthanoic acid 1.5 *n*-propyl gallate	33 Epon 1001 67 Resol phenolic 100 Al powder 1 Cu-quinolate	75 Epon 1004 25 Resol phenolic 1 H_3PO_4

Source: Data from Ref. 18.

life at room temperature—tending to gel on storage unless refrigerated. Note the use in many of these formulations of quinolate or gallate stabilizers. These additives are good chelating agents for iron and increase the bond durability above 200°C, particularly when used to bond stainless steel.

Many manufacturers now offer FPL-type epoxy-phenolic adhesives, including Shell's Epon 422, Narmco's Metlbond 302, Aero's Aerobond E422, Bloomingdale's HT424, and (PPG's) Plymaster ACG1031. In many cases the trend is toward higher epoxy loadings for more toughness and toward more acid accelerator for shorter shelf life but faster cure, as in Epon 422 (Table 3).

Despite the fact that it is now over 30 years since the first FPL formulations were published, and despite the enormous effort which has gone into the development of new polymeric systems designed for use at elevated temperatures, epoxy phenolics are still considered the best adhesives for long-term use in the range from about 350 to 450°F (180 to 250°C). Solvent-based epoxy-phenolic primers are still widely used to prime aluminum and other surfaces to enhance the durability of modern elastomer-modified epoxies.

While developing the epoxy phenolics, Black and Blomquist [16] at the FPL began finding and publishing favorable results using a chromic acid sulfuric acid mixture to etch aluminum surfaces before bonding. This method came to be known as the "FPL etch" and became the standard preparation method specified in ASTM-1002 and in most military and aircraft specifications. It became, for the next 25 years, virtually the universal method for preparing aluminum for adhesive bonding.

C. Nitrile Phenolics

Nitrile-phenolic adhesives are made by blending a nitrile rubber with a novolac-phenolic resin. Although they were previously used for aircraft bonding, their major present use is for bonding automotive brake shoes and clutch discs. They are also used in many other nonmilitary applications because of their low cost, high strengths at temperatures up to about 140°C, and exceptional bond durability on steel and aluminum. Minford [15] has documented the strength retention of nitrile phenolics, such as Narmco's FM-61, on aluminum after extended salt-spray, water-immersion, and other long-term exposure tests. It is probably true that no other adhesive type exceeds the nitrile phenolic's ability to maintain bond strength on steel or aluminum after extended exposure to water, salts, or other corrosive media, and to prevent undercutting via corrosion of the metal substrate.

The nitrile phenolics were developed by the major U.S. rubber companies shortly after World War II. The formulation given by Engel [19] for Bloomingdale's PA-101, typical of the complicated recipes for which rubber compounders are noted, contains a crumb rubber, two novolac phenolics, carbon black, sulfur, hexa, and other additives. The novolac phenolics most often used, such as Hooker's Durez 12968, are made by condensing formaldehyde with an excess of a phenol derived from cashew nut oil to yield a thermoplastic phenolic having unsaturated 15-carbon side chains R, of the formula:

$$(5)$$

Until recently, the nitrile rubbers most widely used in these adhesives were medium molecular-weight grades, of about 30% nitrile content, such as Goodrich's Hycar 1001, Goodyear's Chemigum N5, and Naugatuck's Paracril D. Recent practice tends toward the use of nitrile rubber grades which contain a small proportion of carboxyl groups, such as Goodrich's Hycar 1072. Reynolds [20] has noted the beneficial effects on adhesion to metals obtained by incorporating small amounts of carboxyl groups into nitrile-phenolic adhesives. Many other authors have also shown the importance of carboxyl groups in adhesives and coatings [21–24].

The peak use of the nitrile-phenolic adhesives was during the decade of construction of the B-57 and B-58 bombers. The B-58 "Hustler" design featured extensive and innovative use of adhesives to obtain speed and other performance improvements [24]. But the high cure temperature of 350°F (180°C) and the high clamping pressures required for the nitrile phenolics, while not a serious problem in making automotive brake shoes and clutch discs, caused serious problems in aircraft construction because of the size and wall thickness of the autoclaves required and because of distortion of the aluminum parts at 350°F (180°C).

By the late 1960s, other elastomer-modified adhesive types began to displace the vinyl and nitrile phenolics in the aircraft industry. Crosslinked with epoxy, rather than with phenolic resins, these new epoxy adhesives offered advantages in lower cure temperature and lower cure pressure. They also provided improved toughness and peel strengths.

D. Nylon Epoxies

The nylon-epoxy adhesives, which first came into prominence in the early 1960s, appeared to herald a new age in adhesives. Whereas adhesive technologists had tried for two decades to provide more than about 50 lb/in. in peel strength, these nylon epoxy adhesives provided a quantum jump in toughness—yielding climbing-drum peel strengths above 150 lb/in., tensile shear strengths upwards of 7000 psi, and exceptional impact properties.

The various crystalline polymers which comprise the nylon family—nylon-6; nylon-6,6; nylon-6,10; nylon-11; etc.—have long been noted for their toughness and tensile properties. But because of their high polarity, high melting point, and high degree of crystallinity, the standard nylon types are incompatible with most other resins, notably with epoxies. The key to the development of the nylon epoxies was the availability of noncrystalline nylons which were soluble in alcohols and other solvents, which melted at temperatures below their decomposition temperatures, and which could be blended with epoxy resins to yield curable solid films, tapes, or solvent-based adhesives.

These semicrystalline, adhesive grade nylons were prepared by a variety of techniques. Smith and Sussman [25] describe low-melting nylons prepared by melt blending mixtures of crystalline polymers to obtain random copolymers by chain segment interchange. Scrambled nylons can also be made by polymerizing mixtures of amines and acid chlorides. Smith and Sussman give performance data for adhesives prepared from a variety of mixtures of epoxy resins with such nylons.

Commercial examples of adhesive-grade nylons include BCIs 600- and 800-series nylons and Du Pont's Zytel 61. Bloomingdale's FM-1000, used for many years for laminating helicopter rotor blades and for honeycomb core-to-skin bonding, consists essentially of Zytel 61, resorcinol diglycidyl ether, and an alkyl-substituted melamine catalyst.

Table 4 gives tensile shear strengths and peel values for a series of adhesives made by dissolving various ratios of nylon and epoxy resins in an alcohol/water solvent mixture. Note the dramatic effect on bond values due to the nylon addition—peel strengths increase from 2−3 lb/in. without nylon to 180 lb/in. at room temperature with 85% nylon. Tensile shear strength increases from about 3000 psi to between 6000 and 7000 psi. Although some nylon-epoxy adhesives

Table 4 Nylon-Epoxy Adhesive Compositions

Adhesive composition			Tensile shear strength ASTM D-1002 (psi) at			Peel strength, MIL-A-25463 para 4.6.1 (lb/in.) at		
Nylon	Epoxy	Triazine	-67°F	R.T.	180°F	-67°F	R.T.	180°F
0	100	24	2690	2855	3500	3.0	3.0	2.0
20	80	19	2690	2950	2400	3.7	4.5	6.0
40	60	15	4850	4800	4200	14	49	71
50	50	12	6100	5700	4930	10	78	102
70	30	7.3	7240	6420	5200	26	135	119
85	15	3.6	8250	5740	4850	37	180	143
95	5	1.2	7560	4125	2700	18	145	82

Procedure

Dissolve nylon (Zytel 61), Epoxy (ERL 2774), and curing agent (2,4-dihydrazino-6-methylamino-S-trazine) in methanol/water/furfural alcohol solvent mixture. Apply to clean (acid etched) aluminum, allow solvent to evaporate, close bond, and cure 60 min at 350°F.

Source: Ref. 27.

have been produced by this solution-casting technique, modern practice is to calender dry blends of powdered nylon with a liquid epoxy
resin plus accelerators and other modifying resins directly onto a mesh
support.

Although these exceptional bond values have won for the nylon
epoxies some important applications, this class of adhesives has never
overcome several serious drawbacks which include loss in peel
strength at low temperature, poor creep resistance, and extreme sensitivity to moisture. Nylon-epoxy films have the unfortunate tendency
to pick up substantial amounts of water before cure and tend to lose
bond strength rapidly after cure upon exposure to water or moist air.
De Lollis [26] contrasts the performance of one of the best present-
day nylon-epoxy aircraft adhesives to a conventional nitrile-phenolic
film adhesive. After 18 months of exposure to 95% RH, the nitrile phenolic had lost only a fraction of its initial strength, going from 3000
to 2500 psi tensile shear. In contrast, the nylon epoxy had degraded
from about 5000 to under 1000 psi in just 2 months under the same test
conditions. Similarly, Minford [15] and Sharpe [28] have shown that
nylon-epoxy tapes which give initial strengths above 6000 psi will
fracture at loadings which are a small fraction of the initial dry-failure
load after a few days of exposure to humid air.

The problems associated with the moisture sensitivity of nylon
epoxies were, of course, recognized from the outset. The exciting
properties of these adhesives, however, led dozens of laboratories and
triggered many government-sponsored research programs to try to
formulate around the moisture problem. Most of these efforts involved
primers in an attempt to upgrade the resistance to bond disruption by
water at the interface. Although primers undoubtedly prolong service
life, the present consensus is that it does not appear to be possible to
bring the bond durability of adhesives containing a high proportion
of nylon up to the levels attainable with nitrile-phenolic or nitrile-
epoxy film adhesives.

E. Elastomer–Modified Epoxies

The modern trend in aircraft adhesives is toward tape or film adhesives in which an epoxy-resin mixture is toughened with an elastomer
(most frequently a nitrile rubber) and cured with catalysts which permit cure at low pressures in the shortest possible time and at the lowest possible temperature. The use of adhesives to assemble progressively larger wing, tail, and other honeycomb subassemblies has placed
an increasing premium on cure temperatures in the range of 250°F
(125°C), which permits the use of lower steam pressure and thinner-
walled autoclaves, and on shorter cure cycles, which permit more units
to be assembled per autoclave per day.

Table 5   Comparison of Bond Strengths of Epoxy Adhesive with and without Nitrile Rubber Addition

Test temperature (°F)	Rubber-modified		Without rubber	
	Tensile shear (psi)	"T" peel (lb/in.)	Tensile shear (psi)	"T" peel (lb/in.)
-40	4900	48	2100	2
0	5100	47	2500	3
25	5200	44	2700	3
75	3600	22	2200	4
100	2800	21	3000	5

Source: Data from Ref. 7.

Table 5 summarizes data showing the effect of adding a nitrile rubber to a rigid epoxy formulation. The improvements in tensile shear and peel strength compare favorably to the increments reported in Table 4 for the addition of nylon to rigid epoxy adhesives. A major advantage of the nitrile-epoxy adhesives, however, is that the peel strength does not decrease as abruptly at subzero temperatures as the peel values of the nylon epoxies do.

Bond durability with these nitrile-modified epoxies is satisfactory as measured by most long-term moisture tests—for example, a tape with a cure temperature of 250°F (125°C) was found by Minford to have an initial tensile shear strength of 6700 psi as measured by ASTM D-1002 and to hold useful values after extended aging in air or water. But Minford [29] also warned that the degree of bond retention with these epoxies with curing temperatures of 250°F (125°C), although much better than that of the nylon epoxies, still did not match the permanence of the older adhesive types with curing temperatures of 350°F (180°C).

By the mid 1970s the standard method for bonding honeycomb panels or other bonded-aluminum structures was to use a nitrile-epoxy film adhesive with a curing temperature of 250°F (125°C), such as Bloomingdale's FM-123 or FM-96, or 3M's AF-126 with an FPL-etched aluminum surface. Because the FPL etch gave a thin, hygroscopic, and easily contaminated oxide layer, it was common practice to seal the surface with a thin coating of a "shop primer" very shortly after etching.

In 1974 Bethune, of the Boeing Aircraft Company, gave his landmark paper [30] at the April Society of Aerospace Material and Process Engineering (SAMPE) conference, disclosing that Boeing was beginning to find sporadic disbonding of aluminum honeycomb structures which had been bonded with the new, low-curing-temperature [250°F

(125°C)] epoxies. The failures had begun to occur after about one year in the field and could not be related to climate or service conditions. The mechanism, however, was clearly interfacial failure due to growth and subsequent fracture of an oxide layer on the aluminum surface. This type of failure had never been observed with the older, high-curing-temperature [350°F (180°C)] epoxy-phenolic or nitrile-phenolic adhesives.

Because the failure mechanism appeared to depend on a combination of moisture plus stress, Boeing and the other aircraft companies began to develop tests which could correlate with the field failures. One widely used current test is Boeing's simple but valuable wedge test [31]. Another important acceptance test maintains a bonded joint under a long-term, constant load of 800-1000 psi in a chamber at 140°F (60°C) and 100% RH as in the procedure in Fig. 8. The low-curing-temperature [250°F (125°C)] adhesives generally fracture at the interface after 20—40 days in this accelerated test, whereas the older, high-curing-temperature [350°F (180°C)] adhesives will survive until the 120-day end point.

The solution to these corrosion and moisture problems required changes in the method of cleaning and priming the aluminum as well as changes in the adhesives. The phosphoric acid anodization (PAA) process specified by Boeing [32] proved to yield a more stable oxide than the older FPL-etch method. In place of relatively simple shop primers, special corrosion-inhibiting primers, generally containing strontium chromate or other chromates, were specified and used. And the adhesives were modified to have higher heat-deflection temperatures and lower moisture-permeability rates.

The success of these efforts is seen in the performance of the bonding systems subsequently submitted by the major U.S. adhesive companies for the "PABST" program. PABST stands for the "Primary Adhesive Bonded Structures Technology" contract awarded by the Air Force to McDonnell Douglas to build and evaluate a fuselage section for the YC-15 transport plane, which will use adhesives for all structural bonds. Until now, adhesives, although used in large quantities, have been restricted to control surfaces and other secondary structures or to structures using rivets or other methods of attachment in addition to adhesives. Adhesives were not permitted to be used as the only fastening method for primary structures where failure could mean loss of the aircraft.

The PABST contract has demonstrated that adhesives can be used reliably for primary structures, has validated targets of 20% cost and 15% weight savings, and has shown improvements in structural integrity. The four elastomer-modified epoxy adhesives submitted for the PABST program (Bloomingdale's FM-73, 3M's AF-55, Hysol's EA9628, and Narmco's M133), together with their associated primers, currently represent the best state of the adhesives art. The combination ultimately selected for PABST [33] was FM-73 used with Bloomingdale's

BR-127 primer, applied over PAA-treated aluminum. Readers wishing more information on PABST, the PAA process, or the most recent information on procedures are referred to the 30 papers presented at the 1977 symposium at Picatinny [34] or to the other related papers [35–37] listed at the end of this chapter.

F. High Temperature Adhesives

Over the past 30 years, the Air Force Materials Laboratory at Wright Field, plus other government and industrial laboratories, have made substantial progress in the synthesis of new polymeric materials which can be used as adhesives for progressively higher service temperatures (Fig. 10). These materials include a variety of linear, heterocyclic, aromatic polymers such as polyimides (PI), polybenzimidazoles (PBI), and polybenzothiazoles. Many of these have the ability to retain useful strength levels at temperatures at which the previously best epoxy or phenolic adhesives weaken and fail due to rapid thermal and oxidative degradation. Of all the many different polymers which have been developed and evaluated during the past 30 years, it is the PIs which have emerged as the most useful resins for heat-resistant adhesives and coatings as well as for high-temperature composites. The success of the PIs is largely due to the many important nonmilitary and nonadhesive industrial used for PI resins. These uses include the manufacture of high-temperature wire enamels and dielectric materials (Du Pont's Kapton and "H" film).

Polyimides are synthesized by condensing an aromatic diamine with a tetracarboxylic dianhydride such as PMDA or benzophenone tetracarboxylic dianhydride (BTDA) to produce a linear prepolymer which is soluble in solvents such as N,N-dimethylformamide (DMF) or N-methylpyrol. Final cure then proceeds via solvent evaporation plus internal condensation (ring-closure) reactions during heating above 250–300°C, via Eq. (6).

PI prepolymer (solution) → PI + condensation products

$$+ 2n \ H_2O \qquad (6)$$

where Ar is an aromatic group, such as

$$(7)$$

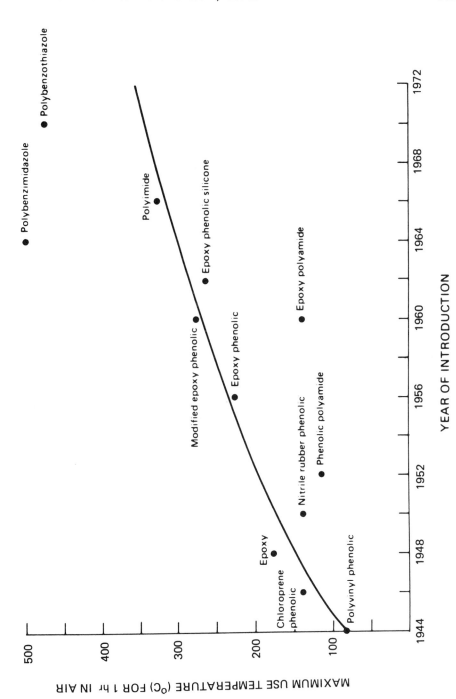

Fig. 10. Sequence of development of heat-resistant adhesives. (From Ref. 38.)

Ikeda, of Du Pont [39], and Twiss [40] have compared bond strengths for a typical PI (Du Ponts' NR-150), a PBI, and an epoxy-adhesive after various periods at elevated temperatures. The PI and PBI adhesives do indeed show better bond strengths above 200°C in air than the best epoxy-based adhesives, although the epoxy phenolic gives better strength retention up to 200°C after long-term immersion in room-temperature water.

The major disadvantages of the PIs, which have limited their use as adhesive systems to date, are their brittleness, their high costs, their high cure temperatures, and the problems involved in the elimination of volatiles during cure to obtain a void-free bond. Best results with the condensation-cured PIs, such as NR-150 or Monsanto's "Skybond", require a long careful series of cure and post-cure steps at progressively increasing temperatures up to 350°C, coupled with intermittent application and release of high clamping pressures.

The NASA-Lewis and NASA-Langley Laboratories have sponsored research programs aimed at reducing the volatiles evolved during cure, which cause strength loss due to void formation. These programs have produced several addition-curing PIs which include NASA Lewis' PMR-15 and Langley's LARC-series resins [41]. LARC-160 is the binder used in the graphite composites on the space shuttle. Other addition-curing PIs include Gulf's acetylene terminated "THERMID" resins and Rhone-Poulenc's "KERIMID" bis-maleimids.

The ladder model discussed in Sec. II leads to the prediction that all of the steps used to increase T_g and long-term hot strength in these heterocyclic aromatic polymers, such as the PIs, would also tend to decrease useful elongation, leading to very brittle adhesives. This is indeed the case, as shown in Fig. 11, in which one of the physical properties associated with toughness (peel strength) is plotted as a function of maximum load-bearing temperature. The Air Force Materials Laboratory and other government agencies have worked for years to develop methods of flexibilizing these high-temperature aromatic polymers. Some of the modifications which have proved to be beneficial in the PIs involve the introduction of short, flexible, polysiloxane or polyamide chain segments, leading to less brittle products, usually referred to as silicone-polyimides or polyamid-imides. These modifications lead to the peel-strength improvements shown in Fig. 11 with relatively slight attendant loss in hot strength. Even the best of these toughened PIs are, however, an order of magnitude lower in peel strength and in other toughness properties than the best current elastomer-modified epoxies.

Despite these brittleness, cost, and processing problems, the use of PI adhesives for bonding steel, nickel alloys, graphite, titanium, silicon, and ceramics is expected to continue to grow. Many of these new bonding applications will be in the electronics industry—for

Key

Number	Adhesive Type	Date Developed	Max Service Temp (°C)	Climbing Drum Peel Strength (lb/in.)
1	Vinyl phenolic	44	79	10-25
2	Nylon phenolic	52	104	5-10
3	Nitrile phenolic	50	143	25-35
4	Chloroprene phenolic	46	132	15-30
5	Epoxy (dicy cure)	48	182	5-10
6	Nylon epoxy	60	132	70-100
7	Nitrile epoxy	66	160	40-60
8	Nitrile epoxy (high temp)	65	193	25-40
9	Amine-diisocyanate	63	82	80-120
10	Modified phenolic	60	293	4-6
11	Epoxy phenolic silicone	63	271	6-10
12	Epoxy phenolic	56	249	6-10
13	Polyimide	66	343	4-8
14	PBI	66	538	3-6
15	Polybenzothiazole	70	510	3-6

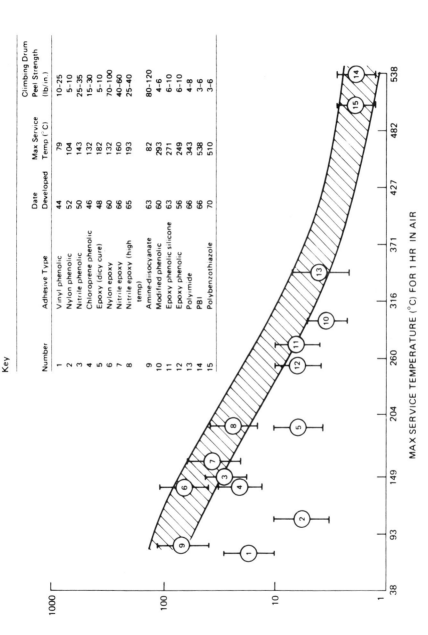

MAX SERVICE TEMPERATURE (°C) FOR 1 HR IN AIR

Fig. 11. Peel strength versus maximum use temperature of tape and film adhesives. (From Ref. 38.)

bonding silicon chips to ceramic substrates or to copper lead frames, and for other assembly or lead-attachment operations. These high-reliability electronic applications can tolerate the high cost of the PIs because of their unique dielectric properties, high T_g, low permeability to oxygen and moisture vapor, and low ionic content, and because of the ability of the PIs to survive assembly operations at the processing temperatures needed to reflow solder or to melt glass.

IV. PASTE AND LIQUID ADHESIVES

Few of the many types of liquid adhesives now in commercial usage can be classified as structural adhesives for the permanent bonding of metals and other rigid adherends—the subject of this chapter. Cyanoacrylate adhesives, for example, are sold in large volumes because of their fast cure capabilities, but they do not provide the strength, toughness, or heat and moisture resistance needed for high-reliability metal-bonding operations. In this section we will consider the five most important types of paste and liquid structural adhesives, based on epoxy, urethane, acrylic, or polyvinyl chloride (PVC) resins.

The distinction between paste and liquid adhesives refers primarily to viscosity and to the method of application. A "liquid" adhesive generally contains no mineral filler. It may contain solvents, and its viscosity is low enough to permit application by pouring, spraying, brushing, or roll coating. Paste adhesives, on the other hand, are generally thixotropic, contain no solvents (100% solids), generate no volatiles during cure, and may contain mineral fillers as well as colloidal thickeners. Paste adhesives are formulated for application by air operated pumps or caulking guns; they are used for gap filling and other bonding applications where "sag resistance"—a very limited amount of drip or flow during heat cure—is essential.

A. Two-Component, Room Temperature Curing Epoxy Adhesives

The two-component epoxy adhesives, which first became available in the early 1950s very quickly gained acceptance in virtually every industry as well as in virtually every household. Almost every adult has, at one time or other, used a 2-tube epoxy kit for some repair job at work or at home. Epoxies have gained this acceptance because they provide major advantages over any other previous adhesive type including:

1. *Useful cure properties.* Previous room temperature setting adhesives had been water- or solvent-based liquids which hardened by drying or by absorption into the adherends. This caused high shrinkage during cure and limited their use to thin films between porous or absorbent surfaces. But

the epoxies evolved no solvents or other volatiles during
cure, gave convenient mixing and working times, cured at
room temperature, and produced very low volumetric shrinkage during cure. They could be cured either in thin films
or cast and cured in large masses.

2. *Excellent physical properties.* Even the simplest epoxy res-
in-hardener mixtures give tough, high-strength bonds; adhere well to metals, glasses, and most other surfaces; and
provide good resistance to heat, water, and most chemicals.
Because these epoxies also give excellent electrical insulation values, they are used to insulate and encapsulate as
well as to assemble a wide variety of electrical and electronic devices. And because of their toughness and ability to
be cast and molded in large masses, epoxies became the
basis for the modern tooling industry.

3. *Versatility.* Epoxies can be modified with pigments, fillers,
and other resins to give a virtually endless range of viscosities, decorative effects, and physical properties. Epoxies can be filled with metals or oxides to conduct heat or
electricity, can be combined with glass fibers for very high
strength or with microballoons for low density and thermal
insulation.

A full discussion of the chemistry and composition of epoxy adhesives is beyond the scope of this chapter, but the interested reader
is referred to several good texts [42—44] on the subject. For present purposes, the most important class of epoxy resins, those derived from bisphenol-A, are listed in Table 2 together with the trade
names and product numbers for each of the molecular-weight grades
offered by the major domestic resin suppliers.

The liquid, bisphenol-A epoxy resins, having an epoxide equivalent of 300 or less, are used to formulate most of the one-part and
two-part paste and liquid adhesives discussed in this section. These
are cured by an addition reaction between the terminal oxirane group
and a compound containing an active (i.e., replaceable) hydrogen
(Table 6). The solid epoxy resins, having epoxide equivalents of
400 or greater, are used in the tape and film adhesives of Sec. III,
and although partial cure of solid epoxy resins may proceed through
the oxirane groups, cure more frequently proceeds through the hydroxyl group as in Eq. (4). Solid resins, such as Epon 1007, are
more accurately described as polyols, rather than as epoxy resins,
since these contain about nine hydroxyl groups versus only about one
oxirane group per molecule, and the important applications in both
coatings and adhesives technology depend almost entirely on the ability of their hydroxyl groups to condense with phenolic or amino resins
via Eq. (4), or to esterify with organic acids.

Table 6   Epoxy-Resin Curing Agents

General Reaction:

$$-CH_2-CH-CH_2 + HXR- \rightarrow -CH_2-CH-CH_2-X-R-$$

	CURING AGENT TYPE	CURE	X-LINKAGE FORMED	COMPARATIVE PROPERTIES
HX–R–				
HS–R–	MERCAPTAN	R.T. OR BELOW	SULFIDE	POOREST CHEMICAL, HEAT STABILITY
HN(R–	ALIPHATIC AMINE	R.T. OR ABOVE	AMINE	
HN–φ–	AROMATIC AMINE	HEAT CURE >80°C	AMINE +ETHER	BEST CHEMICAL, HEAT STABILITY
HN(C=N / C=C)	IMIDAZOLE			
HO–CH / HO–φ–	ALCOHOLIC HYDROXYL PHENOLIC HYDROXYL	HEAT PLUS BASE CATALYST	ETHER	BEST ELECTRICALS, LEAST EXOTHERM. SLOWEST CURES
O=HOC–R–	ORGANIC ACID	→	ESTER	
O=C–O–C=O	ACID ANHYDRIDE			
HN–C–N / H	DICY OR MELAMINE		ETHER +AMINE	HIGH EXOTHERM. GOOD TOUGHNESS

Table 6 summarizes the most important types of curing agents used with the liquid epoxy resins, showing the general reaction by which a compound containing an active hydrogen adds to a terminal oxirane group to yield an amine, an ester, or an ether crosslink. Although a very large number of compounds will ultimately react with epoxy resins at room temperature, most of these (aromatic amines, alcoholic and phenolic hydroxyls, acid anhydrides, Lewis acids, most tertiary amines) require elevated temperatures to yield a satisfactory cure in a reasonable length of time. Most room-temperature–curable epoxy adhesives are, therefore, cured with an aliphatic amine or an aliphatic amine adduct.

Simple aliphatic polyamine curing agents, such as diethylene triamine (DETA) or triethylene tetramine (TETA), are rarely used in two-part adhesives because they are objectionable skin and respiratory irritants, require an inconveniently high A–to–B-component mixing ratio, have a very short pot life, and cure to yield rigid, glassy, and easily fractured adhesives. Most of these objections can be overcome by condensing the simple polyamines with a mono- or polybasic organic acid to yield higher-molecular-weight, lower-vapor-pressure, more easily handled compounds containing amide as well as amine groups. By reducing the density of active hydrogens on the curing agent molecule, these so-called polyamide curing agents can give tougher, less brittle (although sometimes more water sensitive) adhesives.

The most widely used polyamide curing agents of this type are made by reacting an aliphatic polyamine with a dimer acid, thereby introducing a bulky, oil-compatible, C_{36} carbon group between the amine sites. General Mill's Versamid series of curing agents are the original and best known "polyamid" curing agents. The familiar hardware store two-part epoxy, as well as most proprietary adhesives which call for mixing nearly equal parts of A- and B-components, are of this type.

Even the simplest of such mixtures, consisting of equal parts of Epon 815 or Epon 828 and Versamid 125, can provide reasonable bond durability in most environments. By judicious compounding, involving the use of fillers, phenolic or tertiary amine accelerators, silanes or other coupling agents and by optimizing the mixing ratios, the bond durability of some of the proprietary two-part room temperature–curable epoxies can be brought to surprisingly high levels. This good durability under favorable bonding conditions is also evidenced by the many successful uses of these adhesives in the marine and outdoor construction industries.

Yet these two-part, polyamide-cured epoxies are also noted for rapid failure under many other commonly encountered service conditions. There are probably three major reasons for these seeming contradictions, which must be understood if the epoxy polyamides are to be used successfully:

1. *Surface preparation.* Sharpe [28] and Minford [15] have shown that best results, particularly for steel and aluminum bonding, are obtained with carefully cleaned surfaces. Room-temperature—cured epoxies are, in general, much less able to overcome deficiencies in surface preparation than are heat cured epoxies.

2. *Stress levels.* Bond durability of polyamide-epoxy adhesives is unusually dependent on bond stress in a humid environment. An adhesive which will last for months when aged unstressed in a humidity cabinet will frequently fail within hours, sometimes minutes, if placed in the same environment while it is loaded to only a small fraction (10—20%) of its "dry" failure stress via the procedure shown in Fig. 8.

3. *Formulation variables.* Some of the compounding ingredients normally used in epoxy polyamides can adversely affect moisture resistance. The use of reactive diluents can cause the equilibrium swelling of epoxy-polyamide films in water to increase by factors of 2 or more. Alkyl substituted phenols are excellent plasticizers and accelerators but can cause serious degradation in bond retention to steel and aluminum under humid aging conditions.

1. Other curing agents for two-component epoxies

One widely studied amine curing agent is diethyl amino propylamine (DEAPA), which was used in early epoxy adhesives, such as Armstrong's A-2 and Shell's Epon VI. Schornhorn and Sharpe [45] have shown that this amine is surprisingly effective in reducing the surface tension of epoxy resins.

Epoxy	Surface tension
Epon 828 alone	$\gamma = 44$ to 49 dyne/cm
Epon 828 + DETA	$\gamma = 44$ dyne/cm
Epon 828 + DEAPA	$\gamma = 33$ dyne/cm

Schornhorn and Sharpe therefore speculate that the utility of DEAPA may be due in part to better wetting of metals when DEAPA, rather than primary polyamines, is used as the curing agent.

Mercaptan curing agents (Table 6) are used where very fast set is needed, at or below room temperature, in a two-component epoxy adhesive. The offensive mercaptan odor plus relatively low softening temperatures are the major disadvantages of the fast-curing, two-part mercaptan-epoxy adhesives.

Although most two-component, epoxy adhesives are intended for room temperature cure, there are also some specialty adhesive applications which require the high-temperature properties provided by a heat-cured, two-component epoxy adhesive. There is an important difference in morphology between a room-temperature—cured and a heat-cured epoxy adhesive. At ambient temperature, reaction proceeds very slowly between an epoxy group and an alcoholic hydroxyl group. Hence, the hydroxyl group which is formed (Table 6) when the oxirane group reacts with an amine generally cannot enter into further reaction with a second oxirane ring at room temperature, and a diepoxide is ultimately bound into the cured network by just two terminal covalent bonds. In a heat-cured system, particularly when base catalysts and/or acid anhydrides are used as the curing agents, the hydroxyls can, and generally do, react with either epoxy or acid groups. Hence, heat-cured diepoxides can be regarded as being potentially tetrafunctional, rather than difunctional, and epoxy resins can, therefore, yield much more highly crosslinked structures with lower moisture permeability and higher thermal stability after heat cure than after ambient cure.

Barie and Franke [46] have reported impressive thermal stability data for epoxy adhesives cured with BTDA. Other curing agents which give unusually good thermal stability in two-part adhesives include imidazoles and certain aromatic amines—particularly diamino diphenylsulfone (DADPS). Thermal stability is enhanced when such curing agents are used with epoxy resins having a higher functionality than the bisphenol-A types—for example, tri-or tetrafunctional resins of the epoxy-novolac type, such as Dow's DEN 438, or Shell's tetrafunctional Epon 1031. A Shell patent [47] reports that a mixture of 100 parts Epon 1031, 100 parts aluminum powder, and 30 parts DADPS, cured 30 min at 180°C, will yield bond strength in excess of 1000 psi at 260°C.

2. Other resins and additives

In addition to the two principal ingredients—a liquid epoxy resin of the type shown in Table 2 in the A-component plus an aliphatic in the B-component—two-part epoxies generally contain other compounding ingredients. Aluminum, silica, and other fillers, thickeners, and pigments may be added to control the viscosity and color or to reduce shrinkage and thermal expansion during cure. Low-molecular-weight diluent resins may be added to reduce viscosity. Phenol or substituted phenolic compounds are frequently added to speed the cure rate at or below room temperature. Other types of epoxy resins are frequently added to or blended with bisphenol-A diepoxides either to reduce viscosity, improve toughness or flexibility, or for other special purposes. Most of these "diluent" resins are made by reacting epichlorohydrin ("epi") with an alcohol, a phenol, or a polyol to produce a mono- or polyglycidyl ether resin:

$$R(OH)_n \xrightarrow{EPI} R(O-CH_2-\overset{O}{\overset{\diagup \diagdown}{CH}}-CH_2)_n \qquad (8)$$

Simple alcohols, such as butanol or phenol [n = 1 in Eq. (8)], give low-molecular-weight reactive diluents such as butyl glycidyl ether (BGE) or phenyl glycidyl ether (PGE) which are the most effective viscosity reducers. Because these monofunctional diluents reduce crosslink density, they can be tolerated only at low levels to avoid degrading heat and moisture resistance or other properties of the adhesive. The high vapor pressure of the low-molecular-weight diluents can also cause outgassing and worker-irritation problems.

The use of polyols, such as butanediol or trimethylol propane, yields polyglycidyl ethers, such as butanediol diglycidyl ether (BDGE), which are still reasonably effective viscosity reducers but which are less volatile and usually cause less loss in properties which depend on crosslink density. The use of long-chain polyether polyols, such as polypropylene glycol, in Eq. (8) yields a diepoxide resin with an internal polyether chain (Dow's DER 732 and 736) which serves both to decrease viscosity and to increase flexibility.

Other flexibilizing and/or diluent additives are made by reacting epi with an organic acid or polybasic acid to form glycidyl esters:

$$R(COOH)_n \xrightarrow{EPI} R(COOCH_2\overset{O}{\overset{\diagup \diagdown}{CH}}-CH_2)_n \qquad (9)$$

Simple, straight-chain organic acids give low-cost diluents which reduce viscosity and improve wetting. Long-chain polybasic acids, particularly those made by dimerizing or trimerizing unsaturated C_{18} acids, are reacted with epi to yield polyglycidyl ester flexibilizers, such as Shell's Epon 871. Another class of epoxy-ester resins is made by adducting dimer or trimer acids with a bisphenol-A diepoxy to yield the high-molecular-weight, high-viscosity resins, such as Epon 872 and Epi-Rez 5132. These resins retain the fast reactivity of the terminal bis-A epoxy groups but contain a long-chain, internal segment which improves toughness, crack resistance, and peel strength of the adhesive after cure.

Polysulfide resins are sometimes used to flexibilize amine-cured, two-component epoxy adhesives. de Bruyne [48] cites high tensile strength (4500 psi initial) and fairly good bond durability (retention of 1700 psi after 30 days in water) for a highly filled, two-part, amine-cured epoxy adhesive flexibilized via the addition of 100 phr LP-3, a liquid polysulfide resin manufactured by Thiokol. de Bruyne concluded that the polysulfide significantly helped initial bond strength both before and after water immersion although his best results were obtained curing at 250°F (125°C) rather than at ambient cure.

Resorcinol diglycidyl ether (RDGE) and the triglycidyl ether of
p-amino phenol (Ciba's Araldite 0500) are two low-molecular-weight,
polyfunctional epoxy resins used in many high-performance tape and
liquid adhesives. As expected from their compact structure, the resins are able to reduce viscosity, improve hot strength, and increase
reaction rates. Surprisingly, these two high-functionality resins also
increase shock resistance. But Araldite 0500 is relatively expensive
and has a limited shelf life even in the absence of catalysts. Resorcinol
diglycidyl ether is also expensive but, more importantly, it is a toxic
and very dangerous liquid, causing skin burns, nausea, and severe
"poison ivy" reactions to personnel manufacturing or using adhesives
containing even small quantities of this resin.

3. Worker hazards and handling precautions

The most widely used epoxy resins derived from epichlorohydrin and
bisphenol-A (Table 2) are relatively harmless, nonirritating liquids or
solids which pose essentially no health hazard. In part, this is because these resins are insoluble in water (or in body fluids) and because they have essentially zero vapor pressure at room temperature.
Another reason is the recent success of the resin manufacturers in
removing all but trace quantities of potentially hazardous residues of
unreacted epichlorohydrin.

As noted above, however, many of the diluent resins, curing
agents, and other additives used with the epoxy resins of Table 2 may
be volatile or water soluble and may introduce specific health hazards.
Some of the additives used most widely in the past, such as BGE and
PGE, are now cited specifically by the Occupational Safety and Health
Agency (OSHA) and other agencies for suspected carcinogenic or
mutagenic activity.

As a result, there now exists considerable confusion, as well as
genuine concern, regarding either the real or suspected hazards associated with using industrial grades of epoxy adhesives. The user
must remember that the resins themselves are essentially harmless.
Workers hazards will generally arise from an unwise choice, either by
the adhesive manufacturer or by the user, of the additives, curing
agents, or solvents used with these resins, or from poor hygiene, insufficient ventilation, or incorrect handling procedures in the work
place.

Section IV.A.2 listed just a few of the large number, literally
hundreds, of different diluent resins, curing agents, and other additives which can be used in epoxy adhesives. Only a small percentage
of these additives is potentially harmful. Using the latest toxological
information from OSHA and from the resin suppliers, the major adhesive manufacturers can almost always select a combination of raw materials which meets the industrial customer's cost and performance,
as well as worker safety, requirements.

Table 7   Guide for Classifying Epoxy Products According to Their Skin Irritating, Sensitizing, and Carcinogenic (in Animals) Potentialities

Hazard category (Class)	Description	Definition
Class 1 (practically nonirritating)	The product is nonirritating or produces only transient mild skin irritation following prolonged or repeated contact.	The undiluted product produces *no* irritation when applied to the skin of rabbits[a] for 24 hours *and* it produces no irritation (or very slight irritation) when applied to rabbits for 8 hr/day for 5 consecutive days.
Class 2 (mildly irritating)	The product will produce only irritation of the skin following prolonged or frequent contact.	The undiluted product fails to produce severe edema[b] or severe erythema[c] when applied to the skin of rabbits[a] for 24 hr.
Class 3 (moderately irritating)	The product may produce injury with some persons when in contact with the skin for a prolonged period of time following a relatively few short-term contacts.	The undiluted product fails to produce severe edema[b] or severe erythema[c] in 4 hr but does so when applied to skin of rabbits[a] for 24 hr.
Class 4 (strong sensitizer)	The product is a strong sensitizer.	A strong sensitizer is a substance that produces an allergic sensitization in a substantial number of persons who come in contact with it.

| Class 5 (extremely irritating) | The product is capable of producing substantial injury when in contact with the skin for a relatively short period of time. | The undiluted product produces severe edema[b] or severe erythema[c] when applied to the skin of rabbits[a] for 4 hr or the undiluted product produces injury in depth (destruction or irreversible change in structure of tissue) when applied to the skin of rabbits[a] for 24 hr. |
| Class 6 (suspected carcinogen in animals) | The product can cause a carcinogenic response when applied topically to or inhaled by experimental animals. | The product is a material which (1) Causes tumors or cancers in a statistically significant number of mice[a] in a lifetime of skin painting by standard test procedures *or* (2) By animal studies, causes depression of blood-forming organs. |

[a] Or other suitable laboratory animals selected at the discretion of the testing or research team.
[b] Severe edema—swelling 1 mm or more (Draize score of 4).
[c] Severe erythema—bright or beet redness (Draize score of 4).
Source: Ref. 49.

The Society of the Plastics Institute (SPI) publishes a very useful guide [49] to handling, classifying, and labeling epoxy resins which ranks all common resins and additives on a hazard scale from 1 to 6 (Table 7). The bisphenol-A resins themselves are rated SPI Class 2, whereas most diluents, curing agents, solvents, and accelerators are rated Class 4. Most epoxy adhesive manufacturers now follow the SPI guidelines on labeling and identifying their products. Most avoid entirely any ingredient having an SPI classification higher than 4 and are attempting to bring the overall rating for their finished products down to a classification of 2 or 3.

Regardless of how safe an adhesive is when manufactured, the user must follow sensible handling precautions and maintain good cleanliness in work areas. These precautions are normally specified on the manufacturer's data sheet and are also summarized in a helpful booklet available from the National Institute of Occupational Health and Safety (NIOSH) [50]. In this regard, the improper use of solvents can pose particularly serious problems. Although most epoxy adhesives have the advantage of being "100% solids" and are solvent-free, strong, flammable, irritating solvents, such as ketones, are widely used to clean out equipment and to clean up spills in work areas. Individual workers, despite warnings to the contrary, unfortunately also tend to use these solvents to clean their hands and clothing after work. Such solvents can strip oils from the skin and carry chemicals into the body and are, therefore, responsible for many of the problems which are incorrectly blamed on the epoxy resins themselves.

B. One-Component, Heat-Cured Epoxy Adhesives

The previous section dealt with two-component epoxy adhesives. From the standpoint of the large industrial user, the requirement to mix together two or more reactive liquids just before use involves two major disadvantages:

1. Increased production costs due to the labor involved in metering, mixing, and degassing separate resin components plus costs associated with wastage of short-pot-life compounds which have thickened or gelled before use or which harden in, or plug up, dispensing equipment
2. Reduced reliability due to the possibilities of improper or incomplete mixing, errors in weighing out proper ratios of resins and hardeners, and problems in eliminating moisture pickup or air entrapment during mixing

For these reasons the manufacturers of epoxy adhesives have made major efforts over the past 20 years to provide one-component epoxy adhesives which could be used directly as received—eliminating mixing, metering, degassing, and other objectionable operations by the user.

The industrial users of liquid or paste epoxy adhesives in the manufacture of lamps, appliances, motors, speakers, metal furniture, hydraulic oil filters, and other high-volume consumer products also prefer a heat-cured, rather than a room temperature–cured adhesive, for these reasons:

1. *Production economics.* A plant producing a large number per day of almost any hardware item cannot afford to wait any significant length of time for bonded parts to cure before moving on to the next assembly step. This is particularly true if the bonded part must be clamped or fixtured until the adhesive hardens. Heat and the means to provide heat are inexpensive compared to the time and space requirements for curing at ambient temperatures. Industrial adhesives now are expected to cure in times measured in minutes, rather than hours, and this trend to higher speed cures is accelerating as induction, dielectric, infrared, UV, and other fast-cure techniques come into use.

2. *Durability and reliability.* Room temperature–cured epoxies are not capable of matching the best performance features of heat-cured epoxies. Heat-cured epoxies give higher heat-distortion temperatures, higher electrical-service temperature ratings, superior resistance to attack by acids, bases, and swelling solvents and to bond disruption by water. They also provide higher combinations of peel and tensile strengths and better tolerance to oily steel and other improperly cleaned surfaces.

The best, current, one-component epoxy-paste adhesives are relatively complex mixtures of epoxy resins, flexibilizers, and surfactants plus colloidal fillers to provide sag resistance (Fig. 12) and "latent" curing agents which are inactive at normal storage temperatures but which initiate crosslinking via addition reactions when the mixture is heated.

1. Dicyandiamide

The most widely used latent curing agent is a cyano-substituted guanidine known as dicyandiamide or "dicy".

$$N \equiv C - \underset{\underset{H}{|}}{N} - \underset{\overset{\parallel}{NH}}{C} - NH_2 \tag{10}$$

Dicy is a crystalline solid, soluble in water and lower alcohols but relatively insoluble in lower polarity liquids, including liquid epoxy resins. It begins to melt and decompose at about 300°F (150°C), which is a surprisingly high temperature compared to other aliphatic

Fig. 12. For many bonding applications, the adhesive must not flow or "sag" during heat cure. In the test shown, a 3/8-in. diameter bead of epoxy is applied to an aluminum plate and oven cured in the vertical position. The thixotropic adhesive on the left shows the desired ability to stay in place during cure.

amines of comparable molecular weight, such as DETA, which are low-boiling liquids. The melting point, solubility, and other properties of dicy can be explained by noting that it does not really exist as a single molecule, as in Eq. (10), but rather as an ionic salt, internally bonded by strong acid-base forces. Used as a curing agent, dicy is ground to a very fine powder and then blended with liquid or solid epoxy resins. Typically, 6–12 phr are used with liquid resins having an epoxide equivalent of about 180 to give dispersions having a shelf life of 12 months or more at room temperature. When heated to about 140–180°C, dicy decomposes to initiate cure, which typically requires 20–30 min at 180°C.

Dicy has many advantages which account for its importance in the adhesives industry. It is inexpensive. Bond strengths and performance properties are surprisingly good for even the simplest mixtures of dicy, fillers, and liquid epoxy resins. Data sheets of the various manufacturers generally show tensile shear strengths ranging from 3500 to 4500 psi via ASTM D-1002. Recommended long-term use temperatures generally range up to about 160°C. Resistance to attack from phosphate esters, acids, bases, and other fluids which attack many epoxies is generally excellent. Shell's Bulletin SC-54-54 [18], which reports no significant loss in bond strength for Epon IX after 30 days in salt water or tap water, and Minford's data [15] on PPG's Bondmaster M620, are representative of the generally good moisture resistance of well formulated dicy-cured epoxies.

With no elastomers added, these dicy-cured epoxies are hard and brittle, generally giving low peel values. But peel strength and other properties associated with toughness can be increased sharply via the addition of epoxy-ester resins [Eq. (9)], urethane adducts [51], or other polymers [47]. The most important tougheners are carboxylated nitrile rubbers, as in the nitrile-epoxy tape adhesives (described in Sec. III) and in many commercial, one-part adhesives, such as 3M's EC-2086 and EC-2186 [52], and in Amicon's Uniset A-500 types.

The major disadvantage of the early dicy-cured epoxy adhesives was that cure rates were slow and cure temperatures were high. The penalties associated with cure temperatures above 300°F (150°C) are particularly severe in the assembly of large parts. High oven temperatures mean that energy costs are high, that many plastics, electronic components, and other heat-sensitive components cannot be bonded, and that alloys which anneal at temperatures close to 300°F (150°C) cannot be bonded. Dicy has one additional unpleasant characteristic. It can only be cured in thin films between heat-conductive surfaces. If cured in thick sections or at temperatures above about 350°F (180°C) in thin sections, dicy epoxies exotherm excessively, resulting in the evolution of ammonia and other gases and leaving behind a charred, puffed up, resin mass (Fig. 13).

It has been known for some time that a wide variety of nitrogen-containing compounds will reduce the temperature at which dicy decomposes, thereby greatly accelerating the rate of reaction with epoxy resins. These dicy "accelerators" or "promoters" include primary, secondary, tertiary, and quaternary amines; amides and pyridines; substituted ureas, melamines, and guanidines; borinanes, imidazoles, and imidazolines; zinc and cadmium salts of amines, and other amino complexes. Most of these amine promoters are, however, themselves curing agents for epoxy resins and, therefore, give epoxy—dicy-promoter mixtures having shelf lives at room temperature which are too short to be of practical value. Some improvement is obtained by adding the amine promoter as a salt (e.g., of a carboxylic acid), but the most fruitful approach has been to seek individual compounds, from the many classes listed above, which increase cure rate with a minimum loss of shelf life. For example, Nowakowski et al. [53] describe the reaction product of a secondary amine with an isocyanate to form a substituted urea:

Fig. 13. Epoxy adhesives differ in their heats of exotherm. Equal amounts of two types of single-component, aluminum-filled epoxy adhesives were placed on a hot plate at 400°F. The conventional, accelerated dicy epoxy, rear, puffed up, charred, and emitted smoke. The other type, described in the text, cured in 15 sec without charring or fuming.

$$2\left(CH_3\right)_2 NH + \phi \left(NCO \right)_2 \rightarrow \phi \left[\underset{H}{N} - \overset{\overset{O}{\|}}{C} - N \overset{CH_3}{\underset{CH_3}{\diagdown}} \right]_2 \tag{11}$$

Adding 10 phr of this compound to 10 phr of dicy is claimed to give six weeks of shelf life at room temperature and a curing time of 60 min at 110°C in Epon 828.

Most of the low-curing-temperature [250°F (125°C)] elastomer-epoxy tape adhesives described in Sec. III.E are based on one of these accelerated dicy—curing-agent combinations. So are many paste adhesives, such as 3M's EC-2214 series or Amicon's A-400 series which have a shorter shelf life than straight dicy-cured epoxies but which feature the ability to cure at 250°F (125°C).

Copper, because of its unusual oxide characteristics, is normally a difficult metal to bond. A project sponsored by the International Copper Research Association (INCRA) [54] showed the unique ability of properly formulated, dicy-cured epoxy coatings and adhesives to provide bond strength retention to copper after extended exposure to moisture and stress. Long-term peel, boiling-water, and thermal-shock tests were run on steel, aluminum, and copper samples using a variety of successful commercial coatings and adhesives as well as formulations with and without special additives. Most of the adhesives studied showed no loss in adhesion in high-humidity tests up to 6 months long on aluminum or steel, but failed rapidly—frequently in a few days—on copper. Amine- and anhydride-cured liquid epoxy adhesives, as well as several vinyl-phenolic film adhesives, were among the many formulations showing most rapid deterioration on copper.

This INCRA report [54] showed that only those heat-cured epoxies which contained dicy or melamine provided long-term bond durability under humid aging conditions. They performed about as well on copper as on aluminum or steel. Dicy was shown to be beneficial when used as the sole curing agent with epoxy resins, when mixed with imidazoles or other curing agents, or when used to pretreat the copper surface before bonding. Even when simply added to coatings (e.g., phenolic-cured epoxies) which cure by a different mechanism, dicy and a melamine compound both increased the time to adhesive failure significantly on either bare or alkaline permanganate-treated copper.

2. Other latent curing agents

Until mixed with other reactive ingredients, epoxy resins are stable almost indefinitely at ambient temperatures. They are also stable for long periods when mixed with many of the compounds listed as curing agents in Table 6 (alcohols, phenols, and many acid anhydrides) provided that there are no strongly acidic or basic compounds present to initiate cure.

Much of the present art aimed at the development of one-component epoxy adhesives with long shelf lives involves the way in which acidic or basic catalysts can be added. Present techniques involve adsorption in molecular sieves, formation of Lewis-acid salts or other amine salts, microencapsulation of amines, and other segregation techniques.

Lewis-acid salts based on BF_3 or BCl_3 have been used for many years as latent catalysts. The BF_3—monoethyl amine salt, when dissolved in liquid epoxy resins, gives a shelf life of 6 months or more at 25°C in closed containers. Cure rates are slightly faster than with dicy. Lewis acids are widely used to cure many insulation compounds,

but they are rarely used in high-quality adhesives because toughness
and bond strengths are inferior to most dicy-cured epoxies.

A new family of one-component epoxy adhesives, introduced in
1968 by the Amicon Corporation, is based on a latent imidazole complex [55] which provides significant property differences from other
latent curing agents. Cure rates for the UNISET A-300 adhesives are
considerably faster than any reported for dicy-cured epoxies of comparable shelf life. Bond strengths are comparable to those reported
for epoxy phenolics and other adhesives formulated to comply with
MMM A-132, Type III, heat-resistant adhesives.

An additional useful feature of the latent imidazole catalyst used
in the UNISET A-300 series adhesives is that they do not char or burn
out due to exothermic heat during cure (Fig. 13) and can, therefore,
be used where extremely rapid cure rates are required. An article on
new heat-cure techniques [56] describes a variety of high-speed assembly operations where these one-part epoxy adhesives are cured
in times as short as 15 sec, using induction or dielectric heaters or
focused infrared lamps as the high-intensity heat source.

These properties of fast cure and controlled exotherm, plus the
excellent heat and chemical resistance which is characteristic of imidazole cured epoxies [57], account for the major current industrial
uses of the UNISET A-300 series adhesives. UNISET A-316 and A-304
are used in the assembly of hydraulic oil filters for aircraft where
resistance to attack by hot phosphate esters is needed, and in diesel-
engine oil filters, where engine-oil temperatures may exceed 350°F
(180°C). UNISET A-359 has been used for many years to replace
brazing alloys in the assembly of aluminum heat-exchanger tubes in
refrigeration and air-conditioning equipment.

Section IV.A explains why, of the many types of curing agents
shown in Table 6, the most important ones for use in two-component
epoxy adhesives are aliphatic amines or amine adducts. Dicy and imidazoles are the best curing agents for one-part, heat-cured epoxies.
Other curing agents may, however, be used where maximum strength
or moisture resistance of bonded metal parts is not a primary concern.
Acid anhydrides, such as dodecyl succinic anhydrides (DDSA) or adducts of DDSA with polyglycols, give long pot life with epoxy resins
and may be used for heat-cured adhesives which require good electrical properties. Acid- or anhydride-cured epoxies are useful for
bonding many plastics—notably polyesters, such as Du Pont's Mylar,
and polycarbonates.

C. Polyvinylchloride "Plastisol" Adhesives

Only one type of metal-bonding adhesive is currently used in really
large quantities in the passenger car industry and in the manufacture
of appliance and computer cabinets. This is the vinyl-plastisol adhe-

sive used to bond sheet steel to inner stiffener panels and to seal around the crimped panel edges. All passenger-car steel hoods and trunk lids and most steel doors and roofs use these adhesives in order to avoid the higher finishing costs which would be required to obtain a Class A surface after spot welding. Because of the large number of passenger cars and appliance cabinets produced each year, the total usage of these vinyl-plastisol adhesives is very large—currently ranging from 20 to 30 million lb/yr. These adhesives are formulated as high-solids, thixotropic pastes and are applied as discrete dots or droplets to the stiffener surface or panel edge before joining and crimping. These adhesives are, therefore, usually called "Hershey drop" adhesives in the trade because of the characteristic shape of these droplets.

These adhesives are mixtures of a plastisol grade PVC powder; primary plasticizers, such as dioctyl phthalate (DOP); a liquid epoxy resin, such as Epon 828; thickeners, stabilizers, surfactants, and other additives. The epoxy serves as a secondary plasticizer, acts as a stabilizer (acid scavenger), and helps to "fortify" the plastisol by crosslinking during cure. Good thixotropy, "sag resistance", is essential because considerable time may elapse between the time the adhesive is applied and the time the car or cabinet is finally assembled, painted, and baked. Cure, which takes place in the primer and paint bake ovens, depends primarily on the swelling of the PVC powder by the plasticizers. Hence, cure temperatures cannot be reduced below 270°F (130°C) without unacceptable loss of shelf life.

These plastisol adhesives are soft and flexible after cure, giving 50–100% elongation at 25°C and typical tensile shear strengths of about 400–700 psi on steel [40]. Higher bond strengths are not needed because of the large bond areas. The important property requirements are long-term flexibility to accommodate relative motion between the stiffener and the outer panel and the ability to bond to, and retain adhesion to, oily steel. The other important requirement is low cost since these adhesives presently sell in large volumes at a very low margin over direct materials cost.

The major disadvantages of these plastisol adhesives are that their high cure temperatures may limit current efforts to reduce bake-oven temperatures and that the possibility of corrosion exists due to liberation of HCl near hot-spot areas. To date, many epoxy adhesives have been formulated which match the plastisol properties, cure at lower temperatures, and which do not contain halides or cause corrosion. But to date no epoxy has been able to approach the lower selling price of the plastisols.

D. Polyurethane Adhesives

Isocyanate-based "polyurethane" adhesives are most widely used for
bonding elastomers, fabrics, fibers, and fiberglass-reinforced plastics
(FRP). Polyurethane adhesives were first developed in Germany during World War II specifically for making rubber-to-metal bonds on tank
treads. They continue to represent an excellent choice for bonding
most plastic and elastomeric adherends. But the modulus of most
polyurethane adhesives, although high enough to yield bond strengths
which can "tear rubber" or "tear fibers" in FRP during fracture, is
too low to provide the tensile shear strengths and levels of creep resistance which are required in most metal-to-metal bonding.

Polyurethanes also have the disadvantage of poorer chemical and
thermal stability than most epoxy adhesives. They lose bond
strength at temperatures above about 80—100°C. Unless they are applied over a suitable primer, adhesion and adhesion retention to metal
surfaces are generally unsatisfactory. But Twiss [40] and Smith and
Sussman [25] have pointed out that the best of the polyurethane adhesives are superior in performance to virtually any other adhesive type
at cryogenic temperatures—maintaining their impact strength and
toughness at temperatures far below those which cause serious embrittlement of epoxies or other high-strength adhesives.

All polyurethane adhesives are made of a combination of the following three ingredients:

1. A long-chain, flexible polyol, normally of mol/wt. > 1000.
   This can be a polyether, a polyester, or a polybutadiene,
   having terminal OH groups.
2. A diisocyanate, invariably either toluene diisocyanate (TDI)
   [Eq. (14)] or methylene bis-4,4'-phenyldiisocyanate (MDI)
   [Eq. (17)].
3. A curing agent which is either an aromatic diamine if item 2
   is TDI, or a short chain diol or triol if item 2 is MDI.

Although it is conceivable that one could make "one shot" adhesives
by adding the diisocyanate to a mixture of polyols just before use,
the normal practice is to prereact the long-chain polyol with a molar
excess of TDI or MDI to make a prepolymer A-component which is
subsequently cured by addition of the third ingredient B-component.

The prepolymers most widely used in the past for high-strength
urethane adhesives have been made by reacting polytetramethylene
glycol, mol wt. about 1000, with an excess of TDI. Examples include
Du Pont's liquid "Adiprene" prepolymers. These were then cured with
a solid aromatic diamine known as "Moca", which is 4,4'-methylene-bis-
(2-chloroaniline) as in the following reaction sequence,

Step 1. Diisocyanate + long-chain polymer → prepolymer

$$OCN\text{--}\phi\text{--}NCO + HO\text{-}(C_4H_8O)_{\overline{10\text{-}14}}H \longrightarrow OCNR_1NCO \qquad (12)$$

where

$$R_1 = -\phi-\text{NHCOO}\mkern-2mu\left(\mkern-2mu C_4H_8O\mkern-2mu\right)_{\!10\text{-}14}\mkern-6mu\text{CONH}\phi- \tag{13}$$

and

$$\phi = -\!\!\left\langle\!\bigcirc\!\right\rangle\!\!-\text{CH}_3 \tag{14}$$

for TDI prepolymers

Step II. Prepolymer cure with aromatic diamine

$$\text{OCN}-R_1-\text{NCO} + \text{H}_2\text{N}-R_2-\text{NH}_2 \longrightarrow \left[\begin{array}{c} \overset{\text{O}}{\overset{\|}{\text{N}-\text{C}}}-\text{N}-R_1-\text{N}-\overset{\text{O}}{\overset{\|}{\text{C}}}-\text{N}-R_2 \\ \text{H}\quad\ \ \text{H}\qquad\ \ \text{H}\qquad\ \ \text{H} \end{array}\right]_n \tag{15}$$

where, for Moca

$$R_2 = -\!\!\left\langle\!\overset{\text{CL}}{\bigcirc}\!\right\rangle\!\!-\text{CH}_2\!\!-\!\!\left\langle\!\overset{\text{CL}}{\bigcirc}\!\right\rangle\!\!- \tag{16}$$

The long-chain, flexible polyether group [R_1 in Eq. (15)] provides high elongation and low T_g. This long-chain polyether might also be expected to cause low cohesive strengths, but this tendency is balanced by regions of very intensive interchain hydrogen bonding at the urethane linkages and, to an even greater extent, at the urea linkages in Eq. (15). This combination of flexible, weakly bonded chain segments alternating with regions of strong secondary-force attractions accounts for the balance of low-temperature flexibility and good room-temperature tensile strength.

Adiprene L-167, for example, when melt blended with Moca, applied to cleaned and primed aluminum, and heat cured at 100°C gives the bond strengths shown in Fig. 5, a durometer of Shore D = 65 at 25°C, and exceptional tear and abrasion resistance.

Beginning about 1973, however, OSHA took the position that Moca was carcinogenic and began to curtail its use. At the same time OSHA also began to enforce strict limits on airborne concentrations of TDI vapor in the work areas. Although there are many other aromatic diamines which might be considered curing-agent candidates, none can be used as a satisfactory substitute for Moca. The chlorine groups at the Moca 4- and 4'-positions slow the reactivity enough to give a convenient 5−10 min pot life for TDI prepolymers. Other aromatic diamines, such as methylene dianiline (MDA) gel within seconds after mixing with TDI prepolymers. Polyol curing agents, which react much more slowly than amines, give useful pot life but yield softer,

weaker products after cure because substitution of a diol for the diamine in Eq. (17) produces only urethane groups, rather than the more
strongly hydrogen-bonding mixture of urea *plus* urethane groups.

After several years of litigation and unsuccessful efforts to find
a Moca replacement, OSHA has agreed to permit the continued use of
Moca by users who comply with new handling, warning-sign, ventilation, and other worker-safeguard regulations.

Meanwhile, during the past 5 years a very large market for urethane adhesives has developed in the transportation industry for
bonding the new polyester-fiberglass hoods, trunks, deck lids, doors,
and inner and outer panels which are now being used to replace steel
to reduce the weight of passenger cars, trucks, and other vehicles.
Panels of FRP cannot be riveted, so they can only be attached by adhesive bonding. Urethanes are the best current-adhesive choice.
But the automotive industry is unable to comply with the Moca-TDI
handling restrictions. This has resulted in the development of an important new family of polyol-cured urethane "structural adhesives".

Of the two diisocyanates, MDI is much more reactive with any
class of curing agent than TDI is and produces more rigid products
with polyols than TDI does.

$$MDI = OCN-\underset{}{\bigcirc}-CH_2-\underset{}{\bigcirc}-NCO \tag{17}$$

Most "structural urethanes" now sold to the automotive industry,
therefore, consist of MDI-terminated prepolymers cured with short-
chain polyols. Examples include Goodyear's Pliogrip 6000 [58] series
of two-component adhesives with 5—10 min pot life, good flow and gap-
filling properties, and excellent flexural and fatigue strengths. Tensile shear strengths are lower than those of previous Moca-cured
urethanes but are high enough (400—500 psi) to tear fibers in FRP
panels after cure. These "100% solids" urethane adhesives cause essentially no vapor or flammability problems. Satisfactory adhesion to
FRP normally requires only a clean surface, but before these urethane
adhesives can be bonded satisfactorily to metal or glass surfaces,
special solvent-based primers must be applied and heat cured.

One of the factors in the success of this FRP bonding program
has been the development, primarily by the Ford Motor Corp., of a
simple and reliable procedure for monitoring the quality of the bonded
FRP joints. Meyer and Chapman [59] have described a new ultrasonic inspection method and the quality control procedures needed to
locate voids or disbond areas and to determine how many voids can be
tolerated within an acceptable FRP joint.

E. Acrylic Adhesives

Until recently acrylate and methacrylate resins, although widely used for making soft acrylic sealants and for flocking, pressure-sensitive and textile adhesives, have generally not been considered candidates for structural bonding of metals, FRP, or other rigid surfaces.

The most important prior use of acrylic adhesives was the use of the "anaerobic" adhesives, which were discovered and named by V. K. Krieble while a professor at Trinity College in Hartford in the early 1950s. He, together with his son Robert, used these anaerobic adhesives to found and build the Loctite Corporation.

The anaerobics owe their utility to the unique advantage of being one-component liquids which can cure at room temperature without the need for catalyst addition. Their cure depends upon the fact that certain redux catalyst systems are inhibited by oxygen, so they can be mixed with vinyl monomers to form liquids which are stable for long periods in contact with air. But when these liquids are cut off from oxygen, free-radical—initiated polymerization begins, proceeding either slowly, at room tempterature, or very rapidly, at elevated temperatures.

A typical "Loctite-type" anaerobic adhesive [60,61] is based on diester monomers of methacrylic acid, such as tetraethylene glycol dimethacrylate (TEGDMA) cured with 2% cumene hydroperoxide catalyst, 0.3% benzoic sulfimide accelerator plus a stabilizer, such as benzoquinone. Packaged in small, thin-walled polyethylene containers which permit a continuous supply of oxygen to reach the contents, these liquids have a shelf life of over 1 year at room temperature. Cure is initiated when the adhesive is confined between two oxygen-impermeable adherends. Cure time can vary from 1 to 24 hr at room temperature, depending on temperature and the nature of the metal or other adherends since these redux catalysts are sensitive to the identity of the metal ions present. Cure is generally rapid at elevated temperatures, for example, 10 min at 100°C or < 1 min at 150°C on most surfaces. Cure can also be accelerated by the use of primers which contain peroxides or other promoters. Loctite product literature indicates that about half of the current applications for anaerobic adhesives require some sort of surface pretreatment "activation" to provide the desired cure rates.

Bond strengths for the anaerobic adhesives are similar to strengths for two-component, room-temperature-curing epoxies, being relatively good at room temperature but falling off sharply above about 80°C. Because contact with air prevents cure except in close-fitting joints, the anaerobics cannot be used for gap-filling applications where more than a few mils of adhesive are exposed. These adhesives are also quite expensive, selling for an order of magnitude more per unit volume than epoxy adhesives due to both the high cost of the monomers and the requirement that these be made in small batches and packaged and shipped in small, breathable containers. Nevertheless, anaerobic adhesives have secured several important segments of the adhesives market due to their unique cure mechanism and controlled break-loose characteristics. These uses include screw fastening, pipe and thread locking, and the making of formed-in-place gaskets. Higher strength applications include bonding magnets to speaker frames and assembling small electric motors [62].

Other types of acrylic adhesives have been developed for use as dental adhesives and as bone cements in orthopedic surgery. Generally, these adhesives consist of a solution of a low-molecular-weight (PMMA) polymer or copolymer in (MAA) monomer, plus fillers and pigments. Peroxide catalysts are added just before use to provide rapid, body-temperature cure. Applications for such adhesives are limited by their high shrinkage and high exotherm during cure, very short pot life, and handling problems due to monomer odor and flammability. Their use is also limited by the inhibiting effect of oxygen on cure and by the relatively weak and brittle bonds formed after cure.

During the past 5 years adhesive manufacturers have begun to promote "second generation" and "third generation" acrylic structural adhesives [63]. The terminology is meant to imply that these newer adhesives have handling and performance advantages over the earlier adhesives which were based on simple solutions of PMMA in MMA, and to emphasize that these are intended for "structural" bonding applications which were previously not considered suitable for acrylics.

Some of these new acrylics employ new monomer types intended to be less volatile and to have less odor than styrene or MMA and to give better physicals. These new monomers include methacrylic esters made from epoxy or isocyanate resins. For example, two molecules of methacrylic acid reacted with one molecule of a liquid bisphenol-A di-epoxide (Epon 828) give a difunctional vinyl monomer which incorporates aromatic groups and imparts "epoxy-like" properties to acrylic formulations while permitting cure either by peroxide addition or by exposure to UV radiation. Other low-volatility polyfunctional starting resins have been made by esterifying hydroxy-terminated urethane prepolymers with methacrylic acid to add "urethanelike" toughness.

Another technique used to toughen these later generation acrylic adhesives is to substitute a long-chain, nonacrylic backbone polymer for the PMMA component. A widely licensed Du Pont patent [64], for example, discloses a two-component acrylic adhesive made by dissolving a linear chlorosulfonated polyethylene (Du Pont's HYPALON) in a monomer mixture of MMA, methacrylic acid, and ethylene glycol dimethacrylate. This solution is then cured just before use by adding a mixture of a peroxide and an accelerator (dimethyl aniline). The patent claims bond strengths of 2800 psi on grit-blasted and vapor-degreased steel after a 14 day cure at room temperature and good peel strengths on aluminum.

The terminal double bonds in all of these acrylic formulations can be cured in a variety of ways. They can be cured either by adding a peroxide-containing second component or by adding mixtures of peroxides plus inhibitors to make one-component anaerobic adhesives. They can also be cured without catalyst addition by exposure to UV light or by application of the acrylic mixture between two surfaces which have been "activated" via a primer solution which leaves a peroxide residue after drying. A Hughson publication [65] claims, for example, that some of their Versilok acrylic structural adhesives will cure rapidly at room temperature to yield high-strength bonds between steel surfaces which have been pretreated with an activator solution containing peroxides and chlorinated solvents. Such one-component, surface activated, "third generation" structural acrylic adhesives are now being recommended for bonding steel and for making FRP-to-FRP and FRP-to-metal bonds [66,67].

V. SUMMARY: TOUGHNESS VERSUS DURABILITY

The structure-property relationships given in the foregoing sections to explain relative bond durability for various adhesive types agree well with the extensive results reported by Bodnar and Wegman [68] on long-term weathering of aluminum overlap shear specimens bonded with 17 different adhesive types, as well as with Minford's data [15] which summarize several years of testing performed at Alcoa with 27 different epoxy and other adhesive types.

Bodnar and Wegman [68] reported tensile shear strengths measured at -67°F (-55°C), 73°F (23°C), and 160°F (71°C) both after initial cure and after exposure for times up to three years in a very dry climate (Yuma, Arizona), in an Eastern industrial climate (Picatinny Arsenal, New Jersey), and in a humid, subtropical climate (Panama).

As expected, the high-humidity environment (Panama) was the most disruptive. Bodnar and Wegman [68] concluded that only two adhesives (an epoxy-phenolic and a nitrile-phenolic) showed a really high degree of bond strength retention after 3 years in Panama. Five other adhesives (two heat-cured epoxies and three epoxy polyamides) showed large percentage losses in strength but were able to maintain about 2000 psi shear strength after 3 years in Panama. The other 10 adhesives either fell apart on the test racks or gave unacceptably low bond strengths. These 10 included a nylon epoxy; epoxies cured with an acid anhydride, with a polysulfide flexibilizer, and with an aliphatic amine; several polyurethane and polyester adhesives.

All of the many performance properties desired in a structural adhesive can be summarized by the terms toughness and durability. "Toughness" properties include peel and fatigue strengths, crack resistance, and useful elongation. "Durability" parameters include strength at elevated temperatures, long service-life times at high op-

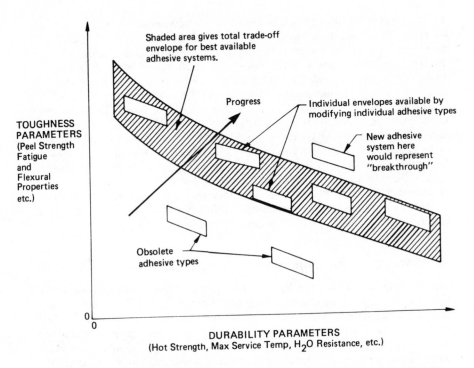

Shaded area gives total trade-off
envelope for best available
adhesive systems.

TOUGHNESS
PARAMETERS
(Peel Strength
Fatigue
and
Flexural
Properties
etc.)

Progress

Individual envelopes available by
modifying individual adhesive types

New adhesive
system here
would represent
"breakthrough"

Obsolete
adhesive types

DURABILITY PARAMETERS
(Hot Strength, Max Service Temp, H_2O Resistance, etc.)

Fig. 14. Basic trade-off between conflicting toughness and durability requirements. Shaded area defines the current "best state of the art". True progress consists of developing structural adhesives with better combinations of toughness plus durability properties and with cost and processing advantages.

perating temperatures, and resistance to water, humid air, salt, hydraulic fluids, and other corrosive media. Gains in any of the toughness parameters can be regarded as a true advance in the state of the art only when they are achieved without major sacrifice in any of the durability parameters. Section II of this chapter explains why almost any formulation change which enhances one of the toughness properties will degrade one of the durability properties and vice versa. This fundamental trade-off is illustrated by the data of Bodnar and Wegman [68] and by Fig. 10, which shows that attempts over the past 30 years to increase hot strength were almost always achieved at the expense of reduced toughness.

Any competent formulator can make an epoxy, phenolic, or polyamide adhesive that is rigid, durable, and heat-resistant. He can make a polyurethane or PVC adhesive that is flexible and tough. The problem is to *combine* toughness and durability in a single adhesive to offer the user a better *balance* of properties. Figure 14 uses this fundamental trade-off to define the current state of the structural-adhesives art. Progress must consist of motion *both* up and to the right in Fig. 14. The best present-day structural adhesives lie at the top edge of the shaded envelope, and true "progress" made over the past 40 years has come by exploiting a small number of technical breakthroughs which permit better physical property combinations *plus* cost and processing advantages. The most important of these breakthroughs, which have been reviewed in this chapter, have been

1. The two-polymer principle of "alloying" a rigid thermoset with a high-molecular-weight linear polymer
2. Development of the epoxy resins based on bisphenol-A and, starting about 1950, their increasing use in structural adhesives
3. Development of isocyanate resins and their use to make the polyurethanes
4. Use of acrylic adhesives for anaerobic, UV or surface-activated cure
5. The synthesis of PIs and other polyaromatic polymers permitting service temperatures above those previously possible for epoxy or phenolic-based adhesives
6. The development of nylon and nitrile polymers whose solubility, functionality, and other factors make these specifically compatible with epoxy resins in adhesives
7. The polyphase principle which recognized that one way to improve toughness is to use a multiphase resin system in which one or more dispersed filler or elastomer phases serve as crack stoppers within a more rigid, continuous resin phase
8. The synthesis of latent curing agents for epoxy tapes and one-component liquids, beginning with dicy for cure temperatures of 350°F (180°C) and continuing through to today's faster and lower temperature (100°C) curing one-component epoxies.

REFERENCES

1. J. C. Bolger, Structural Adhesives for Metal Bonding, in *Treatise on Adhesion and Adhesives*, Vol. 3, (R. Patrick, ed.), Dekker, New York, 1973.
2. J. C. Bolger and A. S. Michaels, Molecular structure and electrostatic interactions at polymer-solid interfaces, in *Interface Conversion for Polymer Coatings* (P. Weiss, ed.), Elsevier, New York, 1969.

3. F. Gehimer, and F. Nieske, *Insulation 39:*   Aug. 1968.

4. L. H. Sharpe, *Machine Design*:2 (1966).

5. J. Gardon, in *Treatise on Adhesion and Adhesives*, Vol. 1 (R. Patrick, ed.), Dekker, New York, 1967.

6. D. Kaelble, in *Treatise on Adhesion and Adhesives*, Vol. 1 (R. Patrick, ed.), Dekker, New York, 1967.

7. A. F. Lewis and R. Saxon, Epoxy resin adhesives, in *Epoxy Resins* (H. Kakuichi, ed.), Dekker, New York, 1969.

8. F. J. McGarry and A. M. Willner, *A. C. S. Div. Org. Coat. Plast. Chem.* 28:512 (1968).

9. L. H. Linebarrier, *Materials Eng.* Sept. 1975.

10. N. A. de Bruyne, U. S. Patent No. 2,499,134 (1952).

11. J. L. Been and M. M. Grover, U. S. Patent No. 2,920,990 (1955).

12. *WADC Tech. Report 53-126*, Wright Patterson AFB, OH.

13. *WACD Tech. Report 57-513*, Wright Patterson AFB, OH.

14. G. Solomon, in *Adhesion and Adhesives*, Vol. 1 (R. Houwink and G. Solomon, eds.), Elsevier, New York, 1965, p. 325.

15. J. D. Minford, Durability of adhesive bonded aluminum joints, in *Treatise on Adhesion and Adhesives*, Vol. 3 (R. Patrick, ed.), Dekker, New York, 1973.

16. J. M. Black and R. F. Blomquist, Development of metal-bonding adhesives with improved heat-resistant properties: progress report on adhesives, *FPL-710* (NACA Rm. 54 D01), May, 1954.

17. J. M. Black and R. F. Blomquist, *Adhesives Age* 5(2):and 5(3):(1962).

18. *Bulletin SC-54-57*, Shell Chemical Co., Houston, TX

19. N. de Lollis, cited in Engel, *WADC Tech. Report 52-156*, Wright Patterson AFB, OH.

20. W. B. Reynolds, to W. Phillips, U. S. Patent No. 2,774,703 (1957).

21. W. H. Smarnook and S. Bonotto, *Polymer Sci. Eng.* 8(1):41 (1968).

22. H. P. Brown and J. F. Anderson, in *Handbook of Adhesives* (I. Skeist, ed.), Reinhold, New York, 1962.

23. A. K. Doolittle and G. M. Powell, *Paint Oil Chem. Rev.* 107(7): 9, 40 (1944).

24. T. J. Reinhart, SAE Paper 800,212, SAE, Detroit, 1980.

25. M. B. Smith and S. E. Sussman, *Development of Adhesives for Very Low Temperature Application*, Narmco Res. and Dev. (NASA Contract NAS-8-1565), May 1963.

26. N. de Lollis, *Adhesives Age* Jan. 1969.

27. R. A. Frigstad, to 3M, U. S. Patent No. 3,449,280 (1970).

28. L. Sharpe, Some aspects of the permanence of adhesive joints, in *Structural Adhesives Bonding* (M. Bodnar, ed.), Interscience, New York, 1966, pp. 353-359.

29. J. D. Minford, *Adhesives Age* March 1978.
30. A. W. Bethune, *SAMPE J.* 2(3): (1975).
31. J. A. Marceau, J. Moji, and J. C. McMillan, *Adhesives Age*, October, 1977.
32. Specification BAC-555, Boeing Aircraft Co.,
33. E. W. Thrall, Jr., *Adhesives Age* (1979).
34. Durability of Adhesive Bonded Joints (M. Bodnar, ed.), *J. Appl. Polymer Sci.*, *Appl. Polymer Symp. No. 32*, Wiley, New York, 1977.
35. R. B. Krieger, *Adhesives Age* June 1978.
36. R. W. Vaughon, *Adhesives Age* December 1976.
37. R. F. Wegman and M. J. Bodnar, *Adhesives Age* July 1978.
38. H. Schwartz, Structural adhesives, in *Treatise on Adhesion and Adhesives*, Vol. 4 (R. Patrick, ed.), Dekker, New York, 1976.
39. C. K. Ikeda, Polyimide high temperature binders, in *du Pont Tech. Bull. PI-1104*, n.p., (March 1964).
40. S. B. Twiss, in *Structural Adhesives Bonding* (M. Bodner, ed.), Interscience, New York, 1966, pp. 455-488.
41. *Materials Eng.* 91:40 (1980).
42. H. Lee and K. Neville, in *Handbook of Epoxy Resins*, McGraw-Hill, New York, 1967.
43. P. F. Bruins, in *Epoxy Resin Technology*, Interscience, New York, 1968.
44. I. Skeist, in *Epoxy Resins*, Reinhold, New York, 1958.
45. H. Schornhorn and L. Sharpe, *Polymer Letters* 719 (19).
46. J. R. Barie and N. W. Franke, *IEC Product Res. Dev.* 8:72 (1969).
47. Hopper and Naps, to Shell Chemical Co., U. S. Patent No. 2,915, 490 (1959).
48. N. A. de Bruyne, *J. Appl. Chem.* (1956).
49. *Guide for Classifying and Labeling Epoxy Products According to Their Hazardous Potentialities*, Epoxy Resin Formulators Div., Society of the Plastics Industry.
50. Epoxy wise is health wise, in *Publ. 76-152*, NIOSH, Cincinnati, 1976.
51. J. M. Hawkins, to Dow Chemical, U. S. Patent No. 3,525, 779 (1970).
52. *"Scotch-Weld" Structural Adhesives Design Manual*, 3M Company, St. Paul, MN.
53. A. C. Nowakowski, A. M. Schiller, and S. Wang, to Am. Cyanamid, U. S. Patent No. 3,386,956 (1968).
54. J. C. Bolger, R. W. Hausslein, and H. E. Molvar, *A New Theory for Improving the Adhesion of Polymers to Copper*, INCRA Project No. 172, INCRA, New York, 1971.
55. J. C. Bolger, to Amicon Corp., U. S. Patent No. 4,066,625 (1978).

56. J. C. Bolger and M. J. Lysaght, *Assembly Eng.* March 1971.

57. *EMI-24 Curing Agent for Epoxy Resins*, Houdry Proc. Chem.,
    Philadelphia, PA, 1966.

58. Goodyear Chem. Bull., 1978.

59. F. Meyer and G. B. Chapman, *Adhesive Bond Strength Determination*, Adhesives and Sealants Council Fall Seminar, Detroit,
    1979.

60. Burnett and Norlander, to General Electric, U. S. Patent No.
    2,628,178 (1955).

61. R. Krieble, to Amer. Sealants, U. S. Patent No. 2,895,950
    (1958).

62. B. Murray, SAE Paper 800,211, Detroit, 1980.

63. Versiloc Acrylic Structural Adhesives, in *Bull. 3019E*, Hughson
    Chem., Erie, PA, 1974.

64. P. C. Criggs to Du Pont, U. S. Patent No. 3,890,407 (1975).

65. *Tech. Bull. DS-103031*, Hughson Chem., Erie, PA, 1978.

66. Modern Plastics October 1978.

67. D. Zalucha, *Adhesives Age* (1979).

68. M. J. Bodnar and R. F. Wegman, *SAMPE J.* Aug./Sept. 1969

8

Urethane Structural Adhesive Systems

Gregory M. MacIver and Douglas P. Thompson, II *Goodyear Tire and Rubber Company, Ashland, Ohio*

I. INTRODUCTION

A. Historical

Urethane structural-adhesive systems are relative newcomers to the marketplace. However, in the area of fiberglass-reinforced plastic (FRP), urethane structural adhesives have captured a very large share of the market. This is partly due to the good specific adhesion of urethanes, their excellent chemical and environmental resistance, and the high lap shear strengths of urethane bonds.

The fiberglass structural bonding market is a rapidly growing one. Projections indicate a growth of 15–20%/year for the next 5 years. Major projects requiring huge amounts of adhesive per part, for example, all-plastic truck cabs or conventional assembly-line parts with large numbers of parts (car doors, hoods, or trunk decks), are major growth areas and help account for the 15–20% annual growth projections. With this in mind, let us proceed to urethane structural-adhesive chemistry.

B. Advantages and Limitations

Urethane adhesives offer significant advantages over other structural
adhesive systems. Urethane polymers can be viewed as a series of in-
terconnecting, soft and hard segments (Fig. 1). While other structural adhesives have one or the other (soft or hard segments), only
urethanes have this unique combination. The ratio of soft to hard
segments may be varied to produce a wide range of physical properties.

Urethane structural adhesives have excellent water and humidity
resistance. The urethane linkage is hydrolytically stable and unaffected by a high concentration of water at elevated temperatures.
This valuable property is important in any bonded part which will, or
could, see environmental exposure. Urethane structural adhesives
are equally resistant to salt water and show little or no loss of bond
strength when exposed to salt spray.

Urethane structural adhesives can be compounded to resist high-
temperature paint bake ovens and to retain structural integrity after
exposure to 400°F (204°C) for short periods of time. This allows
bonded parts to be processed on conventional assembly lines.

The disadvantage of two-component, urethane structural adhesives is that they are moisture sensitive in the uncured state. They
cannot be conveniently hand mixed, which limits the amount of material used to comparatively large quantities. Meter-mix machines should
be used for maximum bond strengths, and their use represents some
type of capital investment. Having machines to meter and mix adhe-

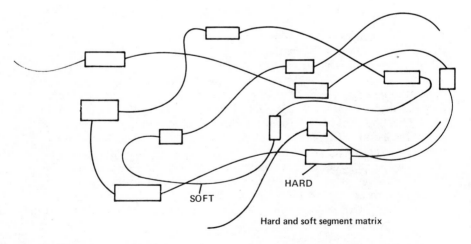

HARD

SOFT

Hard and soft segment matrix

Fig. 1. Urethane polymers are composed of hard and soft molecular
segments.

hesive also means that they may break down, may go off ratio, and, of course, do require periodic maintenance.

II. CHEMISTRY

A. Basic Concepts

Urethanes, as a generic class of organic chemicals, are the reaction product of an alkyl or aromatic diisocyanate and a multifunctional polyol (bi- and trifunctional polyols are the most common, but others are often used for special purposes). The reaction is a nucleophilic attack on the carbonyl of the isocyanate group by one of the lone electron pairs of oxygen of the hydroxyl group of the polyol (Fig. 2).

The two most common diisocyanates are toluene diisocyanate (TDI)—a mixture of 80% 2,4-isomer and 20% 2,6-isomer—and methylene bis-4,4'-phenyldiisocyanate (MDI). These diisocyanates (Fig. 3) can be used separately or in combination to produce the desired physical properties in the cured adhesives.

Two types of polyols are available in the industry, polyester polyols and polyether polyols. Early urethane polymers utilized polyester polyols. Unfortunately, ester linkages are susceptible to hydrolytic cleavage, so these early urethanes degraded in a fairly short period of time, due to moisture. This, of course, gave urethanes a bad name. At the present time, however, the polyether polyols are used exclusively in any urethanes which may see environmental exposure, for example, adhesives.

Diamines are used in urethane adhesives as chain extenders. The reaction between diisocyanates and diamines produces substituted ureas (Fig. 4).

Urea linkages are part of the hard segment of the urethane polymer. They are "harder" than urethane linkages, and this property can be used to the urethane chemists' advantage. These "harder" hard segments lend better tensile properties and higher heat resis-

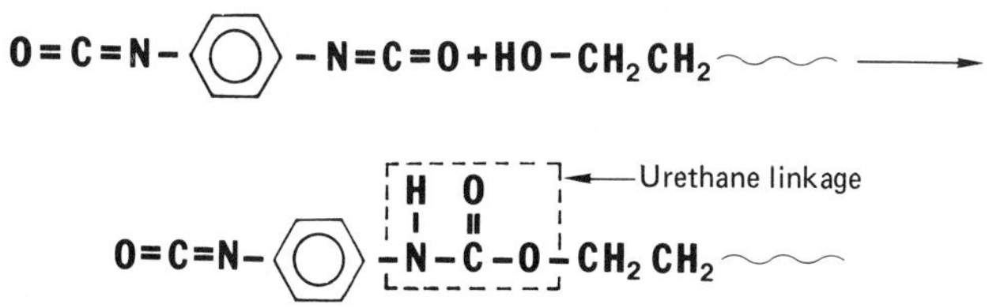

Fig. 2. Urethanes are formed by reacting a polyol with a diisocyanate.

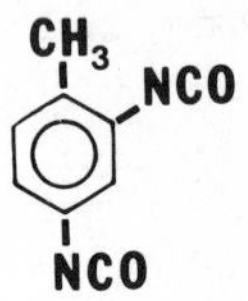

CH₃ NCO ... (structures)

2, 4-toluenediisocyanate
TDI (80%)

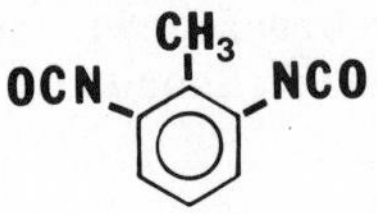

2, 6-toluenediisocyanate
TDI (20%)

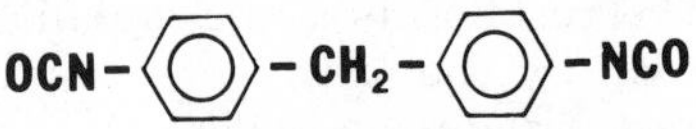

OCN-⟨○⟩-CH₂-⟨○⟩-NCO

methylene bis (4, 4-phenylisocyanate)

MDI

Fig. 3. The two diisocyanates most commonly used in urethane structural adhesives are TDI and MDI.

tance to the polymer. Tertiary amines make good catalysts for the
urethane reactions.

The advantages of urethanes come from their unique polymeric
structure (Fig. 1). The combination of hard and soft segments allows
properties of both rigid and elastomeric polymers. The hard segments
provide good high-temperature properties, good tensile strengths,
and good modulus properties, while the soft segments provide excellent low-temperature properties as well as some elastomeric properties.
The combination of these two sets of characteristics makes a good
structural adhesive.

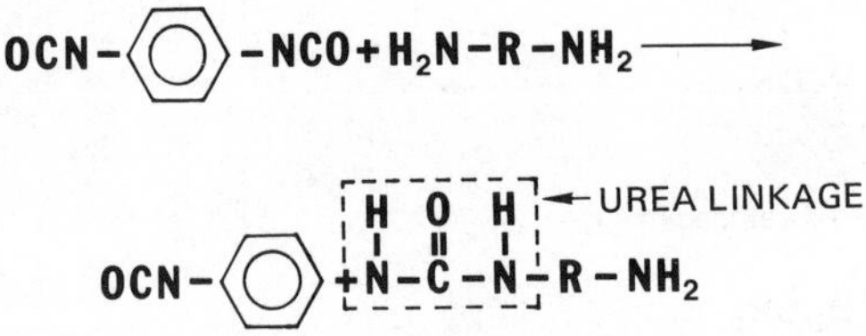

OCN-⟨○⟩-NCO + H₂N-R-NH₂ ⟶

OCN-⟨○⟩ ⌐ H O H ⌐ ← UREA LINKAGE
          ⌊N-C-N⌋ R-NH₂

Fig. 4. Ureas are formed by reacting a diamine with a diisocyanate.

III. APPLICATION—METER-MIX EQUIPMENT

Two-component urethane structural adhesives should be meter-mixed to provide consistent, high quality, mixed adhesive. As the name meter-mix implies, these machines perform two functions:

1. Metering the correct amount of prepolymer and curative
2. Mixing the two components to provide an air-free, complete mix

Several companies currently produce such equipment. Some of the requirements of meter-mix machines are

1. Ability to pump materials of different viscosities (prepolymer usually of higher viscosity than curative)
2. Metering of prepolymer and curative accurately enough to have a consistently high-quality product (usually within 5–10% of the stated ratio)
3. Temperature-controlled material pots, to ensure consistent gel times year round

Several different metering-pump designs are found in the marketplace. The first type of machine utilizes a follower-plate mechanism. This type of machine is normally used for prepolymers and curatives of high (75,000–250,000 cP) viscosity. The follower-plate type of machine can be used with lower-viscosity materials, but the cost of this type machine does not warrant its use.

The second type of machine is a double action, air-driven piston pump. Cylinders of different diameters are used for metering the two components. A recent improvement of the design has been a shorter shaft connecting the air cylinder and the pistons, allowing faster cycle times.

The third type of machine is an impeller-type pump. The impellers, driven by an air motor, drive the pumps which meter the adhesive.

Lastly, there is the gear-driven type of pump. This type of machine, unlike impeller machines, has the advantage of "no surge" operation.

All but the follower-plate machines are air (pneumatically) powered. This is so these machines can be used in plants where flammable liquids might be used in production process.

Two different types of mixers are used in the industry: static and dynamic. The dynamic mixers are grids, inside a mixing chamber powered by an air motor. The problem with dynamic mixers is that the heat they add to the mixed adhesive makes the gel time shorter. The dynamic mixer is, however, easier to clean than the static mixer. Static mixers are simple mechanically and do not add much heat to the mixed adhesive. Cleaning of small [<1 0.75 in. (20mm) I. D.] static mixers can be a problem due to urethanes' excellent solvent resistance. Burning out of static mixers is not recommended by the manufacturers but is often done by the customer.

IV. CURING, TESTING, AND DURABILITY

A. Curing

Two-component urethane structural adhesives have the advantage of room-temperature curing. Some customers do use heated fixtures for their parts to accelerate the gel time of the adhesive. Obviously, in high-speed production operations, the ability to speed up the cure is desirable.

Fixtures are used with bonded parts to maintain dimensional stability (Fig. 5), to ensure adequate contact of adherends, and in the case of heated fixtures, to accelerate the cure. Unheated fixtures

FIG. 5. Pneumatically driven fixture to maintain dimensional stability.

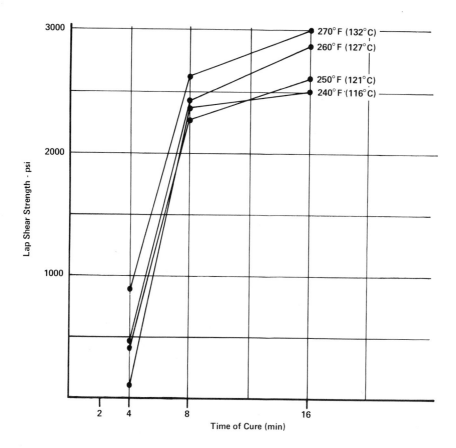

Fig. 6. Effect of cure time at several temperatures on the lap shear strength of a one-component urethane structural adhesive.

should exert 2–5 psi (15–35 kPa) throughout the bondline; heated fixtures must exert 20–40 psi (140–280 kPa). The reason for this difference is that at elevated temperatures [200–225°F (93–107°C)], water preferentially reacts with the isocyanate, producing CO_2. This CO_2 will form bubbles in the cured adhesive, thereby weakening the bond. A pressure of 20–40 psi (140–280 kPa) will force the carbon dioxide into solution and prevent the bubbling. Temperatures over 250°F (121°C) will "blow" the bond due to the same water/isocyanate reaction. At these temperatures, however, not enough force can be applied to the part to prevent "blowing" the bond.

A new one-component structural adhesive requires heat for curing. A time-strength graph for several temperatures is shown in Fig.

6. Generally speaking, 5 min at 200°F (93°C) or 1 min at 300 °F
(149°C) will cure the material totally. The one-component adhesive
has the obvious advantage of not requiring meter-mix equipment.

B. Testing and Durability

Extensive testing of the urethane structural-adhesive systems has
been done in cooperation with the major automotive and truck companies, both in the United States and in Europe. A summary of the results of this testing is found in Tables 1 and 2.

Table 1 contains tensile, percent elongation, Young's modulus,
100% modulus, and shore hardness data for six types of structural adhesives over a temperature range from -40°F (-40°C) to 250°F (121°C).
The first column represents a standard, two-component, urethane
structural adhesive; the second, a version resistant to high-heat
[400°F (204°C) for 60 min]; the third, a sandable, paintable, two-
component urethane; the fourth, a two-component, urethane elastomer
for use in the air- and oil-filter industries; the fifth, a one-component, urethane, elastomeric adhesive/sealant; the sixth, a new, experimental, one-component structural adhesive.

All of the data in Table 1 were obtained from primed 0.060-in.−
(0.15-cm) thick steel. The 1-in. (2.5-cm) overlap bonds were pulled
in tensile. All failures were cohesive within the adhesive.

Urethane structural adhesives do adhere to unprimed metals;
however, to protect the metal surface a primer is highly recommended.

Table 2 shows the effect of substrate and various environmental
conditions on the flex fatigue strength of urethane adhesive. Sheet-
molding compounds (SMC), high-glass sheet-molding compound
(HGSMC), directional-glass sheet-molding compound, cold-rolled steel
(CRS), and aluminum were used as substrates. The flex fatigue test
is an Owens Corning Fiberglass test for durability. The test consists
of placing a bonded panel in the flex machine. One end is fixed,
while the opposite end is flexed 7.5° to either side of the normal plane.

Urethanes also exhibit a useful gap-filling trait when compounded
as adhesives. Table 3 and Fig. 7 illustrate this property of gap-filling. Often with plastic production parts tolerances are ± 0.020 in.
(0.04 cm). If an adhesive works well only with thin bondlines or with
very consistent bondlines, it may not be suitable for production parts.
Figure 7 and Table 3 illustrate that even at 0.110 in. the joint retains
a lap shear strength of 1300 psi (9000 kPa), which is enough to de-
laminate most SMC and HGSMC laminates.

V. HEALTH, SAFETY, AND ENVIRONMENTAL CONSIDERATIONS

Worker safety factors should be considered when designing a system
or specifying a material for use in production. Not only should you

Table 1 Properties of Cured Urethane Structural Adhesives

	Test temperature	Adhesive number[a]					
		1	2	3	4	5	6
Tensile shear strength (psi)	-40°F (-40°C)	2871	6090	7324	1302	2166	
	R.T.[b]	2496	3246	5655	1112	825	3040
	180°F (82°C)	582	622	667	217	825	
	250°F (121°C)	266	607	405	164	340	
Elongation %	-40°F (-40°C)	73	17	7	15	254	
	R.T.	52	39	7	105	200	100%
	180°F (82°C)	15	10	11	24	90	
	250°F (121°C)	12	30	15	18	141	
Young's modulus	-40°F (-40°C)	15,687	73,209	70,936	23,633	8281	
	R.T.	13,834	19,856	66,591	1535	1491	43,600
	180°F (82°C)	4831	5779	11,514	924	922	
	250°F (121°C)	3737	4381	5275	776	629	
100% Modulus	-40°F (-40°C)						
	R.T.				1078	591	3040
	180°F (82°C)					294	
	250°F (121°C)						
Hardness shore		65(D)	80(D)	90(D)	73(A)	72(A)	61(D)

[a]See text for a description of the adhesives.
[b]Room temperature.

Table 2   The Effect of Various Environmental Conditions on the Tensile Strength (psi) of Urethane Bonds Between Various Substrates[a]

Test conditions	SMC[b] (low profile)	HGSMC[c]	Directional-glass SMC	CRS[d]	Al[e]
-40°F	646	956	2500(C)	3025(C)	1638(C)
72°F	740	1232	2744(C)	2550(C)	1050(C)
180°F	480(C)	471(C)	690(C)	794(C)	700(C)
250°F	280(C)	293(C)	304(C)	525(C)	580(C)
14 days 190°F	852	952	2610(C)	2520(C)	1650(C)
14 days H$_2$O immersion, 72°F	834	1158	1958(C)	2395(C)	1624(C)
14 days 100°F, 100% rh	620	1076	1970(C)	2170(C)	1604(C)
500 hr weatherometer	798	1250	2736(C)	2557(C)	1043(C)
Flex fatigue	4×10^6 cycles	2×10^6 cycles	2×10^5 cycles	1×10^5 cycles	
Mode of failure	No fail	Plastic fracture	Plastic fracture	Stress cracks in steel	
Slam test	Passes	Passes	Passes	Passes	Passes

[a]All bonds resulted in substrate failure except where (C) denotes cohesive break.
[b]Sheet-molding compound.
[c]High-glass sheet-molding compound.
[d]Cold-rolled steel.
[e]Aluminum.

Table 3   Effect of Bondline Thickness on Bond Joint Strength

Thickness		Lap Shear Bond Strength	
(in.)	(cm)	(psi)	(kPa)
0.010	0.025	2540	18,000
0.030	0.076	1940	13,300
0.050	0.127	1632	11,000
0.070	0.178	1496	10,500
0.090	0.229	1352	9,100
0.110	0.279	1304	9,100

Tested according to (ASTM) D-1002—0.060-in. cold-rolled steel primed with Pliogrip 6025—except with a 1-in. by 1-in. bond area. (See Fig. 7.)

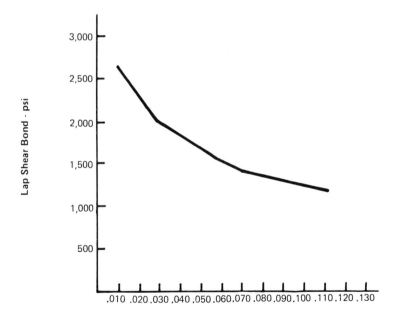

Fig. 7. Effect of bondline thickness on lap shear strength for a ure-thane structural adhesive.

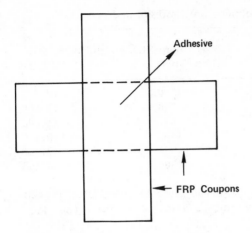

Cross Peel Bond

Fig. 8. Configuration for cross peel testing of urethane adhesives.

ask yourself, "Does it work?" but also, "Is it safe for the people using the product?"

Urethane prepolymers, by their chemical nature, contain isocyanate groups. Both TDI and MDI are materials for which the Occupational Safety and Health Agency (OSHA) has set breathing zone limits. Urethane manufacturers must consistently be aware of the potential hazards of isocyanate exposure. Urethane-adhesive manufacturers must be concerned on two fronts—first, when the prepolymer is manufactured and, secondly, when the prepolymer is used.

Monitoring of the air in workers breathing zones while they were dispensing the material showed a concentration of 0.001 ppm* TDI. This concentration is well below the OSHA standard of 0.002 ppm of TDI or MDI. Because TDI has a higher vapor pressure, it is often of more concern than MDI. The 0.001 ppm value just cited was obtained in a customer's plant with no special ventilation in the workers' area. Of course, we would highly recommend worker monitoring in every plant where urethane adhesives are being used to assure all parties involved that OSHA's standards are being met.

---

*Method of Grim and Linch. Am. Ind. Hyg. Assoc. J., 25, 285 (1964).

VI. QUALITY CONTROL OF URETHANE ADHESIVES

Obviously, quality control is a vital part of any manufacturing process. Management needs to know that both incoming and outgoing materials and products are of the desired quality.

Where urethane structural adhesives are concerned, evaluating incoming materials is comparatively easy. Several simple tests, such as Brookfield viscosity, visual appearance, and gel time will suffice to ensure the quality of incoming material.

Storage of adhesives should also figure into quality control of the material. Urethanes are moisture sensitive, so water contamination should be avoided. Extremely high temperatures will accelerate aging. Indoor storage with a temperature range of 60–90°F (15–30°C) is ideal. Materials should be at room temperature when used.

Visually inspecting mixed adhesives right out of the gun can tell the experienced eye a lot. The material should be smooth and glassy on the top of the adhesive bead, with no bubbles or waviness. Many customers make cross peel bonds (Fig. 8) as a check on the adhesive in production. When these cross peel tests are run, the failure mode should be delamination of the laminate. If it is not, the problem with the adhesive (poor mix, porous adhesive) can easily be seen and corrected.

Nondestructive testing of adhesively bonded joints is still in its infancy. Some work has been done with a sonar-type machine, and work is just starting on acoustic emissions. This nondestructive testing may become important to in-house, adhesively bonded assemblies.

9
Modified Acrylic Structural Adhesives

James A. Graham *Chemical Product Group, Lord Corporation, Erie, Pennsylvania*

I. INTRODUCTION

The most recent, and perhaps the most versatile, generic family of structural adhesive products to be introduced to the parts-assembly industry has been designated modified acrylic structural adhesives. It is the purpose of this chapter to provide the reader with a broad understanding of the technology, history, use, performance, and handling of this new adhesive joining tool. Additional material pertaining to modified acrylic structural adhesives may be found in Refs. 1-3. Important adhesive tests are described in Ref. 4.

209

II. HISTORY

Although "chemical fastening", or adhesive bonding, is as old as recorded history itself, its use has been limited by a multitude of factors. Until only recently the use of adhesives had been essentially a last resort when no other method of joining could be used— from the ancient Egyptians manufacturing paper from papyrus by bonding cross plies of fibers to the assembly of lightweight, durable, aerodynamically designed aircraft of modern day.

With the advent of synthetic polymer chemistry, the adhesive technologist in the early twentieth century could begin to explore resources other than the naturally occurring raw materials of centuries past. The advent of synthetic polymers, ranging from elastomers to rigid plastics, provided the adhesive technologist with a continuing flow of new raw materials: phenolics, urethanes, vinyl resins, epoxies, etc.

The earliest, truly structural adhesives were based on phenolic resins but were found to be limited by rigidity. They were subsequently modified with flexible resins, such as polyvinyl butyral, to provide impact strength and flexibility. The limitations of the phenolic chemistry mothered the invention of the new epoxy-based adhesives of the late 1940s and early 1950s. The epoxies, until very recently, had been the "state of the art" of high-performance structural adhesives. The limitations of the epoxy adhesives—they require clean, well-prepared surfaces; they require heat cures for best performance; they provide only moderate adhesion to the newer engineering plastics now replacing metals in many areas—mothered the invention of the modified acrylic structural adhesives.

In the late 1960s and early 1970s, the adhesive technologists began to exploit the potential of synthetic polymer chemistry "in situ" or *within the bondline*. In other words, they began to build the adhesive polymer and, at the same time, bond the assembly. Very distinct advantages of this concept emerged. Poorly prepared surfaces could be tolerated quite well. Adhesion to metals was retained, and far improved adhesion to engineering plastics was accomplished. Room-temperature curing without physically mixing two components was provided. Far improved handling characteristics were evident because very low viscosities could be applied while high-molecular-weight, crosslinked polymers were formed in the bondline, without volatiles or solvents being emitted.

With the new advantages, the modified acrylic structural adhesives can now begin to compete successfully in many areas. They are considered, not as a last choice when no other method of parts assembly, such as mechanical fastening or welding is adequate, but as a first choice when new assemblies are being designed. In many cases, they are used as high-quality alternatives to mechanical fastening on parts that are already designed and in production.

III. PERFORMANCE PROPERTIES

The performance of any adhesive product in any given assembly is dependent on joint design, application conditions, load distribution, environmental conditions, substrate properties and in-service conditions. The ultimate test of any adhesive performance is the actual in-use history. Regardless, certain standard testing procedures that aid in the selection process, have been adopted to communicate somewhat representative performance values on test configurations.

Tables 1-4 provide representative data on the three families of acrylic adhesives. The advantages of these materials over all other room-temperature—curing structural adhesives in processing and performance are quite significant.

A. Advantages

1. *Wide substrate versatility.* Commonly the same adhesive performs satisfactorily on steel or aluminum as well as on engineering plastics.

2. *Unsurpassed hydrolytic resistance and permanence in various aggressive environments.* Acrylic polymers are well known for resistance to aggressive environments. They are not readily plasticized by moisture and can be highly crosslinked "in situ" during curing reaction.

3. *Versatility in processing variables.* Room-temperature cure, without prior mixing, is possible. Mix-in systems may also be used. Since the adhesives have very low molecular weights prior to curing, low-power pumps and metering devices may be used even though "non-sagging" rheology may be required, even with these 100% reactive systems.

4. *Excellent price-to-performance ratios.* Although they provide improved processing and performance compared to most cyanoacrylates, anaerobics, and specialized epoxies and urethanes, the modified acrylics are commonly priced significantly lower.

5. *Minimal or no surface preparation on metals and plastics.* Commonly, mill-finished steel and aluminum may be bonded as received with little or no deleterious effects on initial strength or long-term durability. In some cases, improved performance is noted on unprepared, as opposed to rigorously prepared, metals.

Table 1 Lap Shear Strength Measured by ASTM D-1002-72 (psi).

Adherend		General-purpose acrylic	Second-generation acrylic	High-performance acrylic
Aluminum 6061-T6	as received	800	2300	4400
	Grit-blasted	4300	1700	5000
Aluminum 2024-T3 clad	as received	700	1300	4700
	Grit-blasted	3800	3400	4500
Steel SAE 1010 cold-rolled	as received	300	2800	5500
	Grit-blasted	4300	3500	6000
Copper	as received	1300	800	3100
Brass	as received	500	900	2400
SMC-Polyester abraded		540 (Adh)[a]	20 (Adh)	600 (F.T.)[b]
ABS	as received	500 (S.B.)[c]	700 (S.B.)	600 (S.B.)
Polycarbonate	as received	1200 (S.B.)	1000 (S.B.)	700 (S.B.)
Polymethylmethacrylate	as received	600 (S.B.)	600 (S.B.)	600 (S.B.)

[a]Adh, adhesive failure.
[b]F.T., fiber tears of substrate.
[c]S.B., Stock break of substrate.
Source: Data for general-purpose acrylics from Refs. 1-7, for second-generation acrylics, from Refs. 8 and 9, for high-performance acrylics from Ref. 12.

Table 2   Peel Strength Measured by ASTM D-1876 (pli)

Adherend		General-purpose acrylic	Second-generation acrylic	High-performance acrylic
Steel CQ 1010 cold-rolled, 24 mil	as received	0	24	32
	Grit-blasted	41	32	54
Aluminum 3003-H14, 20 mil	as received	0	4	11
	Grit-blasted	15	5	30

Table 3   Impact Strength Measured by ASTM D-950 (ft-lb/in.^{2})

Adherend		General-purpose acrylic	Second-generation acrylic	High-performance acrylic
Aluminum 2024-T3 bare	as received	0.4	7.2	7.1
	Grit-blasted	10.7	7.2	8.1
Steel CQ 1018 cold-rolled	as received	2.3	9.7	9.5
	Grit-blasted	12.0	8.1	14.0

Table 4   Heat Resistance Measured by ASTM D-1002-78:   Shear Strength (psi) as a Function of Temperature

Adherend Surface prep.	General-purpose acrylic SAE 1010 cold-rolled Grit-blasted	Second-generation acrylic SAE 1010 cold-rolled None	High-performance acrylic SAE 1010 cold-rolled None
Test temperature °F (°C)			
-40 (-40)	2600	800	2000
0 (-18)	3000	3400	3400
75 (24)	3800	3900	5500
150 (66)	1900	2400	4400
200 (93)	1400	1800	3600
250 (121)	800	1400	1500
300 (149)		600	800

6. *Structural, or load-bearing, physical properties.* The modified acrylic structural adhesives provide extremely high initial bond strength and long-term durability on load-bearing assemblies. Analogous performance with other generic structural adhesives normally requires high-temperature curing and rigorously prepared substrates.

7. *Tolerance for poorly mated surfaces.* Although best performance is realized with 5–10-mil bondlines, adequate and acceptable performance is possible with bondlines of 1-125 mils or thicker if necessary.

B. Disadvantages

1. *Poor adhesion to most unprepared, cured elastomers and low-energy surface plastics.* The modified acrylic structural adhesives do not readily cure on untreated, cured elastomer surfaces. Adhesion to untreated polyethylene, polypropylene, and various fluoropolymers is poor.

2. *Distinctive odor.* Commonly, in poorly ventilated work areas, the distinctive acrylic odor is unfamiliar and considered unpleasant by some application personnel.

3. *Mix-in systems commonly of unequal portions.* Mix ratios of 20:1–4:1 are most commonly used in mix-in systems. One-to-one systems are available for certain applications but have limited storage life and less performance versatility. Tolerance to slightly off-ratio mixing is good.

4. *Flash points.* The uncured adhesive fluids may commonly have flash points slightly below room temperature. Specially developed systems may have flash points above 100°F (38°C).

5. *Limited long-term, "in use" credibility.* Due to relatively recently developed technology, case histories beyond 10 yr in use are rare. Newer systems have very limited "in use" histories.

C. General Performance

Additional performance properties, such as resistance to aggressive environments, fatigue resistance, crack propagation, effect of bond-line thickness, joint design, etc., have been characterized on individual products within the families discussed. The results of such testing have indicated a high level of durability compared to presently known structural adhesives. The few major suppliers of acrylic adhesives can provide specific durability data on these materials to aid in the engineering design of new bonded assemblies.

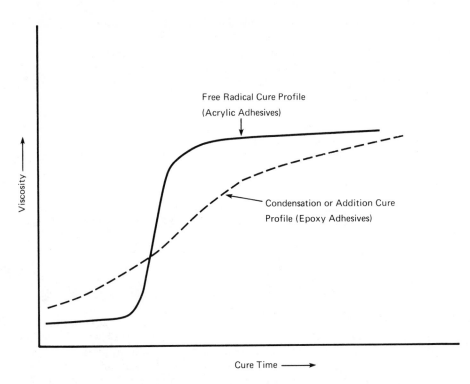

Fig. 1.   Cure time versus viscosity for acrylic and epoxy adhesives.

IV. CURING PROPERTIES

Unlike the urethane or epoxy adhesives, the modified acrylic structural adhesives cure by a free-radical polymerization process rather
than by ring opening or condensation polymerization. Therefore,
the time span between detectable thickening and achievement of full
handling strength of the acrylic adhesives is very short compared to
epoxies or urethanes. This is viewed as another distinct process
advantage of the modified acrylic structural adhesives. A cure profile, such as the one depicted in Fig. 1, provides an unusual combination of usable repositioning time coupled with a relatively short
interval between starting to cure and reaching full handling
strength. In the case of epoxy adhesives, the curing adhesive
thickens to a point where repositioning is not recommended long before full handling strength has developed. In the case of acrylic
adhesives, the parts may be repositioned for a period approaching
80% of the time required to reach full handling strength. The caution recommended in the case of the acrylic adhesives is that once

thickening has become obvious, full handling strength will rapidly be achieved, so repositioning cannot be tolerated or recommended. With epoxies and urethanes some latitude is possible regarding repositioning after the onset of visible thickening.

The majority of acrylic structural adhesives provide cure times in the range of 2 min to 1 hr at room temperature. In contrast to the curing of the epoxies or urethanes, the use of heat to speed the cure is not recommended. The cure speed may be accelerated in certain conditions by heating at only mild temperatures in the range of 130–150°F (55–66°C). In no case should the cure temperature exceed 160°F (72°C) during the initial stages of cure. Also, unlike the epoxies, the acrylic adhesives cannot be "B-staged". Once the free-radical cure has commenced, it will go to completion by a self-propagating mechanism.

V. TECHNOLOGY

The chemistry of the modified acrylic structural adhesive is quite complex and extremely versatile, but the chemical concepts are rather simple and straightforward. Several patents [5–15] have been granted in the technological area and are beyond the scope of this chapter. It is important, however, to describe the concepts in a broad sense.

The adhesive resins are essentially made up of various polymers dissolved or dispersed in reactive, unsaturated monomers. This fluid also contains free-radical initiators, free-radical scavengers, fillers, nonreactive diluents, and, in some cases, unsaturated oligomers. Typical recipes of the adhesive-resin portion would be as follows:

1. *Resin.*
   a. Blend of reactive, unsaturated monomers (methyl methacrylate, methacrylic acid, etc.)---------- 70–85%
   b. Modifying elastomer polymer or oligomer (neoprene, nitrile, etc.)------------------------------ 15–20%
   c. Free-radical initiator (tertiary amine)-------- 0.5–2.0%
   d. Free-radical scavengers----------------- 0.001–0.05%
   e. Fillers, diluents, etc.---------------------- remainder
2. *Catalyst.* The catalysts are commonly free-radical sources, such as peroxides carried in plasticizers or non-reactive diluents.
   a. Reactive organic peroxide--------------------10–25%
   b. Nonreactive diluent or plasticizer-------------75–90%
   c. Filler or flow-control agents may also be present.

A somewhat different case in the second-generation acrylics [8,9] provides a unique, free-radical source in the adhesive resin portion, while the free-radical initiator is carried in the catalyst portion. This is just the reverse of the majority of the newer, high-performance systems [12] and the older general-purpose adhesives [1-7] discussed earlier.

VI. HANDLING PROPERTIES

A. Accelerator Lacquer Method

This application technique utilizes the greatest processing advantage of the acrylic structural adhesive family. The accelerator lacquer is applied to the substrate or surface that is to be bonded by roller, brush, spray, or any conventional coating process. Once the surface has been activated by the accelerator lacquer, it may be bonded immediately or stored for periods up to 12 months prior to application of the adhesive resin. The only exception occurs with the second-generation acrylics [8,9]. This accelerator lacquer cannot be coated and stored for periods longer than several days. In either case, at the time and point of bonding, the adhesive is applied to the activated surfaces, and curing commences. In use, it appears to be a single-component adhesive that cures at room temperature.

It is important to point out several distinct advantages of this handling property. Surfaces that are to be bonded may be treated with the accelerator lacquer at some remote facility or by an outside custom-coating facility in a mass-production technique. The pretreated substrates may then be fed into the actual assembly station or assembly facility and bonded rapidly at room temperature with a stable, single-component adhesive resin. The resin may be fed to the actual bonding or assembly station by comparatively simple metering pumps or caulk tubes and applied by relatively unskilled assembly personnel with relatively simple production equipment (Fig. 2).

An additional benefit of this technique is that on surfaces that have not been activated or areas of surfaces that have not been activated, the adhesive will not cure; thus, cleanup is greatly facilitated. Areas not to be bonded can be controlled by application of the accelerator without close control of the adhesive-dispensing operation.

Adhesive bondlines up to 30 mils will cure with the accelerator applied to only one side of the assembly. If thicker bondlines are desired, both surfaces may be activated. Bondlines of to 60 mils will cure efficiently if this method is used.

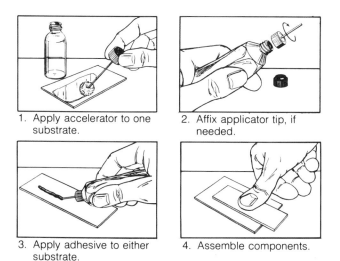

1. Apply accelerator to one substrate.

2. Affix applicator tip, if needed.

3. Apply adhesive to either substrate.

4. Assemble components.

Fig. 2. Handling sequence for acrylic adhesive.

B. Two-Component Mix Method

Should the accelerator-lacquer method not be appropriate, the acrylic structural adhesives can be supplied as the more conventional two-component, mixed systems. The common mix ratio for these products is normally 10 parts of resin to 1 part of catalyst. Equal-mix components are possible in certain circumstances but presently they are not preferred due to package stability of the components and the lack of broad-use, high-strength systems formulated for 1:1 mix. An exception is noted in Sec. VI.C.

C. Two-Component, No-Mix Method

A less commonly known, and less commonly used, method of application can be provided. Most adhesive resins can be made up in two packages. Package A is applied to one surface; package B is premixed with a catalyst up to 12 hr prior to use and applied to the second surface. Upon mating of the two surfaces, the cure commences. This procedure allows both the bonding of very large surface areas (with very long open times) and very rapid curing upon mating of the surfaces. If it is absolutely necessary in specific applications and if a 12-hr pot life of the catalyzed part B is acceptable, it is possible to mix the system in a 1:1 ratio and still obtain high performance.

VII. REPRESENTATIVE CASE HISTORIES

A. Solar Heating Panels

There are currently three major types of solar-energy power plants. One type uses heliostats—large mirrored devices that concentrate the sun's rays onto a central receiver which is often referred to as a power tower (Fig. 3). The sun's rays heat the water in the power tower resulting in steam to drive turbines to produce electricity. Excess heat produced during peak hours is stored in oil and gravel for use during cloudy periods or nighttime.

The construction of these heliostats requires an adhesive to bond very large surfaces at room temperature with high strength, excellent durability, and relatively simple handling properties in less than ideal application environments. The adhesive must bond aluminum to aluminum, galvanized steel to galvanized steel, and painted metal to glass.

A second type of solar-energy power plant utilizes banks of solar cells containing point-focusing "Fresnel lenses". A bank of photovoltaic cells mounted on a mechanical device for following the sun is called a photovoltaic array.

These photovoltaic cells convert sunlight directly to electricity without the use of turbines. A computer system monitors and controls the power system. The conversion of sunlight to direct current takes place within a cell, causing a potential difference which, in turn, causes the flow of electrons. Like in the power tower, excess energy can be stored. Batteries offer a convenient method of storing excess electrical energy.

The construction of the photovoltaic arrays requires an adhesive to bond an acrylic lens to an acrylonitrile butadiene styrene (ABS) engineering plastic housing.

A third form of solar-energy plant converts the sun's rays directly to heat through the use of panel collectors which are usually mounted on the roof of a building (Fig. 4). Water is circulated through the collector panel via a conventional pump and can be used to heat buildings, houses, and swimming pools. This energy can also be stored for use when the sun is not shining.

The construction of these panels requires an adhesive to bond ABS to itself and aluminum to aluminum.

All of the above applications presently utilize modified acrylic structural adhesives because of the excellent performance and ease of application realized with the product. As performance-proven products for the solar industry, these products offer the following benefits:

Competitive prices
Room-temperature cures and handleable bonds in as little as 3 min

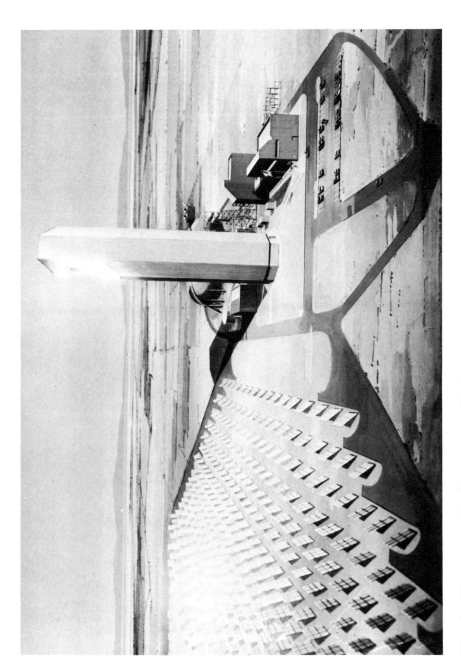

Fig. 3. "Power tower" solar-energy collector.

Fig. 4. A liquid-filled, solar collector panel.

Little or no surface preparation, adding speed and efficiency
to the production cycle

Ability to bond a variety of substrates in many varied combinations

Ability to withstand temperature ranges from -50 to +260°F (-45
to +127°C).

The primary substrates used for the three types of power
plants described in this section are mainly painted and bare metal,
glass, and plastic. Once a cure parameter and viscosity specifications have been determined, a single modified acrylic structural adhesive will usually do the job of bonding all the different generic
substrates. Other options include the long or short pot life of a
mix-in system and the use of a no-mix, wipe-on accelerator method.
Modified acrylic structural adhesives can be cured using the mix or
no-mix method.

B. Ceramic Magnets

"For about 12 years we were bugged by the frustrating problem of
bonding a fragile, expensive ceramic material in the magnet assembly
of our gasoline-driven generators."

This is the way a Vice President of Engineering of a major
manufacturer of small, gasoline-driven, electric generators describes
his introduction to acrylic adhesives and technology [16].

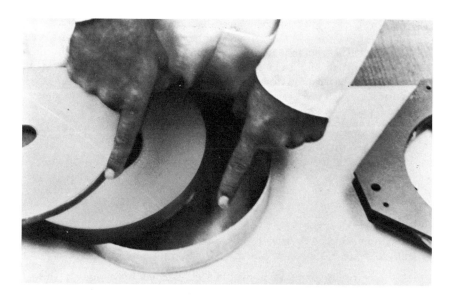

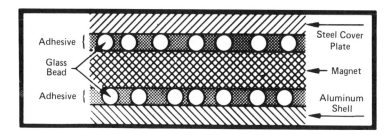

Fig. 5. Ferrite-to-metal bonded assembly used in an electrical generator.

Since 1961 our company has produced gasoline-driven, portable electric generator sets. These are 110 or 220-volt a-c power sources used in a variety of industrial, commercial, and construction applications. Homeowners also use our units as standby power sources in case of municipal power outages.

For the past three years we have been using an acrylic structural adhesive to bond a strontium ferrite magnet disc to a steel structural disc. The same adhesive is used to bond an aluminum stamping to the opposite side of the magnet disc to complete the magnet assembly (Fig. 5).

The acrylic adhesive is doing a real job for us, he continues. With the adhesives we were using before, we were getting a

25% scrap rate due to breakage of the fragile ferrite as a
result of bonding. Our production rate is about 100 units per
day and at $8.00 to $9.00 each, this was costing us a bundle.
I am happy to say that since switching to acrylic adhesives not
one case of damaged ferrite can be traced to the bonding procedure. We have saved thousands of dollars during the past
three years, thanks to these new adhesives.

The magnet assembly is heavy— 26 lb— and spins at 3600 rpm,
subjecting the unit to high centrifugal forces which the adhesive
must withstand. Each generator unit contains two similar magnet
assemblies separated by a critical 0.550-in (13.97 mm) air gap. The
adhesive must hold all components of the magnet assemblies perfectly
rigid to maintain this precise air gap during operation.

We had some problems with oil-canning (distorting) of the
aluminum stamping with previous adhesives. This we could
not tolerate because of the precise air gap, he explains.

But our really expensive problem concerned breakage of the
fragile sintered ferrite discs when bonded to the steel support
ring. We were using a single-component, heat-cure adhesive
system that required a bake at 325°F (162°C) for four hours
to cure. The differential expansion of the steel and ferrite
during this bake set up stresses that were breaking the magnets.

We investigated various air-dry systems to get around the
bake problem but found them unsatisfactory because of poor
adhesion. Also, most suffered from a rather low temperature
capability. Our units normally operate at about 175°F (79°C),
but must be able to withstand a temporary extreme of 275°F
(135°C).

The acrylic adhesive used by this manufacturer is a room-temperature—cure system that possesses a strength necessary to withstand the high forces developed in the magnet assembly. It is also
flexible enough to accommodate the differing amounts of expansion due
to operating temperature changes without inducing dangerous stresses in the magnets.

Prior to bonding, glass beads 0.010-in in diameter are sprink-
led over the part to maintain a uniform bond thickness. Adhesive
is then manually applied, and the parts are clamped for 6 min with a
500-lb clamp load during the room-temperature cure.

C. Shipbuilding

During the past several decades, the Japanese shipbuilding industry
has surpassed many of the great shipbuilding ports of the world in

becoming a leading producer of ocean bulk carriers, particularly the "super" tanker. But fishing boats too are being built in Japan with equal skill and ingenuity and have become a substantial industry in their own right. At Hokkaido, an eastern Japanese seaport, more than 30 shipyards construct mostly fishing boats. At 15 of the shipyards, acrylic adhesive is being used to repair boats and to construct new boats.

Not content with building conventional steel-hull boats, the innovative Japanese boat builders developed a lighter-weight fishing boat by substituting fiberglass reinforced polyesters (FRP) in the upper hull. The reduction in weight is said to increase the cruising speed of 10-ton—class boats by about 4 mph and to improve boat balance by lowering the hull's center of gravity.

First developed at the Unjo and Sakai Shipyards, the two-step technique employs a fiberglass upper hull bonded to a steel lower hull (Fig. 6). After sanding the portion of the steel hull to be bonded, a two-part, mix-in acrylic adhesive is mixed and applied, and the first wet lay-up of the inside section of the upper hull is completed within a 15-min cure time. In all, 13 fiberglass mats are laid over with polyester resin.

Once the FRP has cured, the outside joint area is likewise prepared. The second, or outside, hull lay-up is made so that it is

Fig. 6. A modern Japanese fishing boat which utilized acrylic adhesive to bond an FRP upper hull to a steel lower hull.

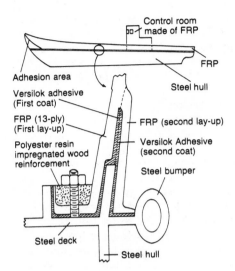

Fig. 7.   Cross section of the bond between the FRP upper hull and the steel lower hull shown in Fig. 6.

bonded to both the inside portion of the FRP hull and to the lower steel hull in a kind of tongue-and-groove fashion (Fig. 7).

The construction method has been approved by local authorities.   Six fishing boats in the 10-ton class have been built with a combination steel and fiberglass hull—five more were completed by the end of 1978.   Previous attempts using urethane primer on steel for wet FRP lay-up had ended in failure when the fishing boat couldn't pass an examination by the port authority and had to be reconstructed entirely of steel as before.

In other fishing-vessel applications, acrylic adhesive is being used for the joining of an FRP pilothouse to the boat deck and for the lay-up of wet FRP to protect the boat's metal fish storage tanks from saltwater corrosion.

D.   Sporting Goods

Can you imagine the stress developed in a tennis racquet when it smacks a ball hard enough to send it on its way at 115 mph?   Shock of this magnitude requires the maximum in racquet frame strength and toughness.

A manufacturer in Chicago has spent over 8 years developing what it believes to be the ultimate racquet material—one that has the ability to stand up to such repeated punishment.   The Tremont MAG-1 racquet is molded from a proprietary composition of ABS and

Fig. 8. A modern tennis racquet which uses an acrylic adhesive in its assembly.

graphite (Fig. 8). The structural foam core is injection molded at 15,000 psi to produce a super strong frame of controlled density. This reduces vibration and provides superior power and control. To achieve a critical balance between strength and stiffness, the Tremont MAG-1 racquet features an aluminum stiffening ring bonded to each side of the graphite composition frame.

Tremont engineers chose acrylic adhesives to bond the 0.032-in.−thick, high-strength aluminum panel to the core material. A general-purpose, acrylic structural adhesive displays just the right amount of flexibility to accommodate the extremely high stresses transmitted through the frame core and ring with no adverse effect on adhesion. These superior characteristics of the adhesive were proven in an exhaustive series of tests conducted during development of the MAG-1 design. These included tests for strength and adhesion and tests to verify its resistance to environmental deterioration.

Production engineers at Tremont have found that hand mixing of resin and the accelerator just prior to bonding is the application technique most adaptable to their production requirements. The selected adhesive's shear-sensitive viscosity permits easy mixing of measured amounts of adhesive and accelerator in a small bag. Its resistance to "flow" makes it easy to apply to the core and to the aluminum ring right from the mixing bag. This formulation has a pot life of 5−8 min and cures in 10 to 20 min.

The MAG-1 racquets have been in production for 2 years, and production rates have been as high as 40,000 (projected) on a yearly basis. The racquets are available throughout this country and are marketed in France, Italy, New Zealand, Japan, and Hong Kong.

Graphite composite racquets are increasing rapidly in popularity, and Tremont's MAG-1 is no exception. Widely distributed, the MAG-1 is responsible for Tremont's dramatic growth from eight employees in 1975 to thirty employees in 1977 and for sales of approximately $1 million.

E. Aircraft

The sleek, trim lines of a sailplane belie its inner strength. To safely withstand the rigors of powerless flight, and yet be light and agile enough to ride on a thermal cusion for hours, requires a high degree of structural strength throughout the craft.

At Schweizer Aircraft Corp., Elmira, New York, acrylic structural adhesives play a critical role in creating a unified glider structure where every major component−even the canopy−adds to the overall structural strength (Fig. 9).

Fig. 9. Acrylic structural adhesives replaced rivets for joining the two-part FRP aerodynamic fairing in Schweizer's 2-33 two-place sailplane. About 400 2-33s have been produced, making it the standard trainer for commercial soaring schools in the United States and Canada.

1. Bonding versus fastening

When designing the new Model 1-35—a single-passenger, high-performance sailplane for the serious glider enthusiast—Schweizer engineers evaluated the pros and cons of either adhesively bonding or mechanically fastening the clear plexiglass canopy to the glider frame.

The superior bond strength and load-carrying capability of acrylic structural adhesives were just what the Schweizer designers needed. They found that bonding with an acrylic adhesive gave a product in which the joint became load bearing. The assembly was twice as strong as when assembled with other techniques and was also completely airtight. Bonding also eliminated any chance of cracking the expensive, Swiss-made, plexiglass canopy during assembly as might have occurred when using mechanical fasteners.

The "blown" 1/8-in.—thick (0.318 cm) plexiglass canopy measures approximately 4½ ft (1.37 m) long, 23 in. (58.4 cm) along its major axis, and 20 in. (50.8 cm) high at its peak. It is joined to the glider frame which is of laminated epoxy lay-up construction.

During production engineering tests of the canopy assembly, an unusual problem developed when performance of an epoxy adhesive was being evaluated. The epoxy generated so much heat during cure that it caused warpage and separation of the plexiglass and laminated frame due to differential expansion of the materials. Clamping the parts during the 18—20-h cure cycle didn't help because the combination of heat and clamping pressure caused the plexiglass to craze. The properties of an acrylic adhesive—fast

cure, low exothermic heat, and excellent bond strength—were the
answer to these problems. Pull tests also pointed out the acrylic
adhesive's superior shear strength compared to other adhesives.

2. Applying acrylic adhesive

Schweizer Aircraft is using a two-part, paint-on formulation with fast
cure and shear-sensitive viscosity.

Guide holes are predrilled in a coordinated pattern in the
plexiglass canopy and the frame. Prior to bonding, plexiglass dowels are inserted into the holes in the frame. The two adhesive components are applied to the bond areas of the canopy and frame, respectively, and the canopy is then set on the dowels. "The adhesive
'kicks over' in about 10 minutes," says Schweizer, "and no clamps
are needed" [17]. This particular system also possesses a good
measure of resilience to accommodate shock and sudden stress loading.

Having proved its mettle on the Model 1-35, the adhesive is
being used on other craft built at Schweizer. The company manu-
factures between 70 and 80 two-passenger, training gliders each
year. On this craft—the Model 2-33—two halves of an FRP lay-up
"nose cone", or aerodynamic fairing, are bonded together with the
same system as used on the Model 1-35. The two-part construction
permits the fairing to be fitted around control elements during assembly.

Previously, the cones had been riveted. When adhesive replaced the rivets, the assembly procedure was switched to the two-
part, paint-on technique. Not only have the structural properties
of the bonded joint been improved, but assembly time has been cut
between 25% and 30%, Schweizer estimates.

Still another job for acrylic adhesives on the Model 2-33 is
bonding two halves of vacuum-formed ABS sheet that make up the
wheel-well covers. The covers, which are about the diameter of a
small aircraft tire are mated by a few tack rivets for proper alignment, then bonded.

VIII. METER, MIX, DISPENSE EQUIPMENT

The modified acrylic structural adhesives provide the unique capability of curing rapidly at room temperature. In applications that
require mass production, it is necessary to provide automatic dispensing equipment in the joining process.

Most equipment suppliers in the list below are capable of designing, manufacturing, and delivering automatic equipment that will
handle nearly all acrylic adhesives.

The primary difference, as regards equipment, between modified acrylic structural adhesives and the more well-known epoxies

and urethanes is that the resins used may be somewhat corrosive to cold-rolled steel and aluminum. Therefore, it is recommended that stainless steel be used where adhesive is in contact with the equipment. Additionally, care must be used in the selection of organic polymer seals and hoses that contact the adhesive resin. Considerable experience has already been developed in meter-mix-dispense units for the modified acrylic adhesives. The equipment or adhesive supplier should be consulted prior to using any acrylic adhesive for the first time in equipment that has been used with other types of adhesives.

Suppliers of mix, meter, and dispense equipment, meter and dispense equipment, and dispense equipment are listed below.

1. Accumetrics, Inc.
   Division of Detrex Chemical
   P.O. Box 843
   Elizabethtown, Kentucky 42701
   Telephone: 502-769-3385
2. Alemite
   Division of Stewart-Warner
   1826 West Diversey Parkway
   Chicago, Illinois 60614
   Telephone: 312-883-6000
3. The Aro Corporation
   One Aro Center
   Bryan, Ohio 43506
   Telephone: 419-636-4242
4. Automatic Process Control
   1123 Morris Avenue
   Union, New Jersey 07083
   Telephone: 201-688-1618
5. Glenmarc Manufacturing, Inc.
   330 Melvin Drive
   Northbrook, Illinois 60062
   Telephone: 312-272-9030
6. Graco, Inc.
   P.O. Box 1441
   Minneapolis, Minnesota 55440
   Telephone: 612-378-6000
7. Hardman, Inc.
   Triplematic Division
   Belleville, New Jersey 07109
   Telephone: 201-757-3000
8. Kenics Corporation
   Kenics Park
   North Andover, Massachusetts 01845
   Telephone: 617-687-0101

9. Liquid Control Corporation
    7576 Freedom Avenue, NW.
    P.O. Box 2747
    North Canton, Ohio 44720
    Telephone: 216-494-1313
10. Otto Engineering, Inc.
    36 Main Street
    Carpentersville, Illinois 60110
    Telephone: 312-428-7171
11. Pyles Industries, Inc.
    28990 Wixom Road
    Department TR
    Wixom, Michigan 48096
    Telephone: 313-349-5500
12. Sealant Equipment & Engineering, Inc.
    2100 Hubbel Road
    Oak Park, Michigan 48237
    Telephone: 313-967-2111
13. Tridak
    Division of Indicon
    Secor Road
    Brookfield Center, Connecticut 06805
    Telephone: 203-775-1287

Suppliers of spray equipment include the following:

1. Binks Manufacturing Company
    9205 West Belmont Avenue
    Franklin Park, Illinois 60131
    Telephone: 312-671-3000
2. The DeVilbiss Company
    P.O. Box 913-T
    Toledo, Ohio 43692
    Telephone: 419-470-2169
3. Graco, Inc.
    P.O.Box 1441
    Minneapolis, Minnesota 55440
    Telephone: 612-378-6000
4. IRC Corporation
    1363 East 286th Street
    Wickliffe, Ohio 44092
    Telephone: 216-944-7500

The range of equipment available to handle the acrylic structural adhesives is quite extensive. Commonly, the delivery systems
for specific applications must be selected based on the joining process selected. It is important to learn the range of equipment capabilities prior to designing the assembly stations in each individual

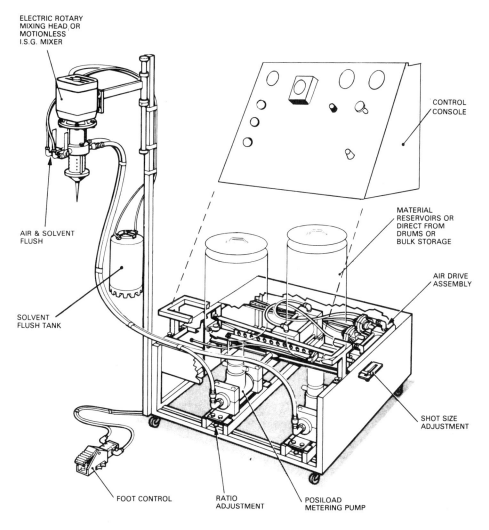

ELECTRIC ROTARY
MIXING HEAD OR
MOTIONLESS
I.S.G. MIXER

CONTROL
CONSOLE

AIR & SOLVENT
FLUSH

MATERIAL
RESERVOIRS OR
DIRECT FROM
DRUMS OR
BULK STORAGE

AIR DRIVE
ASSEMBLY

SOLVENT
FLUSH TANK

SHOT SIZE
ADJUSTMENT

FOOT CONTROL

RATIO
ADJUSTMENT

POSILOAD
METERING PUMP

Fig. 10. Twinflo machine: Typical dispensing set-up for acrylic adhesives. (Courtesy of Liquid Control Corp., North Canton, Ohio.)

joining operation. This is necessary to take full advantage of the
efficiency and capabilities of both the equipment and the acrylic adhesive that is selected based on performance, cure speed, rheology,
and throughput of the assembly operation.

A representative schematic of one of the more versatile pieces
of equipment supplied by Liquid Control Corporation is shown in
Fig. 10.

IX. PRESENT LIMITATIONS AND FUTURE DIRECTIONS OF MODIFIED ACRYLIC STRUCTURAL ADHESIVES

Although this family of structural adhesives is relatively new, the early prototypes have been used successfully for nearly 20 years [2]. The technology has advanced at a rather rapid pace over the past few years and now is becoming more commonly known. The distinct advantages in energy savings (100% reactive, room-temperature curing) and the competitive economics relative to the epoxies and urethanes, along with the advent of engineering plastics as materials of construction, have caused these materials to be recognized as superior products for mass produced joining operations.

As with any new system, there is a reluctance to change to it because these materials are "different" from that which is familiar, that is, "the epoxies". The acrylic adhesives have a distinct and different odor; their cure cycle is not familiar to most people; their affinity for unprepared surfaces is in direct contrast to the "white glove" approach that has been taught in the past.

While research and development continue to provide improved systems, those available today are performing assembly operations that could not be done with adhesives only a few years ago. Lower-odor systems, broader temperature performance, higher flash points, and greater substrate versatility are being, and will continue to be, developed by the major adhesive suppliers. The resin capabilities are challenging the equipment suppliers to design and build totally automated delivery systems that can not only meter, mix, and dispense, but also will be able to provide motion control (Robotics), self-sensing automatic adjustments on fluid flow volumes, and rheological changes in the dispensing operation.

Totally automated assembly operations are now on the horizon in several major manufacturing fields. That these automated operations will replace mechanical fastening and welding is due, nearly entirely, to the capabilities that are inherent in the adhesive technology of the newer modified acrylic structural adhesives.

REFERENCES

1. J. A. Graham, *Machine Design* (8 December 1977).
2. J. A. Graham, *Machine Design* (7 October 1976).
3. D. J. Zalucha, High performance structural bonding of unprepared metals, in *SME Tech. Bull.* SME, September 1978.
4. B. Gould, Acrylic structural adhesives/challenging the epoxies, in *Assembly Engineering* Hitchcock Publishing Company, 1978.
5. *1979 Annual Book of ASTM Standards, Part 22,* ASTM, Philadelphia, 1979.

6. W. J. Owston, to Lord Corporation, U.S. Patent No. 3,832,274 (1974).

7. E. Bäder and O. Schweitzer, to Deutshe Gold und Silber, U.S. Patent No. 2,981,650 (1961).

8. E. Bäder, to Deutshe Gold und Silber, U.S. Patent No. 3,321,351 (1967).

9. W. J. Owston and D. D. Howard, to Lord Corporation, U.S. Patent No. 3,873,640 (1975).

10. W. J. Owston, to Lord Corporation, U.S. Patent No. 3,838,093 (1974).

11. W. J. Owston, to Lord Corporation, U.S. Patent No. 3,962,498 (1976).

12. W. J. Owston, to Lord Corporation, U.S. Patent No. 3,970,709 (1976).

13. P. C. Briggs, to E. I. Du Pont de Nemours & Company, U.S. Patent No. 3,890,407 (1975).

14. A. S. Toback, to Loctite Corporation, U.S. Patent No. 3,616,040 (1971).

15. D. J. Zalucha, to Lord Corporation, U.S. Patent No. 4,223,115 (1980).

10

Phenolic Adhesives and Modifiers

Robert H. Young* and J. M. Tancrede† *Union Carbide Corporation, Bound Brook, New Jersey*

I. INTRODUCTION

In the early 1900s, Leo Baekeland discovered a way to use, for practical applications, a very simple chemical reaction between phenol and formaldehyde (Fig. 1) [1]. The chemistry of this application had been investigated many years before [2]. The initial reaction products are

---

Present affiliations
*Weyerhaeuser Company, Tacoma, Washington
†Exxon Chemical Company, Baton Rouge, Louisiana

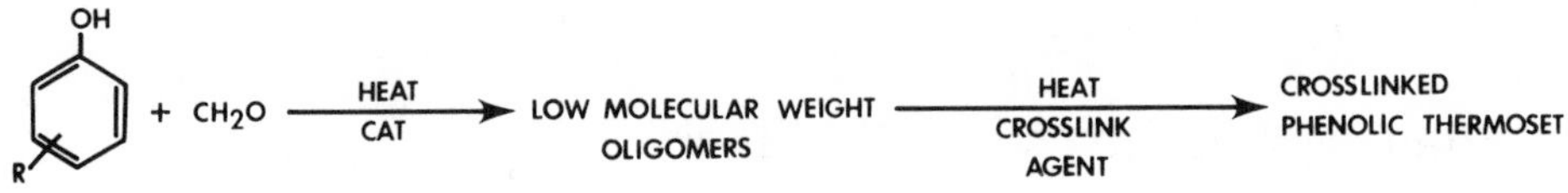

Fig. 1. Formation of phenolic thermosets from phenol and formaldehyde.

a series of relatively low-molecular-weight oligomers with molecular
weights from a few hundred to a few thousand. With additional heat,
and sometimes a catalyst or hardener, these oligomers will chain extend and crosslink to yield a phenolic, thermoset product.

Baekeland discovered a method to make useful products from a
resin which previously had had no special utility. However, it was
only after 5 years of intensive effort and after many failures, that he
succeeded in the development of a useful material called Bakelite.

Leo Baekeland had a varied educational and an interesting occupational career prior to his discovery of Bakelite. His early studies
were in Belgium at the University of Ghent. Later he studied at the
University of London, Oxford University, and the University of Edin-
burgh. After coming to the United States he worked in the area of
photographic materials. It was in this area that he made a major development which, after some hard business times, gave him the finan-
cial freedom to continue investigations into new areas.

After a long and systematic investigation in which Baekeland
tried to study all factors of the reaction between formaldehyde and
phenol, he found that the reaction could be dissected, or separated,
into different steps. He also found that pressure was valuable in controlling the reaction, and that in the presence of ammonia or another
base he could spread the reaction over a longer period and so could
stop it at any stage he wished by cooling [1]. In 1910 the General
Bakelite Company was founded. At this time there were a number of
industries which were in need of this Bakelite material, a plastic which
could be used to mass-produce standardized, interchangeable parts.

The product had excellent dimensional stability and electrical
properties; thus, one of the early uses of the product was an electrical application. Subsequent to this, the phenolic material was used in
the automotive industry because of its dielectric strength and its "im-
munity from temperature, acids, oils, and moisture" [3]. Other areas
Dr. Baekeland pioneered for this new polymeric material were the
abrasive industry, the telephone industry, the radio industry, the
packaging industry, and the protective-coatings industry.

As early as 1912 there were hundreds of uses and applications
for Bakelite materials. These included many for the automotive re-

lated area, such as molded parts, moisture-resistant cements, and timing gears. However, until 1915 most of the uses were confined to lighting and ignition equipment. By 1918 these phenolic-based products were also used as radiator caps, gearshift knobs, battery terminals, door-latch handles, sliding circuit connectors, commutators, as well as in spark plugs, gauges, and as cement for bonding electric-headlight lamp bases. By 1935 the uses had expanded further to instrument panels, steering wheels, magnetic couplings, ignition locks, robe rails, door-lock buttons, ash trays, heaters, and parts of the auto radio. Most of these applications were based on molding materials and on coating resins (e.g., varnishes). In the early 1930s phenolic resins as adhesives began to become more important. The typical "glues" available at that time had limitations which resulted in an inability to produce a uniform product. Staining, lack of moisture resistance, and lack of resistance to bacteria and fungi were all problems which were encountered with adhesives at that time. About 1931 development of the use of a new phenolic resin for plywoods and veneers began. It was recognized that phenolics had an advantage of being chemically inert, and, thus, were free from attack by fungi and bacteria. They were unaffected by heat, cold, and moisture, and did not stain.

Since these early times of the plastic industry, many new plastics have been discovered, but phenolics have remained "as lively and as important as they were in those first formative years" [3].

At the present time phenolic resins are used as the major bonding agent, or contribute to bonding, in a variety of automotive related application areas including foundry, friction, abrasives, fiberbonding, contact adhesives and sealants.

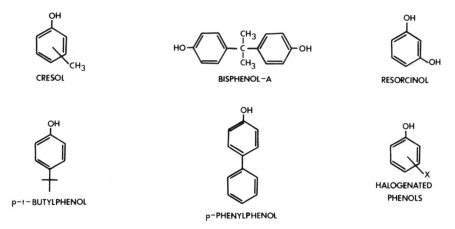

Fig. 2. Examples of substituted phenols used in the manufacture of phenolic resins.

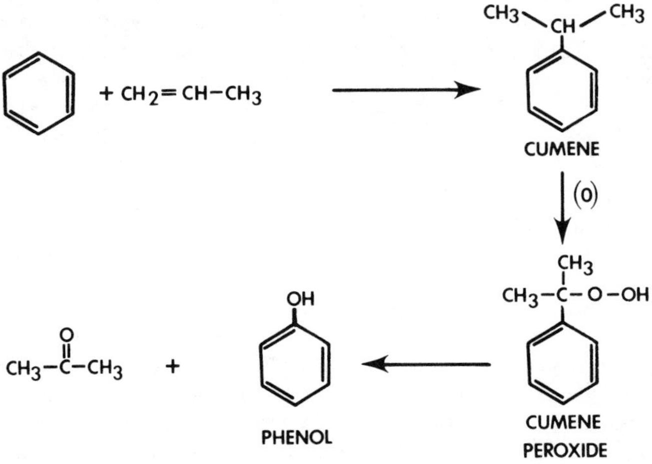

Fig. 3. Synthesis of cumene and phenol from benzene.

II. CHEMISTRY OF PHENOLIC RESINS

Phenolic resins are manufactured from phenol and a large number of
substituted phenols through reaction with an aldehyde, primarily formaldehyde. Some examples of phenols are cresols, bisphenol-A, resorcinol, p-t-butylphenol, p-phenylphenol, xylenols, cardenol (meta-
substituted alkyl phenol from cashew-nut shell liquid), and others
(Fig. 2).

The major chemical route to the most common reactant, phenol, is
outlined in Fig. 3. Benzene is initially reacted with propylene to yield
cumene. Cumene is then oxidized to cumene hydroperoxide which, in
turn, undergoes an acid-catalyzed rearrangement reaction to yield
phenol and acetone.

There are two basic chemical types of phenolic resins, resols
and novolacs. They are differentiated by their phenol-to-formalde-
hyde ratio, the type of catalyst used in manufacture, and the chemical

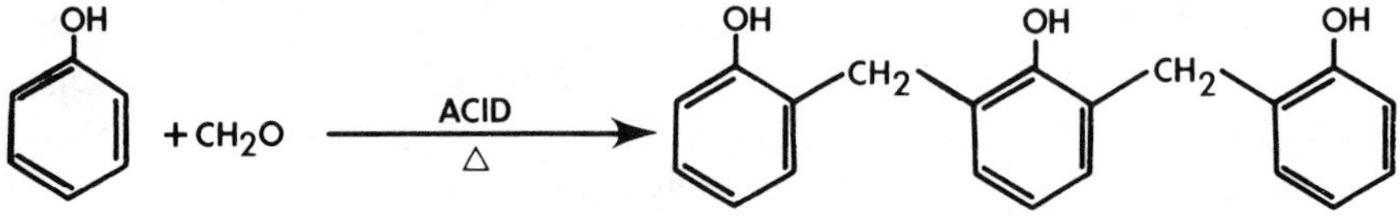

Fig. 4. Novolac-phenolic-resin chemistry.

Table 1   Characteristics of Resol Versus Novolac Phenolic Resins

Characteristic	Resols	Novolacs
Catalyst type	Alkaline	Acid
Mole ratio of CH_2O:phenol	> 1	< 1
Resin structure	High branched	More linear
Reactivity	Cures with heat	Requires both heat and hardener to cure

structure of the resulting resin (Table 1). These chemical differences are further illustrated in Figs. 4 and 5.

A novolac resin is characterized by having no reactive methylol groups but having unsubstituted ortho and/or para reactive sites where a hardener, such as hexamethylenetetramine (hexa), can react to yield a chain-extended and, ultimately, crosslinked polymeric system. A resol resin, on the other hand, contains not only open reactive sites, but also reactive methylol groups. The result is that resols require only heat to effect chain-extension and crosslinking reactions. The cure of both types of resins is dependent on temperature, catalyst type, hardening agents such as hexa, concentration of catalyst and/or hardening agents, and the type of phenol and aldehyde used.

The chemical reactions involved in the cure of phenolic resols, for example, include: substitution reactions of the methylol groups at a reactive site of another phenolic ring, yielding a methylene linkage; a substitution reaction by a methylol hydroxyl group to yield a methylene ether linkage (Figs. 6 and 7).

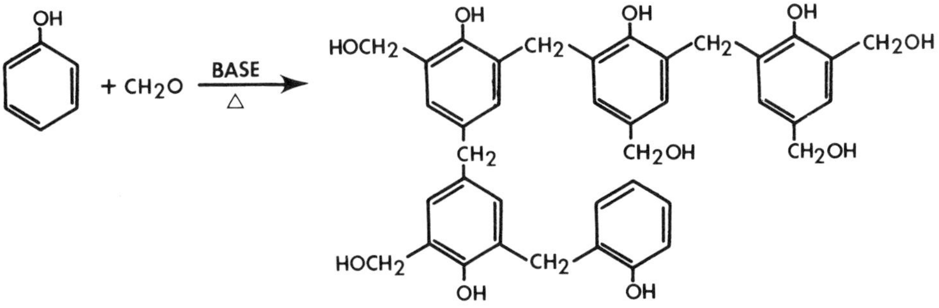

Fig. 5.   Resol-phenolic-resin chemistry.

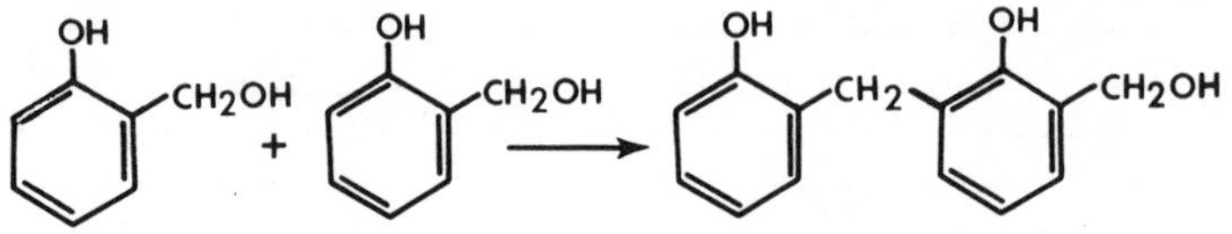

Fig. 6. Chemical reaction of a resol phenolic to form a methylene linkage.

The chain-extension crosslinking reactions of the phenolic resol or novolac result in a fully cured system. Many factors contribute
to the degree of this cure which, in turn, affects the performance
properties of the ultimate product. Leo Baekeland in 1909 described
the curing process as going through three phases of reaction [4].

The first phase results in the formation of low-molecular-weight
oligomers and is designated as A-stage. At ambient temperatures the
phenolic may be a low to high viscosity liquid, paste, or solid. This A-
stage product is soluble in alcohol, acetone, or similar polar solvents
and in sodium hydroxide solution. The solid form will melt on being
heated.

The second phase involves the formation of an intermediate condensation product and is designated as B-stage. In this form the
phenolic is a brittle solid which is slightly harder than a solid in the
A-stage. The B-stage resin is now insoluble in all solvents but may
swell in acetone or similar solvents. Although it will not melt on heating, it will soften and can become somewhat thermoplastic-like. Further heating will take it into the fully cured C-stage. In this stage
the phenolic is infusible and insoluble in all solvents. The cured resin is now resistant to chemicals, thermally stable, and a good insulator
to heat and electricity.

III. ANALYTICAL TEST METHODS

A number of different methods are used in the laboratory to elucidate
the composition and structure of phenolic resins and the chemistry of

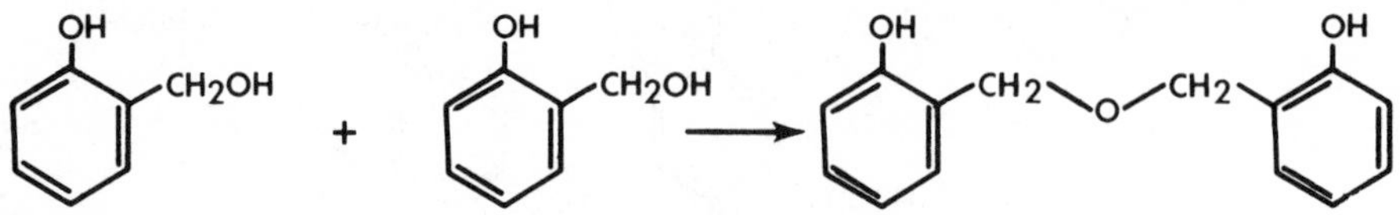

Fig. 7. Chemical reaction of a resol phenolic to form a methylene
ether linkage.

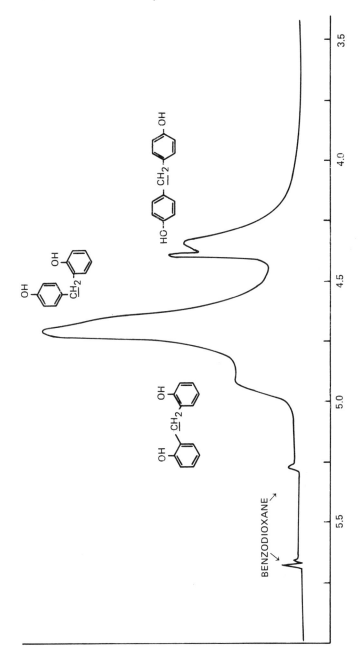

Fig. 8. Nuclear magnetic resonance spectrum of a novolac phenolic.

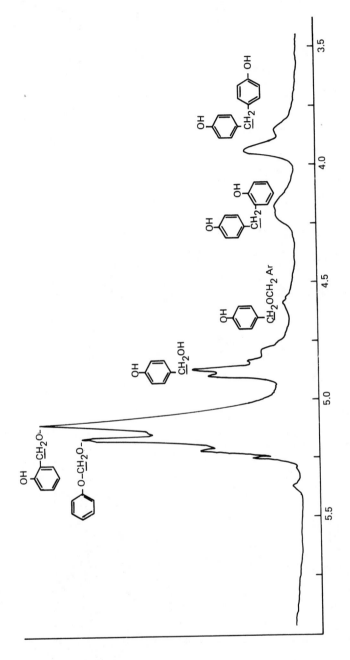

Fig. 9. Nuclear magnetic resonance spectrum of a resol phenolic.

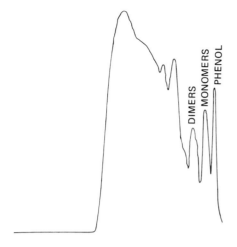

Fig. 10. Gel permeation chromatogram of a resol phenolic.

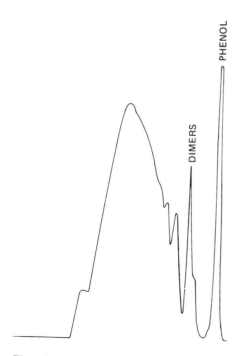

Fig. 11. Gel permeation chromatogram of a novolac phenolic.

the curing reactions. These methods include: infrared spectroscopy (IR); nuclear magnetic resonance spectroscopy (NMR); differential scanning calorimetry (DSC); thermal gravimetric analysis (TGA); gel permeation chromatography (GPC); vapor phase chromatography or gas chromatography (GC); and dynamic mechanical analysis (DMA), among others. Each of these methods offers some unique insight into the chemistry of phenolic resins. These methods are in addition to the normal quality-control techniques that are commonly used. Typical of the latter are plate flow, gel time, viscosity measurement, and other tests that are a function of molecular weight, reactivity, and crosslink density of the phenolic resin.

Both IR and ultraviolet (uv)-visible absorption spectroscopy can yield information on structure and functionality of phenolic resins. However, these methods require careful interpretation and appropriate standards. A more useful analytical technique is proton or ^{13}C NMR. Most of the published work on NMR of phenolic resins discusses proton NMR. In recent years ^{13}C NMR results have become more abundant, and it is predicted that in the next few years solid state NMR will be used more frequently.

Proton NMR quite readily distinguishes the different types of substituents and ring linkages between the phenol groups. Examples of typical NMR spectra for novolacs and resols are shown in Figs. 8 and 9, respectively. Note that these spectra give information, not only on the type and location of substituents on the phenol ring, but also on their relative concentrations.

Gel permeation chromatography is a variation of high-pressure liquid chromatography designed to separate molecules based on molecular-weight differences. Uncured phenolic resins can be easily separated into monomers, dimers, trimers, etc., as shown in Fig. 10 for a typical resol and Fig. 11 for a novolac resin. Note that in Fig. 10 a peak is shown for a methylolated phenol monomer and is clearly indicative of a resol resin. Depending on the analytical equipment and the columns used, even better separation of the peaks can be achieved. Thus, a GPC scan can be an excellent way to characterize a phenolic resin in terms of resin structure and relative component ratios, as well as resin molecular-weight distribution.

Differential scanning calorimetry and TGA are two methods which measure the response of a phenolic resin to increasing temperatures. Differential scanning calorimetry is used to obtain information on the resin softening point T_g and on its cure characteristics. Generally, caution is advised on interpreting the cure information due to complications arising from the formation and emission of by-products, such as water and ammonia (from hexa). To circumvent this problem, a pressure cell is used to prevent the loss of such components, thus yielding somewhat better results.

Thermogravimetric analysis gives an interesting insight into one of the key properties of phenolics, that is, their thermal stability.

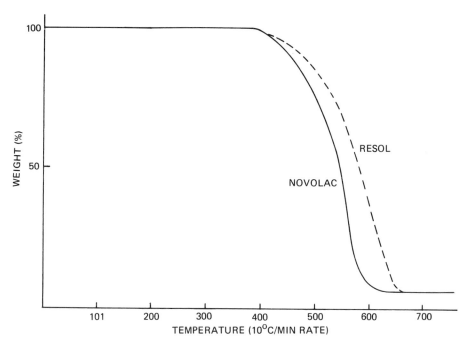

Fig. 12. Example of a TGA of a novolac and a resol phenolic.

This technique measures the weight gain or loss of a material as a function of increasing temperature. It is a useful monitor of the uptake of oxygen and degradation of the phenolic resin. Examples of such measurements are shown in Fig. 12. These TGA scans are obtained on cured materials and show that the oxidation of the methylene linkages does not occur until about 752°F (400°C).

With the exception of TGA, most analytical methods of characterizing phenolic resins involve measurements on phenolic oligomers, that is, A-stage resins. For this reason, they often do not yield all the information required to evaluate the cure characteristics of a given resin.

A more recent technique, DMA, is offering promise for evaluating phenolic resins during simulated cure conditions. One type of instrument, the Du Pont Model 880 DMA, measures the ability of a sample to transmit an applied frequency. The standard method of analysis involves scanning a sample at a programmed rate of temperature increase while simultaneously monitoring frequency. This frequency response is directly related to the modulus or stiffness of the sample which, in turn, is dependent on the molecular weight/crosslink density at a specific point in time. Figure 13 illustrates a typical DMA

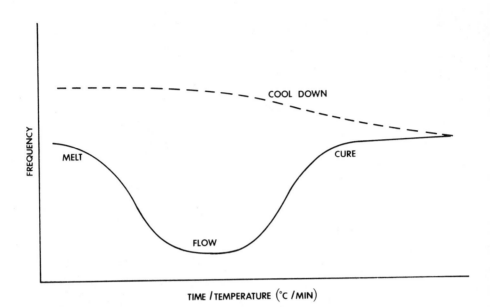

FREQUENCY

COOL DOWN

MELT   CURE

FLOW

TIME / TEMPERATURE $\left(^{\circ}\text{C /MIN}\right)$

Fig. 13. Typical DMA scan of a resol phenolic.

scan of a solid, resol resin. The DMA scan gives information on the
melting range and relative cure rates and, in a cool-down mode, gives
a modulus-temperature profile of the cured resin system.

In addition to monitoring the sample's frequency response, DMA
also provides a mechanical loss scan which yields a unique measurement of the T_g and the gel temperature based on a mechanical property of the resin (Fig. 14). One of the major deficiencies of this scanning method is that there are three variables (time, temperature, and
modulus) in a two-dimensional monitoring system. Thus, the kinetics
of the cure reaction cannot be quantified. In order to quantify the
kinetics of the cure reaction(s), a modified technique has been developed which involves monitoring the frequency response of the phenolic sample in an isothermal mode (Fig. 15) [5]. Rate constants are
obtained as a function of temperature. From these rate constants a
temperature response factor for the overall reaction can be calculated.
This type of data can be used to predict the degree of cure of a phenolic resin under a given set of reaction conditions for a given time
[5].

Phenolic resins are available in a wide variety of forms and can
be used in a large number of applications (Table 2). The methods of
choice for evaluating a phenolic resin are dependent on the application
area. Many times simple quality-control methods are all that is needed
and required. At other times a more detailed analysis is necessary in

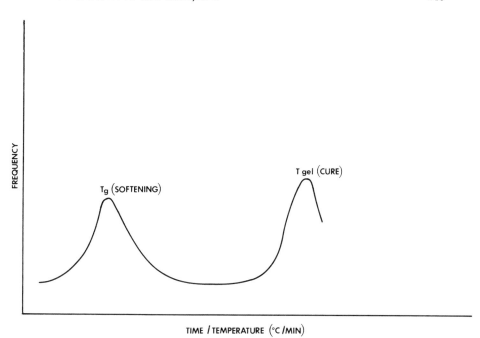

Fig. 14. Mechanical loss scan of a resol phenolic.

order to compare resin structure and properties with actual application performance needs. The following selected section (Sec. IV) will discuss phenolic bonding applications in the automotive area. A more detailed discussion will be presented on phenolics as modifiers for adhesives (Secs. V—XII).

IV. PHENOLIC ADHESIVES

Phenolic resins can be bonding agents as neat resin (adhesive) or as part of a formulation (phenolic modifier). The phenolics have good adhesion to polar substrates, good high-temperature properties, resistance to burning, and high strength. Phenolics are used as bonding agents in fiberbonding, friction, abrasives, and foundry applications, among others, all of which utilize the material as a neat resin.

In the fiberbonding area the phenolic resin is used as a binder in products such as thermal-insulation batting, automotive acoustical padding, and cushioning materials. These products can consist of a variety of fibers—such as glass, mineral, cotton, and polyester—laid down in a randomly oriented, loosely packed array to form a mat. The phenolic resin is used to bond the individual fibers together using

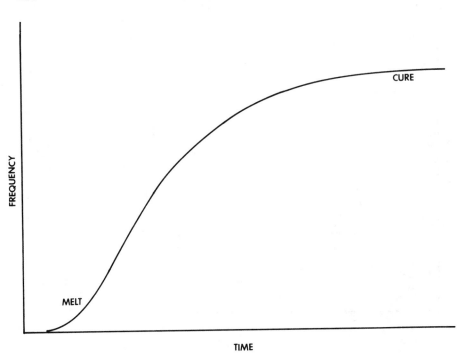

Fig. 15. Isothermal DMA scan of a resol phenolic.

either a dry-bonding or a wet-bonding process. The dry-bonding
process uses a pulverized phenolic resin, either a resol-based or a
novolac-hexa system, to bond reclaimed fibers. Both automotive
acoustic padding, normally involving organic fibers, and low-grade
thermal-insulation batting, utilizing glass fibers, are made using the
dry-bonding process. Resol-based phenolic resins offer some advantages for manufacture of both types of products, especially for the
glass-based materials, but this type of resin requires special handling
and refrigerated storage. In the wet-bonding process virgin spun
glass is bonded with a liquid, low-molecular-weight, water-miscible
resin. Higher-grade thermal insulation for construction applications
is made by this process.
      Phenolics are used in the friction area for a variety of applica-
tions which include automotive brake linings, disk pads, transmission
elements, and clutch facings. They are also sometimes used to bond
the friction element to a metal backing. The resin can directly affect
the coefficient of friction, high-temperature performance, fade, recovery, and wear resistance. Many chemically different resins and
modified resins are required in order to meet the performance criteria
of a wide number of friction applications.

Table 2 Material Forms and Application Areas for Phenolic Resins

Liquids			Phenolic resins			Dispersions
Solutions		Neat	Solids			
Aqueous	Organic		Lumps	Flakes	Pulverized	
	Laminates		Coatings	Foundry	Fiber bonding	Friction
	Foams		Adhesives		Grinding wheels	Adhesives
	Foundry				Wood bonding	Coated Abrasives
	Coated abrasives				Friction	Coatings
	Coatings					
	Fiber bonding					

Powdered phenolic resins are used extensively in the abrasives area for bonding various types of abrasive grains. In general, the manufacturing process involves the initial application of a wetting agent, such as a liquid phenolic resin or furfural, to the abrasive grains which are compounded further with a powdered phenolic resin, normally a novolac-hexa system. These compounded grains are then subjected to either a cold-pressing or a hot-pressing operation followed by a careful, high-temperature cure.

In the foundry industry phenolic resins are used as the bonding agent for the sand particles in the manufacture of shell molds and cores. Two shell halves are joined together by clamps or adhesive agents to form a shell mold into which molten metal is poured to form metal castings. The largest volume of phenolic resin is used in the hot-sand coating process. In this method the sand is preheated to a temperature of 275–338°F (135–170°C), and a solid novolac resin in flake form (2–3% by weight) is charged to the hot sand in a muller where it quickly melts and coats the sand. The batch is cooled by adding water, usually containing hexa. There is only minimal reaction between the hexa and the novolac resin at this point. The shell is then formed by depositing a resin-sand mix on a hot metal pattern plate and, after a predetermined period, inverting the pattern and dumping off the excess resin-sand mix that has not fused. An oven cure permanently bonds the sand particles. The typical flake phenolic resin used is an intermediate-melting-point novolac containing a low concentration of lubricant. The resin must be fast curing with excellent melt flow, good peel-back resistance, and high strength.

Other automotive-related application areas for phenolic resins include air and oil filters, battery separators (involving paper impregnation by liquid resins), and modifiers in formulated coatings, sealants, and adhesives.

V. PHENOLIC MODIFIERS

Phenolic modifiers are a smaller subset of the general category of phenolic adhesives. They refer to those phenolic resins used in conjunction with other adhesive systems. There are two general types of phenolic modifiers distinguished by whether the phenolic resin functions in the adhesive formulation primarily as a tackifier or as a crosslinking agent. As a tackifier, the phenolic resin is used to enhance the adhesion of relatively nonpolar elastomers by improving the wetting property of the adhesive mass. These elastomeric adhesives are to be distinguished from the structural-type systems which rely on the crosslinking ability of the phenolic resin to provide chemical and thermal stability.

VI. PHENOLIC MODIFIERS AS TACKIFIERS

In adhesive technology, tack can be defined as the property of a material which enables it to form a bond of measurable strength upon contact with another surface, usually with low applied pressure. Most synthetic elastomers and natural rubbers used in adhesives have little tack, either for themselves or other surfaces. Accordingly, phenolic tackifiers in the form of resins are added to these systems to increase their tack.

The type of phenolic resin used to tackify elastomers in adhesives is primarily based on the reaction products of alkyl-substituted phenol and formaldehyde. These phenolic tackifiers enhance the ability of the elastomers to wet a surface by increasing the polarity and altering the viscoelastic properties of the adhesive mass. The increase in polarity is largely due to the presence of phenolic hydroxyl groups and, in the case of heat reactive resins, aliphatic hydroxyl and ether groups. In addition, the tackifier facilitates plastic deformation by reducing elastic recovery, which enables the adhesive mass to contact the surface more intimately, resulting in higher bond strengths.

Plastic deformation has been shown to result from the ability of the tackifier resin to dissolve some of the rubber material and form a discrete phase at the surface of the film [6]. The extent of formation of this discrete phase depends on the compatibility of the resin with the rubber. With the more polar elastomers—like neoprene—p-butyl—substituted phenolic resins are found to be very effective tackifiers. On the other hand, with less polar elastomers—like butyl and styrene-butadiene rubbers (SBR)—phenolic resins containing longer-chain octyl and nonyl substituents are preferred.

VII. SOLVENT-BASED CONTACT ADHESIVES

Perhaps the best known and, certainly, the most important use of phenolic tackifiers is in the area of contact adhesives [7-10]. Contact adhesives are characterized by their aggressive tack and excellent peel strength. They are used to join surfaces which must bond on contact and must not slide after assembly. As such, they have found application in the automotive industry for bonding interior upholstery and exterior vinyl roof tops; in the furniture industry for attaching high-pressure plastic laminates; in the construction industry for assembling sandwich panels, metal doors and dry wall, wood, or plastic installation; and in the shoe industry for the attachment of soles to shoes.

The adhesive systems available to the contact adhesive area are outlined in Table 3. The solvent-based, neoprene contact adhesive tackified with a heat-reactive, butylphenol resin is still the most wide-

Table 3 Elastomer-Based Adhesives Modified with Phenolic Tackifiers for Use as Contact Adhesives

Adhesive vehicle	Elastomer	Type of phenolic resin used	Examples of commercial product[a]
Solvent	Neoprene	p-t-Butyl Phenol resol	CK-1634 CK-1636 SP-134
	Neoprene	Terpene phenolic	Durez 12603
	Nitrile	One-step	BKR-2620
Emulsion	Neoprene	Phenolic dispersion	Durez 12603 BKUA-2370
	Acrylic	Phenolic dispersion	BKUA-2370

[a]Trade name suppliers are listed in Sec. XIII.

ly used formulation. This type of formulation is noted for its rapid bond-strength development and for its high ultimate strength. The nitrile-based contact adhesives, which are more effectively tackified with a more polar phenolic resol, have excellent oil and grease resistance. They can be used on substrates that contain migrating materials, such as plasticizers and oils, which can enter the glueline and weaken the adhesive bond.

Water-based formulations are newer members of the contact adhesive family. With the growing concern over the toxicity and flammability of solvent systems, these formulations offer viable alternatives to solvent-based formulations. Phenolic tackifiers in the form of dispersions are being developed to help these latex adhesives achieve the high-performance properties characteristic of solvent-based systems.

A. Neoprene–Phenolic Contact Adhesives

The key components associated with formulating a solvent-based neoprene contact adhesive are the neoprene rubber, phenolic tackifier, and solvent blend. The adhesives are prepared at 18–30% solids and contain, in addition to these key compounds, magnesium and zinc oxides, which function as HCl acceptors and crosslinkers, and an antioxidant. The magnesium oxide plays an additional and important role in that it forms a nonfusible, but soluble, metal complex with heat-reactive, substituted phenolic resins. This complex improves the cohesive strength of the adhesive at both room and elevated temperatures.

Table 4   Typical Neoprene-Phenolic Contact Adhesive

Ingredient	Parts
Mill mix	
Neoprene AC	100
MgO	4
ZnO	5
Solvent blend	333
Phenolic-MgO complex	
Butyl phenol resin	45
MgO	4
H_2O	1
Solvent blend	48

The solvent blend is generally a mixture of toluene, ketone, and aliphatic hydrocarbon. Toluene is the primary solvent for the rubber, and the other solvents are added to reduce cost and to control the evaporation rate of the adhesive. A typical formulation is shown in Table 4.

B. Adhesive Compounding

There are two general methods for compounding a neoprene-phenolic adhesive. The first method involves a two-part mix. One part is prepared by masticating the neoprene rubber in a two-roll mill to improve solubility and adhesive uniformity. Magnesium oxide, zinc oxide, and antioxidant are worked in during the milling operation. The magnesium oxide is added before the zinc oxide to prevent premature curing of the rubber. The homogenized sheet of rubber is then cut and dissolved in a solvent blend.

In the second part of the mix, the phenolic resin is separately dissolved in additional solvent. In the case of the more generally encountered heat-reactive, substituted phenolic resins, additional magnesium oxide is added with a catalytic amount of water to form an organometallic complex. Generally, this reaction is near completion within 2–3 hr. This resin solution is then mixed with the first part of the mix containing the dissolved milled rubber.

If milling equipment is not available, the adhesive is simply prepared by mixing all the components together in an appropriate mixing tank. Adhesives made by this method are usually produced at lower solids because of the higher solution viscosity of the unmilled rubber. Also, the adhesive pigments are not as uniformly mixed with the rubber and are more prone to settle on aging.

C. Adhesive Testing and Performance

Important properties for a contact adhesive are open (tack) time, cohesive strength, and heat and phasing resistance.

Open time, also known as bonding range, is the maximum time two coated surfaces may be allowed to dry before bonding and still attain satisfactory adhesion. A good method for comparing the open times of various formulations is to apply a wet adhesive film to both surfaces of unsized canvas. The surfaces are allowed to dry tack-free and are then assembled using a constant, low pressure load (e.g., a 10-lb roller) at various predetermined times. These bonds are tested about 30 min after assembly for green strength and again after aging several days to a week for ultimate bond strength. Ultimate strength based on canvas-to-canvas bonds using a tensile tester is probably the best measure of the cohesive strength of the adhesive film.

The relationship between open time and green strength is an important feature of a contact adhesive. For an ideal system, a long open time accompanied by a rapid buildup of green strength is desired

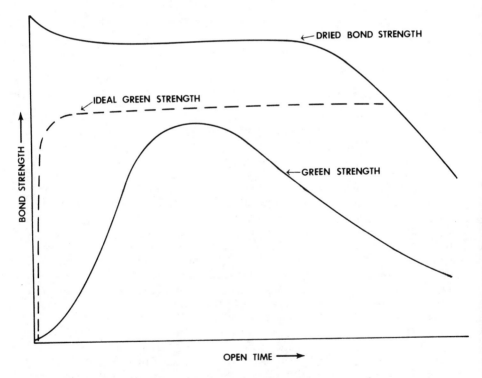

Fig. 16. A generalized illustration of the dependence of green strength and dry bond strength on open time.

(Fig. 16). Normally, green strength increases with open time to a maximum value and then begins to decrease. The dry, or ultimate, bond strength, on the other hand, is greater at shorter open times. The bonding range of neoprene contact adhesives is most readily varied by the grade of neoprene and the type and level of phenolic resin used. The fast-crystallizing grades of neoprene rubbers, such as Neoprene AC, AD, and AF*, have poor tack retention. Addition of slower-crystallizing grades of neoprene rubber and tackifiers in particular will enhance tack. Tackifiers decrease in efficiency in the order of terpene, terpene-phenolic and heat-reactive alkyl phenolics. The complexation of alkyl phenolics with metal oxides further decreases adhesive tack, presumably because of more rapid solvent release. However, this type of tackifier greatly enhances adhesive heat resistance and metal adhesion.

Heat resistance can be determined by several measurements including peel and/or shear strength as a function of temperature, high-pressure laminate lift [11], and dead-load creep at elevated temperature.

The preferred test(s), substrates, and bond-preparation method depend on the intended application. The ability of neoprene rubbers, such as Neoprene AC and AD, to crystallize on aging accounts for their exceptionally high cohesive strength at room temperature. Peel strengths of 35 pli and higher are possible for canvas-to-canvas bonds at room temperature, but they decrease dramatically above 140°F (60°C), the point where the crystallization process is reversed. The addition of a heat- and metal-oxide—reactive, alkyl-substituted phenolic resin to the neoprene adhesive will enable the bond to maintain greater than 15 pli canvas-to-canvas peel strength at 140°F (60°C) and to resist dead-load creep up to 200°F (93°C). Adhesives with higher heat resistance can be made with a carboxyl containing polychloroprene, such as Neoprene AF.

Adhesive phasing, or flocculation, refers to the separation of the adhesive mix into two layers on standing. The separated mix consists of a clear or slightly opaque upper layer and a lower layer of flocculated metal oxide. Formulation variables—such as neoprene type, solvent mixture, solids content, heat history, and milling time—are factors affecting flocculation. The heat-reactive resin is by far the most important factor. Studies have shown that low-molecular-weight fractions [12] and various dialcohols such as p-butylphenol dialcohol [13] in particular, are similarly responsible for the flocculation phenomenon. Based on these studies, "nonphasing" resins, such as CK-1636 and SP-154, have been developed. The thermoplastic, non-heat-reactive phenolics, such as the terpene-phenolic resins (Durez 12603 or Sp-560) and butylphenol novolacs (CK-2103 or CK-2432),

---

*Suppliers of trade-name chemicals are identified in Sec. XIII.

Table 5   Guide Formulation for a Nitrile-
Phenolic Contact Adhesive

Ingredient	Parts
Nitrile rubber (high-nitrile type)	15
Phenolic resin (BKR-2620)	15
Ketone [methyl ethyl ketone	
(MEK), acetone]	70
	100

yield nonphasing adhesives at the expense of peel strength and heat
resistance.

D. Solvent Blend

The choice of solvents is very important for achieving a compromise
in cost, toxicity, viscosity, and volatility in solvent-based adhesives.
The use of the concept of solubility parameters for selecting solvents
has gained wide acceptance. Comparison of solubility parameters with
the solubility characteristics of the base elastomer can be used to predict, with a fair degree of accuracy, the efficiency of a wide range of
solvent blends. Graphic methods make this technique particularly effective [14,15].

E. Nitrile-Phenolic Contact Adhesives

Commercial nitrile rubbers are predominantly copolymers of butadiene
and acrylonitrile. Two most important properties of nitrile rubbers
are excellent oil and plasticizer resistance. These properties improve
as the acrylonitrile content of the copolymer is increased. Increased
acrylonitrile content also improves compatibility with phenolic tackifiers.

An unmodified film of nitrile rubber deposited from solution
tends to skin over the surface, thereby considerably restricting further loss of solvent. Thus, the nitrile rubber exhibits poor tack,
and, consequently, poor autoadhesion on contact. Resol or novolac
resins, prepared from phenol, are added to nitrile rubber to promote
solvent release and to improve tack and heat resistance.

A formulation for a nitrile contact adhesive, modified with a
resol phenolic resin, is shown in Table 5. This adhesive formulation
is less complicated than a solvent-based neoprene system, but is high-

Table 6   Preparation of Phenolic-Resin Emulsions

Ingredient	Parts by weight	
	Method A	Method B
Phenolic resin	100	100
Toluene	54	
Oleic acid	6	
NaOH (25% aq)	6	
Ammonium caseinate (10%)	30	30
Gum arabic (5%)	60	
Daxad 11 (10%)		30
H_2O	77	40
Triethanolamine		

er in raw-material cost. The use of solvent-borne, nitrile contact adhesives appears to be declining, largely due to the trend away from
solvent-based systems.

VIII. PHENOLIC DISPERSIONS

Phenolic tackifiers in the form of aqueous dispersions can be combined
with a number of other thermoplastic resin latices to yield polymer
alloys for use as adhesives. These polymer alloys are finding use in
areas previously dominated by solvent systems, such as contact adhesives. For example, high-solids, water-based contact adhesives
are readily prepared by simply blending the appropriate acrylic latex
and water-dispersed, phenolic tackifier. This is to be contrasted with
solvent systems which involve specialized equipment to compound and
dissolve the rubber component. Also, greater attention is required
for the selection of suitable solvent blends to yield a proper balance of
adhesive solids and viscosity. In the case of water-based systems,
the formulation concerns are primarily the stability of the individual
dispersions and the stability of the resulting blends.
    Several approaches have successfully been used to emulsify phe-
nolic resins. Two of these approaches [8,16,17] are outlined in Table
6. In method A, the resin is first dissolved in a suitable solvent,
such as toluene, along with the emulsifying agent, oleic acid. This
solution is then added, with agitation, to the basic aqueous phase
which also contains other stabilizing ingredients. This yields an
anionically stabilized dispersion with solids content of about 30%.

Table 7   Neoprene-Latex Contact Adhesive Tackified
with a Terpene-Phenolic Resin Emulsion

Ingredient	Dry parts by weight
Neoprene 750	100
Durez 12603	50
Antioxidant	2
ZnO	10

Alternatively, method B results in a solventless dispersion with
higher solids content (about 50%). After the ingredients have been
slurried together, they are added to a pebble mill and rolled for 48 hr.
In both methods oleic acid soap and ammonium caseinate colloid are
utilized with stabilizers for the phenolic dispersions.

Resins emulsified by either method A or method B are compatible
with anionically stabilized adhesives such as the one shown in Table 7.
The combination of neoprene latex and tackifier provides a good balance of cohesive strength, contactability, and heat resistance suitable
for contact adhesive applications [18]. In this formulation, a terpene-
phenolic resin dispersion enhances peel strength and overall specific
adhesion.

A third approach for incorporating a phenolic-resin tackifier
into a water-based adhesive is illustrated by the formulation in Table
8. The phenolic resin, a butylphenol novolac, is first dissolved in a
50:50 blend of ethanol and butyl acetate similar to method A already
described. In contrast to method A, this solution is then added directly to an acrylic latex. In this example, the formulation is further
filled with calcium carbonate to yield a construction mastic. The phenolic resin enhances adhesion to wet, frozen, and dry lumber. This
formulation meets the AFG-01 specifications for bonding plywood to
wood framing [19].

Recently, phenolic dispersion tackifiers based on proprietary
formulations have been commercialized. Union Carbide's BKUA-2370
is described as a heat-reactive, phenolic dispersion stabilized by a
nonionic emulsifier [20]. Typical physical properties of this dispersion are shown in Table 9. It has excellent mechanical, freeze-thaw,
and shelf stability, and, moreover, is compatible with a wide variety
of waterborne vehicles [21]. This phenolic dispersion is recommended
as a tackifier to enhance specific adhesion and bond strength of water-
based adhesives. For example, a starting point formulation of an
acrylic contact adhesive is shown in Table 10. The pH of the phe-
nolic-latex formulation is adjusted to about 6.5 with an amine to ensure
emulsion compatibility.

Table 8   Direct Addition of Phenolic Tackifier to a Latex Mastic Adhesive

Ingredient	Percent by weight
UCAR latex 154	57.41
Cellosize QP 300H	0.50
Daxad 301	3.20
Triethanolamine	0.75
ASP-400P	17.21
Dowicil S-133	0.25
Bakelite resin CK-2103 (50%)	10.34
Ethanol, denatured (25%) solution)	5.17
Butyl acetate (25%)	5.17
	100.00

Substrate	Shear strength[a] (lb)
Wet lumber	335
Frozen lumber	332
Dry lumber	730

[a]AFG-01 test based on 1.5-in.^2 blocks.

The peel strength of the acrylic-phenolic, water-based formulation is somewhat lower than a conventional solvent-neoprene system (Table 11). However, it is comparable in green strength and metal adhesion and is superior in high-temperature shear resistance. This water-based acrylic-phenolic formulation also passes the 140°F (60°C) edge lift test under Federal Specification MMM-A-130B for contact adhesives involved in bonding of wood to high-pressure plastic laminates [11].

IX. WATER-BASED VERSUS SOLVENT-BASED CONTACT ADHESIVES

For nearly 30 years the term "contact adhesive" has implied the use of solvents. Recently, solvent contact-adhesive formulations have come under pressure because of the heightened concerns over the toxicity, flammability, and high cost of solvents. Water-based systems offer safer and more ecologically acceptable contact adhesives. Formulations are being developed which are based on neoprene latexes (Table 7); newer, carboxylated, neoprene latexes [22]; and acrylic latices (Table 10). The acrylic-latex approach represents a departure from the traditional neoprene-based systems. Phenolic tackifiers in the

Table 9   Typical Physical Properties of BKUA-2370
Phenolic Dispersion

Property	Value
Solids (%)	45
Viscosity (cP)	5000
pH	4.5
Stability	
Mechanical	Excellent
Shelf life at 70°F	> 1 year
Freeze-thaw at 0°F	
5 cycles	Passes
Weight per gallon (lb)	9.0

form of dispersion, such as BKUA-2370, are being developed to help
these latex systems meet the demanding adhesive requirements expected of contact adhesives.

Water-based formulations to date, however, have experienced
slow acceptance in the marketplace. Although strides have been
made, water-based formulations are not as forgiving as solvent-based
systems. They are slower drying under ambient conditions, and their
overall performance is more sensitive to changes in humidity and temperature. These problems are being addressed by raw material suppliers and adhesive formulators.

Nevertheless, water-based formulations do offer some significant
advantages. In addition to being nontoxic and nonflammable, water-
based formulations are generally higher in solids content, lower in
viscosity, and afford greater latitude in controlling adhesive rheology.
Less expensive and less critical manufacturing techniques can be used
resulting in the reduced raw-material costs per pound of applied adhesive. It is not surprising, therefore, that water-based contact adhesives are receiving increasing attention.

X. OTHER USES FOR PHENOLIC TACKIFIERS

Phenolic resins and terpene-phenolic resins are used to tackify polymers and elastomers, such as butyl rubber, polybutadiene, SBR,
ethylene-propylene co- and terpolymers, and other relatively nonpolar
adhesives. These polymers have very little tendency to crystallize
and depend on molecular entanglement or crosslinking for their
strength. The completely amorphous character of these polymers imparts flexibility, tack, and impact-shock resistance. However, these

Table 10   Acrylic-Latex Contact Adhesive Formulation

Formulation	Wet parts
UCAR latex 154, 152 (60% N.V.[a])	100
BKUA-2370	44
Triethanolamine	0.5
Typical properties	Values
Latex:phenolic solids ratio	3:1
pH	6—6.5
Solids (%)	55
Freeze-thaw at 0°F	> 2 cycles

[a]N.V., nonvolatiles or % solids.

polymers have little or no polarity and, in spite of their inherent tack, have weak attraction to many surfaces. Commonly employed tackifiers are polyterpene resins, wood rosins, wood-rosin derivatives, hydrocarbon resins, and coumarone-indene resins. Terpene-phenolic resins and substituted, non-heat-reactive phenolic resins are also used to improve specific adhesion, cohesive strength, and heat resistance of adhesives based on these elastomers. These materials are used as general-purpose solvent adhesives, mastics, and hot-melt and pressure-sensitive adhesives.

Phenolic resins also play a small but significant role in the formulation of hot melts based on ethylene vinyl acetate and ethylene ethyl acrylate copolymers. High-melting-point, substituted, non-heat-reactive resins—such as CK-2103, CK-2432, and CK-2400—have been found to be effective modifiers for improving the elevated-temperature bond strength of ethylene-copolymer—based hot-melt formulations.

XI. STRUCTURAL ADHESIVES

Resols or hexa-modified, novolac phenolic resins are used in structural adhesives. They function primarily as crosslinking components rather than as tackifiers, as is the case with most elastomer-based adhesives. For this reason they provide the strength, durability, and heat resistance required for structural adhesives. Phenolic resins also enhance the wetting of and adhesion to metal and oxide surfaces. In order to reduce shrinkage and provide stress relief in a structural

Table 11   Comparison of Water- and Solvent-Based Contact Adhesives

Property	UCAR-154-BKUA-2370 (3:1 solids)	Solvent-based neoprene-phenolic
Canvas-to-canvas peel strength (pli) at 74°F		
Initial	11	7
24 hr	19	20
1 week	20	35
Canvas-to-steel peel strength (pli) at 73°F		
24 h	20	20
1 week	20	25
DLHS[a] (°F)		
5 lb	> 300	170
1000 g	> 300	190
Edge lift MMM-A-130 B	Pass	Pass

[a]Deal Load Hot Strength, 1-in.^{2} overlay canvas-to-steel bond.

bond, phenolic resins are often coreacted with high-molecular-weight elastomers, such as vinyl and nitrile polymers or epoxy resins. Examples of phenolic resins used in structural adhesives are shown in Table 12.

A. Vinyl-Phenolic Structural Adhesives

Polyvinyl acetals of the structure shown in Fig. 17 are used in combination with phenolic resins for structural adhesives in, for example, metal bonding. They are used as bonding agents in the aircraft industry: as copper adhesives for printed circuits, as adhesives for brake linings, and as adhesives for honeycomb construction.

The phenolic resins used in these applications are generally resols catalyzed by sodium hydroxide. The hydroxymethyl groups in the phenolic resol can react with the hydroxyl groups of the polyvinyl-acetal resin. Consequently, the hydroxyl and acetal contents are important factors in the selection of the polyvinyl-acetal component.

The ratio of phenolic resin to vinyl polymer can range from 0.3:1 to 2:1 depending on the desired tensile strength, creep, and temperature resistance. These adhesives are generally applied from alcohols,

Table 12   Structural Adhesive Systems Modified by Phenolic Resins for Strength and Durability

Polymer base	Phenolic resin	Examples of commercial products[a]
Nitrile rubber	Resol	BKR-2600
	Novolac-hexa	BKPA-5864
Epoxy	Novolac	BRWE-5833
	Resol	BLS-2700
Polyvinyl butyral	Resol	BLS-2700

[a]Material suppliers are listed in Sec. XIII.

which are often blended with other solvents to improve polymer solubility. They are cured at temperatures of 284—356°F (140—180°C) and pressures of 100—400 psi. These vinyl-phenolic structural adhesives are suitable for continuous use to about 212°F (100°C).

B. Nitrile-Phenolic Structural Adhesives

Nitrile-phenolic structural adhesives are made by blending a nitrile rubber with a defined phenolic resin. The nitrile component imparts oil resistance and toughness whereas the thermosetting phenolic contributes chemical, solvent, and heat resistance. They are suitable for continuous use to 248°F (120°C) and for short-term use to 175°C (about 300°F). Commercial adhesive formulations also contain hardeners, accelerators, antioxidants, and fillers, and are available in the form of liquid solutions as well as dry films. These adhesives have outstanding humidity and salt spray resistance and, thus, show exceptional durability on steel and aluminum. In comparison to vinyl-

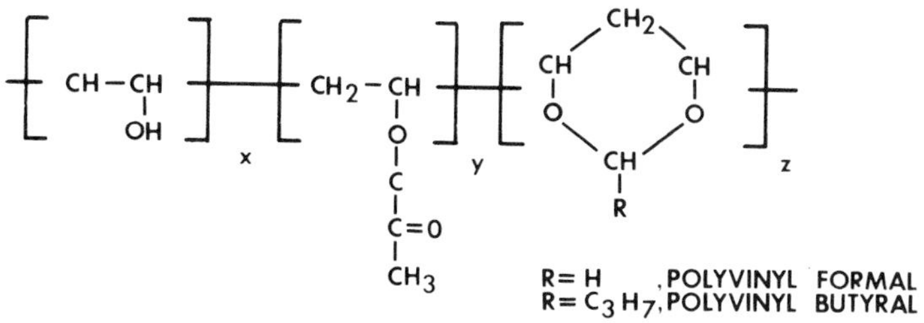

Fig. 17.   Acetal resin used as vinyl-phenolic adhesive.

phenolic adhesives, nitrile-phenolic blends have higher peel strength, impact resistance, and heat resistance.

C. Epoxy-Phenolic Structural Adhesives

Combinations of epoxy and phenolic resins provide structural adhesives with superior high-temperature resistance in comparison to phenolic-modified vinyl or nitrile-based systems. Properly formulated epoxy-phenolic structural adhesives are suitable for continuous use as high as 392°F (200°C), but their peel strengths are low due to the high crosslinked density of the cured films. These blends cure through the reaction of the epoxy with the phenolic hydroxyl, and thus, both novolac- and resol-type phenolics may be used. When resols are used, the epoxy resin may also react with the hydroxymethyl groups on the phenolic rings.

XII. SUMMARY

Phenolic resins have been used widely as modifiers to improve the tack, specific adhesion, bond strength, and heat resistance of elastomeric adhesives. In structural adhesives, the crosslinking ability of phenolic resins is employed to impart strength, durability, and chemical resistance to the adhesive bond. As such, phenolic resins have proven to be one of the most versatile formulating components in adhesives.

Phenolic resins in the form of aqueous dispersions are relative newcomers to this resin family and promise to bring to water-based adhesives the desirable tackifier and heat-resistant properties for which phenolic resins are well known.

XIII. SUPPLIERS OF TRADE-NAME MATERIAL

Union Carbide Corporation, Danbury, Connecticut.
BKR-2620
BKUA-2370
BRWE-5833
BKPA-5864
BLS-2700
CK-1634
CK-1636
CK-2103
CK-2432
CK-2400
UCAR-154,152
Cellosize QP 300H

Durez Division, Hooker Chemical Corporation, North Tonawanda,
New York.
Durez 12603
Schenectady Chemicals, Inc., Schenectady, New York.
SP-134
SP-154
W. R. Grace, New York, New York.
Daxad 11, 301
E. I. Du Pont de Nemours, Wilmington, Delaware.
Neoprene AC, AD, AF
Neoprene 750
Dow Chemical, Midland, Michigan.
Dowicil S-133
Engelhard Minerals & Chemicals Corporation, New York, New York.
ASP-400

REFERENCES

1. C. F. Kettering, in *National Academy of Sciences, Biographical Memoirs*, Vol. 24, 1947.
2. A. Bayer, *Ber. Dtsch. Chem. Ges.* 5:25, 1095 (1872).
3. *Bakelite Review*, General Bakelite, (1935).
4. L. H. Baekeland, *Ind. Eng. Chem.* 1:149 (1909).
5. R. H. Young, P. W. Kopf, and O. Salgado, in TAPPI, *64* No. 4:127 (1980), and TAPPI Paper Synthetics Conference, 1980.
6. C. W. Hick and A. W. Abbott, *Rubber Age 82*:417 (1957).
7. B. P. Barth, Phenolic resin adhesives, in *Handbook of Adhesives*, 2nd ed. (I. Skeist, ed.), Reinhold, New York, 1977.
8. M. Steenfink, Neoprene Adhesives: Solvent and Latex, in *Handbook of Adhesives*, 2nd ed. (I. Skeist, ed.), Reinhold, New York, 1977.
9. A. Knop and W. Scheib, in *Chemistry and Application of Phenolic Resins*, Springer-Verlag, New York, 1979.
10. Phenolic resins for neoprene and nitrile contact adhesives, in *Union Carbide Prod. Bull. F-42997B*, Union Carbide, Danbury, 1978.
11. *Federal Specification*, MMM-130-B for Contact Adhesives, 6 May 1975.
12. J. W. McDonald and R. W. Kearn, Factors affecting phasing in neoprene adhesives, in *Du Pont Elastomers for Adhesives Bull. E-12347*, Du Pont, Wilmington.
13. T. Tanno and I. Shelbuya, Special behavior of p-tert-butylphenol dialcohol in polychloroprene solvent adhesives, Adhesives and Sealants Council, Spring Meeting, 1967.
14. Solvent systems for neoprene-predicting solvent strength, in *Du Pont Elastomers Bull.*, Du Pont, Wilmington, 1964.

15. K. W. Harrison, Solvents in polymer-based adhesives, in *Adhesion*, Vol. 3, (K. Allen, ed.), Allied Science, London, 1978.
16. R. G. Azrak and B. P. Barth, *Adhesive Age 18*:23 (1975).
17. B. P. Barth, to Union Carbide, U. S. Patent No. 3,887,539 (1975).
18. F. W. Doherty, Neoprene latex adhesives, in *Du Pont Elastomer Chemicals Dept.*, *Literature SD-138*, Du Pont, Wilmington.
19. *AFG-01 Performance Specification for Adhesives for Field-Green Plywood to Wood Framing*, American Plywood Association, Portland, OR, September 1974.
20. Bakelite phenolic dispersion BKUA-2370, in *Union Carbide Prod. Bull. F-47050*, Union Carbide, Danbury, June 1978.
21. Bakelite phenolic dispersion BKUA-2370, in *Union Carbide Prod. Bull. F-47415*, Union Carbide, Danbury, October 1979.
22. A. M. Shaw, Jr., *Adhesives Age 23*:35 (1980).

11
Anaerobic Adhesives

Martin Hauser and Girard S. Haviland *Loctite Corporation,*
Newington, Connecticut

I. THE CHEMISTRY OF ANAEROBIC ADHESIVES AND SEALANTS

A. History and Origin

The origins of commercial anaerobic products can be traced to work
done by Burnett and Nordlander [1] at General Electric Corporation
in the 1940s. When diacrylate or dimethacrylate monomers [Structure
(1)] based on diethylene glycol or high homologs were heated at 60–
80°C for some hours while being bubbled with air, oxygen, or ozone,
they were converted to materials capable of spontaneously polymerizing at room temperature to form crosslinked polymers possessing adhesive properties.

$$CH_2=\overset{\overset{\displaystyle CH_3}{|}}{\underset{\underset{\displaystyle O}{\|}}{C}}-C-(OCH_2CH_2OCH_2CH_2)_n-O-\overset{\overset{\displaystyle CH_3}{|}}{\underset{\underset{\displaystyle O}{\|}}{C}}-C=CH_2$$

$$\underline{I}$$

The major problem encountered with these products, named Anaerobic Permafils by General Electric, was that they required either continuous refrigeration or bubbling with air to remain unpolymerized liquids. This property of polymerization in the absence of oxygen was the genesis of the term anaerobic, that is, active in the absence of oxygen. The mechanical difficulties inherent in maintaining these products in a usable form presented problems which eventually precluded their successful commercialization.

In the early 1950s Krieble [2] attacked the problem of obtaining a stable anaerobic system by eliminating the monomer oxygenation step and substituting cumene hydroperoxide (II), a very stable free-radical source, as the polymerization initiator. Due to its high order of stability, cumene hydroperoxide underwent homolytic decomposition to free-radical species (III) at an exceedingly slow rate at room temperature. The free radicals thus produced reacted with the monomer (I) to initiate polymerization, but polymerization did not proceed (propagate) in the presence of small amounts of oxygen which reacted preferentially with the monomer radical to form inactive species.

$$\underline{II} \longrightarrow \underline{III}$$

$$(1)$$

$$I + III \longrightarrow III-I\cdot \underset{O_2}{\overset{I}{\diagup}} \longrightarrow III-I-O_2\cdot$$
(inactive)

When the mixture of I and II was confined so as to exclude further contact with atmospheric oxygen, the oxygen dissolved in the system was consumed by the III-I free radicals generated from II, and polymerization eventually occurred. However, because the rate of spontaneous generation of the free radicals was extremely slow, the time necessary for the onset of polymerization was exceedingly long [2].

The key to a practical anaerobic system is the fact that hydroperoxides undergo a very facile oxidation-reduction reaction with transition-metal ions (such as iron and copper) leading to a rapid heterolytic (as opposed to homolytic) decomposition to free radicals:

$$\text{II} + Fe^{2+} \longrightarrow \text{III} + OH^- + Fe^{3+} \tag{2}$$

When the anaerobic adhesive was placed on a transition-metal surface under conditions which precluded further contact with oxygen, there was a rapid generation of free radicals (III); the dissolved oxygen was quickly consumed; polymerization proceeded at a commercially acceptable rate. Therefore, this first phase of Krieble's work resulted in the development of a shelf-stable, anaerobic system capable of polymerization to a usable adhesive in the absence of oxygen[2].

Formulation of this basic monomer-hydroperoxide system was followed by further work that identified a class of organic compounds known as trialkylamines (IV) as accelerators for the polymerization[3]. It was also found that the presence of oxygen alone was not always adequate to ensure that the product would not polymerize during storage. The addition of small amounts of another class of organic compounds, the quinones (V), produced major improvements in the shelf stability of the products[4].

$$\underline{\text{IV}} \qquad\qquad \underline{\text{V}}$$

The "first generation" of anaerobic products cured fairly rapidly on "active" surfaces, such as iron and copper, where a surplus of transition-metal ions was available. Fifty percent of ultimate strength was obtained in 2—12 hours at room temperature. However, on "inactive" surfaces, such as zinc plate, cadmium plate, and passivated stainless steel, where the concentration of transition-metal ions was very low, cures were often very slow or nonexistent.

Paramount importance was placed on the necessity of increasing the speed of cure and decreasing the dependence of the composition on the surfaces being bonded. Research effort was focused on identifying new accelerators which would render the hydroperoxide initiator much more sensitive to heterolytic decomposition by exceedingly low levels of transition-metal ions. The result of this research program was the discovery that a coaccelerator system based on saccharin (VI) and dialkylarylamines (VII) was capable of producing major increases in cure speed on all surfaces without a consequent negative effect on shelf stability [5].

VI   VII

This new accelerator technology was incorporated into a "second generation" of anaerobic products possessing a far higher degree of tolerance for the variety of surfaces found in actual bonding applications. It should be noted that in this new product line, speed of cure on all surfaces had been increased. There was still a considerable difference in cure speed on "active", as opposed to "inactive", surfaces. The key was that cure times on "inactive" surfaces had been decreased to a point that was generally acceptable from a practical point of view.

Despite the considerable advances in anaerobic technology from 1950 to 1968, there were still significant limitations on formulation capabilities. The majority of attention had been devoted to improving the cure-speed characteristics of the products. Formulation parameters were still basically limited to the thickener-plasticizer combinations of the 1950s. This was primarily due to the fact that the oxygen-quinone—stabilized anaerobic systems were prone to destabilization on addition of the variety of chemicals needed to produce other useful properties.

During the late 1960s and early 1970s, basic discoveries relating to anaerobic chemistry provided strong direction in terms of designing even faster-curing and far more shelf-stable products. The destabilization of anaerobics by the addition of a variety of chemical species was traced to the fact that these chemicals were introducing small amounts of transition-metal ions to the formulations. Incorporation of chelating agents[6] to complex and deactivate these "tramp" metal ions produced major improvements in the ability of anaerobics to tolerate low levels of contamination. Furthermore, it was then possible to utilize certain classes of accelerators, particularly the hydrazides (VIII) [7], which had previously exhibited too strong a destabilizing effect. These new accelerators, when used in combination with the previous accelerator systems, produced anaerobic compositions which were even faster curing, particularly on "inert" surfaces.

$$CH_3-\overset{\overset{\text{O}}{\|}}{C}-NH-NR$$

$$\underline{\text{VIII}}$$

B. Modern Capabilities

This new stabilizer and accelerator technology forms the basis for the current "third generation" of anaerobic products whose high degree of stability and rapid cure on almost all surfaces are combined with a number of previously unattainable properties.

1. Anaerobic products can now be formulated with viscosities ranging up to a pastelike consistency or can be made thixotropic[8] to flow readily onto parts and then remain in place until the bonding operation is performed, eliminating problems with liquid nitrogen.
2. Anaerobics can be formulated which cure through gaps up to 0.8 mm, as contrasted to a previous maximum of 0.1—0.3 mm.
3. Anaerobics can be formulated to cure reliably at temperatures as low as -4°C.
4. It is possible to pigment anaerobics to afford a high degree of on-part visibility.

5. Anaerobics can now tolerate additives which enable them to cure on parts which have not previously been degreased or cleaned.
6. Various lubricants, which allow maintenance of the rated torque-tension relationship of "as-received" fasteners, can be added to anaerobics designed for threadlocking applications.
7. Various additives have increased the maximum use temperature of specially formulated anaerobics from 177°C (340°F) to 232°C (450°F)[9].

Much of the foregoing discussion has centered around those anaerobic products for which the term "machinery adhesives" was coined some years ago. These products have numerous applications for use with cylindrical fitted parts but, in general, lack the tensile strength and impact and peel resistance necessary for structural bonding.

In order to obtain structural adhesives which could use the anaerobic cure mechanism, it was necessary to design and synthesize new types of monomers. The first new monomers (characterized as "urethane methacrylates") were actually prepolymers obtained by reacting a dihydroxyl alcohol (diol) in less than stoichiometric amounts with a diisocyanate and then "capping" the resulting intermediate with a hydroxyalkyl methacrylate[10]. constituents contained both "rigid" and "flexible" segments in the same molecule. This hypothesis formed the basis for an expanded synthesis program in which "rigid" and "flexible" prepolymers were chemically joined prior to "capping" with hydroxyalkyl methacrylates.

The methacrylate-terminated polymers formed by this reaction sequence retained the ability to cure anaerobically and, when formulated with monomeric diluents, gave a new class of structural adhesives with outstanding bonding properties[11]. The cured properties of these adhesives (tensile, impact, peel, thermal, solvent, and environmental resistance) compared favorably to those of the highest performance epoxy adhesives; yet they retained the highly desirable quality of the room-temperature cure of the anaerobics as contrasted to the protracted, high-temperature cure of the epoxies.

Variation in the nature of the level of the "backbone" and diluent constituents led to three basic adhesive types:

1. General-purpose adhesives with an excellent balance of cure speed and cured properties
2. Very high-performance adhesives offering the maximum in cured properties with some sacrifice in cure speed
3. Very high-speed adhesives offering the maximum in cure rate with some sacrifice in cured properties

Introduced in the late 1970s, these adhesives are today at the leading edge of anaerobic technology and, in fact, structural-adhesive technology in general.

These prepolymers, diluted with mono- or dimethacrylates for workability and formulated with anaerobic cure systems, were indeed structural adhesives as measured by shear and tensile properties. However, because of their high degree of crosslinking and their rigid "backbones", they were extremely brittle and had poor impact and peel resistance. In order to obtain more flexible systems, monomers synthesized by the reaction of isocyanate-capped polyethers with hydroxyalkyl methacrylates were introduced into the adhesive formulations. To some extent, these monomers copolymerized with the urethane methacrylates to give adhesives with a more desirable combination of strength and flexibility. By judicious blending of "rigid" and "flexible" components, it was possible to formulate a line of structural adhesives possessing various combinations of hot strength, tensile strength, impact and peel resistance, and so forth.

Despite the fact that these mixed "rigid-flex" formulations were considerably superior to the all "rigid" formulations and far superior to the machinery adhesives in structural-adhesive properties, they still fell short of exhibiting truly outstanding structural-bonding characteristics. One possible explanation for this short-fall in performance was that the various monomer components did not readily copolymerize.

Following this line of thought one step further led to the hypothesis that better performance might be achieved if the prepolymer

II. APPLICATION OF ANAEROBIC ADHESIVES

A. Introduction

The unique properties of anaerobic stabilization have led to the use of these adhesives in closely fitting machinery parts. They are able to form neat bondlines because, due to the presence of oxygen, excess adhesive can be washed away uncured. By the same token, they are unsuitable for large cured-fillet caulking or bonding. The maximum bondlines are generally 1 mm or less. A more normal bondline is 0.1 mm. Anaerobics fit easily into either maintenance or high-volume production because the single-component materials eliminate mixing and pot-life problems occurring with two-component systems. This results in cost savings for the user. The applications are often broadly defined in three functional categories.

1. Sealing

Anaerobic resins can be made with a wide range of viscosities from 0.01 mPa-sec to 500 Pa-sec. The low-viscosity resins are often used to seal powdered-metal parts or porous die castings, and the low-viscosity resins are used as a curing sealant for pipe threads. In each case tailored properties of wetting, cure speed, and lubrication are formulated into the product.

2. Holding or retaining

Anaerobic resins have the ability to fill spaces between machinery parts where no other adhesives can do the job as conveniently or as well. Hence, the term machinery adhesives is often used to further identify them. A press fit may have no more than 30% contact from part to part because of machining roughness. To fill the extra 70% of air space with a liquid which turns into a dense, hard resin has benefits of reliability and cost savings which are substantial. These materials, in viscosities from 10 mPa-sec to 7 Pa-sec are used to

1. Augment or replace press-fitted assemblies of shafts, bearings, bushings, and pins
2. Prevent loosening of threaded assemblies due to sideways movement as caused by differential thermal expansion, bending, and impact
3. Hold studs by replacing an expensive interference fit with an inexpensive, high-strength, anaerobic resin
4. Improve the fit of keys and gibs to prevent backlash
5. Retain the position of pads, feet, labels, and inserts

Important properties of the holding and retaining materials are cure-speed control, good wettability, thixotropy for positive placement, lubricity on threads, and ultimate shear strength. All of these properties have been attained in anaerobic resins. It is worth special note that their lubricity can be equivalent to that of mineral oil. Cures take only minutes over oil. Shear strengths routinely are as high as 27 MPa, or they can be controlled to be as low as 7 MPa.

3. Structural bonding

Structural bonding includes all applications in which adhesives are placed in mechanical assemblies to transmit the primary functional load. In other words, they are critical to the functional integrity of the structure. The so-called retaining compounds are often used structurally, as in bonding shafts to laminations in electric motors. Conversely, the structural adhesives are often used just to hold things like labels and decorative knife handles. A few other typical structural bonding applications are as follows:

Audio speaker magnets
Metal frame to glass
Fuel-pump components
DC-motor magnets
Fiberglass-reinforced automobile, train, and airplane parts

Table 1 Dispensing Equipment Capabilities

Application type	Typical rate	Adhesive quantity	Viscosity range (mPa-sec)	Typical application
Spot or drop				
Hand	10/min	0.01–0.04 ml	1–10,000	Hand assembly, maintenance
Automatic	100/min	0.005–0.1 ml	1–50,000	Nuts, bolts, various assemblies
Extrude	200 in./min	1/16–1/8 in. dia. beads	500–2,000,000	Flanges, bonded assemblies, speakers
Tumble	5 barrels/hr	0.01–0.05 ml/part	500–500,000	Pipe plugs, set screws, screws, bolts
Screen				
Manual	4/min	0.001–0.015 in. thickness	⩾ 50,000	Flanges, flat-bonded assemblies such as speakers
Automatic	30/min	0.001–0.015 in. thickness	⩾ 50,000	
Template	2/min	0.040–0.080 in. thickness	500,000–1,000,000	Flanges
Spray				
Manual	5/min	0.1–1 ml	1–100	Wheel rims, gas containers, die castings, transformers, brazed fittings, welds
Automatic	20/min	0.1–1 ml	1–100	
Roto-spray	6/min	0.01–0.1 ml	1000–50,000	Core plugs, bearings, axle tubes, sleeves
Vacuum or vacuum-pressure impregnation	3–5 baskets/hr	Depends on parts 0.1–50 ml/part	1–100	Powdered metal, aluminum die castings, various castings

The most important characteristics of anaerobic structural adhesives are

1. The ability to cure to high ultimate tensile, tensile shear, and shear strengths.
2. The ability to adhere reliably to clean or moderately clean surfaces of all types of material. This latter ability has only recently been obtained with anaerobics.
3. Curability under preselected conditions such as
   a. The exclusion of air and contact with an active metal at room temperature
   b. Contact with an activator
   c. Application of moderate (100°C) heat
   d. Exposure to ultraviolet (UV) light.
4. Impact, peel, fatigue, and creep resistance. Recent anaerobics exhibit high impact (over 2 J/cm^2) and peel (over 7000 N/m) resistance, whereas in the past they had been known as high-strength, but brittle, formulations.
5. Environmental resistance—including resistance to heat and chemicals. Anaerobics have always been good in this respect although moisture resistance on some substrates has only recently equaled epoxies and some rubber-based adhesives.

B. Equipment and Application Techniques

Because of the unique nature of anaerobic liquids, which become unstable when confined, a special line of containers and applicators proved necessary for efficient shipping and handling in production [12] (see Table 1). The shipping bottles are made of low-density polyethylene which allows free access to oxygen molecules through the sides. In addition, each bottle is only partially filled to allow adequate air access to the liquid surface.

Applicators have been developed which will not cause the material to cure. The simplest is a peristaltic pump, bottle-top applicator which can dispense amounts, variable by a screw adjustment, from 0.01 ml to 0.04 ml[13].

For a user who must apply material to a large number of parts in production, automatic dispensing is accomplished by using a pressure-time system of feeding. These machines can dispense up to 6000 drops per hour in quantities from 0.005 ml per shot to continuous streams. The dispensers can be integrated into assembly machines, or parts can be stored for later assembly. Care must be taken in storage so that "popcorning" of parts does not occur. Storage is usually limited to 3 days for nuts and 1 day for bolts.

Table 2   Quantity of Anaerobic Material Required for
Tumbling (ml resin/1,000 pieces)

Size of screw or pipe	2	4	8	12	1/4	3/8
Slotted, round head, 3/8 long	5	6	10	15	18	32
Pipe plugs					50	80

Tumbling is another very effective technique for applying anaerobics to threaded parts, such as pipe plugs (Table 2). The process
is simple but must be controlled. Parts are put into a tumbling barrel
with a measured amount of resin. A tumble of 5—15 min completes the
process of applying material to thousands of parts.

Screen template application is very practical with anaerobics on
flat surfaces. Since the materials are nondrying and stable in air,
they never cure or dry on the screen. A very precise quantity of
complex geometry can be laid onto a flat surface in less than 1 sec.

Roto-spraying equipment has been highly developed to apply
precise quantities of material to the internal bores of machinery parts.
It works by placing a measured amount of material on a spinning disk.
The material spreads to the periphery of the disk and flies across a
gap (5—300 mm), creating a fairly narrow, even band of material on
the bore of the part.

Direct spray application is often used for low-viscosity products
where weld or local porosity sealing is necessary. Air-type sprays are
most often used because of the difficulty of confining anaerobic materials at high pressure for airless spraying.

Vacuum-impregnation systems are the largest consumers of anaerobic sealing resins[14]. The process is used on both small, powdered-
metal parts and large (0.7 × 0.2 m) castings to seal porosity. The
parts are placed into a vacuum container for air removal and then lowered into a bath of resin. Repressurization forces the material into
pores and cracks. A centrifugal spin-off of excess material conserves
material and prevents entrapment in large holes and recesses. A
warm, activated-water wash completes the cure and cleaning. Pores
up to 0.1 mm can thus be plugged with a dense, nonshrinking material which is impervious to most conventional chemicals.

Design techniques for all this equipment have been developed
over the last decade. It has been necessary to develop special valves,
rotary seals, nozzles, feed lines, and containers. Techniques include
air suck back after dispensing[15] and intermittent or continuous bubbling. The most benign materials in contact with the resin are poly-

ethylene, Teflon, stainless steel, silicone rubber, and Delrin. Even
these materials must be of the highest purity or known formulation.

III. DESIGN CONSIDERATIONS AND DATA

A. Sealing

1. Porosity impregnation

Anaerobic sealing compounds have drastically replaced other sealing
compounds such as linseed oil, varnish, sodium silicate and polyester
(epoxy or styrene). Their unique characteristics of staying liquid in
air and curing without heat when confined in porosity make them ideal
for vacuum impregnation. The materials are thin, 10−100 mPa-sec,
and cure without shrinkage. The process (Fig. 1) is relatively simple,
yielding clean parts which can withstand pressures up to 55 MPa.
Machinability of powdered-metal parts is said to be improved by 500%
because of the elimination of tool chatter (Fig. 2). Impregnation
makes it practical to metal plate porous parts since plating chemicals
cannot be trapped under the plate.

Applications (Figs. 3−5) include iron and brass powdered-metal
parts; zinc, aluminum, and magnesium die castings; steel welds; and
cast iron. Because of the cleanliness of the process, parts are often
finish machined before impregnation.

2. Cylindrical and threaded fits

Plugs of all types and cylindrical, tapered, and straight threaded
parts are sealed with anaerobics. Usually the materials are made thick
or thixotropic to make them hold position (Figs. 6 and 7).

3. Flat surface and gasketing

Cured, crosslinked, anaerobic polymers which cure with little shrinkage offer outstanding resistance to industrial fluids and gases. Thixotropic paste technology and through-volume cures have resulted in
the development of sealing-gasketing compounds which are used extensively in place of cut gaskets (Figs. 8 and 9). They are especially
effective in load-bearing joints where exterior loads often worked the
compressible-type gasket until it leaked. Anaerobic sealants fill only
the voids, allowing metal-to-metal contact for load support. With the
newer, flexible materials it is possible to replace gaskets up to 1 mm
thick. Gaskets thicker than 1 mm respond well to coating. Cure of
the material on a soft, thick gasket stiffens the gasket after it has
been deformed, thus preventing further deformation.

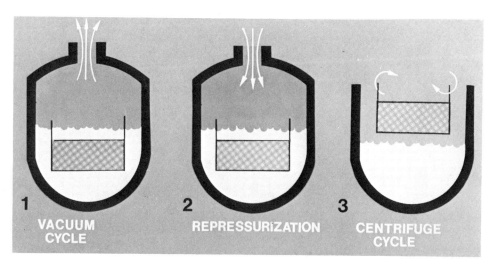

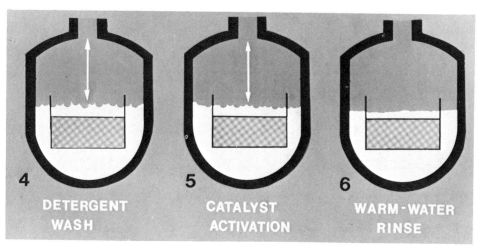

Fig. 1. Typical cycle for an anaerobic impregnation process.

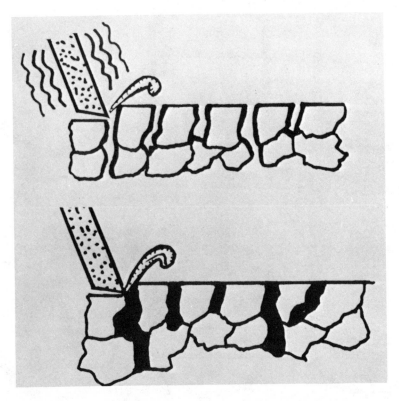

Fig. 2. Powdered-metal components—machinability.

Fig. 3. Automotive air-conditioner compressor housing.

Fig. 4. Die-cast carburetor.

Fig. 5. Gun-sight ramp.

Fig. 6. Hydraulic fittings on a twin-headed, vertical milling machine are sealed with fast-curing paste sealant.

Fig. 7. Automotive core plugs are sealed automatically with roto-spray application equipment.

Fig. 8. Transmission for farm tractor is sealed and unitized with anaerobic, instant-seal paste.

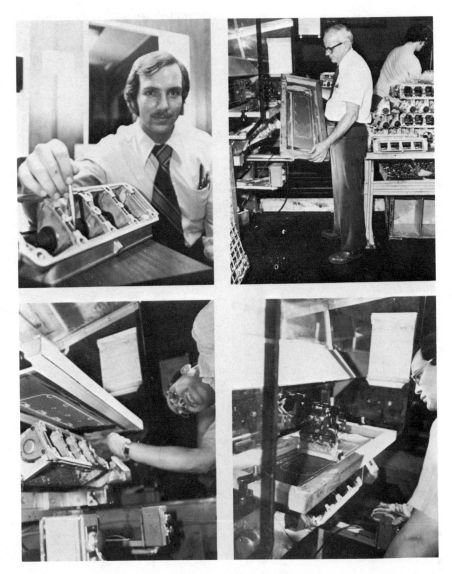

Fig. 9. Overhead cam-shaft housing is screen coated with fast-curing, jelly-like anaerobic in order to seal the housing to the engine head.

Table 3 Typical Properties of Gasket-Type Sealants

Name	Viscosity (Pa-sec)	Tensile strength (MPa)	Maximum gap (mm)	Temp. range (°C)	Time to gel (hr)	Elongation (%)
General purpose flexible	800/200 Thixotropic	5.5	0.3[a]	-54−120 Flexible	2[a]	50
High temp.	950	20	0.08[a]	-57−204	40	2
Unitizing	6.5	38	0.3[a]	-54−149	12	1

[a]Activator or heat cures can quadruple the possible gap and decrease time to gel by a factor of 4.

Because of their anaerobic nature, most of the sealing materials respond very well to screen or stencil printing. Typical properties are shown in Table 3.

B. Holding and Retaining

1. Threadlocking

The first large application of an anaerobic compound was in 1957. A thin resin was used to lock the screws which held the butterfly flapper to its shaft in a Holley carburetor. Staking and other means of keeping the screws from loosening caused shaft bending with consequent sticking of the valve. Failure or loosening of a small screw could have been disastrous to the life of a whole engine. The earliest pure resin formulation did the job and pointed the way for a new industry.

a. How threaded fasteners fail. The failure of threaded fasteners to hold their loads can be attributed to periodic overload conditions which make the threads slide sideways (Fig. 10). If they slide sideways then the clamp load will make them slide "downhill", or unwind. As few as 50 side movements can cause 20% load loss (17). Conditions which can cause side sliding are differential thermal expansion, internal pressure, bending of clamped parts, and breathing or belling of the nut.

Nothing has proven more effective for preventing side motion than filling the threads with a curable, dense, organic material. Anaerobic fluids, which can be made the right consistency to fill the threads, are ideally suited to cure in the confined space between mating threads. The proof of their effectiveness can be obtained in the laboratory on a machine which intentionally slides clamped parts sideways. It is called a transverse shock and vibration (TSV) machine (Figs. 11 and 12). This machine rates various locking methods according to their effectiveness. It is a severe test designed to cause failure in almost any locking system. Field results correlate very well with machine results.

b. Removal torque. All threadlockers are made to augment the removal torque of a fastener. The amount that they do so is dependent on the shear stress of the material and the thread clearance. Some typical shear stresses as found on $3/8 \times 16$ bolts and nuts are shown in Table 4.

Since the average thread clearance is reduced with smaller bolts and screws, it is necessary to adjust the calculation for actual shear stress in accordance with the graph in Fig. 13. This assures that a threadlocker used on a small screw will not give a removal torque unexpectedly higher than screw strength.

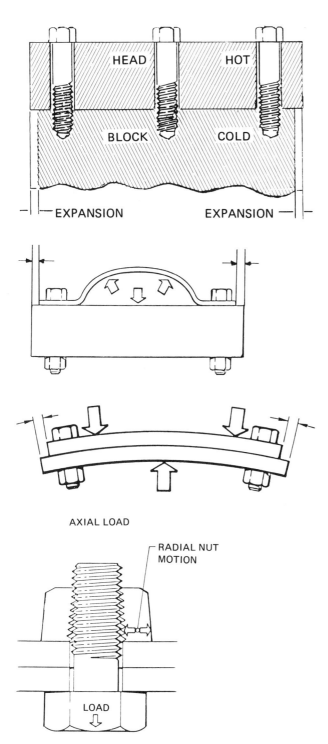

Fig. 10. Causes of thread loosening.

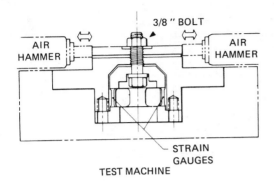

Fig. 11. Transverse shock vibration machine.

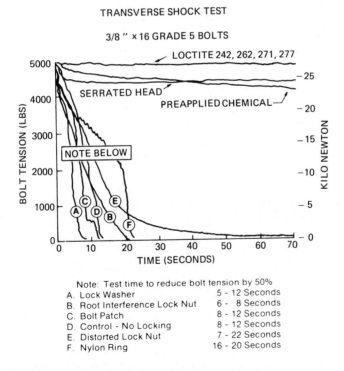

Note: Test time to reduce bolt tension by 50%
A. Lock Washer   5 - 12 Seconds
B. Root Interference Lock Nut   6 - 8 Seconds
C. Bolt Patch   8 - 12 Seconds
D. Control - No Locking   8 - 12 Seconds
E. Distorted Lock Nut   7 - 22 Seconds
F. Nylon Ring   16 - 20 Seconds

Fig. 12. Transverse shock vibration machine.

Table 4   Shear Stress on 3/8 × 16 PHOS and Oil Nuts and Bolts

Product strength MPa					
Low	Medium	High	High and thin	Very high	Wicking
4.1	5.5	14	21	28	5.5

c. Lubrication capability of threadlockers. Since the primary load on a bolt is produced by applying a torque to the assembly, the application torque is usually used to control as well as apply the clamp load. However, 90% of the applied torque is absorbed by friction in the head and threads, so any change in the friction factor is critical to the control of load. Modern threadlockers are made to give the same lubrication as an oiled bolt, thereby maintaining good control of clamp load without change in tightening specifications or production habits.

For any given fastener the torque tension relationship can be expressed as follows:

$$T = KDF \tag{3}$$

where

T = torque, lb-in. (N-m)
D = nominal bolt diameter, in. (m)
F = tension or clamping force, lb (N)
K is a universal constant for all sizes which can be estimated empirically.

Example (3/8 × 16 G5 oily bolt).
T = 0.23 × 0.375 in. × 5000 lb. = 430 lb-in.

or

T = 0.23 × 0.01 m × 22,000 N = 51 N-m

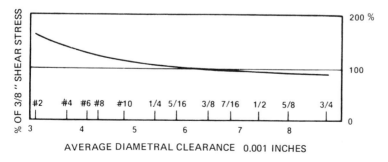

Fig. 13.   Size and clearance effect on shear strength.

Table 5   Some Typical Values of K Found on 3/8 × 16 Nuts and Bolts[a]

Surface	Without anaerobic	Material		
		Low strength	Medium strength	High strength
Oily[b]				
Steel gr. 5	0.15	0.16	0.14	0.13
Cadmium	0.14	0.12	0.13	0.13
Zn phosphate	0.14	0.13	0.11	0.09
Stainless	0.22	0.21	0.17	0.14
Brass	0.16	0.14	0.09	0.10
Al 6262TH	0.17	0.20	0.29	0.18
Silicon bronze	0.19	0.16	0.24	0.17
Zinc	0.18	0.14	0.15	0.13
Dry				
Steel gr. 5	0.20	0.18	0.20	0.18
Nylon	0.05	0.15	0.13	0.15

[a]Accuracy of the above results can vary depending on contact area, thread form, finish, and oxide contamination. Where more precise results are necessary, test the actual parts and locking material.
[b]All oiled specimens were lightly oiled with 5% solution of Lab Oil 72D.

Table 5 gives typical values of K for nuts and bolts made of various materials.

d. Characterization of threadlockers. (1) Thixotropic threadlockers are lubricious and provide a fast, reliable cure even over oily parts. Selection criteria for thixotropic threadlockers are as follows:

Material	Viscosity	Suggested bolt range	Maximum diametral clearance
Low strength	1000 cP (1 Pa-sec)	#2–1/2 in.	0.016 in. (0.4 mm)
Medium strength	1100 cP (1.1 Pa-sec)	1/4–3/4 in.	0.022 in. (0.6 mm)
High strength	1500 cP (1.5 Pa-sec)	3/8–1 in.	0.025 in. (0.6 mm)

(2) Newtonian threadlockers can be made with very high or very low viscosity. The high-viscosity threadlockers are used on studs and large bolts. The low-viscosity threadlockers are used for wicking-in after bolt and nut assembly. Selection criteria for Newtonian thread-lockers are as follows:

Material	Viscosity	Suggested bolt range	Maximum diametral clearance
High strength	750 cP (0.8 Pa-sec)	3/8—1 in.	0.025 in. (0.6 mm)
High strength	7000 cP (7 Pa-sec)	5/8—1 in.	0.025 in. (0.6 mm)
Medium strength	12 cP (12 mPa-sec)	#2—1/2 in.	0.016 in. (0.4 mm)

(3) The availability of preapplied threadlockers which use micro-encapsulated resin is shown by an "X" in the following table. They are easy to inspect and provide reuse per IFI 100 [18]. They are usually applied by a supplier. The water-based slurry is dried on the bolt. Breaking of the capsules during assembly activates the system which causes the cure.

	High production-rate material	Medium production-rate material
Low strength locking and sealing	X	NA
Medium strength, plated fasteners	X	X
Medium strength	X	X
High strength	X	X

Thixotropic and Newtonian threadlockers are covered in Military Specification MIL-S-46163 [19].

2. Retaining or augmenting slip and press fits with anaerobic retaining compounds

Traditional methods of assembling cylindrical parts—such as pulleys, bushings, bearings, and gears—were used to provide an interference fit. This led to two problems: distortion of parts and the cost of machining to small tolerances. Anaerobic adhesives (retaining com-

Table 6 Modern Retaining Compounds—Steel Parts

Name	Viscosity (mPa-sec)	Ultimate shear strength (MPa)	Time to 20% ultimate (Min)	Temp. rating (°C)
General purpose	125	21	15	-54—149
High strength	1250	28	30	-54—149
High temperature	7000	21	45	-54—232

pounds) are now widely used to replace or augment press fits (Table 6). Typically, a heavy press fit will give shear strength of 4 MPa, whereas the anaerobic retaining compounds can readily give 20 MPa. The heaviest of press fits rarely gives more than 30% metal-to-metal contact (Fig. 14). Therefore, even a heavy press fit can be augmented considerably by assembling with an anaerobic resin.

Applications of retaining compounds vary from holding applications, such as bushing or bearing retention, to structural applications, such as shaft and rotor assembly. A variety of anaerobic retaining compounds are specified in ML-R-46082A [20]. Variations occur in strength, viscosity, cure speed, and temperature resistance.

For high rates of production, heat is often used to accelerate the cure of these compounds. Complete cure in 30 sec at 140°C (284 °F) is common.

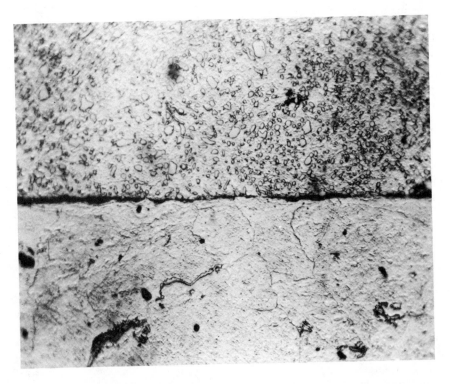

Fig. 14. Photomicrograph of an actual press fit showing inner space where no contact occurs. Approximate magnification is 1000.

Table 7 Typical Properties of Anaerobic Structural Adhesives

Material	Fixture (sec)	Tensile shear (MPa)	Impact (J/cm²)	"T" Peel (kN/m)	Viscosity (Pa-sec)	Gap line (mm)	Temp. range (°C)
Tough, activator-cured, general purpose	180	30	7	8	18	≤ 1	-54-135
Environmentally resistant, activator-cured	300	17	2	1.7	20	≤ 1	-54-177
Fast, activator-cured	15	19	2.7		12	≤ 0.5	-54-107
Surface tolerant, bonds many materials without cleaning or preparation	120	19	2.3	4.8	70	≤ 1	-54-120
Fast, uv light-cured	10	21	2	—ᵃ	20	≤ 1	-54-100

ᵃNot applicable for transparent substrates.

C. Structural Bonding

Anaerobic structural adhesives offer the user benefits unavailable with two-component adhesives, such as epoxies or urethanes. Typical properties of anaerobic structural adhesives are shown in Tables 7 and 8.

Especially useful are the variety of cure systems which make these the true convenience adhesives. Two of the following systems are often used simultaneously.

1. Adhesives can be cured without heat or mixing. Surface activation provides rapid cures and high strength.
2. Heat can be used to accelerate the cure although usually it is the elimination of ovens and multifixtures which produces substantial savings (Fig. 15).

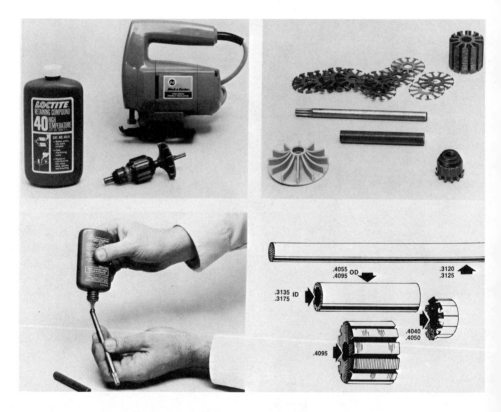

Fig. 15. Four parts are structurally bonded to the shaft of a double-insulated tool: insulator sleeve, armature laminations, communicator, and fan. All operations are done automatically except hand loading of laminations.

Table 8 Physical and Electrical Properties of Anaerobic Structural Adhesives

Property	Value
Physical	
Refractive index (UV)	1.54 ± 0.02
Light transmittance (UV)	95%
Coefficient of thermal	
expansion	$9-13 \times 10^{-5}$ cm/cm/°C
Color	Amber to light brown
Fatigue limit	$3-6.8$ MPa
Electrical	
Dielectric strength	$31-34$ kV/mm
Volume resistivity	$3-20 \times 10^{12}$ Ohm-cm
Dielectric constant	
(100 Hz to 1 MHz)	$3.8-5.5$
Dissipation factor	
(100 Hz to 1 MHz)	$0.024-0.064$

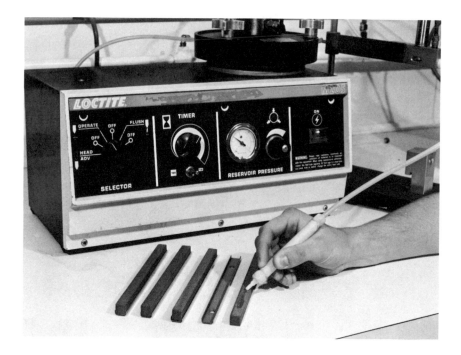

Fig. 16. Honing stones for automatic cylinder-honing machine are loaded to a backing plate. A gap-curing adhesive and activator are used because of the porosity of the stone.

3. Activator cures are very common and convenient. On-part
    life of the activator is usually weeks, and cures are pro-
    duced on activated surfaces in minutes or seconds. (Figs.
    16 and 17).
4. A recent development is a series of anaerobic UV-cured adhe-
    sives. Naturally, the bondline must be visible to the UV
    light, so this technique is limited to parts with at least one
    side transparent to UV light. A side benefit of these resins
    is that cure can often be forced in an exposed fillet over-
    coming the air-induced stability. A light with a predominant
    wavelength of 3660 Å and an intensity of 100 watt/m^2 can
    cure a typical adhesive in seconds (Fig. 18). Bright sun-
    light is similarly effective.

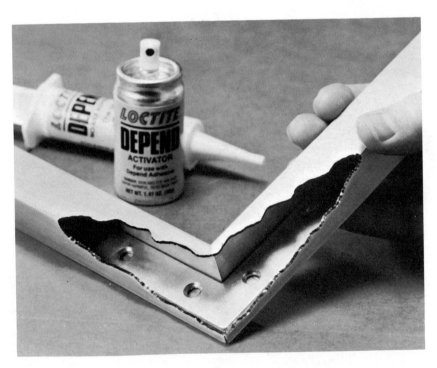

Fig. 17. Activator-cured, tough adhesive is used to bond angle brace
in aluminum furniture extrusion (cut away).

Fig. 18. Ultraviolet adhesive rapidly and permanently bonds the L-shaped stainless channel to the hinged hatch glass.

IV. DURABILITY

Retention of adhesion in the presence of water and industrial chemicals is a function of surface preparation and substrate material, as well as the resin. In general, the anaerobics, as strongly crosslinked materials, are resistant to water, oil, gasoline, alcohols, phosphate esters, and chlorinated solvents. It is recommended that environmental tests be run on the specific parts and surface treatments proposed for assemblies used in questionable ambients. The manufacture of resins often has data from accelerated tests available. Care should be used with an extrapolation of data. Some accelerated tests are too rigorous because no adhesive recovery time is allowed after exposure to destructive materials, such as water and alcohol, whereas a period of drying often partially restores loss of strength.

Anaerobic adhesives perform well in rugged environments—such as exposed automotive glass and metal, the Alaska pipeline, electronics on mortar shells, and motor magnets.

V. SAFETY AND HANDLING

All the anaerobic adhesives are single-component, nonflammable, and nontoxic. They have low odor characteristics. They can be used in production situations with normal ventilation although the activators are sometimes carried in a fast drying, chlorinated solvent which requires extra ventilation during drying. Some adhesives are eye and skin irritants, and proper shop hygiene should be used as with most oils and chemicals. Shelf life is generally at least a year in unopened containers when kept at 20 ± 10°C.

ACKNOWLEDGMENT

All figures and tables are courtesy of Loctite Corporation, Newington, Connecticut.

REFERENCES

1. R. E. Burnett and B. W. Nordlander, to General Electric, U.S. Patent No. 2,628,178 (1953).
2. V. K. Krieble, to Loctite, U.S. Patent No. 2,895,950 (1959).
3. V. K. Krieble, to Loctite, U.S. Patent No. 3,041,322 (1962).
4. R. H. Krieble, to Loctite, U.S. Patent No. 3,043,820 (1962).
5. V. K. Krieble, to Loctite, U.S. Patent No. 3,218,305 (1965).
6. E. Frauenglass and G. Werber, to Loctite, U.S. Patent No. 4,038,475 (1977).
7. R. D. Rich, to Loctite, Patent pending.

8. G. Werber, to Loctite, U. S. Patent No. 3,851,017 (1974).
9. B. M. Malofsky, to Loctite, U.S. Patent No. 3,988,299 (1976).
10. J. W. Gorman and A. S. Toback, to Loctite, U. S. Patent No. 3,425,988 (1969).
11. L. J. Baccei, to Loctite, U. S. Patent No. 4,018,351 (1977), other patents pending.
12. C. Hulstein, *Adhesives Age* (1979).
13. G. S. Haviland, U. S. Patent No. 3,386,630 (1968).
14. T. S. Fulda, SME, Paper FC77-532, SME
15. W. A. Pauwels and R. G. Nystron, U.S. Patent No. 3,638,831 (1972).
16. G. S. Haviland, SME Paper #AD80-329, SME, 29 April 1980.
17. G. K. Junker, SAE Paper 69,005, SAE, Detroit, 13 January 1969.
18. Industrial Fasteners Institute, 1505 East Ohio Blds., Cleveland, OH 44114.
19. *Military Specification MIL-S-46163.*
20. ML-R-46082A

12
Cyanoacrylate Adhesives

William F. Thomsen *Eastman Chemical Products, Inc., Kingsport, Tennessee*

I. INTRODUCTION

Cyanoacrylate adhesives were first offered commercially in 1958 and were used for several years primarily as specialty adhesives in industrial applications. Today, the cyanoacrylates are used for such diverse operations as the assembly of electronic components; bonding of gasket materials; application of strain gauges; repair of antiques and musical instruments; locking of nuts and bolts; and construction of jigs, fixtures, displays, and scale models. In more recent years, these adhesives have been marketed for home-consumer use under a variety of trade names. The public has come to identify the cyano-acrylate adhesives with such adjectives as "instant," "magic," and "miracle." It is understandable that cyanoacrylate adhesives develop-ed these identities, for they have become widely known as unique, one-component adhesives that are able to form strong bonds rapidly at room temperature on a wide variety of materials. They form strong bonds with most rubbers, metals, hardwoods, plastics, and nonporous materials with smooth, close-fitting surfaces, enabling the adhesive joining of these materials to themselves and to each other.

The cyanoacrylates form bonds within a minute or less depending on the substrate being bonded. Bonding rubber and plastic sub-strates requires only a very few seconds, whereas bonding metals may require from a few seconds to a minute for a bond to be strong enough for handling and up to several hours for a bond to reach its ultimate strength. The use of most other types of adhesive usually requires application of heat, addition of a catalyst, or evaporation of a solvent. The cyanoacrylates, when pressed into a thin film between two adhe-rends, undergo a rapid polymerization which occurs at room tempera-ture without the use of a solvent or an added catalyst. This proper-ty is what makes cyanoacrylates unique adhesives.

II. TYPES OF CYANOACRYLATE ADHESIVES

The cyanoacrylate adhesives are alkyl esters of cyanoacrylic acid. Alkyl esters such as propyl, butyl, and isobutyl cyanoacrylates have been evaluated, but methyl and ethyl cyanoacrylates are the principal ones used for the industrial and home-consumer market. Methyl cy-anoacrylate may give slightly stronger bonds than the ethyl ester on certain types of substrates, whereas ethyl cyanoacrylate may be slightly stronger on others; but there is very little difference in bonding performance between methyl and ethyl cyanoacrylate. Typi-cal physical properties of the cyanoacrylate monomers and polymerized adhesives are shown in Tables 1 and 2.

Both methyl 2-cyanoacrylate and ethyl 2-cyanoacrylate adhesives are marketed with and without thickening agents. Without a thicken-ing agent, cyanoacrylates are water-thin liquids with viscosities of approximately 1−3 cP. In some formulations, thickening agents have been added to give viscosities ranging from 20 to 200 cP. These thickened adhesives are easier to use, as they tend to run less on and drip less from the bonding area. However, the unthickened, or water-thin, adhesives will spread more rapidly and will form a thinner film than the thickened formulations, and as a result, they form bonds more rapidly.

Recently, cyanoacrylate-adhesive formulations with additional thickening agents have been developed to give even higher viscosities

Table 1 Typical Physical Properties of Cyanoacrylates

Property	Methyl 2-cyanoacrylate	Ethyl 2-cyanoacrylate
Specific gravity at 68°F (20°C)	1.10	1.05
Boiling point, at 2 mm Hg	122°F (50°C)	138°F (59°C)
Flash point, Cleveland open cup	180°F (82°C)	180°F (82°C)
Autoignition temperature, ASTM D-2155	874°F (468°C)	874°F (468°C)
Vapor pressure at 77°F (25°C)	< 2 mm Hg	< 2 mm Hg
Refractive index η_D^{25}	1.4406	1.4349

Table 2 Typical Physical Properties of Cured (Polymerized) Cyanoacrylates

Property	Methyl 2-cyanoacrylate	Ethyl 2-cyanoacrylate
Softening point	329°F (165°C)	259°F (126°C)
Index of refraction	Approximately same as glass	Approximately same as glass
Flexural modulus (psi) (ASTM D-790)	457,000	300,000
Heat distortion temperature at 264 psi	246°F (119°C)	156°F (69°C)
Temperature of decomposition (temperature at which 10% is decomposed)	388°F (198°C)	415°F (213°C)

(as much as 2,000 cP) for the specific purpose of producing gap-or void-filling bonds. The very viscous formulations are also effective in forming bonds on porous substrates, such as sponges, woods, leathers, and ceramics. These highly viscous formulations usually require a basic catalyst, such as an organic amine, to help initiate the polymerization of the cyanoacrylate in thick gluelines, as would be found in voids or gap areas.

The cyanoacrylates do not usually set up (polymerize) rapidly in thick sections or gap areas unless a surface activator is applied to the bonding surface. The surface activator is applied to the bonding area, never to the adhesive itself. The viscous formulations of cyanoacrylate are only effective in gaps or voids up to about 0.020 in. The surface activator, often referred to as a catalyst or an accelerator or a primer or an initiator, is applied to one surface or both surfaces to be bonded, whereas the adhesive is applied to only one surface. The surface activator is usually a weak organic amine, such as phenylethylethanolamine, dissolved to a concentration of about 1–3% in a fast-evaporating solvent like acetone, methyl ethyl ketone (MEK), naphtha, or isopropyl alcohol. Dilution of the surface activator simply allows a trace amount to be added to the surface, which is all that is needed to initiate polymerization. When the surfaces to be bonded may be acidic, the use of a surface activator is helpful in obtaining a faster-setting bond. It is interesting to observe that when bonding two substrates that may have acidic surfaces (such as metals that have been cleaned with acid and dried without having thoroughly been rinsed), the cyanoacrylate will resist polymerization or form bonds very slowly. The use of a surface activator should not be too strongly emphasized because the bond formation of cyanoacrylate adhesives is usually fast enough for most applications without the use of a surface activator. Application of the surface activator to the bonding area does not produce stronger cyanoacrylate bonds, only faster-setting bonds.

III. MECHANISM OF BOND FORMATION

The liquid alkyl cyanoacrylates must polymerize to function as adhesives. Several theories have been proposed as to the mechanism of the polymerization of cyanoacrylate adhesives, but the most widely accepted theory calls for an anionic mechanism [1]. Polymerization occurs rapidly in the presence of a weakly basic compound, such as an amine, an alcohol, or water. This polymerization reaction is represented by the following equations in which the basic catalyst is represented by A^-, an electron donor:

$$CH_2 = \underset{\underset{\text{CN}}{\displaystyle|}}{C} - COOR \underset{\underset{\delta^+ \quad \delta^-}{}}{\overset{A^-}{\rightleftharpoons}} CH_2 - \underset{\underset{\text{CN}}{\displaystyle|}}{C} - COOR \xrightarrow{A^-} A - CH_2 - \underset{\underset{\theta}{\overset{\text{CN}}{|}}}{C} - COOR \xrightarrow{\quad CH_2 = \underset{\overset{\text{CN}}{|}}{C} - COOR \quad}$$

$$\text{Polymer} \xleftarrow[\text{Reaction}]{\text{Further}} A - CH_2 - \underset{\underset{\text{COOR}}{\displaystyle|}}{\overset{\text{CN}}{|}}{C} - CH_2 - \underset{\overset{\text{CN}}{|}}{C} - COOR$$

$$R = CH_3 ;\ C_2H_5 ;\ \text{etc.}$$

Nearly all substrates have moisture adsorbed on their surfaces, and this moisture provides hydroxyl ions which can serve as initiators for the anionic polymerization. When a drop of the cyanoacrylate is pressed or spread between substrates, the resulting thin film comes into more intimate contact with the traces of moisture adsorbed on the substrate surface, leading to rapid polymerization initiation. The thinner the film of the cyanoacrylate, the more quickly polymerization or bond formation usually occurs. A drop or bead of the cyanoacrylate placed on a normal or nonacidic surface will remain a liquid (unpolymerized) for a long time; however, when it is spread or pressed into a thin film, polymerization will occur. The requirement for thin films also means that very little adhesive is needed to make a bond; hence, cyanoacrylates are inexpensive to use, notwithstanding their high cost per pound.

IV. ADVANTAGES

Among the advantages offered by cyanoacrylate adhesives, some of which have already been mentioned, are the following:

1. *They are easy to use.* The adhesives are ready to use as supplied. No mixing is required. Many users apply the adhesive directly from the container in which it is supplied. The cyanoacrylates usually require no catalyst, no heat, and no solvent. When a drop of adhesive is pressed between two parts, a bond will ordinarily form within a few seconds.
2. *They form strong bonds rapidly.* The cyanoacrylate adhesives form strong and fast-setting bonds on a variety of substrates, such as plastics, rubbers, metals, woods, ceramics, and leathers. No clamps, fixtures, or jigs are normally needed because of their rapid cure. The rapid bonding re-

duces the need for space to store fixtured parts during bond
formation.

3. *A small amount of adhesive is required.* Only very small
    quantities of adhesive are needed to form strong bonds. A
    pound of cyanoacrylate could be expected to form approxi-
    mately 30,000 bonds, each 1 in.^{2}.

4. *They are versatile.* Dissimilar materials can be joined with-
    out the need of special assembly.

5. *They form virtually colorless bonds.* The adhesive bond is
    clear, colorless, and almost invisible.

6. *The adhesive bond is resistant to solvents.* Cyanoacrylate
    bonds have good resistance to most solvents, including alco-
    hol, oil, gasoline, naphtha, and chlorinated solvents.

V. LIMITATIONS

Like all other bonding materials, the cyanoacrylate adhesives have
some limitations. The known limitations are as follows:

1. *They lack gap-filling capability.* There are thickened formu-
    lations that offer some gap-filling ability, but they usually
    require a catalyst, and the ultimate bond strength is usually
    not as great as that obtained with close-mating surfaces.

2. *The bonds have low shock resistance.* The low impact
    strength of cyanoacrylate bonds is particularly noticeable
    when metal is bonded to itself. Bonds on steel samples that
    may give 5000 psi tensile strength and that are capable of
    lifting an automobile can be broken by a sharp blow. How-
    ever, when rubber is bonded to itself or to other materials,
    such as metals or plastics, the impact or shock resistance
    is much greater. Bonding a flexible, rubber interlayer be-
    tween metals will improve impact resistance of the bond as
    well as partially compensate for lack of a close fit between
    the surfaces.

3. *Some bonds lack moisture resistance.* Poor moisture resis-
    tance has been mentioned in a number of publications as one
    of the main limitations of cyanoacrylate adhesives. When
    metals are bonded to themselves or to hard plastics, the lack
    of moisture resistance is especially noticeable. However,
    when cyanoacrylates are used for bonding rubber to itself or
    when rubber is used as an interface between other sub-
    strates, such as plastics or metals, the cyanoacrylate adhe-
    sive bond will have good moisture resistance. Also, when
    some plastics are joined together, the bond usually exhibits
    good resistance to moisture.

4. *Bonds have low peel strength.* Flexible plastics and films
    bonded together with a cyanoacrylate are sometimes easily

peeled apart. The use of cyanoacrylates should be discouraged for applications of this type where the substrates will be subjected to significant peel forces. However, cyanoacrylate bonds between elastomeric substances usually have excellent peel strength.

5. *Bonds lack high-temperature resistance.* Cyanoacrylate bonds do not usually have high-temperature resistance regardless of the type of substrate being bonded. The maximum temperature to which they should be exposed is about 160°F (71°C). Bonds lose strength after prolonged exposure to temperatures higher than 160°F (71°C); the higher the temperature, the shorter the time before the bonds begin to fail.

VI. BONDING CHARACTERISTICS ON VARIOUS SUBSTRATES

Cyanoacrylates form strong bonds on a variety of substrates, such as metals, plastics, rubbers, hardwoods, and other nonporous materials with close-fitting surfaces. Typical bond strengths of low-viscosity, methyl and ethyl cyanoacrylate adhesives are shown in Table 3. Like most adhesives, cyanoacrylates perform best when used to bond surfaces which are thoroughly clean.

A. Metals

Both methyl and ethyl 2-cyanoacrylates form strong bonds on metals such as aluminum, steel, copper, and a variety of alloys, but methyl 2-cyanoacrylate generally forms stronger bonds. Extensive pretreatment is usually not required for most metal surfaces although it is necessary to remove all loose dirt, acidic deposits, grease, and foreign matter from the bonding surfaces. Gently sanding the surfaces to remove loose oxides and to ensure smooth surfaces can be helpful.

B. Plastics

The cyanoacrylates form fast-setting and strong bonds on most plastics, but there are some plastics, such as polyethylene and polypropylene, that are difficult to bond. These plastics can definitely benefit from pretreatment. Satisfactory bonds on these polyolefins can be obtained by oxidizing the surface, by means of, for example, flame treatment or chromic acid treatment. Fluorocarbons, such as Teflon, can benefit from a sodium-etch pretreatment. Polyacetals can benefit from a surface pretreatment, such as sanding or using a chromic acid etch.

Several plastics—particularly the thermoplastic types like vinyl, butyrate, acrylic, polycarbonate, and polystyrene—usually give slightly stronger bonds with ethyl 2-cyanoacrylate than with methyl

Table 3   Typical Shear Strength Values of Cyanoacrylate Adhesive Bonds

		Bond strength (psi) after 48 hr	
Type of bond	Substrate(s)	Methyl 2-cyanoacrylate	Ethyl 2-cyanoacrylate
Metal-metal	Steel	3200	2900
	Aluminum	2400	1700
	Stainless steel	2500	800
Metal-plastic	Aluminum, nylon	900	700
	Aluminum, phenolic	1000[a]	900[a]
Plastic-plastic	ABS	—[a]	—[a]
	Acrylic	—[a]	—[a]
	Nylon	500[a]	700[a]
	Phenolic	—[a]	—[a]
	PVC	—[a]	—[a]
Rubber-rubber	Butyl	—	—
	EPDM	50[a]	60[a]
	Neoprene	—[a]	—[a]
	Nitrile	—	—
Wood-wood	Maple	600	500
	Yellow pine	90	110[a]
	Oak	500[a]	—[a]
	Particle board	—	—

[a]Bond held, specimen broke.

2-cyanoacrylate although the methyl ester normally forms satisfactory
bonds. The results of lap shear tests usually show that the bonds
will hold, but that the plastics will break when either methyl or ethyl
cyanoacrylate adhesives are used. When ethyl 2-cyanoacrylate is used,
the plastics usually exhibit slightly higher tensile break values. When
either adhesive is applied to the bonding surface or thermoplastics,
it shows a solvent characteristic by attacking the plastic surface before polymerization occurs. Presumably, most thermoplastics are attacked more by methyl cyanoacrylate than by ethyl cyanoacrylate,
which results in a slightly weaker substrate when the methyl ester is
used.

C. Rubber

Cyanoacrylate adhesives form outstanding bonds on most rubber substrates, but not on silicone rubber. Preparation of rubber surfaces
requires no more than a wipe with a solvent such as acetone or (MEK).

Polyurethanes sometime require a light sanding followed by a solvent wipe to remove mold release agents that may be embedded in the surface. Methyl and ethyl cyanoacrylates form comparable bonds on rubber.

D. Glass

Both methyl and ethyl cyanoacrylates form fast-setting bonds on glass, but the bonds frequently deteriorate with time. The cause for the bond weakening has not been proved, but it may be related to the alkaline surface of the glass. Therefore, commercial use of cyanoacrylate adhesives with glass should be attempted with caution and only after thorough testing. Glass substrates can usually be prepared for bonding by a simple water wash and thorough drying.

E. Wood and Porous Materials

Wood and other porous materials can be bonded with cyanoacrylate adhesive, but those with excessively porous surfaces tend to have absorbed the adhesive before a bond is formed. A viscous formulation of cyanoacrylate is desirable for such substrates. Many types of wood are frequently acidic, and wood will usually benefit from the use of a surface activator or catalyst.

VII. DISPENSING CYANOACRYLATES

The cyanoacrylate adhesives are easy to apply and can be dispensed directly from the bottle or tube. In some assembly procedures there may be a need for a faster and more precise method of applying the adhesive, and automatic applicators are available for such cases. A popular type of dispenser or dispensing system for small shot applications uses timed air impulses to move the adhesive and employs a vacuum to prevent dripping. Applicator systems that are used for dispensing the adhesive with precision have three basic components: a constantly pressurized reservoir; a valve to control the instantaneous flow of adhesive; and a control module for precise adjustment of the amount of adhesive dispensed. This type of dispensing system uses a variable-stroke pinch valve to control the volume. The dispenser can utilize a variety of tubing sizes to apply very small to large shots. There is no limit to reservoir capacity as long as the adhesive is delivered to the dispenser valve at constant pressure [2].

VIII. REQUIREMENTS FOR SUCCESSFUL USE OF CYANOACRYLATE ADHESIVES

When one of the cyanoacrylate adhesives is used, it ordinarily forms a good bond, or it forms no bond at all. Seldom is a weak bond formed

other than to untreated polyethylene or polypropylene and similar
materials. When the cyanoacrylate fails to bond it is because one
of the requirements described earlier has not been met. Finding
and correcting this deficiency will usually solve the problem. The
following items should be checked.

1. *Bonding surfaces must fit closely together.* If one surface to
   be bonded is flat and the other slightly warped, cyanoacrylate adhesives may fail to bond because they cannot fill the
   gap. When a condition of this type is encountered, it may
   be possible to grind the surfaces together with a fine abrasive to give a suitable fit. The use of very viscous formulations with a surface activator will be helpful, but there is
   a limit (about 0.02 in.) to the size of gap which may be filled. Another possibility is to bond a layer of rubber between
   the two hard surfaces. With adequate pressure during the
   initial bonding operation, the rubber will deform and conform
   closely to both of the other surfaces, enabling the formation
   of a good bond. The rubber serves the purpose of providing close-mating surfaces.

2. *Surfaces must be clean.* Oily materials, mold release agents,
   and acidic contaminants will prevent bond formation. Any
   small piece of grit can hold the surfaces so far apart that
   they will not bond if an unthickened adhesive is used.
   Methods of cleaning dirty surfaces have already been mentioned.

3. *A minimum amount of adhesive is needed.* The smallest
   amount of adhesive that can be spread to cover the area to
   be bonded is the best amount to use. If a considerable
   amount of adhesive squeezes out of the joint when two parts
   are pressed together, too much was applied, and the bond
   forms slowly. Using an excessive amount of adhesive tends
   to give slow-setting bonds. However, if substrates contain
   voids or if they are porous, larger quantities of adhesive
   will be required. Very viscous formulations are suggested
   for these surfaces.

4. *Fresh adhesive is desirable.* On their container, some cyanoacrylate adhesives bear a date by which the material
   should be used for best results. If the adhesive is excessively old, the liquid may have become very thick, and it may
   bond slowly or not at all.

IX. COMMERCIAL APPLICATIONS IN PRODUCT ASSEMBLY

One of the earliest commercial applications of cyanoacrylate adhesive
was the bonding of strain gauges used to measure the stress in metals.
The cyanoacrylate proved to be a very effective adhesive for this pur-

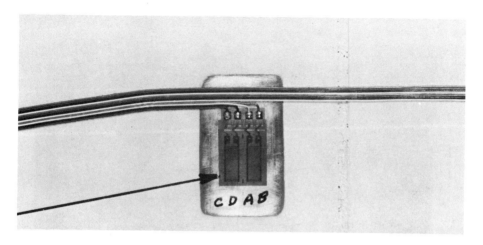

Fig. 1. Plastic-film—backed strain gauge bonded to aluminum.

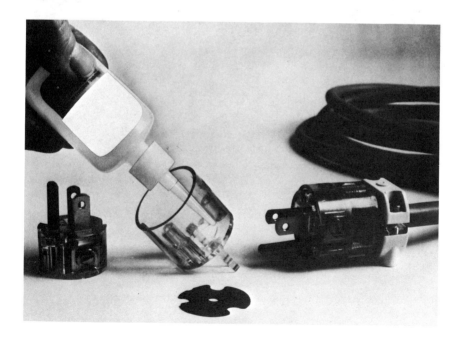

Fig. 2. Neoprene rubber bonded to a polycarbonate plastic.

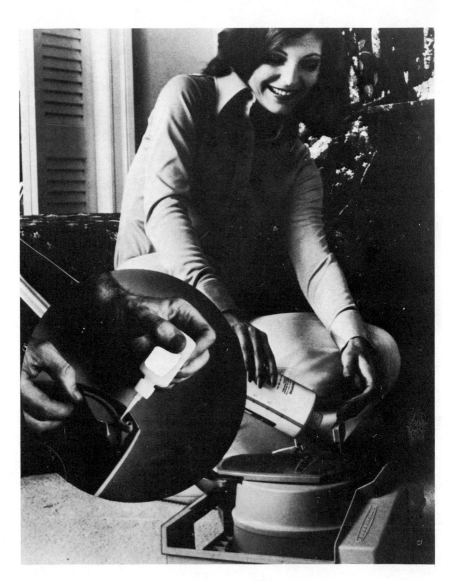

Fig. 3. Neoprene-rubber gasket bonded to ABS plastic.

Fig. 4. Acrylonitrile butadiene styrene plastic bonded to ABS plastic.

pose when the strain gauge had a plastic backing. Figure 1 shows a strain gauge bonded with a cyanoacrylate to a helicopter rotor blade for measuring the stress in the aluminum blade. The ability of cyano- acrylates to bond dissimilar materials has been the basis for utilizing the adhesives in the assembly of hospital-grade electric plugs. A fast-bonding cyanoacrylate adhesive is used to form a long-lasting bond between a neoprene-rubber dust seal and the inside of the trans- parent plug's polycarbonate wiring device (Fig. 2). In Fig. 3, a cy- anoacrylate is being used to bond a neoprene gasket to the acryloni- trile butadiene styrene (ABS) cover on a carpet washer. (Previously, slower setting, solvent-based adhesives were used.) A voltmeter is

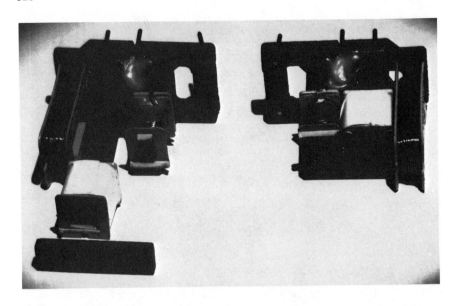

Fig. 5. Iron ferrite bonded to plastic.

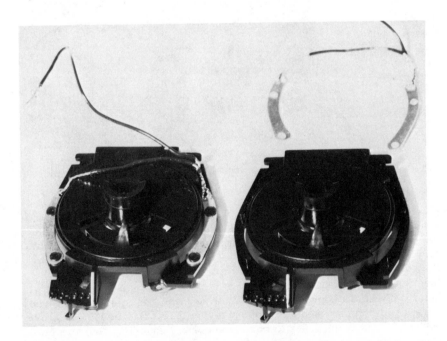

Fig. 6. Beryllium-copper circuit bonded to polyacetal plastic.

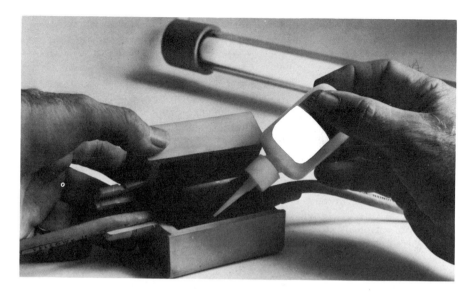

Fig. 7. Hard rubber bonded to hard rubber.

Fig. 8. Nitrile rubber bonded to nitrile rubber.

securely fastened to the cover of a cable locator with only a drop of a
cyanoacrylate adhesive (Fig. 4). The cable locator is used by telephone operating-field personnel to determine the location of buried
telephone cables.

Cyanoacrylates have found increasing use in the electronic industry, mainly because of their ability to form bonds between dissimilar materials rapidly. Most bonding areas of electronic gadgets require very little adhesive. In a magnetic amplifier, an iron ferrite
magnet is bonded to a plastic coil with a cyanoacrylate (Fig. 5). Cyanoacrylate bonds a beryllium-copper circuit to a polyacetal-plastic
control housing (Fig. 6). In Fig. 7, a thin film of a cyanoacrylate
adhesive is all that is required to successfully bond the two halves of
a rubber ballast cover that is part of a portable fluorescent hand-lamp
assembly. One of the earliest uses of cyanoacrylate adhesives was in
the preparation of rubber O-rings (Fig. 8). There are many other
applications for cyanoacrylate adhesives, some of which are listed below.

Industry	Applications and substrates
Aircraft	Neoprene rubber gaskets
Appliance	Components for clothes-dryer lint switch: flexible rubber cap to ABS plastic housing
	Automatic washing machine components: rubber boot to ceramic disk
Automotive	Flexible polyvinyl chloride (PVC) trim strips
	Regulator switch
	Alternator components: nylon to nylon
	Horn assemblies
	Warning buzzer for doors: molded nylon parts
Bicycle	Rubber grips to metal handlebars for bicycles and tricycles
Cosmetic	Eye-shadow applicator: sponges bonded to propionate and polystyrene handles
	Lipstick case: plastic to metal
Electric	Regulator switch cover to switch assembly
Electronic	Components to circuit board
	Magnetic amplifier: iron ferrite magnet to plastic coil
Machinery	Nitrile-rubber O-ring to brass collar
	Rubber gasket to metal end cap
Medical	Syringe: latex rubber to vinyl tube
	Rubber stopper to douche bag
	Blood analysis equipment: nylon to stainless steel, nylon to PVC tubing
Office equipment	Rubber stanps: fabric impregnated with neoprene rubber bonded to itself

Industry	Applications and substrates
Phonographic	Needle assembly: ABS to zinc
Photographic	Film squeegee: neoprene rubber to ABS
Shoe	Tennis shoe: rubber label to rubber shoe heel

X. TOXICITY AND HANDLING PRECAUTIONS

The hazards presented by cyanoacrylates are primarily due to their rapid polymerization and their ability to bond human skin. However, they do have some irritant effects.

A. Toxicity

Determining the acute toxicity of cyanoacrylates is difficult because of rapid polymerization. Polymerized cyanoacrylate, ground and administered as a suspension in corn oil, failed to kill rats at oral doses of 6400 mg/kg. The adhesive caused mild irritation of the skin of guinea pigs after 24 hr of closed contact, but it was not absorbed through the skin and did not cause skin sensitization. In the rabbit eye, the cyanoacrylates caused immediate local irritation which was, in part, due to rapid polymerization which produced adherence of the lids. The cyanoacrylates caused corneal damage. The vapor proved irritating to the eyes and mucous membranes [3].

B. Handling Precautions

Prolonged breathing of vapors of cyanoacrylate adhesives should be avoided. They should be used with adequate ventilation. The cyanoacrylate adhesives have pungent, unpleasant odors and may be mildly lacrymatory (tear-producing) to the eyes and irritating to other mucous membranes. Good general ventilation should be supplied in the working areas, and local exhaust ventilation may be required to maintain vapor concentrations below irritating levels. To minimize the generation of vapors, adhesive should not be poured into an open container for easier access but kept in a container that can be closed when not in use, preferably the original container. If spilled, the liquid should be wiped up immediately with a wet tissue, paper towel, or rag which should then be placed in a closed container.

The threshold limit value (TLV) for methyl 2-cyanoacrylate is 2 ppm (8 mg/m^3). Ventilation should be provided to keep the concentration of vapors in the air below the TLV and below the tentative short-term exposure limit (STEL) of 4 ppm [4]. Contact of the adhesive with eyes and skin should be avoided. Eye contact should be prevented by the use of appropriate protective equipment. If eye contact occurs, the eyelids should be held apart to prevent bonding, and the eye area should immediately be flushed with copious amounts of water. Medical attention should be obtained. Areas of skin contact should

immediately be flushed with water to cause polymerization of the monomer. Repeated soaking in warm water will hasten loosening of the
polymerized material. The cyanoacrylate adhesives are defined as
combustible liquids by the National Fire Protection Association (Std.
No. 321-1973). Both the liquids and the cured adhesives will support
combustion. Cyanoacrylate adhesives in absorbent materials, such as
cloths used to wipe up spills, may undergo polymerization and cause
an increase in temperature that could conceivably result in autoigni-
tion. Contamination of adhesive with polymerization catalyst could
cause rapid heating and rupture of container. Keep cyanoacrylates
away from heat and open flame. The liquid adhesive should be kept
out of the reach of children.

XI. CLEANING UP EXCESS ADHESIVE

Solid (polymerized) cyanoacrylate adhesives are soluble in N,N-di-
methylformamide (DMF) and in nitromethane. Either of these solvents
can be used to remove unwanted polymerized adhesive. *Caution:*
DMF and nitromethane are powerful solvents, and their effects on the
substrate to be cleaned should be determined before they are used.
These solvents are also flammable and should be kept away from
sparks and flames. N,N-Dimethylformamide is irritating to the skin,
eyes, and mucous membranes and is absorbed through the skin. Both
liver and kidney injury may result from skin absorption and repeated
exposure to vapor [5]. Skin and eye contact of these solvents should
be avoided by use of appropriate protective equipment. Rubber
gloves must be changed and discarded frequently as DMF can penetrate rubber. Avoid breathing the vapors of these solvents. Good
general and local exhaust ventilation adequate to keep vapor concentrations of DMF below the TLV of 10 ppm (30 mg/m^3) should be provided. The TLV for nitromethane is 100 ppm (250 mg/m^3) [4]. Avoid
breathing the vapor. Use with ventilation adequate to keep vapor
concentrations below the TLV. Avoid prolonged or repeated skin contact.

XII. HOW TO RELEASE BONDS

Sometimes, there may be a need to break a cyanoacrylate bond. The
following techniques are suggested to release bonded parts. Only
the fourth method combined with washing with large quantities of
water is recommended for the separation of accidentally bonded human
skin. Bonds of the cyanoacrylate adhesives may be broken by

  1. *Solvent.* Immersion in DMF or nitromethane for a minimum
     of 12 hr.
  2. *Impact or shock.* Cyanoacrylate bonds are very strong, but
     parts that cannot be pulled apart can sometimes be separated

by a sharp blow.
3. *Heat.* Heat the bonded part to the softening point T_g of the adhesive bond. [For methyl 2-cyanoacrylate, T_g is about 329°F (165°C); for ethyl 2-cyanoacrylate, about 259°F (126°C)].
4. *Peel.* If one substrate is flexible, the flexible material can be peeled away from the other at as sharp an angle as possible.

XIII. SHELF LIFE OF CYANOACRYLATES

Experience indicates that a cyanoacrylate adhesive is usable as long as its viscosity has not increased noticeably. For maximum shelf life, containers should be stored in a cool, dry place and kept closed when not in use. Contact of a cyanoacrylate adhesive with moisture or alkaline substances causes an increase in viscosity and eventual solidification.

REFERENCES

1. H. W. Coover, F. B. Joyner, N. H. Shearer, and T. H. Wicker, Jr., *Soc. Plast. Eng. J.* *15*(5): (1959).
2. F. C. Herat, *Adhesives Age* *25* (5): (1980).
3. Toxicity information from unpublished data. Health, Safety and Human Factors Laboratory, Eastman Kodak Company, Rochester, New York.
4. TLV from *Threshold Limit Values for Chemical Substances in Workroom Air by the ACGIH for 1975.* Cincinnati, American Conference of Governmental Industrial Hygienists, 1975.
5. *Industrial Hygiene and Toxicology*, Vol. 2, rev. ed. (F. Patty, ed.), Interscience, New York, 1963.

13
Hot-Melt Adhesives

Robert D. Dexheimer* and Leonard R. Vertnik *Henkel Adhesives Company, Minneapolis, Minnesota*

I. INTRODUCTION AND DEFINITION OF HOT-MELT ADHESIVES

We can define hot-melt adhesives as thermoplastic materials, solid at room temperature. When heated above their melting point, they become fluid and are able to wet the surfaces to which they are applied. Generally, a quantity of fluid hot melt is applied to one or both of the surfaces to be joined, and the surfaces are brought together. When the adhesive has cooled below its solidification point, the substrates are bonded. In simple terms, adhesive bonding involves the application of a fluid adhesive which can wet the surfaces to be bonded and the subsequent conversion of the adhesive to a solid. As opposed to

---

*Retired

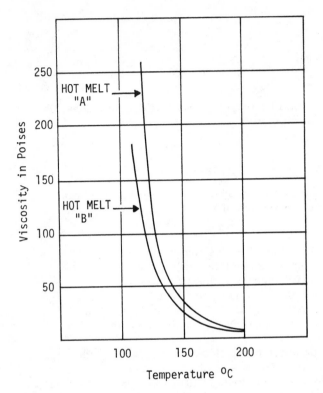

Fig. 1. Viscosity of molten hot-melt adhesives as a function of temperature.

other types of adhesives, which become solid by elimination of solvents and/or water or through polymerization, hot melts convert from a hot fluid to a solid by cooling.

Such a simplistic definition also applies to asphalt, wax, metallic solder, and even water (ice). However, we will limit this discussion to hot-melt adhesives that contain at least one polymer as a major ingredient. These hot melts may contain the backbone polymer either "neat" (uncompounded) or admixed with waxes, tackifiers, plasticizers, fillers, and/or other modifiers.

The viscosity of a hot melt when measured at various temperatures, that is, as it cools from a liquid to a solid, describes a curve which rises very steeply as the solidification temperature is approached (Fig. 1).   This curve is characteristic of hot-melt adhesives, but actual values will differ with different systems.

If one plans to use a hot-melt adhesive it is desirable to choose the optimum type. The optimum may not necessarily be the one that is lowest in cost, nor the one that is easiest to apply, nor even the one that is strongest. Often, the best adhesive is a compromise, which has to be judged on a case-by-case basis, with a weighted value given to cost, performance, and production ease [1].

II. ADVANTAGES AND LIMITATIONS OF HOT-MELT ADHESIVES

An understanding of the viscosity curve (Fig. 1) explains many of the properties of hot-melt adhesives, such as fast bonding, low "wicking-in", ability to control glueline thickness, and void filling. Rapid setting is simply a matter of losing enough heat to achieve a high viscosity or to solidify. This time can be as short as a small fraction of a second, or it can be delayed by design for several minutes [1].

A. Advantages

Some reasons hot melts are used:

1. Form bonds rapidly—resulting in high-speed assembly and short clamp time
2. Invisible bonds—no exposed fastener parts
3. Fill "gaps" between surfaces to be joined
4. Bond most materials
5. Clean, easy handling
6. Load distribution through "area bonding"
7. Cost less per unit assembly than many mechanical fasteners
8. Easy recovery and repair of substandard assemblies
9. No problems with solvent or fume flammability
10. Simpler materials inventory and storage
11. Precise bond control through variations of the temperature and the quantity of adhesive
12. Equipment available for hand or automated assembly
13. Easily maintained equipment
14. Minimal production-line space requirements

B. Limitations

There are several characteristics of hot-melt adhesives, due to their thermoplastic nature, which may limit their use in certain bonding applications. Some of these limitations are

1. Bonds lose strength at elevated temperatures.
2. Some bonds may creep and fail with time.
3. Adhesive may be sensitive to some chemicals and solvents.
4. Some substrates or components may be sensitive to the hot-melt application temperature.
5. Sophisticated application equipment may be required for highly viscous, high-performance hot melts.

6. Some hot melts have a tendency to degrade at application
    temperature unless protected from the air.
7. Some hot melts become brittle at low temperatures.

III. TYPES OF HOT MELTS BASED ON THE BACKBONE POLYMER

One way to typify hot-melt adhesives is to distinguish by the backbone polymer. In some instances the polymer is used neat, but often
other materials are incorporated to form the finished adhesive. We
obviously cannot discuss all of the available backbone polymers, but
Table 1 lists the more commercially significant materials and their
generalized chemical structures.

   Most of the polymers listed in Table 1 are addition-type polymers.
The last three, however, are condensation polymers.

   Addition-type polymers are formed by combining the base, or
repeating units with each other without forming any elimination products; for example, ethylene—vinyl-acetate copolymers are formed by
the combination of ethylene and vinyl-acetate monomers.

   Condensation polymers are formed by combining two or more monomers with the elimination of small molecules; for example, polyamides
are formed by the reaction of organic acids and amines with the elimination of water.

   As a rule the addition-type polymers require compounding to be
useful as hot-melt adhesives. Typically, such a formulated hot-melt
adhesive would contain the backbone polymer plus one or more of the
following: tackifiers, plasticizers, flow modifiers, antioxidants, pigments, extenders, etc.

   The condensation polymers, by contrast, are often used without
compounding.

IV. ELEMENTARY PRINCIPLES OF JOINT DESIGN

Adhesion versus stress—that is what most of adhesive technology is
all about [5].

   Adhesion is the force that holds materials together. Stress, on
the other hand, is the force that pulls materials apart. The basic
types of stress in adhesive technology are illustrated in Fig. 2.

   A joint primarily designed for mechanical fastening may not prove
acceptable for adhesive bonding and may require redesign. Two
prime factors should guide the designer. First, the joint should be
designed so that the stress is distributed over the entire bonded area.
Second, the joint configuration should be designed so that the stress
is largely shear or tensile, with peel and cleavage minimized.

   Joint design is a critical factor in the success or failure of adhesive bonding. This fundamental concept is thoroughly covered in
Chap. 4.

Table 1 Base Polymers for Hot-Melt Adhesives

Backbone polymers	Chemical structure[a],[b]
Ethylene–vinyl-acetate copolymers	$+CH_2\text{-}CH\text{-})_n\,(CH_2\text{-}CH_2)_m$ $\qquad\qquad\;\;$ $O\text{-}C\text{-}CH_3$ $\qquad\qquad\qquad\; \overset{\parallel}{O}$
Polyethylene	$(CH_2\text{-}CH_2)_n$
Polypropylene (atactic)	$(CH_2\text{-}CH_2)_n$ $\qquad\quad CH_3$
Styrene-isoprene-styrene block polymers	$(CH_2\text{-}CH)_n\,(CH_2\text{-}C = CH\text{-}CH_2)_m\,(CH\text{-}CH_2)_\ell$ $\quad\bigcirc\qquad\qquad CH_3\qquad\qquad\quad\bigcirc$
Styrene-butadiene-styrene block polymers	$(CH_2\text{-}CH)_n\,(CH_2\text{-}CH = CH\text{-}CH_2)_m\,(CH\text{-}CH_2)_\ell$ $\quad\bigcirc\qquad\qquad\qquad\qquad\qquad\quad\bigcirc$
Ethylene-acrylic copolymers	$(CH_2\text{-}CX)_n\,(CH_2\text{-}CH_2)_m$ $O = C\text{-}CXH\text{-}CH_3 \qquad X = H \text{ or } CH_3$
Polyvinyl acetate and copolymers	$(CH_2\text{-}CH)_n\,(CH_2CH)_m$ $\;\; R_1\text{-}C\text{-}O \qquad O\text{-}C\text{-}R_2$ $\qquad\; \overset{\parallel}{O} \qquad\qquad\;\; \overset{\parallel}{O}$
Butyl rubbers	$(CH_2\text{-}CH = CH\text{-}CH_2)_n$
Polyurethanes	$\qquad\quad \overset{O}{\overset{\parallel}{}} \qquad\qquad \overset{O}{\overset{\parallel}{}}$ $(R_1\text{-}O\text{-}C\text{-}NH\text{-}R_2\text{-}NH\text{-}C\text{-}O)_n$
Polyamides	$(NH\text{-}R_1\text{-}NH\text{-}C\text{-}R_2\text{-}C)_n$ $\qquad\qquad\quad \overset{\parallel}{O}\;\;\;\, \overset{\parallel}{O}$
Polyesters	$(O\text{-}R_1\text{-}O\text{-}C\text{-}R_2\text{-}C)_n$ $\qquad\qquad\;\; \overset{\parallel}{O}\;\;\; \overset{\parallel}{O}$
Polyesteramides	$(O\text{-}R_1\text{-}O\text{-}C\text{-}R_2\text{-}C\text{-}NH\text{-}R_3\text{-}NH\text{-}C\text{-}R_2\text{-}C)_n$ $\qquad\qquad\;\; \overset{\parallel}{O}\;\;\; \overset{\parallel}{O}\qquad\qquad\quad \overset{\parallel}{O}\;\;\; \overset{\parallel}{O}$

[a]Generalized structure.
[b]R_1, R_2, and R_3 are organic moieties.
Source: Data from Refs. 2–4.

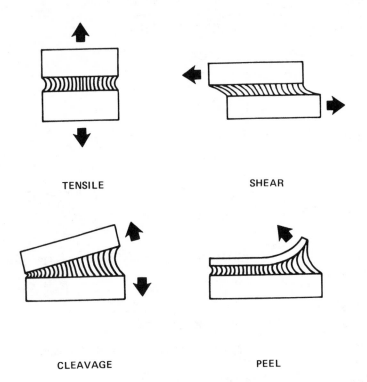

TENSILE   SHEAR

CLEAVAGE   PEEL

Fig. 2. Tensile stress; shear stress; cleavage stress; peel stress.

V. HOT-MELT ADHESIVE USAGE BY INDUSTRY

Table 2 lists the estimated consumption of hot-melt adhesives by industry for previous years and projects consumption for 1981.

The most rapid increases are seen in the areas of pressure-sensitive tapes, textiles, product assembly, and automotive. The overall growth rate is projected to average about 12%/year.

Approximately 60% of the 1979 total hot-melt adhesive use was in
packaging. The next largest volume areas were bookbinding, textiles,
and disposables—each with less than 10% of the total.

Over 50% of the total hot melts are based on ethylene–vinyl-acetate copolymers. Polyolefins (high-pressure polyethylene and atactic
polypropylene) account for over 30% of the balance.

Figure 3 illustrates the data on the use of hot-melt adhesives from
Table 2, showing a growth of approximately 25 million lb/year.

Table 2 Hot-Melt Adhesives Usage by Industry

Industry	Year (usage in millions of lb)			
	1975	1977	1979	1981[a]
Packaging	145	175	200	220
Bookbinding	20	22	24	25
Textile	15	20	25	30
Disposables	15	19	24	28
Carpet seaming tape	4	5	6	7
Pressure-sensitive tape (hot melt)	6	8	12	20
Furniture	10	12	15	17
Footwear	5	6	6	6
Automotive	4	7	8	10
Appliances	5	6	7	8
Filters	2	2	2.5	2.5
Product assembly	10	14	16	19
Consumer	0.5	1	1.5	2
Electronic	1	1.5	2	3
Other	2.5	1.5	1.5	2
Total	245	300	350	400

[a]Assuming no serious disruption in the volume growth curve during 1980 and subsequent years.

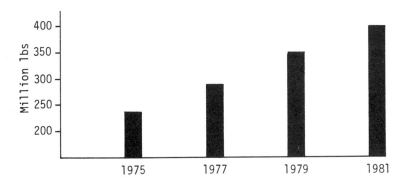

Fig. 3. Hot-melt adhesive usage.

VI. WHERE HOT-MELT ADHESIVES ARE USED

The following list shows some of the application areas that use hot-
melt adhesives. A few examples are illustrated.

Packaging.

Case sealing	Cartons
Bags (Fig. 4)	Composite cans
Trays	Corrugated boards
Labels	Fiber drums
Tapes	Pouches
	Three-piece beverage cans (Fig. 5)

Bookbinding.

Paperbacks (Fig. 6).	Magazines
Catalogs	Directories
Hardcover books	Calendars

Textiles.

Interliners	Aprons
Hemming	Curtains
Waistbands	Towels
Labels	Patches
Cuffing	Basting
Pillows	

Disposables.

Diapers	Sanitary items
Hospital pads	Hospital garments
	Surgical pads

Furniture.

Edge banding	Door panels
Cabinet assembly	Stereo cabinets
Overlay lamination	Window frames (Fig. 7)

Footwear.

Lasting	Counters
Sole attaching	Folding
Box toes	

Automotive and related industries.

Door panels	Trunk and window seals (Fig. 8)
Hand windshield washer	Carpet installation
Tail-light assemblies	Outboard-motor housing (Fig. 9)
Electronic controls	Fire-wall insulation
Vinyl roofs	Headliners
	Seating

Product Assembly.

Vinyl-clad window frames
Envelope batteries (Fig. 10)
Thermal- and noise-insulation pads
Mobile-home components

Mirror assembly
Utensils (Fig. 11)
Electronic assemblies
Flexible ducting (Fig. 12)

Appliances.

Air conditioner components
Coffee pots
Tape-cassette holder

Refrigerator components
Television components (Fig. 13)
Calculator display windows
Slowcook-pot heaters (Fig. 14)

Carpets.

Backing

Seaming tapes (Fig. 15)

Tapes [Hot-melt pressure-sensitive adhesives (HMPSA)].

Electrical
Binding

Sealing
Nylon- or glass-reinforced

Filters.

Oil—automotive
Air—automotive and residential

Exhaust-gas scrubber

Fig. 4. Hot-melt adhesive applied to multiwall paper bags used to export corn-soya-milk products. After being filled with product, the top of the bag is folded over and sealed in a heated clamping device. These bags must withstand export stress and resist mildew and insect infestation.

Fig. 5. Three-piece metal beverage can with the side seam bonded with a high-performance, hot-melt adhesive. Using the hot melt, production rates approach 10 cans per second.

Fig. 6. Paperback book. The hot melt is used to bond the pages and cover to give what is known as "perfect binding".

Fig. 7. A section of window frame, consisting of a vinyl extrusion over wood, fastened to a wood nailing strip. The hot melt bonds the plastic-covered member to the frame.

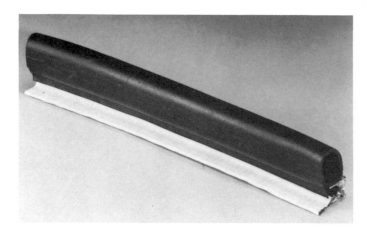

Fig. 8. A section of a trunk or door-seal molding for automotive use. The hollow elastomeric seal is bonded with a hot melt that withstands the severe environment encountered by automobiles. Preapplication of the sealing gasket to the molding facilitates production.

Fig. 9. Plastic cowl for outboard motor. The hot melt is used to bond the urethane-foam sound insulation to the inside of the cowl.

Fig. 10. Envelope-type battery is similar to those used to power special cameras. The battery chemicals are encased in multi-layer laminates which are sealed with hot-melt adhesive. Obviously, leakage of the chemicals cannot be tolerated.

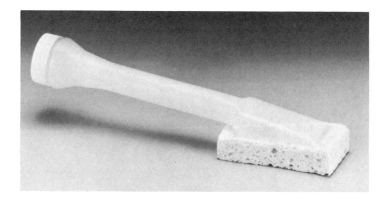

Fig. 11. A kitchen utensil in which a cellulose sponge is bonded to a polyolefin soap-dispenser handle.

Fig. 12. Flexible ducting. An 8-in. diameter, flexible duct made by covering a wire coil with a spiral wound, foil-film laminate. These ducts (of various sizes) are used to carry heated or cooled air throughout buildings. The hot-melt adhesive bonds the laminate to itself, encapsulating the wire coil.

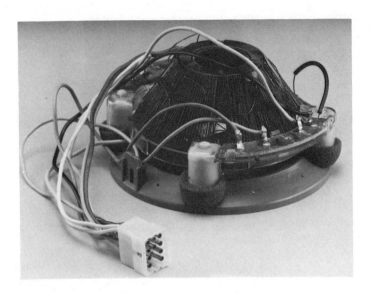

Fig. 13. The focusing device which controls the picture of a TV tube.
The hot melt is used to lock the adjustment and bond the focusing device to the picture tube. Even under the severe temperature conditions which exist during operation no slippage can be tolerated.

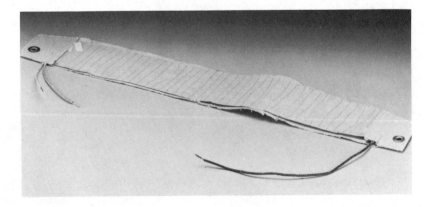

Fig. 14. A heating element for a slowcook-pot heater. The hot melt
is used to bond the position and lead-in wires to the strip heater.

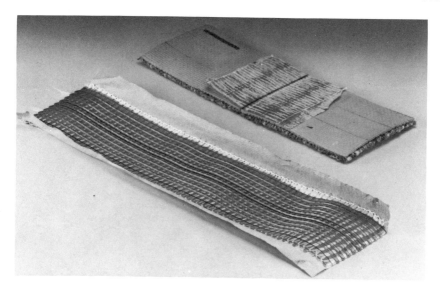

Fig. 15. Carpet seaming tape used to join two sections of carpet. The reinforced tape, with the hot melt side up, is placed under the edges of the carpets to be joined. Then a heated tool is drawn over the tape, melting the adhesive, and the carpet is pressed down immediately behind the tool to give an invisible seam.

VII. SUMMARY OF ADHESIVES BY BASE POLYMER OR USE

Tables 3—7 list a few adhesives by polymer type, illustrating specific examples of supplier, properties, features, and application areas [6].

VIII. WHAT TO DO WHEN PROBLEMS OCCUR WHILE USING HOT-MELT ADHESIVES

Note the wording above is when, not if. Any hot-melt adhesive system used in the commercial manufacture of products will experience problems at times. It is useful to have some guidelines to help locate the trouble and correct it. Hot melts are often used at critical points in production lines, thus difficulties with adhesives exert a leverage effect on overall production costs. These problems can fall into three general categories: poor adhesion, inadequate or uneven volume, and adhesive stability.

Table 8 lists problems, possible causes, and suggested solutions associated with hot-melt adhesive applications. If these suggested solutions do not correct the situation, contact either the equipment or adhesive supplier or both [1,7].

Table 3 Typical Properties and Recommended Applications of Various Hot-Melt Adhesives Based on Ethylene–Vinyl Acetates

Adhesive and manufacturer	Application temperature	Softening point, ball and ring	Viscosity at application temperatures (cP)	Substrates	Features	Applications
Bostik HM899 Bostik Corp.	300°F	—	15– 20M[a]	Metal Plastic Wood	Not for polyvinyl chloride (PVC)	Automotive Electronic
Bostik 9951 Bostik Corp.	375°F	—	16– 19M	Metal Plastic	Quick set	Automotive Electronic
Peter Cooper HW 1320 Peter Cooper Corp.	> 385°F	—	8M	Wood	Fiber tear at 15°F	Edgebanding Furniture
Macromelt 1800 Henkel Adhesives Co.	330– 275°F	186°F	7M	Paper Metal Plastic	Adhesion Hot tack	Bookbinding Product assembly
Macromelt 1966 Henkel Adhesives Co.	330– 360°F	176°F	1100	Aluminum foil Paper	Low viscosity Hot tack	Packaging Product assembly

Product / Company	Application temperature		M[a]	Substrates	Properties	Applications
Amsco Melt 318 Union Oil of CA	330–375°F	190°F	5–30M	Plastic Paper	Adhesion to PE[b] snf PET[c]	PET bottle assembly Packaging
Amsco Melt 293 Union Oil of CA	225–350°F	160°F	600–2600	Textile	Bonds PE films and textiles	Textile Upholstery
Amsco Melt 230 Union Oil of CA	325–375°F	225°F	6M	Paper	Very flexible	Bookbinding
Compo 1200A Compo Industries	350–430°F	165°F	25–35M	Leather Elastomer Metal	Very flex-ible	Apparel Footwear
Tufmelt 3400 Goodyear Corp.	350°F	210°F	9M	Wood Metal Elastomer	Water re-sistance	General industrial
GAC 64X137 General Adhesives & Chemicals Co.	350–375°F	200°F	9M	Wood Metal Plastic	Superior adhesion	Woodworking General industrial

[a]M = thousands.
[b]PE = polyethylene.
[c]PET = polyethyleneterephthalate.

Table 4 Typical Properties and Recommended Applications of Various Hot-Melt Adhesives Based on Polyolefins

Adhesive and manufacturer	Application temperature	Softening point, ball and ring	Viscosity at application temperatures (cP)	Substrates	Features	Applications
Bostik 19-GA-176 Bostik Corp.	355°F	—	29M^a	Plastic Metal Wood	Not for PVC	Automotive Electronic
Thermogrip 6367 Bostik Corp.	350–450°F	> 210°F	7M	Paper Wood	Bonds porous materials	General industrial
Thermogrip 6368 Bostik Corp.	350–450°F	> 210°F	7M	Paper	Bonds porous materials	Toys and novelties
Glu-Beads 77-S Burtonite Corp.	130°F	—	—	Textile Wood Paper	Bonds porous materials	Upholstery Packaging Binding
Eastobond A-3 Eastman	325–350°F	210°F	2.2M	Paper Composites	Low cost	Packaging binding
Eastobond A-381 Eastman	400–425°F	310°F	7.5M	Plastic Metal	Bonds to Polypropylene	Auto filters and batteries
Betastay 52-510 Essex Chem.	325–375°F	300°F	1.5M	Paper	Extended tack time	Laminant for kraft tops
Igetabond A Sumitomo Chem.	205–500°F	175°F	—	Metal Textile Plastics	Resists oil, acids, and solvents	Laminates Composites
Numel 600 Gulf Adhesives	300–350°F	205°F	1.5M	Plastic Wood Composites	Bonds polyolefins	Product assembly Packaging

Table 5 Typical Properties and Recommended Applications of Various Hot-Melt Adhesives Based on Polyamides

Adhesive and manufacturer	Application temperature	Softening point, ball and ring	Viscosity at application temperatures (cP)	Substrates	Features	Applications
Bostik HM 765 Bostik Corp.	355°F	—	2.5M[a]	Metal Plastic	High perf. Not for PVC	Automotive Electronic
Bostik 7232 Bostik Corp.	245—275°F	—	25—40M	Textile	Resists washing and dry cleaning	Textile industry
Platamid H103 Rilsan Corp.	194—212 F	—	88M	Textile Leather Plastic	Readily pigmented	Heat-sensitive substrates
Macromelt 6903 Henkel Adhesives Co.	450 F	275 F	10M	Textile	Flexible, resists washing and dry cleaning	Textile
Macromelt 6239 Henkel Adhesives Co.	> 475°F	285°F	6.5M	Metal Glass Plastic	Adhesion to plasticized vinyl	Automotive Electronic General industrial

343

Table 5 (Continued)

Adhesive and manufacturer	Application temperature	Softening point, ball and ring	Viscosity at application temperatures (cP)	Substrates	Features	Applications
Macromelt 6071 Henkel Adhesives Co.	> 250°F	200°F	.7M	Wood Metal Glass	Resists oil, acids, and solvents	Insulation General industrial
Macromelt 6902 Henkel Adhesives Co.	> 500°F	392°F	60M	Metal Ceramic	Resists acids and solvents	Can making seams Metal bonding
Grill-Tex 1 Emser Werke AG	230—250°F	—	8M	Textile	Fusible interlinings	Textile
Compo 4457 Compo Industries	300—375°F	—	5M	Leather Wood Plastic	Flexible Moderate setting time	Apparel and leather
Compo 4455 Compo Industries	350°F	315°F	70M	Leather Wood Plastic	Fast set	Footwear
Terlan 1531 Terrell Corp.	> 325°F	240°F	4.5M	Elastomer Plastic	Tough Flexible	Chloroprene rubber

[a]M = thousands

Table 6 Typical Properties and Recommended Applications of Various Hot-Melt Adhesives Based on Polyesters

Adhesive and manufacturer	Application temperature	Softening point, ball and ring	Viscosity at application temperatures (cP)	Substrates	Features	Applications
Bostik 7996 Bostik Corp.	—	—	—	Wood Textile	Resistance to wash cycles	Textile Upholstery Furniture
Thermogrip 4175 Bostik Corp.	425—475°F	375°F	20M[a]	Leather Plastic Textile	Solvent resistance	Automotive Electronic
Terlan 6000 Terrell Corp.	450—475°F	230°F	30—40M	Plastic Wood Metal	Thermal stability	Plastic-elastomer General industrial
Pibiter A-18F Montedison	410—465°F	345°F	23.5—31.5M	Leather Plastic Wood	Flexibility Chemical resistance	Footwear Wood Printing
Pibiter A-200 Montedison	410—430°F	390°F	23—31M	Leather Plastic	Rigid	Apparel Leather
Vitel VAR5125C Goodyear Corp.	410°F	240°F	13M	Metal Plastic Wood	Flexible	Preheated metal General industrial

Table 6 (Continued)

Adhesive and manufacturer	Application temperature	Softening point, ball and ring	Viscosity at application temperatures (cP)	Substrates	Features	Applications
Flexclad 5825 Goodyear Corp.	430°F	320°F	40M	Plastic Elastomer Textile	Tough Flexible	Electronic Textile
Grilesta AH-00-5.60 Emser Werke AG	320—375°F	—	140M	Metal Paper Wood	Partly crystal-line	Automotive Electronic
Grilesta 2718 Emser Werke AG	250—320°F	—	—	Wood Paper Metal	Amorphous	General industrial Automotive
Compo 4459 Compo Industries	380—460°F	240°F	200—300M	Metal Wood Plastic	Flexible Slow set	Footwear
Compo 4475 Compo Industries	450—500°F	400°F	4—8M	Leather Plastic	Fast set	Footwear

[a]M = thousands

Table 7 Typical Properties and Recommended Applications of Various Pressure-Sensitive Hot-Melt Adhesives

Adhesive and manufacturer	Application temperature	Viscosity at application temperatures (cP)	Substrates	Features	Applications
GAC 64X275 General Adhesives & Chemical	300–350°F	2.5M^a	Plastic Metal	Bonds to polyethylene	General industrial
Tufmelt 3001 Goodyear Corp.	400°F	47M	Textile	Good retack	Textile Upholstery
Tufmelt 3003 Goodyear Corp.	315–335°F	12M	Plastic Paper	Low temp. flexibility	Packaging Binding
Gulf XH-100 Gulf Adhesives	350°F	8M	Metal Plastic	Quick melting	Stainless steel PE labels
Nicomelt P-1549 Malcom Nicol Co.	325—350°F	10M	Paper Plastic	Bleed resistance	Packaging Binding
Amsco Melt 323 Union Oil of CA	275—325°F	15M	Paper Plastic	Removable labels	Tapes on plastics
Amsco Melt 339 Union Oil of CA	275—325°F	15M	Plastic	Plasticizer resistance	Floor and wall coverings
Amsco Melt 345 Union Oil of CA	275—325°F	20M	Paper	High shear	Tapes for corrugated
Williamson 3625 Williamson Adhesives	350—400°F	100M	Plastic Paper Textile	Automotive Floor covers	Latex-backed carpets
Williamson 8051 Williamson Adhesives	325—375°F	19M	Paper	Permanent applications	General industrial

^{a}M = thousands

Table 8 Troubleshooting Guide

Problem	Possible causes	Suggested solutions
a Bond failure— adhesive on one side only, ad-hesive surface glossy.	Inadequate wetout of secondary surface. Not enough open time.	1. Apply more adhesive. 2. Apply adhesive at higher temperature. 3. Preheat substrate. 4. Shorten time between adhesive application and compression. 5. Try a slower-setting adhesive. 6. Treat substrate sur-face.
b Bond failure— adhesive on both surfaces with leggy strings.	Movement of the joint while adhesive is solidifying.	1. Apply adhesive at lower temperature 2. Increase compression time. 3. Try a faster-setting adhesive.
c Bond failure— adhesive mostly on one side.	Condition of the sub-strate surface or type of substrate.	1. Treat substrate surface. 2. Apply adhesive at higher temperature. 3. Try another adhesive.
d Bond failure— insufficient ad-hesive in bond area.	Absorption of adhe-sive by porous sub-strate. Adhesive squeeze out.	1. Apply adhesive at lower temperature. 2. Apply more adhesive. 3. Use high-viscosity adhesive.
e Bond failure— bond opening when compres-sion is released.	Adhesive not cooling fast enough. Too much adhesive. Substrate too hot. Improper adhesive.	1. Cool substrate and/or adhesive. 2. Decrease amount of adhesive. 3. Increase compression time.

Table 8 (Continued)

Problem	Possible causes	Suggested solutions
		4. Change to faster-setting adhesive.
f Charring or gelling of adhesive in application equipment	Adhesive temperature too high.	1. Lower adhesive temperature.
	Faulty heat control.	2. Check temperature indicator with another thermometer.
	Adhesive heat stability inadequate.	3. Try a more heat-stable adhesive or clean out applicator more often.
		4. Use inert gas blanket.
g Stringing of adhesive at applicator tip.	Adhesive temperature too low.	1. Raise adhesive temperature.
	Adhesive viscosity too high.	*Caution*: Do not exceed recommended application temperature.
	Substrate too far from applicator tip.	2. Position substrate closer to applicator tip.
h Adhesive dripping from applicator tip.	Worn applicator valve.	1. Replace valve.
	Dirty applicator tip.	2. Clean applicator tip.
i Bubbles in adhesive.	Reservoir out of adhesive.	1. Refill reservoir.
	Moisture in adhesive.	2. Check adhesive for moisture.
		3. Try a supply of fresh, dry adhesive.
		4. Consult with the adhesive supplier.
j Excessive adhesive squeeze out.	Too much adhesive.	1. Use less adhesive.
	Adhesive too hot.	2. Lower adhesive temperature.
	Compression too high.	3. Decrease compression.
		4. Use high-viscosity adhesive.

IX. SAFETY SUGGESTIONS FOR USING HOT-MELT ADHESIVES

Working with hot-melt adhesives should not be much more hazardous than cooking the evening meal. However, because of their characteristic rapid solidification at relatively high temperature, they present a potential hazard. Even after solidifying, they are still hot. Therefore, care should be used when working with hot-melt adhesives in the molten state. Severe burns can occur if the molten adhesive comes in contact with the skin. When handling molten adhesive or working near a machine containing molten adhesive under pressure, the use of eye protection and protective clothing is strongly recommended [1].

Should molten adhesive come in contact with the skin, the following actions are recommended.

1. Do *not* try to remove the *molten adhesive* from the skin.
2. Immediately immerse the affected area in water, preferably clean and cold. A supply of water should be readily available at all times near the adhesive equipment. Keep the affected area immersed until the adhesive has cooled.
3. Do *not* try to remove the *cooled adhesive* from the skin.
4. Cover the affected area with a clean, wet compress.
5. In cases of severe burns, be alert for signs of shock. If observed, have patient lie down and use blankets to preserve body heat. Call a physician immediately.

X. HOT-MELT ADHESIVES—FORMS AND SHAPES

Since hot melts are applied in a variety of methods from many different types of application equipment, the adhesive suppliers provide a wide variety of forms. These include

1. Powders (various mesh sizes)
2. Pellets (chopped, strand, or beads)
3. Films (various thicknesses)
4. Webs (open mesh "films")
5. Cylinders (tubes, candles, slugs of various sizes)
6. Crushed (irregular chunks of various sizes)
7. Slats (flat pieces of various sizes)
8. Trays (release-coat—lined boxes of various sizes)
9. Metal pails (filled molten, then solidified)
10. Drums (filled molten, then solidified)

The application equipment for hot melts will be discussed in Chap. 20.

XI. HOT-MELT ADHESIVES—ANTICIPATED FUTURE DEVELOPMENTS

As we move into the decade of the 1980s, we can anticipate at least three areas for future development. Serendipity will probably bring forth others that cannot be anticipated on the basis of today's knowledge.

A. Thermoplastic-Thermoset

A step in the direction of thermoplastic-thermoset, hot-melt adhesives is the radiation crosslinking of HMPSA tapes [8].

Another route being considered is combining two flows of hot-melt material which can react with each other. This concept may also include melting-fusing of an adhesive rope woven from two or more different polymer strands that have some chemical reactivity with each other.

The expected advantage would be improved resistance to elevated temperatures and chemical attack.

B. Foamable Hot Melts

Putting molten, hot-melt adhesives under pressure and adding an inert gas causes a foaming action as the material is dispensed [9].

Some of the expected advantages are

1. Control of glueline deposition
2. Extended "open time"
3. Void filling for poorly mated parts
4. Reduced adhesive consumption.

C. Exotic Polymers

Aerospace and related programs are developing exotic, high performance polymers. Some may be adaptable for use as adhesives [3,4].

REFERENCES

1. R. D. Dexheimer and L. R. Vertnik, The why, how, what, when and where of hot melts, in *Adhesives Ind. Conf. Proc.*, Tech. Conf. Assoc., El Segundo, 1974, pp. 63–96.
2. R. D. Dexheimer and L. R. Vertnik, in *Handbook of Adhesives*, 2d ed. (I. Skeist, ed.), New York, Reinhold, 1977, pp. 581–591.
3. S. R. Eddy, *Adhesives Age*:18 (1980).
4. R. Hinterwaldner, *Adhesives Age*:34 (1978).
5. *Adhesives in Modern Manufacturing* (D. Bruno, ed.), SME, Dearborn, 1970, pp. 27–35.

6. *Adhesives 1979—80 Desk Top Data Bank*. Cordura, LaJolla,
    p. 197.
7. I. Kay, *Adhesives Age*:27 (1978).
8. D. J. St. Clair, *Adhesives Age*:30 (1980).
9. W. H. Cobbs, *Adhesives Age*:31 (1979).

14
Pressure-Sensitive Adhesives

J. W. Hagan and Kenneth C. Stueben   *Union Carbide Corporation, Bound Brook, New Jersey*

I. INTRODUCTION

The term "pressure-sensitive adhesive" (PSA) refers to that type of adhesive which, when in a dry state, will adhere to a variety of surfaces merely by application of light hand pressure. Such compositions are inherently soft, permanently tacky materials which exhibit a balance of adhesive and cohesive strength depending on the viscoelastic nature of the adhesive and the performance requirements of the particular end use. Pressure-sensitive adhesives were first developed in the mid-1800s for use on surgical tapes. Their use has since expanded to include masking tapes for spray painting of automobiles, as well as a myriad of other uses in the consumer-tape and graphic-arts fields. Today PSAs find major applications in tapes and labels. Other uses include wall covering, floor tiles, and protective maskings.

II. THEORY

From a performance point of view, the properties most important to the formation of a satisfactory PSA bond are surface tack, peel adhesion, and shear resistance.

A. Surface Tack

Surface tack, commonly known as quick stick, is defined as that property of a PSA which enables it to spontaneously adhere to a surface under very light pressure and to resist removal almost instantly. Softer polymers generally exhibit a higher degree of tack, but at the same time lower cohesive shear strength, than harder polymers. Tack is also a function of the thickness of the adhesive film with higher film thickness contributing to higher levels of tack. Bates [1], in a study on the nature of adhesive tack, devised a method for measuring tack by determining the energy dissipated during debonding (at constant strain rate) of a substrate from a PSA. Several other techniques have also been developed to measure the tack of PSAs and are widely used in industry today. These will be discussed in more detail in Sec. IX.

B. Peel Adhesion

Peel adhesion is determined by measuring the force required to remove a pressure-sensitive material (tape or label) by peeling at a constant rate, usually at an angle of 180° to the substrate. In addition to the viscoelastic properties of the adhesive, other factors affecting peel adhesion are polarity of the adhesive, thickness of the bond, temperature, the length of time that the adhesive has been in contact with the surface, rate of removal, and the nature of both the substrate and the backing material.

C. Shear Resistance

Shear resistance is the force required to remove a pressure-sensitive item (tape, label, etc.) from a surface in a direction parallel to the surface to which it is affixed. It is a measure of the cohesive strength of the adhesive. Shear resistance can be enhanced by increasing the molecular weight or the crosslink density of the polymer. Either measure will simultaneously result in some sacrifice in tack and peel adhesion. Aubrey [2] maintains that it is necessary to increase the concentration of network or gel so that the adhesive can sustain high shearing forces. Gel concentrations of 30−50% are usually adequate for high-shear PSAs.

D. The Influence of Polymer Structure on Performance Properties

As a first approximation for product design, the tack, adhesive, and cohesive characteristics of a pressure-sensitive polymer can be correlated with three basic properties in addition to the crosslink density, namely, molecular weight, glass transition temperature T_g, and polarity [3]. (See also Chap. 3.)

1. Molecular weight

Molecular weight and molecular-weight distribution will influence polymer viscosity. The presence of lower molecular weight species in an adhesive contributes to lower viscosity. The neat-polymer viscosity at bonding temperatures is a prime factor in how well the adhesive wets a substrate and forms a bond. When the viscosity is lower, the mobility of the polymer molecules is greater; hence, wetting is improved. However, as the molecular weight is decreased, so is the cohesive strength of the bond. Consequently, polymer viscosity must be balanced with the desired physical and mechanical properties.

The influence of the molecular weight on the performance of a PSA is shown by the idealized representation in Fig. 1. Starting with a low molecular weight polymer, an increase in molecular weight re-

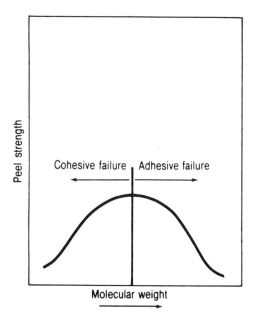

Fig. 1. Idealized representation of relationship between average molecular weight and peel strength of a PSA.

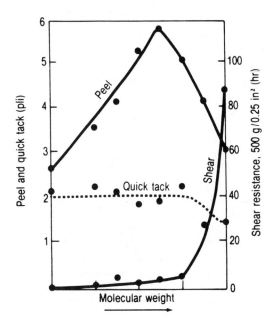

Fig. 2. Effect of molecular weight on PSA properties.

sults in higher peel values due to an increase in cohesive strength.
Wetting of the substrate remains excellent in this range. As the molecular weight is further increased, a decrease in peel strength is observed, and failure, on a macroscopic scale, takes place at the interface between the adhesive and the substrate.

This relationship is borne out by the results of a series of experiments with an acrylic latex adhesive shown in Fig. 2. The molecular weight of the polymer was varied with a chain transfer agent
while keeping the polymer composition constant. Increasing the molecular weight enhanced the cohesive (shear) strength of the polymer but, above a point, adversely affected the adhesive (peel and
tack) properties.

To improve the cohesive strength of PSAs without adversely affecting peel, light crosslinking is often advantageously employed.
The beneficial effects of incorporating very low levels of multifunctional monomers into a pressure-sensitive polymer system are shown
in Fig. 3. In this case, the shear resistance of the polymer increased
markedly, while the peel strength was relatively unaffected until
higher levels of the multifunctional monomer were employed.

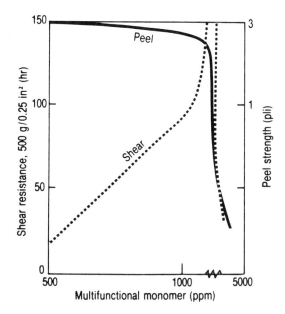

Fig. 3. Effect of multifunctional monomer on pressure-sensitive polymer.

2. Glass transition

The effect of T_g on the performance of acrylic PSAs is illustrated in Fig. 4. The bonding temperatures of PSAs are above the T_gs for the systems. The greater the difference between the use temperature and the T_g, the lower the viscosity and the better the wetting will be. However, an optimum T_g must be sought in order to provide a balance between polymer mobility and cohesive properties. Figure 4 displays the divergent influence of acrylic-polymer T_g on adhesive and cohesive characteristics.

In synthesizing a polymer for adhesive applications, special attention must be given to T_g and molecular weight and their interplay with the mobility of the polymer molecules. Glass transition and molecular weight must be considered simultaneously when correlating the adhesive performance of polymers with physicochemical properties.

3. Polar effects

The polarity of an adhesive affects its adhesion to specific substrates [2]. In general, polar adhesives have a tendency to adhere to polar (high-surface-energy) substrates by virtue of dipole-dipole interac-

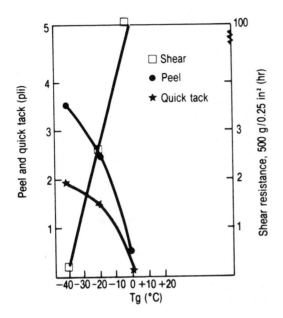

Fig. 4. Effect of T_g of polymers on performance of adhesives.

tions and/or hydrogen bonding. Compositions with lower polarity tend
to adhere to lower-surface-energy substrates [4].

III. MARKET AND TRENDS

A. Introduction

The volume of PSAs consumed annually in the United States is estimated at greater than 274 million dry pounds and valued at $150 million
[5]. These adhesives consist of a combination of elastomers, tackifying resins, plasticizers, and fillers. Pressure-sensitive adhesives
are usually identified by the chemical nature of the elastomer [e.g.,
natural rubber, styrene-butadiene rubber (SBR), acrylic, etc.].
Natural rubber was the first elastomer used in PSAs and is the most
commonly used today, followed closely by acrylates. A recent market
study of the PSA industry by Frost and Sullivan [5] showed the following breakdown by chemical type of PSAs in the United States.

Type of elastomer	Percent of use
Natural rubber	46
Acrylates	34
SBR	10
Block copolymers	6
Others	4

Natural rubber is the most widely used because of its cost/performance relationship. Most PSAs in use today are applied from organic-solvent solutions. However, because of rapidly rising solvent prices, soaring energy costs, and increasing legislation aimed at reducing environmental pollution, the industry is developing alternatives to solvent-based systems [6]. Emulsion technology has progressed to the point where it is finding use in such diverse applications as decorative polyvinyl chloride (PVC), paper labels, polypropylene packaging tape, and many other products [7]. Hot-melt pressure-sensitive adhesives (HMPSAs), which were introduced during the 1970s, are now being used successfully by a variety of industries in, for example, floor tiles, labels, and tapes [8]. Recent estimates indicate that both of these technologies will grow at the expense of solvent-based pressure-sensitive adhesives and by 1985, consumption of water-based and hot-melt PSAs will be three times that of solvent systems (Table 1).

Other technologies, such as the use of 100% reactive, radiation-cured PSAs, are emerging but do not represent a significant portion of the market today.

B. End Uses

The two major markets for PSAs are the tape and the label industries. There is also a variety of other products which utilize PSAs. These include

Wall and shelf covering
Imitation wood grain
Ceiling and floor tiles
Disposable diaper tabs
Medical and sanitary products
Graphic artwork
Protective maskings

1. Tapes

Pressure-sensitive tapes, first introduced for surgical applications, are now finding wide use in both consumer and industrial fields. A variety of backing materials are used, including paper, fabric, plas-

Table 1   Pressure-Sensitive Adhesive Trends

	Percent of total U.S. consumption		
Technology	1974	1980	1985
Solvent-based	95	70	25
Waterborne	3	18	45
Hot-melt	2	12	30

Source: Data from Ref. 9.

tic films (e.g., PVC, cellophane, matte acetate, polyester, polyolefin), and aluminum foil.

A pressure-sensitive tape consists of backing material to which a PSA has been applied on one side. A thin primer is sometimes required in order to anchor the adhesive to the backing material. The primer is applied to ensure that the adhesive layer will remain in place on the backing when the roll is unwound and later when the tape is removed from the surface to which it has been applied. Nitrile rubber, chlorinated rubber, and acrylates are currently used as primers for rubber-based PSAs. Acrylic PSAs do not generally require a primer coat.

In order to facilitate unwinding without splitting the adhesive, a release coating is often applied to the back of the web. The release coat is applied at a very low coat weight (approximately 1-5 g/m^2). A controlled degree of release is required, and several release materials are used—including acrylic acid esters of long-chain fatty alcohols; polyurethanes incorporating long aliphatic chains; and cellulose esters. The release layer should be a coating of relatively low surface free energy [10]. Some plastic films, such as biaxially oriented polypropylene, do not require the use of a primer or a release coat. Adhesion to one side is enhanced by corona treatment to oxidize the polyolefin surface.

Some pressure-sensitive tapes, such as metal-foil tapes for heating ducts, will utilize a separate release liner, usually silicone coated, which is laminated to the adhesive and removed immediately before use. There are also double-sided tapes which have both sides of the backing material coated with a PSA. The tape is rolled with a separate, silicone-coated release liner. These tapes are useful for mounting and splicing.

Paper-backed pressure-sensitive tapes are used for a variety of applications, the most important being masking tape. Masking tape generally utilizes a crepe-paper backing which has been impregnated with a polymeric binder to improve the strength of the paper. Since a major use for masking tape is to protect surfaces during certain paint-

Table 2 Typical Adhesive Performance Requirements

Product	Peel adhesion (pli)	Quick tack (pli)	shear (hr)
General purpose masking tape	2.3	1.1	> 167[a]
Glass-reinforced polyester tape	3.2	1.8	> 500[a]
Aluminum-foil duct tape	> 3.8	1.8	> 6[b]
High-performance polyester strapping tape	2.3	1.1	> 200[c]
Polypropylene carton sealing tape	2.2	1.6	> 80[a]

[a]1 in.^{2}/1000 g.
[b]1 in.^{2}/4540 g.
[c]$1/4$ in.^{2}/500 g.
Source: Data from Ref. 9.

ing operations, the adhesive must resist attack by the organic solvents used in paints and be able to withstand the high baking temperatures used in automobile paint bake ovens.

Plastic films are also widely used in pressure-sensitive tapes for a multitude of applications. Cellophane, one of the original products used in clear pressure-sensitive tapes, is being displaced by cellulose acetate for office and household mending applications. Plasticized PVC has replaced a significant portion of the cloth "friction" tape in electric applications and is used extensively in pipe wrapping and certain medical applications. Recently, biaxially oriented polypropylene film with acrylic latex PSAs has been making significant inroads into the packaging-tape market. It is used for sealing corrugated boxboard containers.

Fiber-reinforced tapes in which fiberglass, polyester, or nylon filaments have been imbedded in the PSA to increase overall tensile strength and tear resistance are being used for high-performance packaging, palletizing, and reinforcing applications.

The tape industry requires a great deal of versatility in PSAs in order to satisfy the broad range of properties required for various products. Table 2 shows typical peel, shear, and quick-tack values required for various pressure-sensitive tapes.

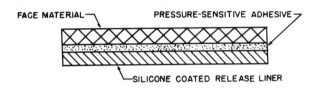

FACE MATERIAL — PRESSURE-SENSITIVE ADHESIVE —

— SILICONE COATED RELEASE LINER

Fig. 5. Pressure-sensitive label stock.

2. Labels

Pressure-sensitive adhesives are gaining increasingly wide acceptance by the label industry at the expense of wet adhesives, pregummed remoistenable adhesives, and heat-activated adhesives. Despite their higher cost, pressure-sensitive, or self-adhesive labels offer a number of advantages over conventional labels. These advantages include cleanliness of operation, speed of label application, and simplicity of application equipment, as well as versatility and ease of application to irregularly shaped surfaces and containers [11]. The growth rate of PSAs in labels is nearly double that of the overall label market.

Pressure-sensitive labels and decals are constructed of three components (face material, PSA, and release liner) as illustrated in Fig. 5.

In the manufacture of label stock the PSA is generally applied to a silicone-coated release liner and dried. In the case of hot-melt adhesives, no drying is required. The coated release liner is then laminated under light pressure to the face material. This process is commonly referred to as "transfer coating". The adhesive will transfer to the face material when the release liner is removed. The label stock is then either rolled or stacked in sheets for subsequent printing and die cutting into individual labels.

Paper, typically a 60-lb coated kraft, is most commonly used as the face material. However, other materials such as plasticized PVC, metal foils, polyesters, and fabric, are also used for specialty labels.

Labels may be classified according to their intended use (i.e., permanent, removable, or repositionable, or low-temperature). Each will require an adhesive with different performance properties. Permanent labels exhibit a sufficiently high degree of adhesion that the label will be damaged upon removal from the surface. A removable or repositionable label requires an adhesive with a balance between adhesive and cohesive strength to allow it to be removed cleanly without destroying the label or leaving a residue of the adhesive on the surface. Low-temperature labels are used primarily for pricing of frozen-food packages and, therefore, require an adhesive that will retain a high degree of tack at extremely low temperatures.

An estimate of the distribution according to the various types is as follows [5]:

Type of label	Percent of total label volume
Permanent	80
Low-temperature	10
Removable	5
Miscellaneous	5

At present, the majority of label-stock manufacturers are using solvent-based PSAs. However, ecological and economic considerations have resulted in a trend toward aqueous and hot-melt systems.

3. Other uses

Pressure-sensitive adhesives find application in a wide variety of other end uses, each with its own specific set of performance criteria. Pressure-sensitive films and papers are used to temporarily protect surfaces from scratches, corrosion, or other forms of damage. For example, Lucite or Plexiglass sheeting is usually covered with a protective masking to prevent scratching until it is used by the consumer. Highly polished or precision-machined metals are often protected with a pressure-sensitive masking during shipping and storage. The adhesive must exhibit a low to moderate degree of adhesion to the surface to allow the masking to be easily removed without leaving a residue or stain on the surface.

Shelf and wall coverings frequently use a PSA on a plasticized PVC backing. The adhesive must have a high degree of adhesion initially and not exhibit a deterioration of the adhesive bond on aging due to migration of the plasticizer from the PVC into the adhesive.

Sanitary and medical products also utilize PSAs. These products include such things as disposable diaper tabs, bandages, and hospital supplies. The adhesives must exhibit a degree of moisture resistance and must not cause skin irritation after long periods of contact.

IV. SOLVENT-BASED PRESSURE-SENSITIVE ADHESIVES

The manufacture of PSA articles by deposition from solvents has a relatively long history. As a result, solvent-based formulations have been developed to such a highly advanced state that their broad range of performance and formulating latitude has generated widespread market acceptance. They represent the largest single market for solvent-based adhesives.

Some general characteristics of solvent-based adhesive systems are summarized in Table 3. Typically, these adhesives consist of

Table 3 Characteristics of Solvent PSAs

Property	Rubber	Acrylic
Adhesive cost	Low	High
Formulating latitude	Excellent	Excellent
Chemical resistance	Good	High
Quick stick	High	Good
Fire hazard	High	High
Aging	Good	Excellent
UV resistance	Poor	Excellent
Clarity	Poor	Excellent
Energy for application	High	High
Space requirement	High	High
Cleanup/changeover	Good	Good

Source: Ref. 12.

various elastomers in conjunction with tackifiers, plasticizers, fillers, and antioxidants [13]. Common elastomers include natural, reclaimed, and styrene-butadiene rubbers, polyisobutylene or butyl rubber; polyvinyl alkyl ethers; acrylics; or thermoplastic elastomers [styrene-isoprene-styrene (S-I-S) block copolymers]. It should be pointed out, however, that acrylics generally do not require the use of tackifiers or antioxidants.

The thermoplastic elastomers appear to have advantages over natural rubber in that they can be used at higher solids levels and tend to be lower in price. These factors, coupled with higher solvent prices, are working to increase the presently small share of solvent-based adhesives held by these elastomers. Acrylic systems tend to require less formulating than the rubber systems and have little need for tackifiers. The bulk of the PSA market (over 3/4) is satisfied by rubbers with most of the remainder using acrylics. Less commonly, products based on polyurethanes, polyesters, and polysiloxanes may be used. The siloxanes, in particular, have an exceptional range of high- and low-temperature properties [14,15].

In order to provide high-temperature properties, most polymers need to be crosslinked either by thermal or chemical means. It is imperative to control the extent of crosslinking carefully since an insufficient amount may result in shear failure at elevated temperatures, while excessive amounts will lead to a loss of peel and tack. Examples of recipes for both rubber- and acrylic-based PSAs are shown in Tables 4 and 5 [12,15].

Among the solvents most commonly used in formulating are aliphatic hydrocarbons (e.g., hexane, heptane) and toluene, for rubber systems, and esters or ketones for acrylic systems. A thorough dis-

Table 4 Rubber-Based PSA for High-
Temperature Masking Tape

Formulation	Parts
Crude rubber	50
GRS	50
Octyl phenol formaldehyde heat-curing resin	20
ZnO	60
Diethylene glycol ester of dihydroabietic acid	110
Antioxidants	2

Source: Based in part on Ref. 6.

cussion of the technical aspects of solvent-polymer interactions in adhesives recently appeared [10].

In recent years, the use of solvents has come under a barrage of
legislative pressure [17] concerned with environmental pollution.
Other negative pressures bearing on this approach are the high energy costs required for evaporation, concern over flammability and
explosion hazards, and the consumption of sometimes scarce and costly solvents. The prices of many solvents have escalated more than
fourfold in the past few years. These factors have combined to spur
the development of alternate methods of manufacture for PSAs. Thus,
while virtually all such products produced in 1970 were solvent-based,

Table 5 Solvent-Based Acrylic PSA
Formulation

Formulation	Parts
Copolymer based on	
Methyl acrylate	100
n-Butyl acrylate	290
2-Ethylhexyl acrylate	590
Glycidyl methacrylate	20
Acetone	300
Benzene	130
p-Toluene sulfonic acid	5
Dried, cured 3 min at 100°C	

Source: Based in part on Ref. 6.

Table 6 Comparative Economics of Improved Solvent Systems

	System		
	Adsorption	Incineration	
		Without heat recovery	With heat recovery
Capital cost (Installed)[a] (thousands of $)	420	65	105
Operating cost[b] (thousands of $/year)	160	207	97
Value of recovered solvent (thousands of $/year)	1,241	—	—

[a]Adsorption capital cost included steam generation and cooling tower. Prices as of January 1980.
[b]Operating cost—fuel gas at $3.708/million Btus.
Source: Based in part on Ref. 6.

ten years later only about three-quarters were made in this manner. In yet another decade, this fraction is expected to decline further, to under half of today's volume. One study on this subject has forecast that solvent-based adhesives will decline to near extinction by the year 2000 in the world's industrialized countries [18].

The increasing cost of energy will be a strong driving force toward systems which are more efficient—such as latices, hot melts, and, possibly, radiation curing. Essentially all new capacity is likely to be dedicated to these nonsolvent systems, and in addition, many solvent ovens will be adapted to latex systems. Solvent systems generally require more energy because of the lower solids levels employed and because of the limits on allowable solvent vapor concentrations in drying ovens [19]. Recent equipment innovations may change this outlook, however [20]. Economic analyses of the factors involved in operating a solvent line have been published which compare their cost with the costs of alternative approaches [6].

Several alternatives exist for those who wish to maintain their present solvent operation, however. These include

Use of "exempt (photochemically nonreactive) solvents. The emission of such solvents is limited, however, and subject to certain requirements.
Solvent incineration with or without heat recovery.
Solvent recovery [21,22].

An economic comparison and a discussion of incineration versus recovery were presented at a recent symposium [23]. The systems compared were based on flows of 10,000 standard cubic feet per minute (SCFM) of air containing toluene at 25% lower explosive limit (LEL) (Table 6).

Each of the three approaches shown requires the outlay of capital investment. Although the solvent adsorption route involves the greatest expenditure of capital by far, the value of solvent recovered results in a net savings of $1,081,000. However, a number of additional factors must be considered before selecting one of these processes, and the decision must be made with regard to the overall plant operation. The combination of hot-air flow, solvent concentration, solvent-laden air collection, and heat recovery which produces the best return on investment consistent with a company's capital capabilities is the best choice for that company [23].

Another approach which has been described involves the use of recycled air depleted of most of its oxygen. This is achieved with a new oven design (Inertair). Substantial reductions in fuel usage and pollution are clamed for this unit [20].

V. WATER-BASED SYSTEMS

The deposition of PSA films from a water medium has a number of advantages when compared to deposition from solvents. Many of the troublesome problems apparent with the use of solvents simply do not exist when water is used. To begin with, water is the cheapest possible carrier obtainable and presents no supply problem. Furthermore, it is odorless, nontoxic, nonpolluting, and nonflammable. These combined characteristics make water-based systems a sensible and viable approach when changing from existing solvent-based coating lines.

Table 7 A Comparison of the Oven Energy Consumed in a Label-Coating Operation

Process	% LEL	Oven air flow (thousand SCFM)	Fuel energy required[a]
Solvent	25	34.4	25.9
system	40	21.5	17.5
Latex PSA	—	6.0	13.4
	—	12.0	17.8

[a]Million Btus/million ft^2 substrate.
Source: Based in part on Ref. 6.

It is well known that the amount of heat necessary to evaporate
water is substantially more than that required to evaporate most organic solvents. However, when account is taken of the nonexplosive
nature of the aqueous systems, a net saving in energy cost accrues
to the latex coating process over the solvent-based line [6,19]. This
occurs because of the vastly reduced need to supply fresh air to the
ovens in order to maintain a safe operating level below the LEL.
Table 7, taken from a recent computer simulation of various PSA coating lines [6], illustrates these points.

In latices, the water acts as a carrier, rather than as a solvent.
Thus, a PSA latex is a suspension of an insoluble liquid polymer in
water, prepared so that it is stable for long periods of time. The
average diameter of the suspended, spherical polymer particles is in
the range of 0.1-0.8 μm. In order to obtain this particle size and
long-term stability, other components are usually present besides the
dispersed resin and water. It is often necessary to have present suitable emulsifying agents (or surfactants). These surfactants function
by their surface activity at the interface of the two phases. They
are essentially balanced molecules with a "hydrophilic" or "water-lov-
ing" part and a "hydrophobic" or "water-repellent" portion, such as
a long-chain hydrocarbon. Another type of emulsifier, or, to use a
more general term, *stabilizer*, is a water-soluble colloid. This class
includes high molecular weight, water-soluble compounds—for example, natural gums and synthetic or semisynthetic polymers, such as
polyvinyl alcohol and hydroxyethylcellulose.

This discussion will focus on latices resulting from emulsion polymerization (i.e., conversion of monomers to polymers). Because of
the insolubility of the polymer in water, the viscosities of latices are
not affected by molecular weight and are substantially lower than
those of solvent-based products. Furthermore, the solids levels of
the latices can be as high as 70%, much higher than it is for solvent
products. The polymer produced by emulsion polymerization normally has a much higher molecular weight than that employed with solution PSAs.

Many latex compositions having the requisite characteristics described earlier have been found to have pressure-sensitive properties
on drying. A summary of the materials first developed for this application during the 1960s has appeared [24]. Illustrative of the various types are the following:

Copolymers of vinyl acetate or, optionally, a higher vinyl ester
with an ester of an alpha, beta-unsaturated dicarboxylic acid, such
as maleic or fumaric acid.

A homopolymer of the vinyl ester of a carboxylic acid. Also
copolymers of mixed vinyl esters provided that

$$\frac{\Sigma M_n}{\Sigma M} \geqslant 2 \, ,$$

Table 8 Recipe for Pressure-Sensitive Emulsion[a]

Ingredient	Parts by weight
Nonylphenol ethylene oxide condensate (97% by weight oxyethylene units)	22.4
Sodium acetate	1.0
Water	465.0
Sodium metabisulfite (in 10 parts water)	0.9
Ammonium persulfate (in 10 parts water)	0.45
Vinyl acetate	83.6
2-Ethylhexyl acrylate	352.0
Acrylic acid	4.4

[a]The polymerization follows a conventional procedure for 3 hr at 60–75°C after which 0.45 parts by weight ammonium persulfite (in 10 parts water) is added and heating continued for 1 hr more. The final emulsion is stable and contains 46% by weight solids.

where M is the total number of molecular units in the polymer and M_n is the total number of carbon atoms in the alkyl groups of the aliphatic fatty acid in each ester unit.

Compositions in which the main constituent is an acrylic ester to which 5–10% of an acid, such as (meth)acrylic acid or crotonic acid, and 3–25% of a "hardening" monomer, such as methyl methacrylate or acrylonitrile, are added.

Formulations that use up to 5% methylolacrylamide or its nitrogen ethers (e.g., N-butoxymethyl acrylamide) in acrylate ester polymers to improve properties.

Combinations of mixed acrylates with small amounts (about 3%) of (meth)acrylic acid and a glycol ester, such as 2-hydroxypropyl acrylate.

Blends of two acrylic emulsions or a blend of natural-rubber latex containing a noncrosslinked acrylic latex or a styrene polymer.

Terpolymers of vinyl acetate, an acrylate ester which is at least 50% of the monomers, and 1–5% of an unsaturated acid, such as (meth)acrylic acid. A recipe illustrating such a formulation is given in Table 8.

Although refinements [25–27] have taken place since the aforementioned review [24], many of the compositions being used commercially today are believed to be comparable to those examples mentioned. The properties of two different types of PSA latex, based on acrylics [3,9] and SBR [28,29], have been described recently and are reproduced in Tables 9 and 10.

Table 9   Rubber Latex PSA Properties

Latex[c]	Properties[a,b]			
	Polyken tack (g)	90° Quick stick (pli)	180° Peel (pli)	Shear (hr)
NR[d]:Foral 85 (30:70)	1420	10	7.3	1.5
SBR latex-K (carboxylated): Foral 85 (50:50)	1320	1.5	4.5	> 50
[1:1 (NR:SBR) Latex-K]: Foral 85 (30:70)	2200	4.5	7.5	15.7

[a]Coated on 2-mil Mylar film.
[b]Tests: Polyken tack (ASTM D-2979-1); 90° Quick stick (PSTC-5);
          180° Peel (PSTC-1); 178° Shear [PSTC-7 modified,
          1000 g/(1/2 × 1 in.)].
[c]Ratios of dry parts of tackifier resin:rubber latex.
[d]NR, natural rubber.

Natural rubber and SBR have often been used for the manufacture of pressure-sensitive tapes and labels. These materials cannot be used as is, however, because the surface tack is extremely low. Furthermore, they require compounding to protect against oxidative and ultraviolet-induced aging. When properly formulated, they do possess good PSA properties, as illustrated in Table 9. Neoprene-latex PSAs have also been introduced recently [30].

Acrylic latices are particularly attractive alternatives to solvent-based rubber adhesives because of their excellent stability to heat and ultra-violet (UV) light and resistance to plasticizer migration. Correlations between the adhesive and cohesive characteristics of acrylic polymers have been made [3,31] with fundamental molecular properties such as T_g, molecular weight distribution, polarity, and polymer branching and crosslinking. Acrylic pressure-sensitive latices can easily be formulated to modify their properties [3,32,33]. Functional acrylic resins can be crosslinked to increase shear resistance. Thus, polymers with carboxylic acid functionality can be crosslinked with urea-formaldehyde at elevated temperature or with zinc acetate at room temperature. Acrylic latices are believed to offer the most widely applicable, versatile alternative to solvent-based PSAs [7]. Despite the many improvements made recently, some deficiencies need to be corrected for full industry acceptance as outlined in Table 11.

Table 10 Typical Acrylic Latex PSA Properties[a]

Property	Ucar latex 173	Ucar latex 174	Ucar latex 175
T_g (°C)	-35	-45	-60
Peel adhesion[b]			
(pli)	3	5	6
Quick tack (pli)[c]	1.5	1.7	1.8
Shear (hr)[d]	50– 100	2– 5	< 0.5
Solvent resistance			
MEK	Poor	Poor	Poor
Toluene	Fair	Fair	Fair
Recommended applications	Industrial tapes General purpose labels Wall coverings	General purpose tapes, labels, floor tiles	Refrig./freezer tapes and labels Other low-temp. applications

[a]Properties measured on adhesives applied directly to 2-mil Mylar film. The dry adhesive coat weight was approximately 2.4 gm/100 in.^{2}.
[b]180° Peel adhesion—PSTC 1.
[c]90° Quick tack—PSTC 5.
[d]Vertical shear, 1/4 in.^{2}/500 g—PSTC 7.

Table 11 Emulsion PSAs: Improvements Needed

Peel strength
  Generally not above 3.5 pli;
  Solution acrylics available at up
  to 6 pli, rubber/resin even higher.

Water, high relative-humidity resistance
  Plasticizing effect reduces strength
  and limits exterior use.
Specific adhesion to nonpolar substrates
  Share this characteristic with solution
  acrylics; rubber/resin types better.
Freeze/thaw stability
  Possible problems in transit, storage.

Coating quality
  Comparable to solution casting possible, but only
  with more care.

Source: From Refs. 12 and 34.

Table 12 Some General Characteristics of HMPSAs

Advantages	Disadvantages
100% solids, no solvent hazard. Simple to ship, handle, and store.	Overall properties not equal to solvent- or water-based adhesives.
Low energy consumption.	Service temperature is limited.
Faster machine speeds—up to 500—1000 fpm.	Limited resistance to attack by solvents or plasticizer
Ability to apply higher coating weights, control penetration into porous stock.	migration.
	Coating equipment and conditions extremely important.
	Decomposition of the resin in melt a potential problem.
	Cannot be applied to heat-sensitive substrates.
	Dirt or contaminants cause streaking of coatings.

Source: Refs. 35, 41, and 51.

VI. HOT-MELT PRESSURE-SENSITIVE ADHESIVES

As their name implies, hot-melt adhesives are solid at normal temperature but become molten on heating. When they are applied to a surface in the form of a hot liquid, wetting occurs, heat is lost to the surface, and a bond is formed. The process of forming a bond with a hot-melt adhesive is one of the most rapid of all techniques because neither solvent loss nor chemical reaction is involved. As a direct result of their thermal characteristics, however, hot-melt adhesives have poor resistance to heat. In many instances, solvent resistance is poor as well (see Table 12). Improvements in these properties can be achieved in some cases by post crosslinking with irradiation [27, 35—39] (Sec. VII) or by blending with other polymers [40].

In comparison with other methods of applying PSAs, the hot-melt technique is the most energy efficient method available at present [41]. Furthermore, it offers the highest known production rates, the lowest capital costs for new lines, and because solvents are absent, no pollution problems. Some odor may be apparent during hot-melt

Table 13 Projected Demand for Hot-Melt Adhesive Compounds in Pressure-Sensitive Tapes and Labels, 1976—1986, North America

Year	Millions of lb	Annual % increase	Consumption ratios, labels:tapes
1976	20	Base year	50:50
1978	36	33	47:53
1980	61	15	62:38
1982	108	21	54:46
1984	153	17	49:51
1986	197	12	47:53

Source: Data from Ref. 42.

operations due to thermal decomposition of the molten polymer. Other advantages inherent in this technique are enumerated in Table 12. The concept of applying an adhesive as a hot melt also has some short-comings, including the sensitivity of certain substrates to heat and streaking caused by contaminants in the coating. The most serious disadvantages, however, are product deficiencies. To quote a recent industry survey,

> It is irrefutable that today's hot melt pressure-sensitive adhesive technology is inferior to the wider spectrum of end-use adhesive properties that can be achieved on tapes and labels from solvent-based and water-based pressure-sensitive adhesives. Hot melt pressure-sensitive adhesives are inferior to the best liquid adhesives in light color, high- and low-temperature strength, durability, electrical properties, and outdoor weatherability.

> While today's hot melt pressure-sensitive technology cannot possibly duplicate the end-use adhesive properties of high-performance PS tape and labels, they can and will duplicate the end-use adhesive properties of low- and medium-performance PS tapes and labels. [42]

Some estimates from this same survey for the demand for hot-melt adhesives in pressure-sensitive tapes and labels over the ten-year period 1976—1986 are given in Table 13.

A concise review of the history of HMPSA development, with emphasis on the basic polymers and compositions employed, has been published [43].

For all intents and purposes, commercial HMPSA compositions are limited to block copolymers, such as S-I-S or styrene-butadiene-styrene (S-B-S), which account for about 90% of usage [5,44]. These products go primarily into tapes and labels. An excellent review of

Table 14 Hot-Melt Pressure-Sensitive Adhesive Based on S-I-S

Ingredient	Percent
S-I-S (Kraton 1107 rubber)	32.8
Midblock resin (Wingtack 95)	32.8
Plasticizing oil (Shellflex 371)	13.1
Endblock resin (Cumar LX-509)	19.7
Stabilizer (zinc dibutyldithiocarbamate)	1.6

Adhesive characteristic	Value
Shear adhesion failure temperature (°F)	220
Rolling ball tack (PSTC-6) (cm)	1.8
Probe tack (g)	1100
180° peel adhesion (PSTC-1) (pli)	3.7
Melt viscosity at 350°F (cP)	40,000

Source: Data from Ref. 44.

the characteristics of the thermoplastic A-B-A block copolymers in adhesives has been published [44].

One important characteristic of such copolymers is that the polystyrene and polybutadiene (or polyisoprene) segments are in separate
physical phases. When other components, such as plasticizers and
tackifier resins, are present they may tend to associate preferentially
with one phase or the other thereby influencing the adhesive properties. An example of a formulation of an HMPSA based on an S-I-S
polymer is given in Table 14. This composition was achieved by bal-
ancing the use of components which interact with either the two polystyrene end blocks or the polyisoprene midblock.

Other products which are used to a lesser degree in HMPSAs include ethylene vinyl acetate (EVA) [45,46], ethylene ethyl acrylate
(EEA) [47], and amorphous polypropylene [48]. These materials find
use in applications which have lower performance requirements, such
as floor tiles and wall coverings.

Acrylics have long been viewed as superior PSAs, but attempts
to develop hot-melt versions of these products have not met with appreciable commercial success thus far. However, newer products are
being made available which may overcome some of the earlier deficiencies. Outstanding color, clarity, resistance to degradation by UV
light, and an ability to be melt processed in air are claimed to be
available in one such product [49,50]. Further improvements in this
area could lead to acrylics capturing a significant share of the
HMPSA market. A comparison of the overall properties of acrylics
with the widely used rubber resins is given in Table 15.

Table 15 Comparative Properties of HMPSAs

Property	Rubber resin[a] (unsaturated)	Thermoplastic acrylic[b]
Peel	High	High
Tack	High	Moderate
Shear	High	Low
O_2 stability	Poor	Excellent
UV stability	Poor	Excellent
Heat resistance	Poor	Poor
Plasticizer resistance	Poor	Excellent
Solvent resistance		
Hydrocarbons	Poor	Good
Oxygenated	Good	Poor
Bleed tendency	Moderate— low	None
Melt thermal stability	Good	Good—poor
Cost	Low	Moderate

[a]Styrene-isoprene-styrene block copolymer tackified with rosin ester and hydrocarbon.
[b]Low-molecular-weight polymer containing a large portion of 2-ethylhexyl acrylate.
Source: Ref. 39.

VII. RADIATION CURING

Curing by UV or electron beam (EB) radiation is the second of two techniques directed at preparing PSAs without evaporating solvents or carriers. In making a PSA, a properly formulated fluid composition will be applied to an appropriate substrate, such as Mylar, polymerized by a few seconds exposure to a radiation source, and thus converted from a low-strength composition to a tacky, higher-strength adhesive tape with little loss of volatiles. This approach has several potential advantages to offer,

No disposal of solvents into the atmosphere.
Extremely low energy requirements.
High line speeds.
Compact operation.
Heat-sensitive substrates may be used.
Improved adhesion to substrate due to grafting.

Furthermore, radiation curing has certain advantages over hot-melt technology including greater formulating versatility, inherent thermal stability through chemical crosslinking, and ease of coating due to

Table 16 Comparison of EB and UV Cure

Equipment cost is an order of mangitude larger for EB as
compared to UV.
Operating skill requirements higher for EB than UV.
Maintenance costs may be higher for EB.
EB generally utilizes more nitrogen inserting than UV.
Residual monomer tends to be less with EB.
Adhesion to substrates is better with EB.
EB can penetrate pigmented or opaque materials including
some laminates.
Energy costs less for EB.
Faster line speeds possible with EB.
EB will cure a broader array of functional groups than UV.

---

lower viscosities [52]. In fact, radiation is often cited as a means of
improving properties of hot melts, as will be discussed later in this
section.

In its simplest form, the concept of preparing PSAs by radiation
curing is potentially superior to all other methods because the conversion from monomer to polymer can take place directly on the substrate
when it is irradiated, whereas solution, emulsion, or hot-melt techniques require the preparation or isolation of the polymer prior to coating. Although examples of direct conversion from monomer to PSA
polymer have appeared in the literature, in practice neither this nor
other variants of radiation curing for PSAs have been much utilized,
and only a small percentage of such products is being manufactured
this way. A recent study indicates that the total sales of radiation-
curable PSAs were only $400,000 in 1979 [53].

Radiation curing has not captured a larger share of this market
for complex reasons that involve the following elements:

Technology less well developed than for alternative techniques—
a broad array of materials and properties not available.

Raw materials costs tend to be high.

Unreacted monomer residues may remain after cure.

Highly reactive monomers (acrylates) used may be toxic and can
lead to sensitization and skin burns.

Inert-gas blanketing required in some instances—adds to cost.

Yet, the advantages of radiation curing are undeniable, and research is actively continuing on several facets of this technique, including the crosslinking of hot melts.

Both UV light and EB radiation have been widely used for the
cure of PSAs. Each of these approaches has its advantages and disadvantages as elucidated in Table 16.

The simplest means of preparing a PSA, irradiating monomer alone, is rarely employed due to coating problems resulting from the low viscosity of the system, incomplete polymerization, and inadequate properties. Irradiation usually results in a low degee of polymerization due to the simultaneous generation of numerous initiating sites. To overcome these problems, other components are utilized to provide increased molecular weight at the start or sites for grafting or cross-linking. Thus, most radiation-curable adhesive systems are made up of monofunctional monomers, polyfunctional monomers, oligomers, and reactive or nonreactive resins which, on irradiation, develop desirable adhesive properties through rapid polymerization. Systems that are to be photocured also require a photoinitiator or photosensitizer, which absorbs the impinging light and induces polymerization. Electron-beam—cured formulations do not require these initiators, however.

The literature on radiation-curable PSA compositions was reviewed in 1977 [54] and again more recently [27]. The majority of compositions utilized for radiation-curable PSAs has contained acrylate components in the form of monomers or polymers.

An example is known in which PSAs were prepared by radiation of monomers alone. In this case, a blend of butyl acrylate containing 9% acrylic acid was applied at a thickness of 50 μm to an aluminum foil and subjected to 5 Mrads of EB radiation. The resulting PSA displayed a 180° peel strength of 4.2 lb/in. and a dead load shear time of 10 hr.

The preparation of PSAs by UV cure of solutions comprised of acrylate monomers containing a small amount (5%) of a polyvinyl alkyl ether has been described. More recent work has discussed the advantages of using higher levels of polyvinyl alkyl ethers. Excellent PSA properties have been obtained with compositions based on both mono-functional [55] and multifunctional [56] acrylates with polyvinyl alkyl ethers. Excellent properties were also observed when using other polyalkyl ethers, such as poly(alkylene oxides) [57].

More commonly, acrylic polymers have been utilized as the sole or major components of radiation-curable adhesive systems. These acrylics have included a homopolymer and many examples of copolymers. One UV-cured acrylic ester copolymer system cited was claimed to have 180° peel strengths of 4.8—5.5 lb/in.

Radiation-curable systems based on copolymers of vinyl acetate and 2-ethylhexyl acrylate have been reported more frequently than other compositions and have sometimes been applied by hot-melt techniques. These PSAs have shown 180° peel strengths ranging from about 1.5 to 3 lb/in. when cured by either UV or EB radiation.

Normally, a small amount of a photoinitiator is required to cure UV systems efficiently. Several patents have described the use of specially prepared monomers or polymers which contain built-in photoinitiators.

Table 17 Hot-Melt Pressure-Sensitive Adhesives Cross-linked by EB Radiation

Composition[a]	Percent
S-I-S (Kraton 1107 rubber)	39.6
Resin (Wingtack 95)	39.6
Plasticizing oil (Shellflex 371)	9.9
Stabilizer (butyl zimate)	0.8
Trimethylolpropane trimethacrylate	9.9

	Radiation dose (Mrads)		
Property	0	5	10
Peel temperature, limit (°C)	110	> 200	> 200
Gel	0	tight	tight
Rolling ball tack (cm)	2	5	15
Polyken probe tack (g)	200	400	400
Holding power to steel (min)	5	50	200

[a]Coated on Kimberly Clark crepe paper at approximately 90 gm/m^2. Cured with Energy Sciences Electrocurtain equipment.

Not surprisingly, solutions of acrylic polymers in acrylic and vinyl monomers have also been used as radiation-curable PSAs.

Although acrylates have received the most attention, a number of other polymer types have been reported in radiation-curable PSAs. Several systems have contained polybutadiene and its copolymers, particularly SBR. Other rubber bases utilized have included those derived from natural sources, polyisobutylene, and silicones.

In addition, components such as tackifiers, fillers, and (meth)-acrylate monomers have been employed in these rubber-based systems.

Ethylene copolymers, specifically EVA and ethylene-5-ethylidene-2-norbornene-propylene, have been used with tackifiers. Pressure-sensitive adhesives based on an entirely different type of chemistry, namely polythiol-ene adducts, have also been patented. A UV-cured system of this type has been reported to exhibit a 180° peel strength of 3.5 lb/in.

Radiation curing is also being used increasingly to upgrade the PSA properties of compositions which contain substantial amounts of polymer [35,39,54]. This technique is of particular interest for correcting some of the deficiencies of hot-melt adhesives based on the thermoplastic A-B-A copolymers. In contrast to other types of PSAs, which can be chemically crosslinked by post heating, hot-melt products must be largely free of such bonds in order to be melt processable. However, the hot-melt processing of an adhesive which contains a radiation-initiated crosslinking component would be a desirable approach to this problem. Because radiation-curable crosslinking agents may also gel when heated, it is necessary to minimize the thermal abuse of such compositions during processing. A successful demonstration of such a process was recently described [58]. On being exposed under nitrogen to a high energy EB, the coated product crosslinked with resultant improvements in both solvent and temperature resistance (Table 17). Without the use of the multifunctional acrylate, relatively little change in properties was observed on irradiation. Rolling ball tack was reduced on radiation, but the resultant properties are claimed to be comparable to the performance of commercial, high-temperature masking tape based on crosslinked natural rubber.

As desirable as the foregoing property improvements may be, they suffer from the disadvantage of requiring the installation of a separate radiation-cure device in addition to a hot-melt coater. Despite this, however, the use of radiation curing is expected to grow appreciably in the years ahead.

VIII. COATING METHODS

There are a number of coating techniques being used to apply PSAs. The selection of the type of equipment best suited to a specific situation will depend on several factors:

Type and thickness of the web
Nature of the adhesive: solvent, aqueous, hot-melt
Viscosity of the adhesive
Adhesive thickness and thickness uniformity required

Jacobs [59], in a discussion of "Fundamentals to Consider in Selecting Coating Methods," points out that to select the coating method most suitable for a specific application, it is first necessary to know the physical properties of the web and coating to be used and to correlate these with the fundamental characteristics of the coating method.

Grant [60] classifies coating equipment as follows:

Equipment that applies an excess of coating to the web and wipes off the excess leaving a desired amount on the surface
Equipment that applies a predetermined amount of coating to the surface of the web
Equipment that saturates the web in a bath and removes the excess coating from the surface by some metering method.

The following are common types of equipment used for the application of PSAs today:

Reverse roll coaters. As the name implies, the reverse roll coaters apply the coating with a roll rotating in the reverse direction to the travel of the web. They vary both as to the number of rolls and the roll configuration. The adhesive is fed either from a pan, via a pickup roll, or from a dam between the metering roll and applicator roll. Reverse roll coaters are capable of applying adhesives of low viscosity using a pan-fed pickup and adhesives of viscosities > 10,000 cP using a dam-fed arrangement. Reverse roll coaters are particularly suited to film thicknesses > 1 mil wet and are not recommended for application of a film < 1 mil thick. The principal limitation of reverse roll coaters is the degree of accuracy with which they are built and, in turn, provide a coating thickness with a minimum variation in coat weight.

Gravure coaters. Gravure coaters are built in either a two-roll or three-roll vertical design. The bottom roll, which is generally the engraved roll, rotates in the coating pan. An oscillating doctor blade removes the excess, leaving the coating material only in the recesses of the engraved roll. This premetered quantity of coating is then transferred directly to the web in a two-roll design or first to an intermediate roll in a three-roll design. Gravure coating is used where relatively thin coatings (1–3 mils) are required and is best suited to smooth substrates. Gravure coating may be used to apply solvent, aqueous, and hot-melt adhesives at viscosities up to 10,000 cP. The major limitation is in applying heavy coat weights.

Wire-wound rod coating. In the wire-wound rod coating technique, the coating is applied directly to the web from the pickup roll and then metered to the desired film thickness by passing over one or a series of wire-wound rods to remove the excess coating. The rods may be rotated in either direction relative to direction of web travel. Wire-wound rod coaters are relatively inexpensive and are used to apply solvent, aqueous, and hot-melt adhesives.

Knife over roll coaters. The knife over roll coater arrangement uses a driven roll to support the web under the knife, which is usually a steel bar with one contoured edge that contacts the web. The coating material is held behind the knife blade and contained by dams which are set for the width to be coated. As the web traverses over the roll and under the knife, the amount of coating applied is determined by the gap between the knife edge and the web. Knife over roll coaters are relatively inexpensive and are particularly useful for application of high viscosity adhesives at relatively thick films. Their major disadvantage is the inability to accurately control film thickness, and their use is generally limited to relatively slow line speeds, up to 100 fpm.

Air knife. In the air knife technique, an excess of the coating is applied to a web by an applicator roll, and the excess is removed with an air knife by a uniform jet of air impinging on the surface. The air knife is used on all types of webs at speeds exceeding 1200 fpm [60]. The major limitation to the air knife is in the solids and viscosities it can handle. Viscosities range from 100 to 1000 cP, and its use is generally restricted to water-based systems [59].

Hot-melt pressure-sensitive adhesives are generally applied by either roll coaters, slot die coaters, or extrusion coaters. The adhesive is fed to the coating head in the molten state from either a jacketed melt-mix tank or continuously through a single or twin screw extruder [61].

IX. TEST METHODS

A comprehensive series of test procedures widely used by the industry has been published by the Pressure-Sensitive Tape Council (PSTC) [62].

A brief description of the most common tests is as follows:

Peel Adhesion (PSTC-1). The peel adhesion test is a measure of the force required to remove a pressure-sensitive tape from a standard stainless steel surface when pulled at a rate of 12 in./min at an angle of 180°. A 1-in. wide tape is applied to a specified stainless steel surface under the weight of a 4.5-lb rubber roller. The force required to remove the tape is expressed in pli or oz/in. of tape width.

Shear or holding power (PSTC-7). The shear, or holding power, test is a measure of the cohesive strength of a PSA to resist a force applied in a direction parallel to the surface to which it has been applied. It is customarily measured in terms of the time required to pull a standard area of tape from a test panel under a standard load, or in terms of a distance the tape has been displaced in a given time on a test panel under a standard load. The procedure involves the application of a tape to a stainless steel surface under the weight of a 4.5-lb rubber roller. The tape is then cut so that a specified area (usually 1 × 1 in.) remains in contact with the steel panel. The panel is then mounted in a jig at an angle of 2° from vertical to ensure that no peeling forces are acting on the tape when a specified constant load (usually 1 kg) is applied. The amount of time required for the tape to completely separate from the panel or the amount of slippage measured to the nearest 1/16 in. during a prespecified period of time is reported.

Quick stick (PSTC-5). The quick stick test is designed to measure the ability of a pressure-sensitive tape to adhere under minimum pressure. In this test, the tape is placed in contact with a standard

stainless steel panel under no pressure other than the weight of the
tape itself. The panel is then placed in a special jig which allows the
panel to move and maintain a constant 90° angle with the tape during
the test. The force required to remove the tape at a rate of 12 in./
min at a 90° angle is reported in pli or oz/in. of width.

Rolling ball tack (PSTC-6). The rolling ball tack test is designed to measure the surface tack of a PSA. It consists of allowing a 7/
16-in.–diameter steel ball to roll from an inclined plane onto the adhesive surface. The distance the ball travels on the surface is a function of the surface tack. The greater the degree of tack, the shorter
the distance the ball will travel.

Probe tack. The probe tack test [63] consists of bringing a metal probe of specified diameter into contact with the adhesive surface
under a given load for a specified period of time. The force required
to separate the probe from the adhesive is a measure of surface tack.
A device for measuring probe tack called the Polyken Probe Tack
Tester has been described [64].

Test methods for measuring properties of pressure-sensitive
coated tapes for electrical insulation are also described in (ASTM) D-
1000.

REFERENCES

1. R. Bates, *J. Appl. Polymer Sci. 20*:2941 (1976).
2. D. W. Aubrey, *Devel. in Adhesives 1*:127 (1977).
3. J. W. Hagan, C. B. Mallon, and M. R. Rifi, *Adhesives Age 22*:29 (1979).
4. M. Yoshii and S. Narusawa, to Sumitomo Chemical Co., Japanese Patent Application—Kokai 54-61284/79 (1979).
5. Frost and Sullivan, Inc., Pressure-sensitive products and adhesives
   market, in *Report No. 614* (1978).
6. A. D. Hamer, J. W. Hagan, and S. G. Krumenaker, *Adhesives Age 23*:23 (1980).
7. J. A. Fries, Emulsion PSAs: Commercial application, in *Procs: Adhesive Coating Tech.*, PSTC Technical Seminar, Rosemont, Illinois, 18-19 June 1980.
8. A. Maletsky and J. Villa, Hot melt pressure-sensitive adhe-
   sives, the 1970s—a look back, the 1980s—a look ahead, in
   Proc.: Adhesive Coating Tech., PSTC Technical Seminar, Rosemont, Illinois, 18–19 June 1980.
9. J. P. Carolan, Water-borne pressure-sensitive adhesives—a
   market in transition, in *Proc.: Adhesive Coating Tech.*, PSTC Technical Seminar, Rosemont, Illinois, 18–19 June 1980.
10. M. E. Hodgson, Pressure-sensitive adhesives and their appli-
    cations, in *Adhesion 3* (K. Allen, ed.), Appl. Sci. Publ. Ltd., London, 1979, p. 207.

11. R. Cosslett, *Labels and Labeling July*: (1979).
12. M. R. Rifi, Water-based pressure-sensitive adhesives structure vs performance, in *PSTC Technical Meeting on Water-Based Systems*, Chicago, 21–22 June 1978.
13. C. W. Bemmels, Pressure-sensitive tapes and labels, in *Handbook of Adhesives*, 2d. ed. (I. Skeist, ed.) Reinhold, New York, 1977.
14. D. F. Merrill, *Adhesives Age* 22:39 (1979).
15. D. F. Merrill, Silicone pressure-sensitive adhesives, in *Proc.: Adhesive Coating Tech.*, *PSTC Technical Seminar* Rosemont, Illinois, 18–19 June 1980.
16. G. W. H. Lehmann and H. A. J. Curts, to Beiersdorf AG, U.S. Patent No. 3,563,953 (1971).
17. J. A. Fries, Regulatory, environmental and safety aspects of pressure-sensitive adhesives, in *PSTC Technical Meeting on Water-Based Systems* Chicago, 21–22 June 1978.
18. Debell and Richardson, Inc., *The Decline and Fall of Solvent-Based Adhesives*, (1974), Enfield, Conn.
19. T. P. Carter, Emulsion pressure-sensitive adhesives: a route to improved oven energy utilization, in *PSTC Technical Meeting on Water-Based Systems*, Chicago, 21–22 June 1978.
20. T. W. Grenfell, What's new in oven design? Interair systems, in *PSTC Technical Meeting on Water-Based Systems*, Chicago, 21–22 June 1978.
21. *Adhesives Age* 23:47 (1980).
22. R. P. Ruud, Solvent recovery with activated carbon, in *Proc.: Adhesive Coating Tech.*, *PSTC Technical Seminar*, Rosemont, Illinois, 18–19 June 1980.
23. T. B. Michaelis, Minimizing volatile organic emission control costs, *Proc.: Adhesive Coating Tech.*, *PSTC Technical Seminar*, Rosemont, Illinois, 18–19 June 1980.
24. H. Warson, *The Application of Synthetic Resin Emulsions*, 310 Ernest Benn Ltd., London (1972).
25. L. F. Martin, Pressure-sensitive adhesives—formulations and technology, in *Chem. Tech. Rev. No. 34*, Noyes Data, New Jersey, 1974.
26. H. R. Dunning, Pressure-sensitive Adhesives—Formulations and technology, in *Chem. Tech. Rev. No. 95*, Noyes Data. New Jersey, 1977.
27. S. Torrey, Adhesive technology development since 1977, in *Chem. Tech. Rev. No. 148*, Noyes Data, New Jersey, 1980.
28. R. C. Oldack, PSAs from natural and carboxylated SBR latexes, in *PSTC Technical Meeting on Water-Based Systems*, Chicago, 13–14 June 1979.
29. J. Y. Penn, Resin emulsions for water-based pressure-sensitive adhesives, in *Adhesive and Sealant Council Fall Seminar*, Washington, D. C., 23–25 October 1978.

30. C. M. Matulewicz and A. M. Snow, Jr., Neoprene latex pres-
     sure-sensitive adhesives, in *Proc.: Adhesive Coating Tech.*,
     PSTC Technical Seminar, Rosemont, Illinois, 18—19 June 1980.

31. J. G. Martins and L. W. McKenna, Status of water-based acry-
     lic pressure-sensitive adhesives, in *Adhesive and Sealant
     Council Fall Seminar*, Washington, D. C., 23—25 October 1978.

32. G. J. Antlfinger and D. M. Yingling, Compounding of acrylic
     pressure-sensitive latexes, in *Proc.: Adhesive Coating Tech.*,
     PSTC Technical Seminar, Rosemont, Illinois, 18—19 June 1980.

33. R. W. Wherry, Resin dispersions for water-based pressure-
     sensitive adhesives, in *PSTC Technical Meeting on Water-
     Based Systems*, Chicago, 13—14 June 1979.

34. J. A. Fries, Pressure-sensitive adhesives: an overview of
     solutions, emulsions, hot melts and radiation curable systems,
     in *PSTC Technical Meeting on Water-Based Systems*, Chicago,
     13—14 June 1979.

35. D. J. St. Clair, Radiation curing of pressure-sensitive adhe-
     sives based on thermoplastic rubbers, in *1979 Course Notes,
     Hot Melts—An Overview for Management*, TAPPI, South Caro-
     lina, 4—7 June 1979.

36. D. R. Hansen and D. J. St. Clair, to Shell Oil Co., U.S.
     Patent No. 4,133,731 (1979).

37. D. J. St. Clair and D. R. Hansen, to Shell Oil Co., U. S.
     Patent No. 4,151,057 (1979).

38. R. M. Christenson and C. C. Anderson, to PPG Industries,
     U.S. Patent No. 3,725,115 (1973).

39. S. H. Ganslaw, Hot melt pressure-sensitive adhesives, alter-
     natives to rubber—resin systems, in *1979 Course Notes, Hot
     Melts—An Overview for Management*, TAPPI, South Carolina,
     4—7 June 1979.

40. D. R. Hansen, to Shell Oil Co., U.S. Patent No. 4,104,323
     (1978).

41. D. de Jager, Economic apparaisal of hot melt produced PSAs,
     in *1976 Course Notes, International Hot Melt Pressure-Sensitive
     Adhesives*, TAPPI, Amsterdam, 3—5 November 1976.

42. *The Rapid Growth and Maturing Future Markets of Hot Melt
     Adhesives, Sealants and Coatings 1976—1986*, A confidential
     study by Springborn Laboratories, Inc., in collaboration with
     H. S. Holappa and Associates, July 1977.

43. A. Maletsky and J. Villa, *Paper, Film, and Foil Converter
     September*:55 (1976).

44. J. T. Harland and L. A. Petershagen, in *Handbook of Adhe-
     sives*, 2d ed. (I Skeist, ed.) Reinhold, New York, 1977.

45. K. C. Brinker, Raw materials for hot melt pressure-sensitive
     adhesives—EVA Copolymers, in *1976 Course Notes, Hot Melts*,
     TAPPI, Boxborough, Massachusetts, 5 May 1976.

46. R. E. Duncan and J. E. Bergerhouse, *Adhesives Age* March: 37 (1980).
47. M. R. Rifi, Pressure-sensitive adhesives—structure vs performance, in *1977 Course Notes, Hot Melts—A Western Vista*, TAPPI, San Diego, 7–10 November 1977.
48. R. R. Schmidt III and J. D. Holmes, to Eastman Kodak Co., U.S. Patent No. 4,143,858 (1979).
49. J. E. Fellerman, Acrylic hot melt pressure-sensitive adhesives, raw materials in a changing adhesive and sealant market in *Adhesive and Sealant Council Fall Seminar*, Washington, D. C., 23–25 October 1978.
50. F. T. Sanderson and D. R. Gehman, Acrylic hot melt pressure-sensitive adhesives: emerging technology for the 80s, in *Proc.: Adhesive Coating Tech.*, *PSTC Technical Seminar*, Rosemont, Illinois, 18–19 June 1980.
51. P. J. DeMarzio, a review of hot melt pressure-sensitive adhesives—advantages and disadvantages, in *1977 Course Notes, Cavalcade of Hot Melts—Their Future and Their Problems*, TAPPI, South Carolina, 18–24 April 1977.
52. A. Schwarz, Radiation curing of pressure-sensitive adhesive products, in *SME Tech. Paper FC 76-497*, SME, Dearborn, 1976.
53. *Chemical Marketing Reporter*, 10 March 1980, p. 5.
54. K. C. Stueben, *Adhesives Age* June:16 (1977).
55. K. C. Stueben, R. G. Azrak, and M. F. Patrylow, to Union Carbide, U.S. Patent No. 4,165,266 (1979).
56. K. C. Stueben, R. G. Azrak, and M. F. Patrylow, to Union Carbide, U.S. Patent No. 4,151,055 (1979).
57. K. C. Stueben, U.S. Patent No. 4,111,769 (1978).
58. D. J. St. Clair, *Adhesives Age* March:30 (1980).
59. R. J. Jacobs, *Paper, Film, and Foil Converter* February through July: (1963).
60. O. W. Grant, Coating equipment, in *PSTC Technical Meeting on Water-Based Systems*, Chicago, 21–22 June 1978.
61. F. C. Palermo and R. Korpman, An alternate method to hot melt coating of pressure-sensitive adhesives, in *1977 Course Notes, Hot Melts—A Western Vista*, TAPPI, San Diego, 7–10 November 1977.
62. *test Methods for Pressure-Sensitive Tapes*, 7th ed., Pressure-Sensitive Tape Council, Glenview, Illinois, 1976.
63. F. H. Wetzel, *ASTM Bull.* 221:64 (1957).
64. *Catalog and Register of Testing Equipment*, 5th ed., Testing Machines Inc., Amityville, New York, 1973, p. 240.

15
RTV Silicone Adhesive Sealants

Jan V. Lindyberg *General Electric Company, Silicone Products Division, Waterford, New York*

I. INTRODUCTION

RTV is a term which, when applied to silicone materials, refers to their ability to room temperature vulcanize or to cure under ambient temperature conditions.

RTV silicone adhesive sealants are available as liquid or paste, one-component and two-component materials. When fully cured these materials are elastomeric and possess a unique profile of physical and performance properties.

This chapter will present a brief insight into the chemistry of silicones beginning with the manufacture of silane chemicals, the production of silicone polymers from those silane chemicals, and the formulation of silicone adhesive sealants using those polymers. The chemistry of the various cure mechanisms will also be discussed.

The subjects of one-component and two-component silicone adhesive sealants will be treated separately. The treatment will cover performance properties, handling techniques, and selection criteria for these materials. A short discussion of the use of primers and the availability and selection of dispensing equipment will also be presented.

II. CHEMISTRY

All RTV silicone adhesive sealants, whether they be one-component or two-component systems, are formulated using the same three basic ingredients [1]. These ingredients are silicone polymer, fillers, and organoreactive silane crosslinker/catalyst systems.

III. SYNTHESIS OF SILICONE POLYMERS

The synthesis of silicone polymers is a multistep chemical process. The first step is the reaction of elemental silicon with methyl chloride in the presence of a copper catalyst, under high-temperature conditions, to form a crude mixture of methyl chlorosilanes [Eq. (1)]

$$Si + CH_3Cl \xrightarrow{\ Cu\ }$$

$(CH_3)_2SiCl_2$	Dimethyl dichlorosilane
CH_3SiCl_3	Methyl trichlorosilane
$(CH_3)_3SiCl$	Trimethyl chlorosilane
CH_3SiHCl_2	Methyl dichlorosilane
	Other organochlorosilanes (1)

This reaction is carried out commercially in a fluid-bed reactor. Elemental silicon is ground to a fine powder, mixed with a powdered copper catalyst, continuously fed into a fluid-bed reactor, and fluidized with methyl chloride. The composition of the crude chlorosilane output stream can be controlled to maximize the concentration of dimethyl dichlorosilane. While the other chlorosilanes have value as inputs into other specialty silicone polymers, dimethyl dichlorosilane is the prime building block for linear dimethyl polysiloxane polymer, which is the prime ingredient for making RTV adhesive sealants. After manufacture of the crude chlorosilane mixture, it is fractionally distilled to produce dimethyl dichlorosilane with a high level of purity. This high-purity dimethyl dichlorosilane is then reacted with water to form a crude hydrolyxate which, when appropriately equilibrated and neutralized, yields a high-purity, linear dimethyl polysiloxane polymer of controlled average molecular weight and molecular weight distributions [Eq. (2)]

$$(CH_3)_2SiCl_2 + H_2O \longrightarrow$$

$$HO \left[\begin{array}{c} CH_3 \\ | \\ -Si-O- \\ | \\ CH_3 \end{array} \right]_n H + HCl \tag{2}$$

The by-product hydrochloric acid is recycled by reacting it with methanol to form methyl chloride, which is the input for the manufacture of the chlorosilanes described in Eq. (1).

The silanol (—OH)-terminated polymer is the basis for the formulation of one-component, moisture-curing RTV silicone adhesive sealants and two-component, condensation-curing RTV silicone compounds.

Silicones are neither organic nor inorganic, but rather a unique combination of the two chemistries. The silicon-oxygen repeating backbone of dimethyl silicone polymer provides the inherent chemical stability, environmental stability, and extremely good high-temperature performance characteristics of these inorganic structures. The methyl side groups provide the flexibility characteristics of organic materials, but because the methyl group is the smallest organic unit, it also contributes to the thermal stability of the structure as a whole.

Silicone polymers have relatively long bond lengths within the polymer and low intermolecular forces between adjacent polymer chains. These characteristics result in relatively low viscosity and relatively high vapor-transmission properties for a given molecular weight in comparison to organic polymers. The high moisture-vapor transmission characteristic, in particular, provides for the ability to produce moisture-vapor—curing compounds.

IV. FILLERS

RTV silicone adhesive sealants contain fillers. Fillers contribute to the development of the desired property profile in the finished compound.

Silicon dioxide in the form of fumed silica, a filler with high surface area, is typically used to improve the physical properties of RTV silicone adhesive sealants. A cured, unfilled silicone polymer may have a tensile strength of only 50–75 psi (350–525 kPa) and virtually no tear strength. The use of reinforcing filler, such as high-surface-area fumed silica, enables the formulation of compounds with cured tensile strengths in the range of 400–800 psi (2800–5600 kPa).

Lower-surface-area fillers—such as ground quartz, micron-sized silica, and metal oxides—are used for improving the thermal conductivity of a compound. Calcium carbonate and carbon black can be used to improve fluid resistance or to serve as extending materials. Metals, such as aluminum, silver, or copper powders, have been used to produce electrically conductive compounds. These compounds, however, are generally classified as specialty materials and are not readily available.

All of these fillers have a high degree of thermal stability and are selected, not only for the properties mentioned previously, but also for their tendency to complement the high-temperature performances characteristics of the silicone polymer.

V. CATALYST/CROSSLINKER

The catalyst/crosslinker used in RTV silicone adhesive sealant systems consists of a species which has the structure described in Eq. (3)

$$R \; Si \; X_3 \; or \; Si \; X_4 \qquad\qquad (3)$$

The R stands for an organic group which could be, but is not necessarily limited to, a methyl radical. The X represents some moisture-hydrolyzable species. RTV silicone adhesive sealant systems typically make use of the species containing three reactive sites, while two-component, condensation-cure systems typically make use of crosslinkers having four reactive sites.

These reactive species may be any of the following:

Acetoxy
Alkoxy
Amine
Octoate
Ketoxime

All of these reactive species are currently being used in one-component sealant systems. The acetoxy system is by far the most available. The technology of this system is about 25 years old, and a broad selection of acetoxy products exists. The by-product of cure in RTV silicone adhesive sealant materials using this crosslinker system is acetic acid. Acetic acid has a relatively high vapor pressure which causes it to evaporate quickly, leaving a cured product free of any odor. The advantages of the acetoxy-cure system are relatively fast cure and short tack-free time. The alkoxy crosslinker systems were developed about 10 years ago and feature a somewhat longer tack-free time and slower cure-through capability. Their advantage is that they produce a by-product that is noncorrosive and has a mild, inoffensive odor while curing.

One-component sealants using the octoate crosslinker system combine the advantage of rapid cure of the acetoxy sealants with the low odor characteristics of the alkoxy materials by using octanoic acid as the by-product of cure. The amine and ketoxime systems possess the benefit of extremely low corrosion behavior but have somewhat longer tack-free and cure times.

All two-component, condensation-cure systems make use of a crosslinker system with four reactive sites and are currently using alkoxy technology, which results in the evolution of an alcohol, usually ethanol or methanol, as a cure by-product.

In one-component formulations the crosslinker/catalyst is added to the filled polymer and immediately reacts with the polymer as described in Eq. (4)

$$R\ Si\ X_3 + HO-\underset{\underset{CH_3}{|}}{\overset{\overset{CH_3}{|}}{Si}}-O-\ldots\longrightarrow$$

$$R-\underset{\underset{X}{|}}{\overset{\overset{X}{|}}{Si}}-O-\underset{\underset{CH_3}{|}}{\overset{\overset{CH_3}{|}}{Si}}-O-\ldots + HX \tag{4}$$

The reaction occurs at each end of each polymer chain and results in the positioning of two moisture-hydrolyzable reactive sites at each end of each polymer chain. The compound at this point is ready for use and must be kept away from moisture or moisture vapor to avoid curing as shown in Eq. (5)

$$\ldots-O-\underset{\underset{CH_3}{|}}{\overset{\overset{CH_3}{|}}{Si}}-O-\underset{\underset{R}{|}}{\overset{\overset{X}{|}}{Si}}-X + X-\underset{\underset{R}{|}}{\overset{\overset{X}{|}}{Si}}-O-\underset{\underset{CH_3}{|}}{\overset{\overset{CH_3}{|}}{Si}}-O-\ldots + H_2O\longrightarrow$$

$$
\ldots -O-\overset{\overset{\displaystyle CH_3}{|}}{\underset{\underset{\displaystyle CH_3}{|}}{Si}}-O-\overset{\overset{\displaystyle X}{|}}{\underset{\underset{\displaystyle R}{|}}{Si}}-O-\overset{\overset{\displaystyle X}{|}}{\underset{\underset{\displaystyle R}{|}}{Si}}-O-\overset{\overset{\displaystyle CH_3}{|}}{\underset{\underset{\displaystyle CH_3}{|}}{Si}}-O-\ldots + 2\,HX \qquad (5)
$$

Equation (5) shows only the crosslinking of two adjacent polymer chains through the primary reactive sites. The secondary reactive sites will continue to react with moisture vapor and to crosslink adjacent polymer chains until all crosslink sites have been completely consumed. The resulting matrix will be a three-dimensionally crosslinked, elastomeric silicone rubber.

The formulation and cure of a condensation, two-component silicone material is similar to that of a one-component system. In this case the filler, polymer, crosslinker, and a controlled amount of moisture are compounded. The crosslinker in this case is generally an alkoxy active species as described in Eq. (3). To begin the curing reaction, a small amount of an active metal catalyst, such as dibutyl tin dilaurate or tin octoate, is added. The moisture contained in the compound then reacts with the alkoxy groups to form active silanol sites which, in turn, condense with the silanol sites at the ends of the polymer chain to form a crosslinked, elastomeric silicone rubber. An alcohol is evolved as a by-product of cure. Equation (6) shows the chemistry involved.

$$
\ldots-O-\overset{\overset{\displaystyle CH_3}{|}}{\underset{\underset{\displaystyle CH_3}{|}}{Si}}-OH + (C_2H_5O-)_4Si + H_2O \xrightarrow{\begin{array}{c} DIBUTYL\ TIN \\ DILAURATE \end{array}}
$$

$$
\left[\ldots-O-\overset{\overset{\displaystyle CH_3}{|}}{\underset{\underset{\displaystyle CH_3}{|}}{Si}}-O- \right]_4 Si + 4\,C_2H_5OH \qquad (6)
$$

Equations (5) and (6) depend on the reaction of moisture with a hydrolyzable species to form a silanol, followed by a condensation of adjacent silanol groups to develop a crosslink. Once the hydrolysis reactions have been completed, crosslinking may be accelerated by exposure to elevated temperatures. It is important to avoid premature application of heat, particularly with lack of humidity control, because these reactions do require moisture for hydrolysis, and elevated temperatures may tend to reduce the relative humidity and, thus, prevent the hydrolysis reaction from taking place.

VI. PERFORMANCE PROPERTIES OF ONE-COMPONENT RTV SILICONE SEALANTS

The one-component RTV silicone adhesive sealants under discussion all cure by reacting with the moisture in the air as discussed earlier. The materials cure from the surface toward the center as atmospheric moisture contacts the material and diffuses inward. As the material cures, the by-product of cure forms and diffuses toward the surface where it evaporates into the atmosphere. A 1/8-in.–diameter bead (as illustrated in Fig. 1) of a typical acetoxy-curing, one-component RTV will take approximately 12 hr to convert completely from a paste to a solid elastomer at 25°C (77°F) and 50% rh.

A 1/4-in.–diameter bead of the same material will take approximately 24–30 hr to cure to a solid elastomer under the same conditions. Since the curing reaction depends on the diffusion of moisture vapor in one direction and cure by-product in the other direction, the thickness of the section to be cured and the distance of any point within the RTV section from an air interface become limiting factors in the utilization of the material. It is not recommended that one-component RTVs be used in sections where any material is more than 1/4 in. from an air interface. Material which is greater than 1/4 in. from an air interface will take an extremely long time to cure. Cure times up to 3–4 weeks for acetic-acid-curing materials at a depth of 1/2 in. have been observed.

Table 1 is a simplified guide for comparing the tack-free times and the cure-through times, at 25°C (77°F) and 50% rh, of the various moisture-vapor–curing RTV silicone sealants which are currently available. The times given reflect merely the period necessary to convert the materials from a paste to a solid for a bead from 1/8 to 1/4 in. in diameter. There may be exceptions to this information, and the reader is cautioned to check with the product manufacturer or to run

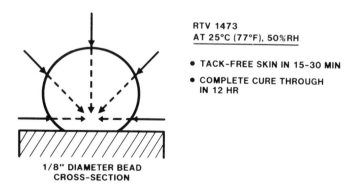

RTV 1473
AT 25°C (77°F), 50%RH

• TACK-FREE SKIN IN 15-30 MIN
• COMPLETE CURE THROUGH IN 12 HR

1/8" DIAMETER BEAD
CROSS-SECTION

Fig. 1. One-component RTV sealants cure from the surface inward.

Table 1 Tack-Free and Cure-Through Times for Moisture-Vapor—
Curing RTV Silicone Sealants (25°C, 50% rh)

	Acetoxy	Octoate	Alkoxy	Amine
Tack-free time (hr)	$\frac{1}{4} - \frac{1}{2}$	$\frac{1}{4} - \frac{3}{4}$	2−3	2−3
Cure time for 1/8−1/4-in. diameter bead (hr)	12−24	12−24	48−72	48−72

tests to determine differences in tack-free time and cure time between
specific products.

VII. ACCELERATING CURE

It logically follows from the previous discussion that the key variables
affecting the rate of cure of one-component, moisture-curing RTVs
are humidity, temperature, and geometry of the bondline, with particular emphasis on the surface-area−to−volume ratio of the material
and the distance of the deepest section from an air interface.

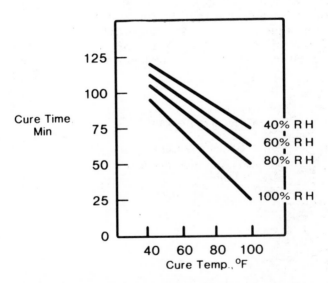

Fig. 2. RTV 1473 cure time versus temperature at four humidities for
1/16-in.−diameter beads.

It is possible to accelerate the cure of one-component RTV sealants. The effect appears to be most dramatic where the surface-area—to-volume ratio is high and where the maximum distance of any section of RTV sealant to an air interface is 1/8 in. or less.

The data presented in Fig. 2 illustrate the cure time for a 1/16-in.—diameter bead of RTV 1473, a typical acetoxy-cure silicone sealant at different levels of temperature and humidity. The data were generated using 1/16-in.—diameter beads extruded on a flat surface and placed in a temperature-humidity chamber at various conditions. At specified periods of time the specimens were withdrawn and sectioned to determine if they had completely cured. The cure times plotted in the figure illustrate the times necessary to convert the material from a liquid to a solid through the entire section. They make no claims for optimum development of material properties.

As can be seen from Fig. 2, the material in a 1/16-in.—diameter bead configuration would take about 1 1/2–2 hr to cure at 25°C (77 °F) and 50% rh. The cure time was accelerated to 1/2 hr at conditions of about 38°C (100°F) and 100% rh.

Because of the thin cross section and the relatively short distances for moisture vapor and by-products of cure to travel, the effect of increased temperature and humidity was quite dramatic in this case. In thicker cross sections (from 1/8 to 1/4 in.) and in configurations, such as fillet seals, where the surface-area—to—volume ratio is somewhat less, the effect of increasing temperature and humidity will be less marked.

Experience has shown that the acetoxy and octoate materials respond most favorably to accelerated-cure conditions. The alkoxy materials do not respond quite as well and, furthermore, may actually be inhibited in their rate of cure if the humidity is high enough to cause condensation on the surface of the RTV.

Because of the difficulty of obtaining sufficiently high levels of relative humidity and because of the sensitivity of the catalyst systems to thermal degradation, curing temperatures in excess of 65°C (150°F) should be used with extreme caution. Exposure of uncured material to a temperature in excess of 65°C (150°F) with insufficient humidity for the cure reaction to take place can result in destruction of the material's ability to cure. For thick sections it is possible to use a two-stage cure consisting of moderate temperatures of approximately 50–55°C (120–130°F) with very high humidities for a short period of time, followed by an increase to 70–80°C (160–180°F) without humidity control to finish the curing reaction.

The information presented in the previous sections is only a guide for the development of accurate process data for each individual application. The following points may be made in summary.

1. Acetoxy and octoate sealants respond most favorably to accelerated-cure conditions.

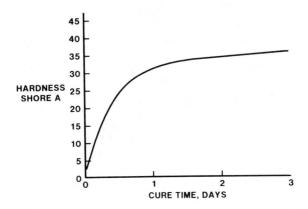

Fig. 3. Shore A hardness versus cure time for a typical one-component, acetoxy silicone sealant cured at 25°C (77°F) and 50% rh.

2. Avoid temperatures in excess of 65°C (150°F), at least for the initial stages of accelerated cure.
3. Humidity is necessary, even under accelerated conditions, to effect cure.
4. Geometry, with respect to the distance of uncured materials from an air interface and the surface-area—to—volume ratio, is the most important variable in determining the impact of accelerated-cure conditions on cure time.
5. In the previous discussion complete cure has been defined as the conversion of the material from a paste to a solid elastomer. Development of design properties requires additional cure time at room temperature.

VIII. DEVELOPMENT OF CURED PROPERTIES WITH TIME

In earlier sections we have discussed the time necessary to convert one-component, moisture-curing RTV sealants from their liquid form to their solid, elastomeric form. Depending on either the material thickness or the cure system and the temperature-humidity conditions, cure times (for conversion of the material from a liquid to a solid) may range from about 1/2 to 72 hr. Development of optimum properties, however, usually requires additional time at room temperature.

Figure 3 illustrates the development of Shore A durometer hardness as a function of time for a typical acetoxy sealant cured at 25°C (77°F) and 50% rh. Even though the material has converted completely from a paste to a solid within 24 hr, it takes an additional 2–3 days for the Shore A durometer hardness to build to its maximum level.

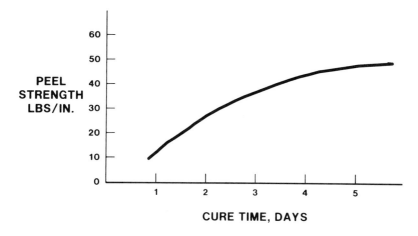

CURE TIME, DAYS

Fig. 4. Peel strength versus cure time for RTV 102 applied to 2024 alclad aluminum.

The additional 2–3 days are necessary to complete the crosslinking process.

Figure 4 illustrates the development of peel adhesion with time for RTV 102 applied to 2024 alclad aluminum. Although the material has cured completely within 24 hr, the data indicate that peel adhesion at that time is virtually nil. Three to four days later the adhesion has reached its maximum design level.

IX. ADHESION OF ONE-COMPONENT SILICONE SEALANTS TO VARIOUS SUBSTRATES

Although one-component RTV silicone sealants have generally been found to bond well to various substrates [2], it is important to evaluate a particular RTV sealant, made by a given manufacturer, with a given cure system on the surfaces to be bonded. It should not be assumed that one-component products having the same cure system, but made by different manufacturers or bearing different product numbers, will perform similarly on a given surface.

Adhesion of one-component, moisture-curing sealants is enhanced by the presence of surface hydroxyl groups, which are usually present in abundance on glass and, to a somewhat lesser extent, on aluminum. Metals which have a high degree of reactivity with the acid by-products of these sealants will generally produce poor bonds because of the formation of salts.

In general, the ability of these materials to bond to olefenic plastics is poor. The presence of silicaceous or glass-fiber fillers, if available in sufficient quantity on the surface to be bonded, will enhance adhesion. Other plastics, such as acrylics and polycarbonates in their virgin form, will adhere well to many of the sealants. Molded plastics having an uncontrolled amount of rework, various slip additives and process aids, and surfaces contaminated with mold release agents form uncontrolled and, therefore, unreliable surfaces for adhesion. Flexible vinyls and other plasticized polymers are also extremely difficult to bond because of the release-agent effect of the migrating plasticizers.

The user is cautioned to control the quality of the surface being bonded and to develop specific adhesion information for a given sealant/surface combination.

X. SURFACE PREPARATION

One of the attractive features of silicone sealants is the relative lack of surface preparation required for developing effective bonds. On glass, aluminum, and steel materials all that is necessary is to render the surface structurally sound by removal of loose particles or corrosion and to remove by means of a solvent wipe all traces of grease or oil contamination. The surface, of course, should be dry and free of any moisture.

Experience has shown that lap shear bonds to glass or aluminum in excess of 200 psi (1400 kPa) can be obtained with some silicone sealants using only the surface preparation procedure described above. Peel strength bonds in the range of 20—40 lb/in. and with certain products in the range of 60—80 lb/in. and above, may also be achieved on these surfaces under the same conditions. Bond failures in either of these modes are in the range of 90—100% cohesive failure.

XI. PRIMERS

If, after having evaluated a certain silicone adhesive sealant/substrate, combination for adhesion, you find that the adhesion levels are not satisfactory, it may be possible to improve adhesion by the use of a primer.

Primers currently available for use with silicone sealants are generally low-solids, solvent dispersion of: silicone resins; silanes; organic resins; or some combination of these materials. There are many types available, and they are generally suggested for evaluation on specific types of substrates. Primers recommended for use on metals or ceramics generally differ from those suggested for use on plastics.

Most available primers are made to be applied by spraying, brushing, or dipcoating structurally sound, contamination-free sur-

faces. Curing of the primers is generally accomplished at room temperature by permitting sufficient time for the solvent to evaporate and
for the remaining resins to hydrolize by reacting with the moisture in
the air. It may be possible to improve the adhesion of certain primers
with a postcure operation at elevated temperatures.

Because of the sensitivity of most of these primers to moisture
contamination, it is important to minimize atmospheric moisture in the
solvent solutions. Some primers may give visual evidence of atmo-
spheric-moisture spoilage by the development of a precipitate. Others
may give no visual indication at all but merely not perform as designed.

XII. TWO-COMPONENT RTV SILICONE ADHESIVE SEALANTS

Any discussion of RTV silicone adhesives sealants would be incomplete without some reference to the technology of two-component RTV
silicone rubber adhesive systems [3].

While not true adhesives in themselves since virtually all products available today require primers to develop adhesion, two-component RTV silicone rubbers combined with the appropriate primer
do offer some distinct differences in performance as adhesives compared to one-component products. The two-component systems offer
a wider range of pot-life and cure times than the one-component formulations do. These differences result from the control of cure-tem-
perature and the use of specific levels of catalysts or curing agents.
Some two-component systems can be cured in sections of unlimited
depth or in totally enclosed spaces since no by-product of cure is
given off.

Two cure systems are available. The older is the condensation-
cure system which is referred to in Sec. II. It is similar in its actual
chemical characteristics to the cure for a one-component system in
that it requires moisture to cure and gives off a by-product of cure.
The other cure system commonly used in two-component RTV compounds is called the platinum-catalyzed, vinyl-addition cure system.
In this particular technology the polymer and crosslinkers are contained in separate components. One of the components contains a silicone polymer with reactive vinyl groups. The other contains a silicone polymer with a reactive hydride species. When the two components are mixed, the hydride combines with the vinyls in a free-radical addition reaction to cause a coupling between the two silicone
polymers. This reaction is catalyzed by the addition of a trace amount
of a platinum catalyst, which is contained in one of the components.
Cure time is generally controlled by raising or lowering the temperature.

Two-component silicone rubber materials are available in a wide
range of viscosities and physical properties [4]. Some two-component

RTV silicone rubbers use silicone resins which have a low percentage of phenyl (for methyl) substitution groups on the linear polymer. Phenyl substitution has the effect of increasing the irregularity of the linear polymer chains, thus depressing the glass transition point T_g. RTV materials formulated around dimethyl silicone polymers customarily have a T_g of approximately $-55°C$ ($-70°F$). RTV materials formulated around phenyl-substituted polymers may have T_gs as low as $-115°C$ ($-175°F$). The improved versatility and variety of choice that two-component systems offer is somewhat offset by the increased complexity of use dictated by the necessity of using primers for virtually all adhesive applications and the need for mixing equipment and degassing equipment. Indeed, most two-component RTV compounds have been used for nonadhesive applications, such as potting and encapsulating or the fabrication of molds.

XIII. HEAT-AGING CHARACTERISTICS OF RTV ADHESIVE SEALANTS

Most silicone adhesive sealants possess the property of extraordinary heat-aging characteristics.

A properly cured silicone adhesive sealant will experience a reduction of approximately 50% in elongation and tensile characteristics after 2000–3000 hr of continuous exposure to a temperature of 200°C (390°F). The same materials will survive for short times (up to 200 hr) at temperatures as high as 260°C (500°F).

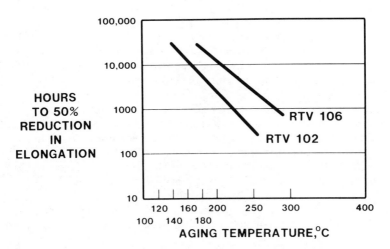

Fig. 5. Heat-aging characteristics of one-component RTV silicone sealants.

The characteristic ability of certain types of red iron oxide to improve the thermal endurance of silicone polymers is utilized in the formulation of compounds resistant to extra high temperatures. These red compounds will survive 2000–3000 hr of continuous exposure to temperatures as high as 260°C (500°F) and short periods (150–200 hr) of exposure to temperatures as high as 300°C (575°F). The data presented in Fig. 5 show the heat-aging characteristics of a standard silicone RTV adhesive sealant and a red, high-temperature silicone RTV adhesive sealant. The data were generated using Underwriters Laboratories procedure UL 746 [5].

These data show the time necessary for the elongation of these materials to decrease to 50% of the original, unaged elongation at each of four temperatures. The lines drawn are least-square fits to the available data.

It should be noted that elongation is the property most sensitive to thermal degradation in a cured silicone adhesive sealant. Lap shear specimens aged at the temperatures described earlier in this section retained at least 90% of their original adhesion characteristics at the time that the elongation of that particular material had decreased 50%. Electrical properties—as measured by dielectric strength, dissipation factor, dielectric constant, and volume resistivity—also remain substantially unchanged, and in some cases improve slightly, after heat aging.

XIV. ELECTRICAL PROPERTIES

Most silicone adhesive sealants have been formulated to be dielectrics. Properly cured silicone adhesive sealants will generally have dielectric strengths in the range of 400 volts/mil or greater, at thicknesses of 50–60 mils. Dielectric constants of 10^{13}–10^{15} are typical, and dissipation factors tend to be in the range of 0.01 or less or in the range of 10–10^5 Hz.

As described in Sec. XIII, silicone adhesive sealants maintain their dielectric properties for long periods of time after continuous exposure to high temperatures.

XV. CHEMICAL RESISTANCE

Cured silicone rubber offers a wide range of chemical and environmental resistance. Continous exposure to ultraviolet light, ozone, salt spray, or any combination of these three things has little or no effect on the physical properties of properly cured silicone rubber adhesives. Intermittent exposure to acids or bases will have little or no effect although continous immersion is not recommended. Silicone rubber adhesive sealants are oleophilic and swell when immersed in various kinds of oil. Differences in formulation can improve the oil-swell characteristics, and certain silicone adhesive sealant materials have been formulated to control swell to 25—30% in typical lubricating oil, such as 10-W-30 SE- and SF-grade engine oils. The user is cautioned, however, to carefully evaluate the particular sealant for compatibility with the oil or fluid with which it is designed to be used since there is a wide variation in the performance of these materials both in oils and among oils of different types.

For the ultimate in chemical and gasoline or solvent resistance, certain RTV silicone adhesive sealants have been formulated around silicone polymers with fluorine substitution. Fluorosilicones, as they are called, have extremely good resistance to hydrocarbon fuels containing high levels of aromatic compounds. Such fuels normally result in the complete destruction of standard dimethyl silicone adhesive sealants.

XVI. DISPENSING AND EQUIPMENT

One-component RTV silicone adhesive sealants are typically supplied in convenient packages consisting of collapsible, aluminum "toothpaste" tubes, 10-12 fl. oz. caulking cartridges, and 6 fl. oz. Semco caulking cartridges.

The collapsible aluminum tubes may be used to apply the material by hand, or the material may be extruded from the appropriate caulking cartridge using a hand-operated caulking gun or an air-

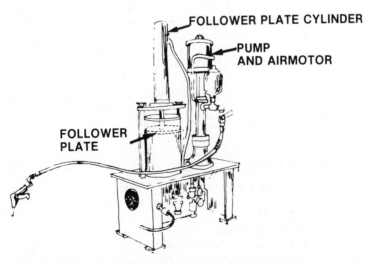

Fig. 6. Bulk-container extrusion pump.

pressure-operated caulking gun. It is advantageous to use nozzles which are made of polyethylene since one-component RTV silicone adhesive sealants, when cured, will not adhere to this material. Cleaning up after use may consist of letting the material set overnight in a nozzle and then pulling the cured plug out of the nozzle immediately prior to the next use.

For larger-volume applications, one-component silicone sealants are available in 5-gal pails or 55-gal drums and may be dispensed from these containers using divorced style, positive-displacement extrusion pumps of a type typically used for dispensing grease materials [6] (Fig. 6).

Because one-component materials cure spontaneously upon contact with atmospheric moisture, certain precautions and changes in materials and design must be made to render extrusion pumps satisfactory for dispensing RTV silicone sealant.

These modifications include replacement of all seals between the RTV material and the atmosphere with polyethylene or Teflon. Leather or rubber seals are not acceptable because they transmit atmospheric moisture which will cause premature curing of the RTV sealant. The follower plate should have a Teflon or polyethylene perimeter seal. All flexible hoses should be of the hydraulic type, rated for service at a minimum of 1000 psi (7000 kPa), and should be Teflon-lined. Absence of a Teflon lining will cause the RTV material in the hoses to gradually cure on the inside of the hose, thus restricting the inside diameter, increasing the pressure drop, and, ultimately, rendering the equipment inoperable. All couplings and joints should be sealed with Teflon pipe tape or pipe dope to prevent the transmission of atmospheric moisture. There should be no piping in the system from the outlet of the pump to the dispensing nozzle with a diameter of less than 0.5 in. (1.25 cm) with the exception of the nozzle. For runs greater than 25 ft (8 m) or where manifolding is necessary to service more than one station, pipe diameters of 1 in. (2.5 cm) or more should be used to avoid excessive pressure drop. A pump with a chopcheck-type inlet, instead of the usual ball inlet commonly employed for greases and other low-viscosity material, should be used. A chopcheck inlet contains a "foot" which mechanically pulls the sealant paste into the pump rather than relying on the operation of a partial vacuum near the inlet to cause material to flow into the pump. Failure to use this type of inlet system may result in cavitation and unsatisfactory performance. A pump with an air-motor—driven piston and a 20:1 ratio between the air-motor drive and the fluid-pumping piston should be used for satisfactory performance.

The same types of pumps, combined with static mixing equipment and proportioning devices, have successfully been used to mix and dispense two-component RTV systems. The use of a proportioning pump system and a static mixer has the additional advantage of eliminating the degassing step since the mixing is conducted in an air free atmosphere.

Since most one-component RTV silicone adhesive sealants are paste-type materials, their viscosities are customarily measured by means of an application-rate technique. A widely used test consists of extruding the material through a 0.125-in.--diameter (0.3 cm) nozzle at a pressure of 90 psi (630 kPa)—typically from a Semco caulking cartridge—for a period of 1 min and expressing the application rate in g/min. Silicone adhesive sealants are formulated with application rates ranging from 40 to 50 g/min to as high as from 700 to 800 g/min when tested under these conditions.

XVII. STORAGE CONDITIONS

One-component and two-component silicone adhesive sealants in their uncured state are sensitive, reactive chemicals and, as such, should be protected from extreme environmental conditions. The one-component, moisture-hydrolyzable materials should, of course, be protected from exposure to atmospheric moisture and be kept in airtight containers. Storage at temperatures in excess of 30°C (80°F) should be strictly avoided. The alkoxy-cure systems in particular are sensitive to elevated temperatures and can benefit from storage at freezer conditions of −20°C (0°F) or below. Two-component RTV compounds should also be stored at temperatures below 30°C (80°F) and will benefit from freezer storage at −20°C (0°F).

XVIII. SAFETY AND TOXICOLOGY

Uncured RTV silicone materials are reactive chemicals, and their by-products of cure may, under certain conditions cause skin irritation or irritation of other sensitive membranes. Cured RTV silicone rubber has been shown to be relatively bio-inert. The by-products of thermal decomposition are typically carbon dioxide, water, and silicon dioxide along with traces of other metal oxides used in the catalyst systems and traces of colorants.

REFERENCES

1. J. M. Klosowski and G. A. Grant, The chemistry of silicone room temperature vulcanizing sealants, in *ACS Symposium Series No. 113*, ACS, Washington, D.C., 1979.
2. RTV silicone rubber, in *Technical Data Book S-35*, General Electric, Schenectady,
3. M. D. Beers, Silicone adhesive sealants, in *Handbook of Adhesives* (I. Skeist, ed.) Reinhold, New York, 1977.

4. *Silicone Elastomers Technical Data Sheets 61-358-77, 61-006C-76, 61-068A-77, 61-349-76, and 61=020C-77*, Dow Corning, Midland, Michigan.
5. Long term heat aging characteristics of polymeric materials, in *Underwriters Laboratory Procedure UL 746*,
6. P. C. Miller, *Tooling and Production April*: (1975).

16

Water-Based Adhesives

W. A. Pletcher and Edmund J. Yaroch   *3M Company, St. Paul, Minnesota*

I. INTRODUCTION

Continuing environmental and economic pressures on solvent-based adhesives will greatly increase the market potential for water-based adhesives. Quite a variety of water-based adhesives exists today. They range from pressure-sensitive adhesives to high-strength, contact-bond adhesives.

Some of the current uses of water-based adhesives are labels, packaging, wood bonding, fabric bonding, panel lamination, high-pressure laminate bonding, construction mastics, bonding of floor and wall tile, and tapes.

There are several types of water-based adhesives. This chapter deals mainly with water-dispersed synthetic rubber adhesives although other types of systems are discussed. Chemistry and formulating are discussed. Information is provided on applying, drying, and bonding of water-based adhesives. Physical and performance properties of water- and solvent-based adhesives are compared, and end uses for adhesives based on several types of polymer are given. Characteristics of water-based adhesives are discussed, and techniques for adhesive selection are reviewed.

II. TYPES OF WATER-BASED ADHESIVES

In general, the area of water-based adhesives can be divided into two broad categories: water-soluble and water-dispersed products. Water-soluble adhesives include the animal glues, vegetable glues, and other (natural) systems [1]. Water-dispersed adhesives include those based on natural-rubber latices, synthetic-rubber (polymer) latices, and postformed-rubber (polymer) latices.

Natural latex refers to material obtained primarily from the rubber tree. Synthetic latices are aqueous dispersions of polymers obtained by the process of emulsion polymerization. These include polymers of chloroprene, butadiene-styrene, butadiene-acrylonitrile, vinyl acetate, acrylate, vinyl chloride, etc. Molecular weights range from 10,000 to 1,000,000, and particle sizes range from 0.001 to 10 μm. Artificial latices are made by dispersing solid polymers. These latices include dispersions of reclaim rubber, butyl rubber, styrene-butadiene rubber (SBR), rosin, rosin derivatives, asphalt, coal tar, and a large number of synethetic resins derived from coal tar and petroleum.

Since the majority of the solution-type adhesives remain water-soluble after drying, their use is restricted to those areas that do not require water resistance in the final bond. Most water-dispersed adhesives cannot be redispersed in water after they have once dried and can, therefore, be used in applications that require water resistance. It is expected that of the two types of adhesive, the water-dispersed types are the ones which will replace solvent-based adhesives.

III. CHEMISTRY AND FORMULATING OF WATER-DISPERSED ADHESIVES

In addition to the base polymer, all three types of water-dispersed adhesives contain a variety of other materials designed to impart specific properties to the adhesive. These materials can include emulsifiers, surfactants, tackifiers, antioxidants, fillers, thickeners, antifoams, freeze-thaw stabilizers, corrosion inhibitors, etc.

A. Natural-Rubber Latices

The natural rubbers are combined with tackifying resins and antioxidants to obtain pressure-sensitive adhesives. The tackifiers are anionic or nonionic resin emulsions that cause the rubber particles to swell and soften by virtue of their mutual solubility. These resins include polyterpenes, aliphatics based on petroleum feedstock, esterified-hydrogenated rosins, and others.

B. Synthetic-Rubber (Polymer) Latices

The majority of today's water-dispersed adhesives fall into the category of synthetic-rubber latices.

1. *Neoprene latices.* The polychloroprene materials, produced by E. I. Du Pont and others, are obtained by emulsion-polymerization techniques and have predominantly the trans-1,4-configuration. Solids usually range from 40—60% with molecular weight, chain branching, gel, and crystallinity differing from product to product. Copolymers of 2-chloro-1,3-butadiene and monomers like 2,3-dichloro-1,3-butadiene, acrylonitrile, methacrylic acid, and others are commercially available able and provide specific property modifications. The neoprenes are generally combined with rosin modifiers, adhesion promoters, crosslinking agents, antioxidants, and other materials to obtain high-performance adhesives. Resins used to modify neoprenes include the terpenes, terpene phenolics, coumarone-indenes, *t*-butylphenolics, rosin esters, pentaerythritol, and resorcinol-formaldehyde types [2]. These modifiers can provide adhesion to specific substrates, increase autoadhesion, increase open times, and reduce raw-material costs. Adhesion promoters are used commonly and include colloidal silicas for bonding glass, metals, and leather; blocked isocyanates for bonding elastomers, and synthetic fabrics; proteins, and other materials for obtaining specific adhesion. The varied crosslinking agents used include metal oxides, methyl ureas, melamines, epoxy resins, and others.

2. *Styrene-butadiene rubber latices.* The SBR materials are obtained by emulsion-polymerization techniques first developed

by Bayer in 1912 and improved for large-scale production in
later years. Carboxylated SBRs provide an excellent range
of polymer properties. Their high tensile and elongation,
excellent water resistance, excellent aging stability, and outstanding adhesion to fibers make them the materials of choice
in many synthetic-fiber adhesive applications. As with the
carboxylated neoprenes, the carboxylated SBRs provide a
mechanism for self-curing and are, therefore, useful in
many high-performance adhesive applications.

3. *Acrylic latices.* The acrylic materials are also obtained by
   emulsion-polymerization techniques and usually consist of copolymers or terpolymers. Many types of properties are possible, depending on the monomers utilized, ranging from
   pressure-sensitive adhesives (PSAs) to high-performance
   adhesives. Typically, the waterborne acrylic latices have
   provided PSAs with unusually varied properties. With the
   appropriate monomer-to-charge ratios, crosslinking mechanisms may be synthetically incorporated to regulate tack,
   shear, and cohesive strength. In these systems the molecular weight, glass transition temperature T_g, polarity, and
   functionality are also very important in providing the final
   required PSA properties.
   Acrylics have fundamental properties in the area of ultraviolet (UV) resistance, thermal stability, and plasticizer resistance. Enhancement of these basic properties may be obtained using a variety of crosslinking or curing mechanisms.
   Monomers, such as acrylamide, N-methylolacrylamide, glyci-
   dylmethacrylate, and dimethylaminoethylacrylate, are commonly used to provide internal curing sites. Zinc acetate,
   urea-formaldehyde, and sodium carbonate or sodium sesqui-
   carbonate may also be considered for use as external crosslinkers.
   More recently acrylic latices have been considered the basis
   for higher-performance adhesives [3]. Particular acrylic
   polymers with T_gs in the range of $0-10°C$ have been tackified to give green strength, open time, adhesion to nonporous substrates, and high-temperature resistance.

4. *Other latices.* Other synthetic-rubber (polymer) latices that
   are used as water-based adhesives include the butadiene-
   acrylonitrile copolymers, polyvinyl acetates, vinyl-acetate–
   vinyl-chloride copolymers, ethylene–vinyl-acetate copolymers, ethylene–acrylic-acid copolymers, polyurethanes, carboxylated butadiene–vinylidene-chloride copolymers, and
   many others.

Postformed-Rubber (Polymer) Latices

Reclaim rubber is recovered from fabricated rubber products, such as automobile tires and tubes. These materials are broken down, devulcanized, and dispersed in water to give high-solids, latex or mastic adhesives. Often they are combined with asphalt for use in the automotive and construction markets as sealers as well as adhesives.

Another type of postformed rubber latex is the butyl-rubber system. Here the monomer is solution polymerized and dispersed in water. The organic solvent is removed. The latex is then generally formulated in a manner similar to those latices derived from emulsion polymerization.

IV. FILM FORMATION OF WATER-DISPERSED ADHESIVES

In the wet state, the particles of polymer, resin, and other modifiers in water-dispersed adhesives exist as individual particles. These particles coalesce into a homogeneous film during the drying operation (Fig. 1).

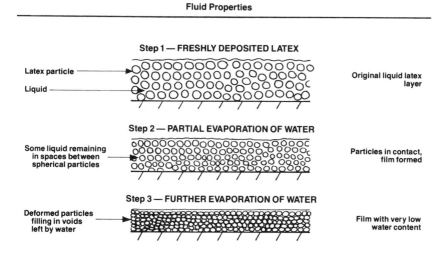

Fig. 1. Film formation of water-dispersed adhesives.

The most consistent theory on film formation of water-dispersed adhesives is the one proposed by Vanderhofft et al. in 1973 [4]. Initially, there is a uniform rate of water loss until particles begin to contact each other and coalesce. This phase is followed by a period of decreasing rate of water loss during which particle coalescence becomes irreversible. Finally, an exponentially decreasing rate of water loss occurs through capillary action and/or diffusion. Heat treating an air-dried film of water-dispersed adhesive will allow the polymer-rubber particles to intermix and will provide improved adhesive properties. This is normally accomplished in the usual process of force drying.

V. APPLICATION OF LATEX ADHESIVES

The normal methods of adhesive application—spray, roll, curtain, flow, brush, and knife coating—can be used with latex adhesives [5]. The allowable application methods are governed by certain properties of the adhesives, including viscosity, mechanical shear stability, degree of foaming, size and shape of parts to be coated, and the required production rate (Fig. 2).

A. Spraying

Spraying is one of the fastest, most economical methods of applying adhesive to large areas. Three means of application are

1. *Air spraying.* The most common spray method, air spraying, is usually equated with a paint sprayer. Adhesive is atomized by high air velocity, while relatively low pressure is used to pump adhesive to the spray tip. Air spraying provides a uniform adhesive film of varying texture.
2. *Airless spraying.* When airless spraying is used, adhesive is atomized by high adhesive velocity and ejected under pressures ranging from 1500 to 3000 psi. No air pressure is directly involved. The high-velocity adhesive is atomized when it strikes the air surrounding the spray tip. Spray pattern depends on the type of spray tip on the gun.
3. *Automatic spray equipment.* Large and fast production runs, where the parts are brought to the equipment, can make effective use of automatic spray equipment.

Some latex adhesives can be coagulated by high shear, certain metals, and solvents. Pumps that do not subject the adhesive to a shearing action, such as a diaphragm pump, are normally used. Hoses should be nylon- or polyethylene-lined. Packings and glands in contact with the adhesive should be made of Teflon. All pipe fittings should be plastic, stainless steel, or nickel plated.

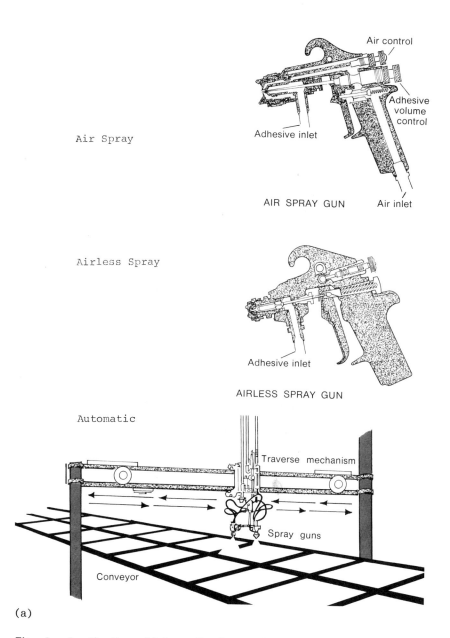

Air Spray

Airless Spray

Automatic

(a)

Fig. 2. Application of latex adhesives. (a) Spraying; (b) Roll coating; (c) Curtain coating; (d) Flow coating; (e) Knife coating.

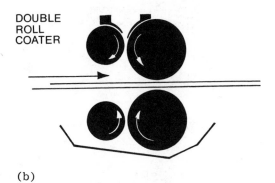

(b)

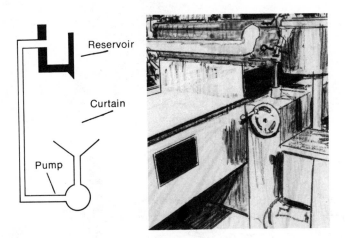

(c)

Fig. 2. (Continued)

A collapsible tube of adhesive is the simplest piece of flow equipment.

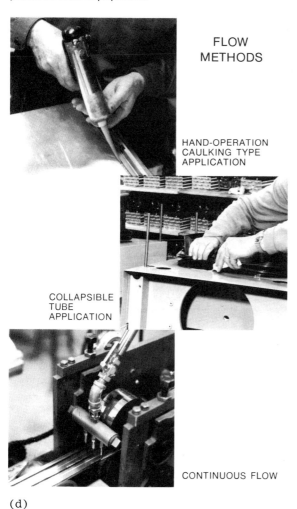

FLOW
METHODS

HAND-OPERATION
CAULKING TYPE
APPLICATION

COLLAPSIBLE
TUBE
APPLICATION

CONTINUOUS FLOW

(d)

Fig. 2. (Continued)

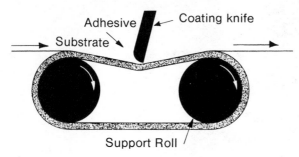

(e)

Fig. 2. (Continued)

If the equipment was previously used for handling solvent-based products, the hoses should be replaced before using a latex adhesive since solvent absorption into the hose could cause coagulation of the adhesive.

B. Roll Coating

For low-volume production or small jobs, paint rollers are convenient applicators. For high-volume or multistation production, pressure-fed hand rollers are available.

Roll coating is used on large, flat areas when large volume and fast production are required. Line speeds of 50–60 ft/min are common. The adhesive is transferred from the surface of the roll to the substrate. Double roll coaters coat from both sides of the substrate simultaneously.

The quantity of adhesive applied is usually controlled by using an engraved roll. The latex adhesive must be designed to prevent excessive foaming during the coating operation since foaming will prevent the roll from depositing the desired quantity of adhesive. Excessive foaming can also present a housekeeping problem.

C. Curtain Coating

Curtain coating is faster than roll coating, but only one side of the substrate can be coated at a time. A wide and continuous waterfall of specially formulated adhesive coats the substrate as it passes underneath. Film thickness is controlled by the adhesive solids content and the speed of the substrate. This method is most useful with broad surfaces.

D. Flow Coating

Equipment for flow coating varies with production requirements. For high-speed production, a continuous flow gun is used. A remotely operated pump maintains constant pressure and adhesive flow. For intermediate- or low-volume applications, hand operated caulking-type guns are satisfactory. A collapsible tube filled with adhesive is the simplest piece of flow equipment.

E. Brushing

The use of a paint brush is common for small-area applications. For higher production rate, a flow brush is recommended. A flow brush is pressure-fed and is usually recommended for continuous application. Intermittent application permits the adhesive to dry on the bristles.

F. Knife Coating

Knife coating, with a mechanically driven unit, is used to apply high-viscosity adhesives at high production rates. Adhesive is applied to the substrate in excess. The substrate then passes under the knife, and all adhesive, except the required amount, is removed. The knife spreads and meters the adhesive.

The simplest version of knife coating is a putty knife used with a mastic adhesive for small jobs. Trowels are used for larger-area applications, such as floors.

VI. BONDING TECHNIQUES

The bonding techniques used for latex adhesives are about the same as those used for solvent adhesives: wet bonding, open-time bonding, contact bonding, and solvent reactivation.

The bonding technique used depends on the parts being joined, the adhesive itself, the immediate bond-strength requirements, and temperature and humidity conditions. All methods cannot be used with all types of latex adhesives.

A. Wet Bonding

When the wet bonding technique is used, at least one of the bonding materials must be porous. Appropriate materials include wood, mason ry, fabric, most flexible foams, leather, cardboard, fiberglass, felt, and cork.

It is usually necessary to apply adhesive to only one surface. The surfaces to be bonded are assembled while the adhesive is still wet. Immediate bond strength is usually low. In some cases it may be necessary to keep the parts being joined under pressure until enough strength has developed for subsequent handling.

B. Open-Time Bonding

Open-time bonding is a technique in which the adhesive is applied to both surfaces and allowed to stand (open) until suitable tack or sticki- ness has been achieved. This technique works best when at least one surface material is porous or semiporous. It is probably the most widely used method of making small- or medium-sized bonds. The simplest test to determine the right tack is to touch the adhesive with the knuckle. If the adhesive feels sticky and does not transfer to knuckles, it is ready. This method cannot be used with some types of latex adhesives that go directly from the wet to the dry state without passing through a tacky range.

C. Contact Bonding

Both surfaces are coated for contact bonding, and the adhesive is permitted to become dry to the touch. Within a given time, these sur- faces can be pressed together, and nearly ultimate bond strength is immediately achieved. Contact bonding is preferred when bonding two nonporous surfaces. Contact-bond adhesives are usually based on specially compounded neoprene latices. Drying times prior to bonding are about 30 min under ordinary conditions of temperature and humi- dity. This time can be reduced to seconds by force drying with heat.

D. Solvent Reactivation

In bonding by solvent reactivation, the adhesive is applied to the sur- face of the part and let dry. To prepare for bonding, the adhesive is reactivated by wiping with solvent or placing the part on a solvent- impregnated pad. The surface of the adhesive tackifies, and the parts to be bonded are pressed together. The tack range is short, so the technique is useful for relatively small-sized applications. It is not generally applicable on large surface areas since the short bonding range gives inconsistent results.

E. Heat Reactivation

Heat reactivation involves the application of a thermoplastic adhesive to one or both surfaces. After the adhesive has dried, coated parts may be stored several weeks or longer before bonding. To prepare a part for bonding, it is heated until the adhesive is soft and tacky. The bond is made under pressure while the adhesive is hot. On cooling, bond strength is as good as the adhesive will provide. This method is used with nonporous, heat-resistant materials when ultimate bond strength is quickly needed.

Heat reactivation can also be a continuous, in-line operation. The adhesive is applied in liquid form to a film or sheet, force dried to remove the water, and then laminated to a second surface while still hot. Temperatures are usually in the range of 250–350°F.

VII. FORCED DRYING OF LATEX ADHESIVES

The heat of vaporization of water is 540 cal/g. This compares to about 100 cal/g for the common solvents used in adhesives (Table 1) [6].

When bonding high-pressure laminate to plywood, a water-based latex contact adhesive requires about a 30-min drying time at 75°F and 50% rh. Under the same conditions, a solvent-based contact adhesive requires a drying time of about 10 min. Under very high hu-

Table 1 Table of Solvent Properties

Solvent	Heat of vaporization (cal/g)	Evaporation rate (ethyl ether = 1)[a]
Hexane	88	1.9
Toluene	87	4.5
Xylene	82	9.2
1,1,1-Trichloroethane	59	2.6
Acetone	125	1.9
Methyl ethyl ketone	106	2.7
Ethyl alcohol	204	7.0
Ethyl acetate	102	2.7
Water	540	45.0

[a]Higher numbers indicate slower drying.

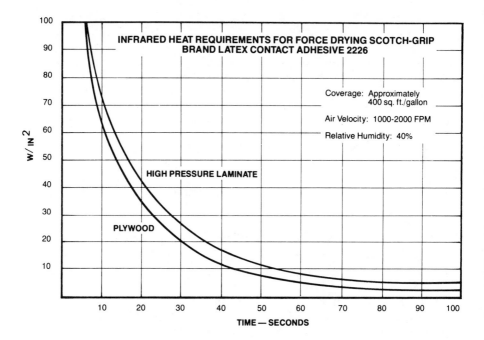

Fig. 3. Infrared heat requirements for force drying Scotch-Grip
Brand Latex Contact Adhesive 2226.

midity conditions, the latex adhesive will not dry at all. The drying
time for the solvent adhesive will be extended, but it will eventually
dry.

When the contact-bonding technique is used and high-humidity
conditions exist or a high production rate is required, it is necessary
to force dry the adhesive prior to bonding.

One of the most efficient methods of drying latex adhesives is
through the use of an infrared heat source in conjunction with air
flow over the adhesive. As the water evaporates, air movement reduces the high humidity directly above the adhesive surface and
speeds the drying.

Infrared heat sources are usually specified by a certain wattage
per unit. Total wattage is expressed as kW/ft^2 or $W/in.^2$. Units are
commercially available with capacities up to 100 $W/in.^2$. With equipment of this type, neoprene latex contact adhesives can be dried in a
few seconds even on heat-sensitive substrates, such as polystyrene
foam (Fig. 3). The rate of drying enables line speeds for latex adhesives to equal those normally used for solvent adhesives.

VIII. PROPERTIES OF LATEX ADHESIVES VERSUS SOLVENT-BASED ADHESIVES

The overall strength properties of a latex adhesive based on a given polymer will be similar to those of a solvent-based adhesive containing a solution polymer of like composition.

The solids content of a latex adhesive is normally in the 40—50% range. This range compares to about 20—30% for solvent-based adhesives.

Since latex adhesives contain little, if any, organic solvent, they are usually cheaper than a comparable solvent system. The cost differential becomes even greater, in favor of the latex adhesive, if the comparison is made to a nonflammable solvent adhesive that contains expensive chlorinated solvents.

Latex adhesives have good brushability and usually require less pressure for pumping or spraying than solvent-based adhesives. Prior to drying they can be cleaned up with water. They are non-flammable in the wet state, and the vapors are nontoxic.

In most cases, latex adhesives must be protected from freezing during shipment and storage. Some materials will recover after freezing, but most of them will coagulate and become unusable.

Solvent-based adhesives tend to have better overall adhesion to substrates such as oily surfaces, metal, rubber, and certain plastics.

IX. APPLICATIONS FOR VARIOUS TYPES OF LATEX ADHESIVES

Latex adhesives, based on neoprene rubber, are popular as general adhesives because they offer an excellent balance of properties. Products based on neoprenes were the original contact adhesives. Some of their properties are

1. The contact-bond characteristic which allows two nonporous surfaces to be joined without trapping water
2. Very high strength immediately after bond
3. Excellent resistance to continuous load stress
4. Good elevated-temperature strengths
5. Excellent film aging

Neoprene contact adhesives are used to bond foamed plastic, plastic laminate, wood, plywood, wallboard, wood veneer, plaster, canvas, leather, and many other substrates. They are not normally recommended for use on metal substrates. Neoprene latex adhesives with high elevated-temperature strength and with metal adhesion are under development. Typical uses include desk tops, hollow-core wood doors, countertops, etc.

Adhesives based on styrene-butadiene latices are among the lowest cost rubber-based adhesives, but they are not as strong as neoprene latex adhesives. They are usually used to bond lightweight material. These materials include synthetic and natural fabrics, felt, plastic foams, fiberglass insulation, canvas, linoleum, cardboard, cork, sponge, paper, foam rubber, and metal. Many of the latex PSAs are based on styrene-butadiene latices.

Adhesives based on butadiene-acrylonitrile latex can be formulated to have good resistance to oil, plasticizers, and gasoline. Applications include the bonding of polyvinyl chloride (PVC) film, aluminum foil, leather, wood, and metals.

Polyvinyl-acetate latex adhesives are the "white glue" of the industry. They are very versatile materials and are used extensively in the bonding of wood and paper products.

Quite a variety of latex adhesives based on acrylate polymers are available. The end uses are varied. They include PSAs, laminated aluminum foil, paper, fabrics, foams, and PVC film. Other applications include adhesives for flocking and installing various types of floor coverings.

X. CHARACTERIZATION OF LATEX ADHESIVES

Latex adhesives are characterized by measuring quite a variety of physical, application, and performance properties. Properties that are measured for any particular adhesive depend on the designed end uses for the adhesive. A few examples of some of the many properties that are measured are shown below.

A. Physical Properties

1. Solids content—weight and volume
2. Viscosity
3. Weight per gallon
4. Color

B. Application Properties

1. Brushability, sprayability, roll coatability, etc.
2. Mechanical stability
3. Chemical stability
4. Ease of cleanup

C. Performance Properties

1. Wet strength
2. Bonding range
3. Required bonding pressure
4. Rate of strength buildup
5. Adhesion
6. Dynamic and static strength in peel, shear, tensile, and cleavage measured over a temperature range

7. Freeze-thaw stability
8. Bacterial resistance
9. Bond durability tests: resistance to
   a. oxidation
   b. UV light
   c. chemicals
   d. temperature-humidity cycling
   e. heat aging
10. Flame spread
11. Smoke emission
12. Self-extinguishing properties

XI. ADHESIVE SELECTION

The object of adhesive selection is to find the most economical adhesive that has the desired application and performance properties.

A good place to start is with the adhesive supplier. If sufficient information about the bonding operation is provided, the supplier can submit several products for evaluation. The quality of the adhesive recommendation will be directly related to the quality of the information provided. This information should include

1. Problem description
2. Allowable economics
3. Desired application method
4. Desired bonding method
5. Substrates to be bonded
6. Operating-temperature range
7. Environmental conditions (humidity, water, fuel, UV, etc,)
8. Strength required (peel, shear, etc.)
9. Special requirements

REFERENCES

1. I. R. Hohwink and G. Salomon, in *Adhesion and Adhesives,* Vol. 1, Elsevier, New York, 1965.
2. R. A. Martin and T. F. Vanier, *Rubber World* 155(1):105 (1966).
3. R. L. Oldack and R. E. Bloss, *Adhesives Age* 22:38 (1979).
4. J. W. Vanderhoff, E. B. Bradford, and W. K. Carrington, *J. Polymer Sci., Polymer Symp.* 41:155 (1973).
5. *Adhesives Answer Book For Product Assembly,* ACS Division, 3M, November, 1976.
6. *Handbook of Chemistry and Physics,* 52d ed., Chem. Rubber, 1971--1972.

BONDING PRACTICE

17
Metal Cleaning

William K. Westray *Economics Laboratory, Inc., St. Paul, Minnesota*

I. INTRODUCTION

Surface preparation is like a neglected headache; it only hurts when you have to think about it. Nowhere does one learn about surface preparation until there is a problem, and then the individual is faced with trying to learn how to troubleshoot the problem, talk to suppliers, understand the terminology, and solve the problem as rejects continue to pile up.

At the most, any current manufacturing textbook discusses surface preparation in one or two pages and assumes that this manufacturing operation need not be discussed in any detail. However, consider that if a surface has not been properly prepared, any application of an organic adhesive coating will be reduced in effectiveness.

427

Over the long term, inattention to surface preparation will lead to product-quality problems.

Another consideration with regard to surface preparation is that the chemicals used do have varying degrees of potential hazards, and care must be used in selecting a chemical system to ensure compliance with government regulations.

II. WHY SURFACE PREPARATION?

Good surface preparation will give better adhesion, resulting in better product quality. On a long-term basis the proper surface preparation will give better corrosion protection to metal surfaces. The goal of any cleaning or conversion technique is to produce a firm, dry, inert surface of known, reproducible behavior. Only then can confidence exist in the performance and durability of a bond.

III. SOILS

Soils are defined as unwanted surface contaminants that interfere with the performance of the following processing steps of the bond. Surface contaminants can be both organic and inorganic in nature [1].

The many different types of metals have varied types of contaminants. Aluminum-mill marks, soils, greases, surface oxides, steel-rolling oils, rust-preventative oils, and heat-treat scales are typical of soils that must be removed before an adhesive should be applied if proper performance is expected. Soil removal will be discussed based on the different cleaning processes available and the optimum performance each one gives for different types of soils.

IV. TYPES OF CLEANING

A. Physical Cleaning

Solvent wiping, shotblasting, sanding, and wire brushing are common physical methods of preparing metal surfaces for adhesion. Where production volumes are involved, physical methods are too costly because of high material and labor requirements. In addition, because abrasion exposes bare metal, no corrosion-resistant residue remains for storage or service protection.

Solvent wiping is a widely used technique for low volume production or for large or heavy objects which can not be easily covered. Physical cleaning methods may subject the surface to contamination by grease, oil, or dirt. Wipe rags and shot should be changed or cleaned regularly.

B. Alkaline Cleaning

Alkaline cleaning is probably the most common and versatile of all methods of surface preparation [2]. It can be used in a variety of processes (spray or soak) and on different types of metals (aluminum, steel, etc.).

Alkaline cleaners are typically blends of caustic, soda ash, silicates, phosphates, and wetting agents. The particular formulations are based on the application for which they are intended and the types of metal to be processed. Suppliers of cleaning compounds should be consulted for assistance in choosing specific formulations.

For aluminum cleaners a formulation of low pH (8–9), with minimum caustic content and with chromate or silicates for etching protection, is recommended. Depending on the effectiveness of the rinse, a nonsilicated cleaner might be a better choice since silicates are somewhat more difficult to rinse and may reduce the effectiveness of the adhesive. Where hard water is used, the metal ions present often form insoluble silicate compounds which may coat the surface of the metal and interfere with adhesion.

For adhesion to steel, a highly caustic (pH 11--12) cleaner is acceptable and gives excellent cleaning capability. A rinse following the cleaning state is essential in order to remove all alkaline residues which would affect the adhesion after bonding. The rinse should be changed often to reduce buildup of alkaline salts. Normally, the rinse should be below 110°F (about 45°C) to ensure complete removal of wetting agents. Strangely, higher temperatures may result in less effective rinsing.

"When do we dump an alkaline cleaning bath?" is a common question asked by operating personnel. Aside from the smart answer of, "When it stops working," a rule-of-thumb answer suggests dumping when chemical additions equal 1.5–2 times the initial charge. Another common solution is to compare the free alkalinity and total alkalinity. When the total alkalinity is double the free alkalinity, dump the bath. The limiting factor for an alkaline bath (which should be established independently for each cleaning system) is its capacity for both oil and particle soil.

C. Vapor Degreasing

Vapor degreasing is used where heavy oil and grease loads are encountered. It is effective where further cleaning is to be done in order to reduce the soil loading of the next cleaning step. It is important to be aware that solvent cleaning will leave an organic film on the surface of the work. This film may affect the following bonding operation, thereby reducing adhesion. In addition, particularly with vapor degreasing, particle soil, such as metal fines or smut, cannot be readily removed [3]. The most common solvents used are trichloroethylene, perchloroethylene, and 1,1,1-trichloroethane. Selection of the proper solvent is dependent on the design factors of the vapor-degreasing equipment and on toxicity. Considerable attention must be paid to the changing Occupational Safety and Health Agency (OSHA) regulations dealing with individual solvents. New Environmental Protection Agency (EPA) regulations tentatively scheduled for publication in 1982 will also have a major effect on the use of vapor-degreasing equipment. The EPA Resource Conservation and Recovery Act requires very strong controls on the disposal of solvent wastes. At the time of writing, the EPA was proposing specific regulations for the design of vapor degreasers and their operations. Though vapor degreasing is a viable cleaning process, a review of current and proposed government regulations would be prudent when considering it as an option.

V. SURFACE PROTECTION

Corrosion of a metal surface begins immediately after removal of contamination (soil, smut, metal fines, oxides, etc.). Adhesives, like any other surface coating, lose their effectiveness when corrosion begins to attack the adherend. Corrosion attack is caused not only by environmental factors, but also by internal factors. An example of internal corrosion of a steel surface is galvanic attack (Fig. 1). Because the surface of steel is not smooth, the peaks and valleys of the steel surface establish minute, galvanic corrosion cells which may lead to adhesion failure caused by internal corrosive attack. A number of techniques have been developed to convert corrosion-prone, clean surfaces to less reactive ones. Three of the more common conversion processes are phosphating, anodizing, and chromating.

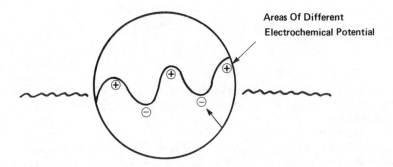

Fig. 1. Galvanic corrosion may occur on the surface of clean metal because of differences in the electrochemical potential of adjacent surface regions.

A. Phospating

The formation of a nonconductive coating on a steel surface is an established method of minimizing the effect of the galvanic corrosion. The crystalline nature of a phosphate coating will normally increase the adhesive properties of a surface. The two most common types of phosphate coatings are iron and zinc. They are produced by treating the surface with acid solutions of iron or zinc phosphate [4].

While iron phosphate coatings are easier to apply and are more environmentally acceptable, zinc phosphate coatings provide better corrosion protection. An important consideration when using these conversion coatings is establishing how much coating weight (thickness) is permissible before adhesive failure occurs between the conversion coating and the adhesive, and not between the adhesive and the adherend. Zinc and iron phosphate coating weights are usually expressed as milligrams per square foot (mg/ft^2). Values from 10 to 200 mg/ft^2 ($0.012-0.24$ mg/cm^2) are typical values, with lower weights giving less protection. An infrared technique has recently been developed to measure phosphate coating weights [5]. Phosphating a metal surface usually involves a cleaning step to remove oil and dirt, a phosphate formation step using zinc or iron salts in phosphoric, and a final rinse step. Sometimes the cleaning and phosphating steps are combined. Other times two or more rinses (of various temperatures) may be used between the other steps of the phosphating process. Typical three- and six-stage phosphate sequences are shown below:

Three-stage iron phosphating—spray
1. Combined cleaning and phosphating, 1.5 min, 38—66°C (100—150°F)
2. Hot water rinse, 0.5 min, 38°C (100°F)
3. Chromic acid rinse, 0.5 min, 38—66°C (100—150°F)

Six-stage zinc phosphating—spray
1. Alkaline precleaning, 1.0 min, 38—66°C (100—150°F)
2. Water rinse, 0.5 min, 38°C (100°F)
3. Titanium activator rinse, 0.5 min, 38°C (100°F)
4. Phosphating stage, 1.0 min, 38—66°C (100-150°F)
5. Water rinse, 0.5 min, 38°C (100°F)
6. Chromic acid rinse, 0.5 min, 38—66°C (100—150°F)

The most important parameters for good control of a phosphate operation are
1. Temperature and time of treatment
2. Concentration of chemicals
3. Ratio of free to combined phosphoric acid

A typical phosphating bath [4] may contain (per liter) 15 grams of zinc oxide and 0.04 liters of phosphoric acid (75%), and 0.2 grams of sodium nitrate. While iron phosphate coatings are easier to apply and are more environmentally acceptable, zinc phosphate coatings provide

better corrosion protection. An important consideration when using
these conversion coatings is establishing how much coating weight is
required. This is best done by actual testing of finished parts.

B. Anodizing

For aluminum, internal corrosion protection is provided by anodizing
(oxidizing) the clean deoxidized aluminum surface in either sulfuric
or phosphoric acid electrolytic baths. The purpose is to create an oxide under controlled voltage and temperature conditions, thereby creating a better protective surface [4].

As with phosphating, care should be used to ensure that the
coating thickness does not reduce joint properties. A common practice today is to use an iron phosphate solution containing fluoride to
anodize aluminum. It is reported that an aluminum phosphate is formed in this process; however, the phosphate coating is very thin [less
than 5 mg/ft^2 (0.006 mg/cm^2)], thereby giving minimum protection.
The principle advantage appears to be that the iron phosphate with
fluoride does an excellent job of cleaning and slightly etching the surface, allowing for better adhesion properties.

Sulfuric acid anodizing is also widely used. It is rapid, economical, and requires low operating voltage. Coating thickness, hardness,
and porosity may be controlled by adjustment of time, temperature,
and concentration. A typical anodizing cell operates at 22 V dc at
room temperature with a sulfuric acid concentration of 15—20% and a
current density of 230 A/m^2 for 15—60 min.

C. Chromating

Chromate conversion coatings are used with some metals to enhance
adhesion. They are tough, hydrated gel structures which are usually
applied by immersion of aluminum, zinc, magnesium, or copper alloys
in heated solutions of proprietary, chromium-containing compounds.
Chromate films are soft when removed from the processing tank but
are hardened by heating to about 150°F (65°C). Immersion times vary
from seconds to minutes, and application temperatures, from ambient
to near boiling. Chromate film thickness rarely exceeds 50 µin.
(0.0001 cm). Care should be taken to not heat chromate films much
over 150°F (65°C) for extended periods because of possible decomposition of the coating.

Chromium compounds have been electrolytically applied to steel to
given discontinuous films which may contain chromium plate. They
apparently increase adhesion—probably by protecting the steel from
corrosion.

VI. CONTROL

Attention to the control of chemical solution strength cannot be over-emphasized. One of the most *common reasons* for poor cleaning is poor control of the cleaning bath concentration. A daily quality-control check of the cleaning-bath concentration is a must for good cleaning. Today, most chemical-cleaning suppliers provide titration kits that permit simple control of a bath. A word of caution—chemical control of a bath is not possible if different chemical suppliers' formulations are mixed in the same tank.

VII. OXIDE REMOVAL

The removal of loose oxide from the surface of the metal will increase the strength of an adhesive bond. For steel, two chemical methods are possible [6]. Acid descaling (pickling), which uses hydrochloric, sulfuric, or some other acid, is very common. Removal of surface oxides is rapid and effective. It is important to use inhibited acid cleaners to avoid corrosion of cleaned surfaces. An acid rinse is used, followed by a neutralizing stage and a water rinse. A limitation of acid descaling is that some dimensional change will occur in addition to oxide removal. Alkaline derusting of steel is possible where the removal of oxides is required, but no dimensional change can be tolerated. An alkaline deruster is used at high concentrations, with temperatures of 200°F (90°C) or higher. Though slower than acid derusting, it is an effective method for removing oxides from steel without changing the shape of the work piece.

VIII. ENERGY SAVINGS

The cost of energy has risen to the point that it must be considered a significant part of the total cost of surface preparation. With a continuous-spray washer system, annual energy costs of $30,000—40,000/year are not uncommon. The old formula for effective cleaning—time, temperature, and concentration—has changed and will continue to change as new products are developed. The cost interrelationship of each parameter has been altered, so that increased chemical cost can, in many cases, realize a greater savings in energy cost. Within the last few years, many suppliers of proprietary, surface-cleaning chemicals have developed formulations that will allow reduced temperature operations without sacrificing performance standards [7]. The needs of industry have created a revolution in the development of new products for energy savings and environmental safety. This revolution has resulted in the opportunity for production personnel to try new products, which have been formulated for the new needs of industrial applications. It is recommended that testing of new products which

have the potential of delivering the performance standards required
be reviewed when a cost benefit has been documented elsewhere.

Many studies have been conducted to determine the energy requirements of a continuous-spray washer system. A rule-of-thumb
method for calculating heating requirements was determined through
test data established by Eclipse Combustion Division of Eclipse, Inc.
and by Caterpillar Tractor. Results from those test data indicate that
evaporative losses due to spray operation can be determined by plotting the solution temperature versus time while the solution is cooling
from 180°F (82°C) with the heat off. The slope of this curve at any
temperature is directly proportional to the rate of heat loss. Heat
losses typical of a spray washer operating at 140°F (60°C) are estimated as follows:

Form of loss	Percent of total
Water evaporation	63
Water heating	12
Heating parts and conveyor	12
Radiation losses	7
Electrical losses	6

The most precise method for determining the energy requirements of a washing system is to measure the amount of energy delivered to the system over a period of time and determine the cost of gen-
erating the energy.

Soak-tank operations have energy-loss characteristics similar to
spray cleaning. The evaporative losses from the surface of the tank
constitute the major demand for energy. With the increased requirement for proper exhaust ventilation of a soak tank, the airflow across
the surface becomes an important factor in determining the sources of
heat loss. If feasible, measurement of the energy demand of the system is the preferred method of determining costs.

IX. SPECIFICATION CONTROL

The cost of surface preparation can be measured in direct and indirect
terms. While direct costs can be based on the pounds or square feet
of metal processed, indirect costs are much harder to measure. Indirect costs are product failures, customer rejection, in-field repairs,
etc., all resulting in poor sales and lost business. The degree of surface preparation should be directly related to the application of the
industrial product. Establishing specifications that relate directly to
the useful life of the product will allow determination of what type of
surface-preparation system is necessary.

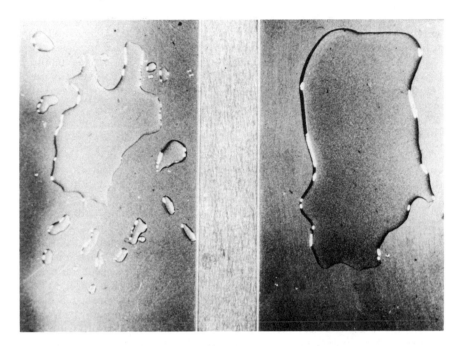

Fig. 2. Water poured on a clean metal surface does not break into droplets. The contaminated panel on the left shows water break; the clean panel on the right does not.

Specification standards allow the evaluation of new processes and the determination of operating costs based on operating data and troubleshooting when product-quality problems occur. Periodic review of specification standards is important since changes do occur. By reviewing specification standards, the relevance of a particular standard can be redetermined.

The proper specification standard will allow optimum operating cost. A standard that requires a higher performance than is required for the end use of the work piece will greatly increase the operating cost of this manufacturing stage.

Interim product inspection is important since identification of poor surface preparation will allow immediate corrective action and eliminate product failures later when more value has been added to the work piece.

What makes a good process-control specification? First, it must relate to the end use of the product. Secondly, a good process test must allow quick identification of problems. Visible soil, water spots, and an uneven coating on parts are quick indications that the cleaning process may not be working correctly. The water-break—free test is widely used to demonstrate a clean metal surface (Fig. 2). A test for adhesive bond strength should certainly be a part of the process control, as should some sort of accelerated-corrosion test. In developing a process control program do not be limited to standard tests but develop your own tests which relate as closely as possible to the end use of the product being produced.

X. TROUBLESHOOTING

Just because the pumps are pumping, the spray nozzles are spraying, the parts are getting wet, and the conveyor is moving, do not assume that the cleaning process is being accomplished.

When your current system is not producing results as desired, the first key step is to determine what has changed, what is wrong. About adhesives you might ask, "Where is the bond failing?" Patience is the most important factor in troubleshooting. Do not make changes just because things are not working correctly. First examine the process. Are the parts actually being cleaned? Is there water break on the parts? Is the concentration right? Is the temperature okay?

When you start to take corrective measures, make one change at a time and wait to determine its effect before making any other changes. Use your vendors; ask for technical support. A good vendor will be able to provide competent technical service to troubleshoot your problem.

XI. SUMMARY

Attention to surface preparation can increase the performance of the adhesive bond. Better adhesion and corrosion protection are possible when the surface preparation is correct.

A proper surface-preparation program includes the selection of the right method for the use life of the work piece and the development of quality-control methods to ensure consistent quality.

REFERENCES

1. R. C. Snogen, *Handbook of Surface Preparation*, Palmention Press, New York, 1974.
2. F. Mazia, Chemical removal of scale oxides, *Metal Finishing*, June 1980.
3. Heat treating and finishing, in *Metals Handbook*, Volume II, eighth edition, American Society for Metals, Metals Park, Ohio, 1964.

4. G. L. Schneberger, Chemical and electrochemical conversion treatments, *Encyclopedia of Chemical Technology*, third edition, Volume 15, Wiley, New York, 1981, p. 304.
5. G. D. Cheener, *American Paint and Coating Journal 62*:14 (1977).
6. G. L. Schneberger, Cleaning, pickling, and related processes, *Encyclopedia of Chemical Technology*, third edition, Volume 15, Wiley, New York, 1981, p. 296.
7. *Metal Finishing Guidebook and Directory*, an annual publication of Metal and Plastic Publications, Inc., 99 Kinderkemack Road, Westwood, New Jersey.

18
Spraying Adhesives

John R. Adams *Binks Manufacturing Company, Franklin Park, Illinois*

439

I. INTRODUCTION

Each of the many methods of applying adhesives have certain advantages and certain limitations. A knowledge of these factors will simplify the selection of the best application method for your particular set of conditions.

The advantages of the spray application methods are speed, versatility, and serviceability.

A. Speed

Spraying permits the rapid coverage of surfaces. In addition, drying time is usually reduced.

B. Versatility

Material can be applied to a wide variety of shapes and sizes of parts either manually or automatically.

C. Serviceability

Spray methods use standard, readily available, time-proven components.

The major limitation of spray application is the material itself. Adhesives, by design, want to stick together. To spray apply an adhesive requires the material to be "atomized," or converted into tiny droplets. Only materials that can be atomized can be spray applied.

Adhesive manufacturers will indicate in their technical data whether the material can be sprayed and what the proper viscosity and the proper conditions for spray application are.

It is important to note that the same processes and equipment used to apply decorative and protective coatings are used to apply adhesives.

II. COMPRESSED-AIR ATOMIZATION

Compressed-air atomization, also commonly referred to as conventional spraying, is the most widely used spray method for adhesive application.

The adhesive is supplied to the spray gun by a pressure feed device (either a pressure tank or a pump). When the gun has been activated, the adhesive is discharged through the fluid nozzle in the form of a liquid stream (Fig. 1). It is immediately surrounded by a hollow column of compressed air emitted from the center orifice of the air nozzle. The action of this column of air on the fluid stream converts the stream into small droplets of adhesive (first-stage atomization), at the same time imparting forward velocity to them. Additional

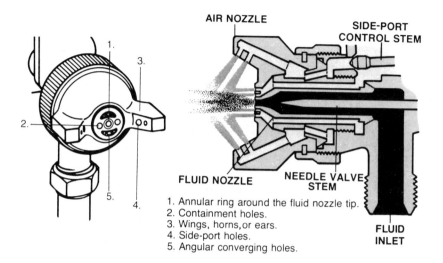

AIR NOZZLE

SIDE-PORT
CONTROL STEM

FLUID NOZZLE

NEEDLE VALVE
STEM

FLUID
INLET

1. Annular ring around the fluid nozzle tip.
2. Containment holes.
3. Wings, horns,or ears.
4. Side-port holes.
5. Angular converging holes.

Fig. 1. Compressed-air atomization.

jets of compressed air are directed into the droplets, forming even
smaller droplets (second-stage atomization). Still other jets of air,
under control, exert force on the sides of the spray pattern to form
the fan-shaped spray pattern necessary for a uniform application.

The ability to control the forces at work at the head of the spray
gun is the key to a successful spray application.

A. Equipment Components

A spray system is really an assembly of components selected to satisfy your needs. It is good practice to identify your needs before proceeding with the selection of equipment. Some typical considerations
include

Frequency of use
Material delivery rate
Portability
Manual or automatic operation
Size and shape of product
Spray characteristics of the material
Price

For example, the equipment components required to apply insulation adhesive to ductwork in the field would be quite different from
those necessary to apply a plastic laminate to particle board in a production installation, although in both cases the same application principle would be used.

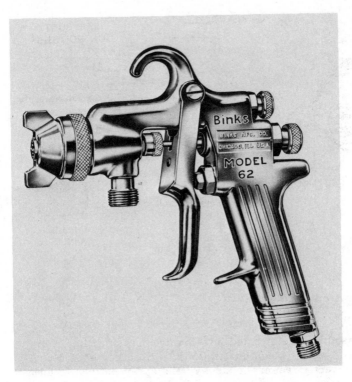

Fig. 2. Heavy-duty—production air spray gun.

B. Spray Guns

Spray guns are manufactured in a broad range of costs and performance capabilities. They are usually cataloged according to their
performance rating, that is, heavy-duty production, standard production, lightweight, special purpose, etc. (Fig. 2).

The "production" category is the most expensive but at the same
time has a much wider range of nozzle assemblies and accessory equipment available. It is false economy to purchase lightweight, low-cost
equipment for production work.

A spray gun—equipped with the proper nozzle combination, properly used, and properly maintained—is the core of the system around
which everything else revolves.

C. Air and Fluid Nozzle Combinations

The spray gun is the application tool, but the nozzle combination used
on the gun actually performs the spray function. The nozzle combination consists of two parts, the fluid nozzle (or fluid tip) and the air
nozzle (or air cap) (Fig. 3).

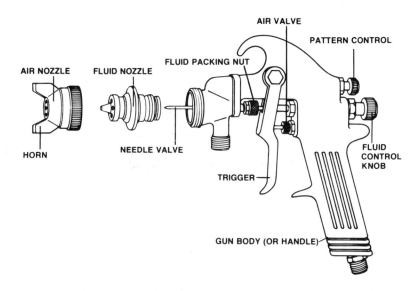

Fig. 3. Air spray gun components.

1. Fluid nozzle

The fluid nozzle serves several functions. The size of the fluid nozzle determines the amount of material that can be dispensed by the gun. "Size" is an indication of the diameter of the center orifice of the fluid nozzle in thousandths of an inch. Nozzle sizes range from 0.022 to 0.500 in. (0.06 to 1.27 cm). Table 1 shows the orifice sizes related to typical adhesive viscosities.

The most widely used orifice range is 0.028–0.040 in. (0.07–0.10 cm), which will deliver 2–6 oz (60–180 ml) of material per minute. Where a higher delivery rate is necessary, as in some automatic applications, additional guns should be used.

Table 1 Fluid-Nozzle Orifice Chart

Orifice size		
(in. dia.)	(cm dia.)	Viscosity
0.022–0.028	0.055–0.070	Very thin
0.040–0.052	0.100–0.130	Thin
0.058–0.070	0.148–0.175	Medium

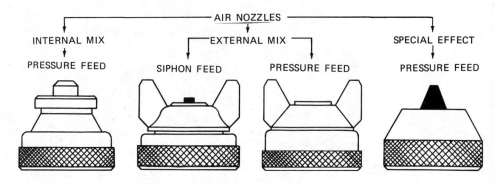

Fig. 4. The three basic types of air nozzles.

Fluid nozzles are available in hardened steel or stainless steel. Stainless steel should be specified wherever corrosion is a possibility, such as with water-based adhesives.

The position of the air nozzle relative to the fluid nozzle is of prime importance when producing a spray pattern. This relationship is determined by the mating of surfaces on each of the nozzles. How this is accomplished differs with each equipment manufacturer. One manufacturer uses matching tapers; another, a method called the "ball and cone". Both methods work; however, since they are different, the nozzles from different manufacturers' products are *not* interchangeable. When servicing the spray guns, care must be taken not to damage these locating surfaces.

2. Air nozzle

The function of an air nozzle is to atomize the material being sprayed and to form this quantity into a usable spray pattern. There are three basic categories of air nozzles (Fig. 4): internal mix, in which the air and material mix inside the air nozzle; external mix, in which the air and material come together after they have left the gun; special effect, in which nozzles are designed to create special decorative effects.

External-mix nozzles are the most widely used in adhesive applications. They may be either the siphon or the pressure type. Siphon nozzles are designed to "draw" air materials out of a container mounted on the spray gun. They may also be used with fluids that are delivered under pressure. Pressure nozzles, however, *cannot* siphon. They must be supplied under pressure. The hole pattern drilled into an air nozzle can be one of hundreds, ranging in complexity from a single 3-hole design to one of 25 holes. The simpler nozzles are the

best for adhesives which have a tendency to stick to the exterior surfaces of the air nozzle. Equipment manufacturers provide a chart
which specifies the most popular nozzles for a given application.

Air consumption is another consideration when selecting an air
nozzle. One must stay within the capacity of the air supply. Each
nozzle will require a given amount of air (measured in ft^3 min or c.f.m.)
depending on the hole pattern in the nozzle and the air pressure used.
A simple nozzle may require 5 ft^3/min (0.15 m^3/min) of air at 50 psi
(350 kPa), whereas a more complex nozzle may require 20 ft^3/min (0.6
m^3/min) at 50 psi (350 kPa). As a rule of thumb, an electric-powered
compressor will deliver 4 ft^3/min (0.12 m^3/min) per horsepower;
therefore, if the demand of the nozzle is 20 ft^3/min (0.6 m^3/min), the
air supply must be a minimum of 5 hp.

D. Hose and Fittings

There are several factors to consider in selecting hose and fittings to
deliver either air or material.

1. Rated working pressure

The rated working pressure of each component in the system must always be higher than the maximum pressure that will be encountered.
This pressure rating will be found in the supplier's literature. Never
use components with a low working-pressure rating in a high-pressure system.

2. Chemical resistance

The materials of construction determine the resistance to attack by the
solvents and/or chemical components being transported. Check your
needs against the solvent resistance charts supplied by the manufacturer. If the material is corrosive (e.g., a water-based material),
stainless-steel fittings should be used.

3. Temperature

Some adhesives spray best when heated and require hose specifically
made for this purpose.

4. Capacity

Hose and fittings must have sufficient capacity to deliver the required
amount of air or fluid with minimum pressure drop. Air-hose capacity
is influenced by the volume of airflow required, the pressure used,
and the length and inside diameter (I.D.) of the hose. The minimum
size air hose for production spraying should be 5/16 in. (0.8 cm) I.D.

Similarly, fluid-hose capacity (measured in fluid ounces per minute) is influenced by the pressure used, the viscosity of the material,

Fig. 5. Horizontal, tank-mounted compressor provides power.

and the I.D. and length of the hose. When a system is not in use, it is good practice to keep the fluid hoses filled with compatible solvent to prevent any residual material from hardening and causing problems.

E. Compressed-Air Supply

The basic source of power for the compressed-air atomization method is compressed air that is clean, dry, and of sufficient volume and pressure to perform the work.

1. Air source

The air source is a machine that takes a quantity of air at atmospheric pressure and mechanically "squeezes" it into a smaller volume (Fig. 5). As the volume decreases, the pressure increases. The controlled release of the compressed air to the atmosphere permits us to use the resulting pressure drop to power spray guns, pumps, pressure tanks, agitators, etc.

Time does not permit us to discuss the many factors involved in the selection of a compressor, so be aware that

1. The compressor must have sufficient capacity to provide enough air volume and sufficient pressure to satisfy the demands of all of the air-operated equipment.
2. Lubrication of the compressor is essential. A maintenance chart should be affixed to the compressor, identifying the areas requiring service and the desired frequency of service.
3. Compressors do not manufacture air, but convert air that is available to them. Be sure that the compressor has a supply of clean, dry air available at the intake, and that the intake filter is cleaned regularly.
4. The compressor should be mounted as near as possible to the point where the compressed air is to be used and should be mounted on vibration isolators or on a firm base to prevent any structural damage due to vibration.

2. Cleaning the air

The removal of contaminants from compressed air is important in the interests of safety and protection against damage to the spray equipment and to the end product. It is important for compressed air to be cooled, dried, and cleaned before it is distributed for use. When air is compressed, it becomes very hot, and in this condition it is potentially explosive. Danger of a combustion explosion, in either the air receiver or pipeline, will always be present if the air has not sufficiently been cooled after leaving the compressor. In addition, heated air will result in less air being stored in your air receiver. It will hold more water and will cause the pipelines to expand and, subsequently, contract when the compressor has been shut down and the pipes have cooled. This cooling will cause condensation and the obvious contamination of the air system.

Vaporized oil can present operating and maintenance problems. Oil should be removed before it enters the distribution system. Once there, it will creep through the system and, most likely, contaminate the product or the process. Oil is introduced into the air supply because of compressor wear or because of an excessively high operating temperature.

If solid particles or other impurities—such as rust, dust, solvent vapors, overspray, etc.—are present in the immediate atmosphere, they may be drawn through the compressor intake and find their way into the air-distribution system. Impurities entering the compressor will contribute to excessive wear and premature parts replacement. They will also contaminate the spraying operation.

Fig. 6. Oil and water extractor.

Aftercoolers are widely used as a means of lowering the air temperature and removing most of the oil and water. The aftercooler is
an essential part of a complete and properly functioning compressed-
air supply system. The larger the air compressor and the higher the
operating pressure, the greater is the need for an aftercooler.

The air in a compressed-air system is never completely dry or
free of moisture. Filters are a normal, mechanical means by which
solid particles and fine water droplets that may still exist in the air
stream are removed from the air supply (Fig. 6). Filters, commonly
referred to as oil and water extractors, are placed close to the actual

Fig. 7. Air regulator provides air control.

spraying operation. The more effective filters have a combination of baffle-type elements to remove the solids and a filter element (of either porous bronze or an absorbant material) to collect the finer particles. Periodic maintenance is required to either clean or replace these filter elements as they load up with impurities. Daily draining of the filters should be standard procedure, for it prevents oil and water from accumulating in the filter and feeding back into the system. Caution must be exercised during this draining procedure to prevent physical injury to the operator since the materials are under pressure.

3. Control

To obtain the best results from the air-atomizing process, we must control it. Air regulators (Fig. 7) are the primary control devices in a compressed-air system. An air regulator is a mechanical device which reduces the main-line air pressure to a specific operating pressure. Once the operating pressure has been set, the regulator will automatically maintain this setting. It could be said that an air regulator is a "floating", or compensating, valve which will automatically compensate for any change in input pressure versus the desired output pressure. However, a regulator will never provide higher air pressure than the air distribution system has to offer.

Periodic maintenance of an air regulator is necessary to ensure free movement of all of the working parts; however, these parts should not be lubricated, merely cleaned.

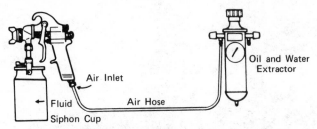

Siphon Feed Hookup:
For limited spraying and touch-up; especially involving many color changes.

Atomization air pressure is regulated at the oil and water extractor. Amount of fluid is adjusted by fluid control screw on gun, as required by the consistency of paint and air pressure.

Fig. 8.   Types of fluid containers and fluid handling systems.   Siphon spray system.

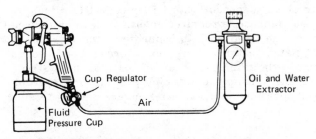

Pressure Feed Cup Hookup:
For fine finishing with limited spraying. It is more efficient than siphon feed systems.

Atomization air pressure is regulated at the extractor, and fluid pressure at the cup regulator. Atomization air bypasses through the cup regulator.

Pressure cups also are available without a cup regulator, in which case fluid pressure equals atomization air pressure. For heavy fluids and internal mix nozzle spraying, fluid flow is adjusted by the fluid control screw on the spray gun.

Fig. 9.   Pressure feed cup system.

F. Material Supply

The material supply portion of a spraying system consists of a container, a means of agitation, a pump (if used), and the auxiliary components that filter and regulate the flow of the adhesive.

1. Siphon feed spraying

Use of the siphon feed method of supplying adhesive to the spray gun (Fig. 8) is limited by the viscosity of the material being sprayed. Very thin, low-viscosity materials may work with this process, but the tendency is toward "cobwebbing" and excessive air pressure. Don't rule out the process, however, until you have first tried it with the material in question. It is the simplest and lowest-cost method of spraying.

2. Pressure feed spraying

Pressure feed spraying delivers material under pressure to the spray gun either by a pressure container or by a material-handling pump. Figure 9 shows a pressure cup, containing the material that is to be sprayed, attached to the gun as a common assembly. This arrangement is ideal for small volumes of material that require low pressure to force them out of the spray gun where they will be atomized by the jets of compressed air.

When larger quantities are to be dispensed, a pressure tank may be used. These tanks are available in a wide range of sizes. It is suggested that the proper size for your installation is equal to the quantity of material used in 1 day. For example, if your spray station is applying 5 gal (20 liters) of adhesive per day, select a 5-gal (20-liter) pressure tank. A typical pressure-tank installation is shown in Fig. 10.

Caution: All pressure vessels are rated as to their maximum safe working pressure. Do not under any circumstances exceed this pressure.

In operation the pressure tank is opened and filled to the appropriate level with the adhesive to be dispensed. The tank cover is then replaced and clamped uniformly. Air pressure, controlled by an air regulator, is exerted on the adhesive, forcing it to flow up the fluid tube, out of the tank, and to the spray gun. The volume of adhesive being dispensed is controlled by air pressure.

Pressure tanks offer a simple, low-cost, controlled way to force feed material to a spray gun. Since there are no moving parts, with the exception of the agitator, maintenance is minimal assuming, of course, that the components are kept clean.

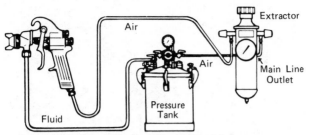

Pressure Feed Tank Hookup (Single Regulator):
For medium production spraying. Atomization air pressure is
regulated at the extractor, and fluid pressure by the tank regulator.

Pressure Feed Tank Hookup (Double Regulator):
This is proven highly efficient for portable painting operations.
Atomization air and fluid pressures are regulated by two individual
air regulators on tank.

Pressure Feed Circulating System Hookup:
Recommended for heavy production spraying. Atomization
air pressure is regulated at the extractor, and fluid pressure is
regulated by fluid regulator.

Fig. 10. Pressure tank system.

3. Material-handling pumps

The reciprocating piston pump is the type most commonly used in an
adhesive spray system. The pump consists of two sections: the power section and the fluid section. The power section can be air driven,
or it can have a piston, driven by a hydraulic motor, that is directly
connected to the liquid-handling fluid piston. The fluid section of the
pump determines the volume capability, usually expressed in gallons
per minute (gpm). The power section of the pump determines the
pressure capability, expressed in psi.

   The action of a reciprocating pump is described in Fig. 11. An
air valve located on the power section alternately pressurizes the top
and the bottom of the piston, causing the piston to reciprocate. The
fluid section is connected directly in line with the power section, so
that as the air-motor piston reciprocates, the fluid-section piston reciprocates at the same speed and stroke length.

   Adhesive may be supplied to the pump by using a suction hose
connected directly to the bottom of the pump or by immersing the fluid
section of the pump in the liquid. Flow through the pump is the result
of the action of two check valves and a displacement plunger. On the
upstroke of the pump, material is drawn into the chamber through the
lower check valve, called the foot valve. On the downstroke, the lower check valve closes, and the fluid-section exerts pressure on the li-

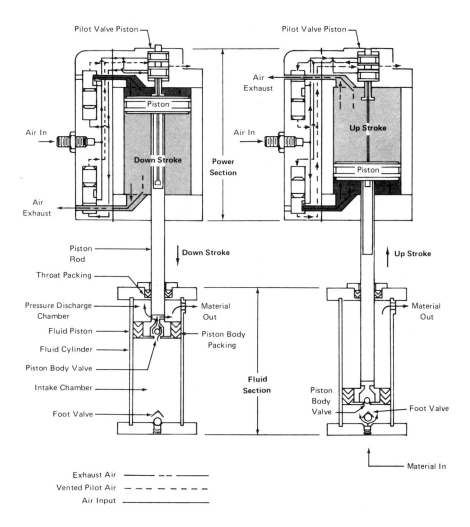

Fig. 11. Air-motor—driven reciprocating piston fluid pump.

quid in the pump chamber. The pressurized liquid then flows up
through the center of the piston, through the body check valve, and
into the upper chamber; however, only 50% of this material can fit in
the chamber because of the position of the piston rod. The material
displaced by the piston rod flows through the discharge fitting on the
pump fluid section. The cycle is then repeated. On the upstroke the
lower chamber is charged, and the balance of the material on top of
the piston discharged.

Fig. 12. Circulating system—Mix Room.

4. Circulating systems

The circulating system moves material continuously through a closed
loop of hose or pipe, to one or more spray stations, and back to the
container. Most circulating systems pump directly out of the shipping
container which minimizes any downtime due to refilling containers, as
in the case of pressure tanks. Material handling is improved when adhesives and coatings are dispensed from one central location generally
referred to as a mix room (Fig. 12).

5. Material heaters

There are significant benefits to be derived from the installation of a
material heater in an adhesive application system (Fig. 13). There
are many designs of material heaters on the market; however, their
functions are the same. As the adhesive is forced to flow through the
heater, it flows over a thermostatically controlled heat-transfer surface where it picks up the desired amount of heat. The adhesive then
flows from the heater to the spray gun where it is spray applied. The

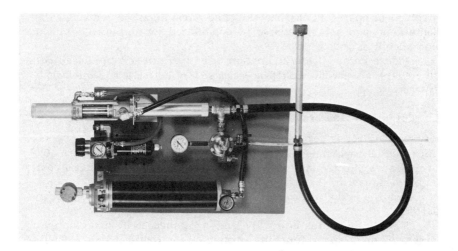

Fig. 13. Material heater.

usual arrangement of components permits circulation of the unused adhesive back into the pumping unit and through the heater to maintain
the desired temperature.

The introduction of heat to the adhesive reduces the viscosity
which will provide the following advantages:

1. *Greater coverage with one pass.* Viscosity is lowered, and
    the "gel" normally associated with rubber-based products is
    broken by the heat, thus providing a better-flowing materi-
    al.
2. *Faster application.* The spray gun can be moved over the
    work at a higher rate of speed than with cold spray.
3. *An improved spray pattern.* The viscosity of the material
    at the spray gun is always the same, thereby eliminating the
    need to constantly readjust air and fluid pressures due to
    changes in ambient temperature and humidity.
4. *Less "soaking in" on porous surfaces.* At temperatures of
    150°F (66°C), the normal temperature obtainable at the tip of
    the gun in a hot-spray setup, 70—90% of the solvent has dis-
    sipated before striking the panel.
5. *Faster drying.* A large percentage of solvent has been driv-
    en off due to the heat before the material strikes the surface.
6. *Less chance of blush.* The fast-evaporating solvents, which
    cause blushing, have already dissipated in the air at high-
    er temperature.

The one limitation of a material heater is the consumption of energy. The source of heat is usually electric, and thus, there is a cost

involved in operating the heater. The advantages far outweigh the
limitations, and heaters should be considered for any spray application except as follows:
    *Please note: Heaters function effectively in solvent-based adhe-
sive systems; however, they are almost ineffective with waterborne
systems.*

G. Equipment Operation

Having the proper components in your spraying system is important,
but of equal importance is the proper use of these components. Improper spraying techniques can make the spraying operation a costly
liability instead of the cost-efficient operation it should be.
    *Use the lowest pressure that will give you a satisfactory spray
pattern.*
    This basic rule is really the secret of success in spray applying
adhesives or in spray finishing products, for that matter. A fluid
nozzle should be selected for your spray gun that will deliver sufficient material to do the job you must do; however, the fluid pressure
necessary to accomplish this delivery rate should not exceed 18 psi
at the head of the spray gun. Excessive pressure will cause the adhesive stream to be emitted from the fluid nozzle at an excessive velocity. When this occurs, the air jets of the air nozzle cannot properly
attack the stream, and a poor pattern results. In pressure feed systems the fluid volume should be controlled by the fluid pressure, the
orifice size of the fluid nozzle, and the fluid control on the spray gun.
    Select an air nozzle for your gun from the list of nozzles avail-
able from the manufacturer. Air nozzles are categorized by the material that they are to spray. Be certain, however, that the air nozzle is within the capacity of your compressed air supply. Since all
air nozzles do not physically fit over all fluid nozzles, be sure to select a matched pair.
    Adjust the atomizing air pressure to approximately 25 psi (175
kPa). Then test for a proper spray pattern by holding the spray gun
perpendicular to the surface you are testing (a piece of cardboard or
newspaper) at a distance of approximately 10—12 in. (25–30 cm). Pull
the trigger on the spray gun and observe the spray pattern that develops. If the pattern is very coarsely atomized, that is, if the droplets of adhesive are large and "splattery" in appearance, slowly increase the air pressure until proper atomization occurs. Incidentally,
adhesives do not spray in quite the same way as paint or coatings.
The pattern will always be somewhat textured in effect depending on
the solvents remaining in the system to level the material. Figure 14
shows a well-balanced spray pattern. Spraying with a faulty or defective spray pattern can cause production problems. Figure 15
shows examples of faulty patterns and how they may be corrected.

Fig. 14. Correctly balanced spray pattern.

SPITTING

Jerky or fluttering pattern. Check for:
- Air leaking into fluid line or passageway.
- Lack of paint.
- Loose or cracked fluid siphon tube.
- Loose fluid nozzle.
- Loose fluid packing nut or worn packing.

Faulty Patterns and How to Correct Them

CAUSE

Dried material in side port "A" restricts passage of air through port on one side. Results in full pressure of air from clean side of port in a fan pattern in direction of clogged side.

REMEDY

Dissolve material in side port with thinner. Do not use metal devices to probe into air nozzle openings.

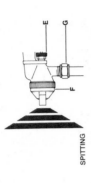

CAUSE

Dried material around the outside of the fluid nozzle tip at position "B" restricts the passage of atomizing air at one point through the center ring opening of the air nozzle. This faulty pattern can also be caused by loose air nozzle, or a bent fluid nozzle or needle tip.

REMEDY

If dried material is causing the trouble remove air nozzle and wipe off fluid tip, using rag wet with thinner. Tighten air nozzle. Replace fluid nozzle or needle if bent.

CAUSE

A split spray pattern (heavy on each end of a fan pattern and weak in the middle) is usually caused by: (1) atomizing air pressure, too high (2) attempting to get too wide a spray with thin material, (3) not enough material available.

REMEDY

(1) Reduce air pressure. (2) Open material control "D" to full position by turning to left. At the same time turn spray width adjustment "C" to right. This reduces width of spray but will correct split spray pattern.

CAUSE
- Air entering the fluid supply.
- Dried packing or missing packing around the material needle valve which permits air to get into fluid passageway.
- Dirt between the fluid nozzle seat and body or a loosely installed fluid nozzle.
- A loose or defective swivel nut, siphon cup or material hose.

REMEDY
- Be sure all fittings and connections are tight.
- Back up knurled nut "E", place two drops of machine oil on packing, replace nut and finger tighten. In aggravated cases, replace packing.
- Remove air and fluid nozzles "F" and clean back of fluid nozzle and nozzle seat in the gun body, using a rag wet with thinner. Replace and tighten fluid nozzle using wrench supplied with the gun. Replace air nozzle.
- Tighten or replace swivel nut "G".

CAUSE

A fan spray pattern that is heavy in the middle, or a pattern that has an unatomized "salt-and-pepper" effect indicates that the atomizing air pressure is not sufficiently high, or there is too much material being fed to the gun.

REMEDY

Increase pressure from air supply. Correct air pressures as discussed elsewhere in this manual.

Fig. 15.   Faulty spray patterns and how to correct them.

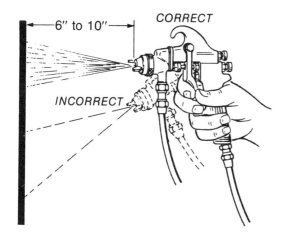

Fig. 16. Hold gun perpendicular to surface being sprayed.

A common experience in developing a spray pattern with adhesives is that they tend to "cobweb". This effect is caused by insufficient solvent to keep the material wet until it reaches the surface. If solvent must be added, it should be kept to a minimum. Add only enough solvent to keep the material wet between the gun and the sprayed surface.

H. Operator Techniques

The application of adhesive with a spray gun can be compared, to an extent, to application with a brush. We can consider a spray gun a brush without bristles and liken the spray pattern itself to the bristles. The brush is moved into an attitude perpendicular to the surface of the work in order to get the benefit of the full width of the brush. Likewise, the body of the spray gun must be held as nearly perpendicular as possible to deposit a uniform swath of material on the work (Fig. 16). If an operator tilts the gun to one side or the other, so that he either "pushes" or "pulls" the spray pattern across the work, a good deal of the adhesive will be deflected from the surface and lost into the air. If an operator tilts the gun in the other direction, only a portion of its width will be utilized, and again, the major portion will be lost through bounce back and blow-by (overspray).

Arcing the gun (Fig. 17) causes uneven film thickness and excessive overspray. In arcing, the adhesive deposit is heaviest when the gun is perpendicular to the work, or in the center of the arc, and is lightest at the ends of the arc. Arcing is caused by using only the wrist to move the gun in a swinging motion instead of using the wrist,

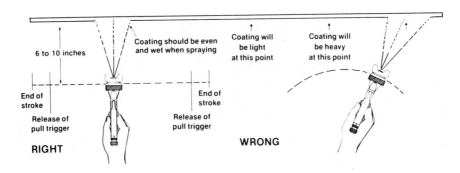

Fig. 17. Arcing.

elbow, and shoulder to maintain a uniform distance between the gun and the work. The only time arcing is permissible is when the work being coated is so large that the operator cannot reach across the entire span. If the work requires sectionalized spraying on a given large surface, a wrist motion causing arcing away from the surface should be used at the end of the stroke. This technique allows feathering of the end of the stroke. When the next section is sprayed, the operator uses the same arcing procedure at the beginning of the stroke to blend into the previous spray pattern.

Spray movement should proceed at a comfortable rate. If the operator must move quickly in order not to flood the work, then either the fluid nozzle is too large, or the fluid pressure is too high. Conversely, if the operator must move slowly in order to apply full, wet coats, then either the pressure is too low, or the fluid nozzle is too small.

A "lap" is the distance between strokes and is generally measured in percentage. For example, if the fan pattern is 12 in. wide (30 cm) and the spray gun is moved down 6 in. (15 cm) before making the next stroke, the lap distance is 50%. Proper lapping is essential in producing uniform film thickness. The lap should not exceed 50% but can be anything less. The closer the lap points, the more uniform the film will be; however, in order to conserve material and operator energy, no more strokes should be used than are absolutely necessary to give the degree of uniformity that will meet the quality requirements.

Lapping should be done in a zigzag manner (Fig. 18). The first spray stroke may start from left to right. At the right position the gun is dropped the lap distance, and the next stroke is made from right to left. If consistent lapping is practiced, any given piece will require a given number of strokes.

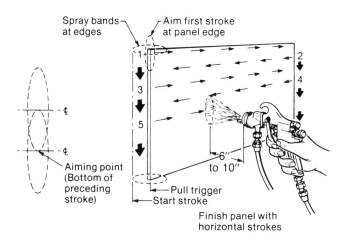

Fig. 18. Lapping.

Cross-hatching is similar to lapping except that it is done in both directions, vertically as well as horizontally. If proper lapping is practiced in either vertical or horizontal spraying, cross-hatching is wasted effort unless extremely close tolerance of film thickness must be maintained.

Proper triggering of the gun is of considerable importance for several reasons. The most important reason is conservation of material; another is maintenance of a clear fluid nozzle.

If the spray-gun trigger is engaged for too long a period, the rim of the orifice on the fluid nozzle will become "plated" with material. The flow rate will diminish; eventually, the orifice will become completely clogged. The needle valve, actuated by the trigger at the end of each stroke, prevents this buildup from occurring and ensures that the same volume of material is consistently delivered.

In considering the proper method of triggering, let us use the spraying of a flat panel as an example. The stroke is started off the work, and the trigger is pulled when the gun is opposite the edge of the panel. The trigger is released at the other edge of the panel, but the stroke is continued for a few inches before reversing for the second stroke. Triggering is really the heart of spray technique.

To develop proficiency in the techniques we have discussed requires much practice; however, as the skill develops, you will find it a relatively simple task to obtain full coverage of the surface with minimum overspray and minimum operator fatigue.

III. AIRLESS ATOMIZATION

The term "airless" has been applied to this process to indicate that no air is used to atomize the material. Atomization occurs when the material being sprayed is forced through a small orifice, or opening, at speed sufficient to cause atomization. A good example of this principle is the fine mist that occurs when you initially open a garden-hose nozzle. If there is sufficient water pressure, the water will be released to the atmosphere through a small orifice at speed sufficient for atomization to occur. As the viscosity of the material being sprayed increases, the pressure necessary to atomize it will also increase. In fact, the number of adhesive formulations that can be applied by this process is limited by this characteristic.

When a material can be applied with the airless process, we can expect certain benefits such as the following:

1. The amount of "spray fog", or rebound, is greatly reduced, thus making it a practical method of application in areas where overspray or fog would be prohibitive. Airless atomization also permits us to spray into recesses and cavities without bounce back.
2. Airless equipment produces a higher rate of fluid delivery to the object being coated; therefore, the surface can normally be covered in one coat.
3. When used properly, the airless process is considerably cleaner than the conventional method.

Warning: A potential hazard exists with this process.

To achieve atomization it is necessary to use relatively high fluid pressures. These pressures may range up to 3000 psi (21,000 kPa). This pressure will create a fluid velocity sufficient to penetrate the skin. This problem, known as hydraulic injection, can occur when the airless equipment is improperly handled. *Operator training is imperative.* All operators and service personnel should be trained in the proper use and safety considerations of the equipment.

A. Equipment Components

A typical airless spray system (Fig. 19) can be described as follows:

1. It must have a fluid pump capable of delivering the amount of material required to do the job at a pressure sufficient for atomization. These pumps are generally referred to as high-ratio pumps and are powered by either compressed air or an electric motor. The pump is mounted in such a manner as to draw material either from the original shipping container through a suction tube, or by immersion of the fluid section of the pump.

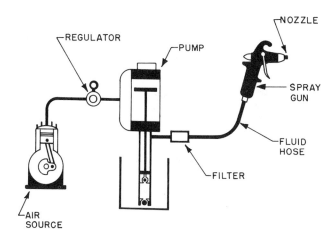

Fig. 19. Airless spray system.

2. As the adhesive leaves the pump, it is under high fluid pressure; therefore, all components in the delivery system— which includes a fluid filter, valves, and the necessary hose fittings—must be rated to withstand the high pressure.
3. The spray gun (Fig. 20) is really a high-pressure valve which can turn the fluid flow on and off. Unlike the spray pattern from air-atomized spray guns, the spray pattern from airless guns cannot be feathered. The guns are either fully on or fully off. Spray guns are available in various models depending on the application, that is, normal cold spraying, heated spraying, or high-volume spraying.
4. A spray tip attached to the spray gun forms the spray pattern. The size and shape of the orifice in the tip determines the output capacity and the area that will be covered. Tips referred to as "flat tips" will provide only one spray pattern. They are not adjustable. To change the pattern one must change the tip. There are several designs of adjustable tip available that permit the operator to change, within limits, the configuration of the spray pattern. A tip guard is provided by the manufacturer of the spray gun for personnel protection. The guard is a reminder to the operator that the spray equipment should not be brought in close proximity to the body.

B. Setting Pressures

When the pattern is right, the pressure is right. This basic rule applies to all forms of spray application. *Use the lowest pressure that*

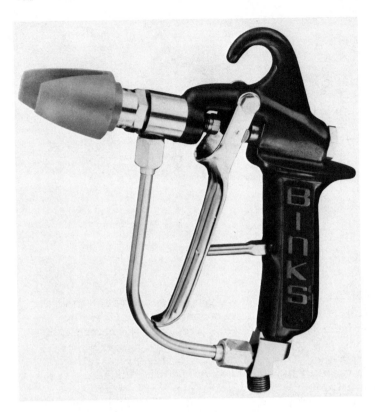

Fig. 20. Airless spray gun. Note the tip guard.

will provide a satisfactory spray pattern.

1. Be sure that the pressure-adjusting knob (if electrically powered) or the air regulator (if air powered) is "backed off" to the lowest setting before you apply power to the unit.
2. Prime the unit with the appropriate solvent to wet out the system according to the manufacturer's instructions.
3. Replace the solvent container with the container of material you intend to spray.
4. Test the spray pattern and gradually increase the pressure until the pattern is well balanced.

This procedure will provide the sprayer with the proper amount of material under control with the least amount of bounce back and off-spray.

C. Operator Technique

The technique for using an airless spray gun is very similar to the technique for using conventional, or air-atomizing equipment with the exception that the distance from the gun to the work should be approximately 12--14 in. (30--35 cm) for airless spraying.

IV. AUTOMATIC EQUIPMENT

Automatic spray machines can be designed to coat almost any product regardless of the shape, size, or material of construction. The basic design of such a machine conforms to one of the motions described later in this section, with minor modifications to suit the particular product. In the final analysis, the automatic machine merely performs the same functions as the sprayer, but it does so under considerably greater control and, usually, at high speeds.

A. Horizontal Reciprocating Machines

A horizontal reciprocating machine (Fig. 21) coats flat objects, such as particle board and other sheet stock, which can be carried on a

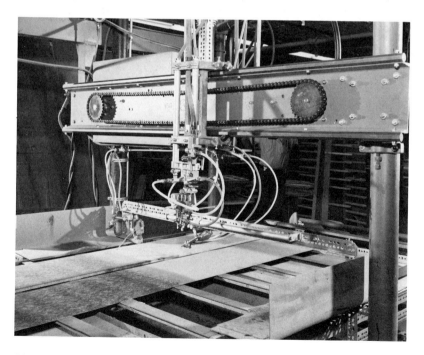

Fig. 21. Horizontal reciprocating machine applying contact cement to particle board and laminates.

lay-down conveyor under the spray gun. The spray guns are fixed
at right angles to the surface of the product. When one spray gun
with a 6-in. (15-cm) spray pattern is used, the product will move 3
in. (7.6 cm) for every stroke of the machine. For a more uniform coat
an overlap is provided to cover one-half of the spray pattern of the
previous stroke. The spray gun is triggered at the beginning and
end of the stroke as in manual spraying. This motion is recommended
for conveyor speeds up to 25 linear ft/min (9 m/min).

B. Vertical Reciprocating Machines

Vertical reciprocating machines are designed to produce a vertical, or
up-and-down stroke. These machines are normally used in conjunction with overhead conveyor systems but can be adapted to other
types. The track that carried the spray-gun carriage in a reciprocating machine can be curved to follow, within limits, the general
contour of a product.

C. Rotary Spray Machines

Rotary spray machines perform the same basic operation as horizontal
reciprocating machines but are able to operate at much higher convey-
or speeds. The arms rotate at approximately 20 rpm, and the spray
pattern, in conjunction with the conveyor, covers the object with a
series of overlapping arcs. The flatter the arc, the more uniform the
coating will be. Four-arm rotary spray machines are recommended for
conveyor speeds up to 50 ft/min (18 m/min); eight-arm rotary spray
machines are recommended for conveyor speeds over 50 ft/min (18 m/
min).

D. Spindle Machines

Spherical or cylindrical items are best coated on a spindle machine.
The object is placed on a specially designed work holder and rotated,
or spun, in front of one or more spray guns. Depending on the
shape and size of the product, the minimum spacing of the work holder is normally every other pin of the chain conveying the product.
If the product lends itself to automatic handling, automatic loading
and unloading can be included in the design of the system.

    Automatic machines are in reality "gun movers" and, as such,
are adaptable to any of the finishing processes we have discussed,
that is, conventional or airless spraying, cold or heated. When used
properly, automatic equipment can provide

    1. Material savings
    2. Increased production
    3. Improved product quality
    4. Finish uniformity
    5. Labor savings

V. RESPIRATORY PROTECTION

No matter how efficient your spray operation, there will always be some air contamination resulting from it. This contamination will be in two forms: solid particulate matter, such as dust, and vapors, such as adhesive overspray and solvents that are being driven out of the adhesive during spray and cure cycles. The human respiratory system can filter out some of the solid materials; however, it is completely defenseless against gases and vapors. Solvent vapors have the same consistency as air, so they are breathed in unchecked. They are easily absorbed into the bloodstream. Harmful vapors, if inhaled, can affect the body in different ways, depending on the solvent type and concentration. Some may just be a nuisance while you are working—have a bad odor, make your eyes water, or give you a headache. Others may make you dizzy or sick to your stomach. Some can irritate the mucous membrane or burn the lungs.

A. Respirators

A family of air purifying devices, which includes dust respirators and chemical cartridge respirators, is available for protection against dust and fumes, including organic vapors. The devices are designed for different working conditions, but in the end they all have a single job to do: to make sure the air you breathe is clean.

Basically, dust respirators have filters that remove particles from the air by a simple, mechanical process. The finer the filter is, the smaller the dust and mist particles that can be trapped and filtered out, and the cleaner the air you breathe will be.

Respirators that protect against toxic gases and vapors have a cartridge filled with chemicals. The purifying chemicals absorb the harmful gases so that only clean air enters your lungs.

For best overall protection, cartridge respirators equipped with paint prefilters may be used. These respirators are equipped with dust covers which snap over the chemical cartridge so that dust has been removed before the contaminated air enters the cartridge for chemical cleaning.

The need for proper care of a respirator cannot be overemphasized. When odors or irritating vapors become noticeable or when breathing resistance becomes excessive, the cartridges or filters should be replaced. Remove and discard the used cartridges or filters. Make sure that the gaskets are in place in the cartridge holders. *In short, read and understand the instruction material supplied by the manufacturer. Your life may depend on it.*

B. Spray Booths

When the spray application is being accomplished in a fixed location, it is desirable, and often mandatory, for a spray booth to be part of

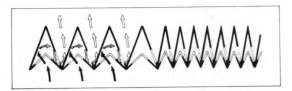

Fig. 22. Dry-type spray booth.

your application system. Spray booths are fire- and health-protection enclosures scientifically designed to provide a positive movement of air through the spray area. This air movement causes the solvent fumes to be carried to the atmosphere and most solid particles to be eliminated in the filter system. Most spray booth designs incorporate a method of filtering or washing the solid particles from the air before discharge. When properly designed and installed, the spray booth can provide the following advantages:

1. Safety through the removal of volatile solvent fumes from the spray area.
2. Lower insurance rates. Some states even provide a tax advantage to firms using spray booths.
3. Improved application quality when overspray is removed from the spray area because unnecessary contamination of other work in the same area is prevented.
4. Working environment of the spray operators and others in the immediate area is improved.

Fig. 23. Water-wash spray booth.

5. Better community relations. Proper filtration of exhausted air will minimize air pollution and damage to neighboring property.

When you are selecting a spray booth, it is advisable to check the safety, fire, insurance, underwriter's, and building codes which affect your area to determine the type of spray booths that are acceptable. Work closely with a qualified equipment supplier to determine the actual size and type that are best for your needs.

Generally, spray booths can be subdivided into two categories: dry-type spray booths and water-wash spray booths.

1. Dry-type spray booths

Dry-type spray booths (Fig. 22) use a mechanical means to distribute the air movement evenly and provide proper filtration when required. No water is required for filtration; however, water may be required in the area for fire protection or cleaning up. The mechanical filter used in this type of booth can take several forms, the most popular of which is called the paint arrestor filter. Paint arrestor filters are made of a fire-retardant, treated paper, formed into a "honeycomb" configuration. The filtering action is accomplished by the rapid back-and-forth movement of the air as it passes through filter media. Centrifugal force throws the solids against the treated paper where they will stick. As the material builds up on the filters, resistance to airflow develops, and eventually, the filters must be changed.

A measuring device mounted on the spray booth called a "draft gauge" is provided to indicate the time to change filters.

2. Water-wash spray booths

Water-wash spray booths (Fig. 23) use chemically treated water as a means of washing the solids out of the exhaust air. A quantity of water is contained in a tank at the base of the booth structure. This water is circulated and discharged through nozzles so that a water curtain develops. The contaminated air is passed through the water curtain at several different points which, in effect, washes the air two to four times to remove the solids. The chemical compound in the water causes the adhesive particles to coagulate and float on the surface. In the standard pump booths, the contaminants are skimmed off the top from time to time.

Properly engineered and sized water-wash spray booths are the most acceptable means of removal of solid contaminants from the exhaust air. It must be emphasized, however, that these booths will not remove solvent vapors.

3. Air replacement

An air makeup or replacement system is necessary because of the large volume of air that is exhausted from the spray booths. This exhaust is sufficient to produce two or more complete changes of air within the building every hour. This is especially difficult during winter months when building heat is being lost. The demand for air also creates a negative pressure within the building which can cause drafts, dirt problems, etc. An air makeup system will provide even temperatures and clean filtered air in addition to ensuring proper booth performance.

ACKNOWLEDGMENTS

All of the material, figures, and tables used in this chapter are courtesy of Binks Manufacturing Company, Franklin Park, Illinois.

19

Dispensing High-Viscosity
Adhesives and Sealants

Barry R. Killick *Pyles Division, Kent-Moore Corporation, Wixom, Michigan*

I. INTRODUCTION

In this chapter we are going to define and explore the equipment and methods required to pump, meter, and mix high-viscosity materials. As the spectrum of materials includes both single-component and plural-, or multicomponent adhesives and sealants, we will cover two types of systems: basic pumping systems and pumping-metering-mixing systems.

Each element of these systems will be discussed, thereby allowing an understanding of the relationship between elements. The discussion will explain how components fit together to give an efficient system for dispensing any of today's high-viscosity fluids. Figures 1 and 2 show typical elements of the two systems.

What is a high-viscosity fluid? Generally, we are talking of materials that are non—level seeking, fibrous, or stringy in nature. This puts us into a viscosity range extending from 60,000 centipoise (cP) to 4,000,000 cP. (Note that highly thixotropic materials appear to be non—level seeking but are, in fact, very easily pumped.)

II. PUMPING SYSTEMS

What does the word "pump" bring to mind? An oilwell? The device used for recirculating your aquarium water or squirting your windshield washer fluid? Maybe your own heart? All of these devices are true liquid pumps, albeit low-viscosity ones. One definition states that a pump is "a device or machine that raises, transfers, or compresses fluids " [1]. Notice the three functions mentioned, any one of which when applied to the description of a device, will cause that device to be termed a pump. However, for the purposes of our discussion—that of moving high-viscosity fluids from the point of supply to a point of application—we require all three functions to be achieved. Let us immediately beware of that old law of physics that states that fluids are incompressible. In our practical applications we find that fluids, especially high-viscosity ones, appear, by intent or otherwise, to be compressible to a small degree.

In considering the transfer or pumping of a high-viscosity, single-component fluid, usually a sealant or adhesive, from a point of supply to a point of application, we require a rather complex device to perform a very simple task. After all, most high-viscosity fluids could be removed from their containers with a spoon and even applied with this same utensil, although the process would be crude and inefficient. What we require, then, is a device that transfers fluids in a controlled manner by variable conditions of pressure in the system. The pressure is necessary to achieve the desired flow rate. The requirements for moving high-viscosity fluids happen to fall into the difficult end of the pump-requirements

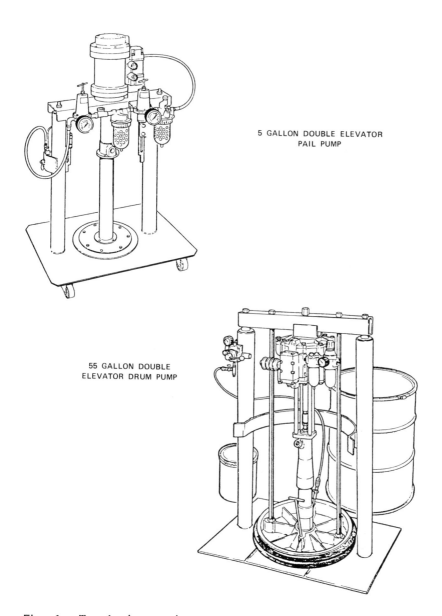

Fig. 1. Two basic pumping systems.

spectrum, especially when precise volumes must be dispensed in precise times at precise pressures— often on automated production lines. It is in these applications that knowledge of critical pump parameters becomes important in putting an efficient system together.

Many of the fluids we may want to pump are single-component materials, such as silicone, plastisol mastic, vinyl neoprene, and butyl. Others are two- (plural-) component materials such as epoxies, urethanes, polysulfides, and acrylics. There are also available, for very critical end-product specifications, three- or four-component materials. However, no matter how many components the final mixed material is composed of, each component is dispensed through its own pump and can therefore be considered as a single-component material. Also, note from Fig. 2 that the pump is the first device in a chain of devices required to transfer a material from a container to a point of application.

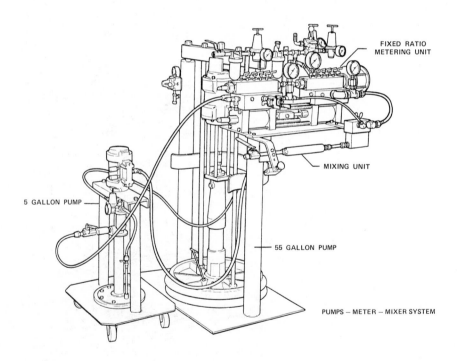

Fig. 2. A typical pumping-metering-mixing system.

III. PUMP TYPES

The industrial pump family includes single- and double-acting ball, rotary, and diaphragm pumps as well as single- and double-acting chop-and-check types for high-viscosity materials.

We are going to expand on the idea of the double-acting, reciprocating chop-and-check pump for high-viscosity dispensing systems. It is the most commonly used device for moving industrial sealants and adhesives. This reciprocating pump is considered a positive-displacement device, that is, its material output volume is directly proportional to its piston stroke. However, false compressibility, caused by minute (sometimes large) air bubbles that are trapped in the material and that compress under internal pump pressure, plus internal leakage will cause some deviations from the positive-displacement theory. Fortunately, these deviations are almost unnoticeable during the operation of modern pumps. Note that ball-type pumps should not be used for high-viscosity materials due to the inability of the balls, used as internal valving, to move rapidly within these heavy materials.

The major components of a dispensing system for high-viscosity materials will now be discussed.

A. Air Motor

The air motor provides the power for the supply pump. It is a cylinder with a double-acting piston driven by air. Various models are available, usually 3, 4, 6, 8, and 10 in. (7.5, 10, 15, 20, and 25 cm) in diameter. All will operate on air pressures from 8 psi (56 kPa) to 120 psi (840 kPa). Most are designed to be efficient with normal American production plant supplies, that is, a minimum of 30 standard cubic feet per minute (SCFM) [80 standard liters per minute (SLM)] at 80–90 psi (560–630 kPa).

B. Elevators and Follower Plates

An elevator consists of one or more simple air cylinders having double-acting pistons. The single-post elevator is primarily used to conveniently lift the pump out of a 55-gal drum. The double-post elevator is used with a follower plate whose purpose is to exert pressure on the material which otherwise would not flow into the pump tube. The follower plate also wipes the barrel clean.

C. Pump-Tube Assemblies

The pump tube is the material-handling section of the pump. It is driven by the air motor. The difference in area between the air-motor piston and the pump piston creates the "power ratio" of the pump. A balanced double-acting pump loads a full charge of material on its upstroke. It pumps out half of this charge on the downstroke

while transferring the other half internally, such that it will be
pumped out during the next upstroke.

IV. PUMP OPERATION

A double-acting piston pump will provide a continuous flow of a
fluid as it is cycled or reciprocated. The rate and smoothness of
this flow varies with the stroke velocity and directional changeover
speed. It is also proportional to the cycle rate. The displacement
rod of a pump can be sized diametrically such that the same volume
of fluid is dispensed on the upstroke as on the downstroke. A
pump with such a piston is called a balanced pump. Balance is essential for accurate and consistent flow, especially in automated

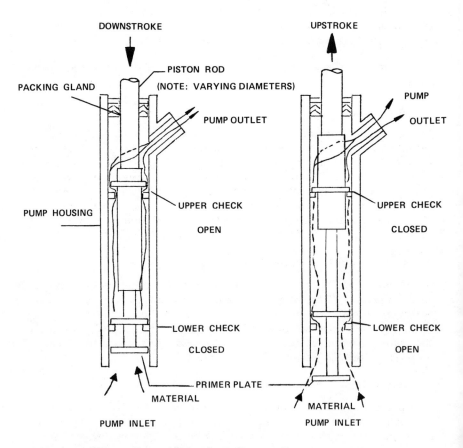

Fig. 3.   Double-acting, high-viscosity pump.

production-line equipment. In manual applications, it is possible
for the operator to anticipate an unbalanced pump and to adjust
for minor volumetric variations in output.

Referring to the pump in Fig. 3, consider the upstroke. The
pump is bringing in, through a combination of pump suction and
elevator down pressure, twice the volume of material that it can
force out on the downstroke. In actual operation half of that incoming volume is dispensed while the other half is stored within the
pump. On the downstroke, that stored half is in turn dispensed.
This cycle is repeated, creating a continuous material output as the
pump reciprocates.

Note that the pump has two internal checks, upper and lower,
and a primer plate. The function of the primer plate is to assist
the elevator in forcing the material into the pump throat. As we
have seen, the elevator, via the follower plate, is attempting to get
the material to flow into the pump throat. The primer plate, during
the upstroke, is also assisting by lifting and compressing the material into the pump throat. During the downstroke, we would prefer that the primer plate not be there at all. It is, therefore, designed carefully as a compromise unit to give positive assistance on
the upstroke and neutral assistance on the downstroke.

V. THE METERING UNIT

A metering unit is essential for accurate volumetric dispensing of
single- or multicomponent adhesives or sealants. In its simplest
form, a metering unit may consist of a double-acting pump with a
stroke counter to indicate the volume dispensed. However, due to
inherent variations in the output of reciprocating pumps this type
of meter is not considered very accurate. In a more complex form
a meter may consist of two or more cylinders coupled by a variable
ratio linkage. Each cylinder is supplied with material from a standard, reciprocating supply pump. We can break metering units
down in the following ways.

A. Single Component

1. A standard reciprocating supply pump that has a means
   to count either total strokes or increments of strokes
   (Fig. 4a). Due to losses at pump changeover, accuracies of better than ±5% should not be expected.
2. As above, but using a time control (Fig. 4b) to determine the volume of material dispensed. Due to variations
   in pump rates, which the timer cannot detect and correct
   for, this method of metering may show errors of even
   greater than ±5% of required volumetric output.

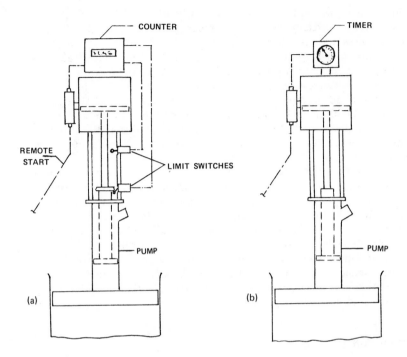

Fig. 4. (a) Pump with volumetric output controlled by an air or electric stroke counter. (b) Pump with volumetric output controlled by an air or electric timer. (c) Pumps with volumetric output controlled by a displacement rod.

3. A positive displacement rod or piston which dispenses on the downstroke and is primed by an external pressure feed on the upstroke (Fig. 4c). The stroke of the rod may be determined in increments down to thousandths of an inch using standard, limit-switching or other types of position sensors. The dispensing-loading cycle is achieved by switching material valve(s) on the bottom of the metering cylinder. If at all possible, use this type of unit for accurate metering of single-component, high-viscosity materials. The limiting factor for this type of unit is the volume that is required per shot. For volumes up to 60 in.^{3} (984 cm^3), the design remains practical. Above this volume, the size of the cylinder becomes cumbersome, though feasible, for single-stroke metering, and reciprocating-type metering should be utilized.

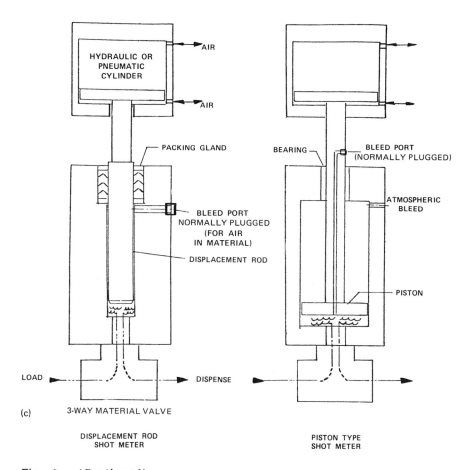

Fig. 4. (Continued).

B. Plural Component

1. A fixed-ratio unit (Fig. 5a) in which the size (area) and stroke of the major and minor volume cylinders remain constant so that the volume of each component dispensed remains constant with respect to the other, irrespective of changes in dispensing rate or pressure.

2. A variable-ratio unit (Fig. 5b) in which we are able by various means to alter the ratio of the major volume material dispensed to the minor volume material.

The more common metering unit will be a reciprocating type, driven either pneumatically or hydraulically. Metering cylinders

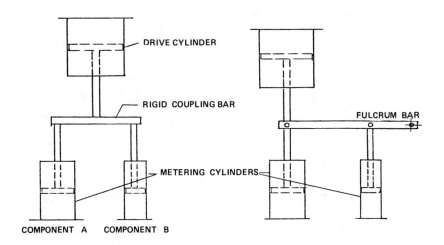

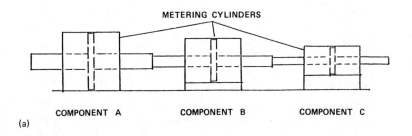

(a)

Fig. 5. (a) Fixed-ratio meters; (b) variable-ratio meters.

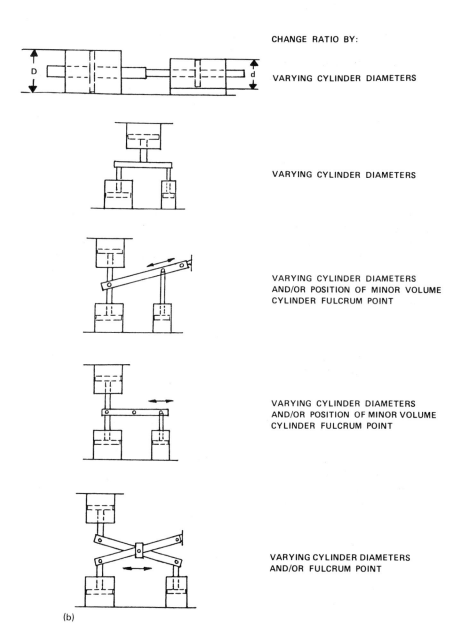

(b)

Fig. 5. (Continued).

may be set horizontally and coupled in tandem or laterally and coupled through a linkage bar. The horizontal method is probably less complicated, but neither enjoys a major advantage. Output ratios are determined by either altering the distance of each cylinder from a pivot point of a linkage, varying cylinder diameter(s), or doing both. Ratios from 1:1 up to 100:1 are possible with accuracies of 0.5–1.5%. However, to achieve these accuracies, it is imperative that the material be completely air free and that the meter be bled correctly. Due to its compressibility, any air in the material will cause ratios to fluctuate widely. Remember to ensure that your materials are delivered air free by the supplier, or add a degassing system to the metering-unit supply source.

C. Meter Ratio

Consider a two-component sealant. In order for the desired properties of the mixed material to be achieved when cured, the two components must be metered "on-ratio" at all times. For example, a 4:1-ratio material means that we require four units of base material to every one unit of catalyst material to give us the required properties of the material when mixed. The units may be defined in two ways:

    1. 4:1 parts by volume (PBV)
    2. 4:1 parts by weight (PBW)

If the specific gravity of either material is not exactly 1, then PBV does not equal PBW.

If the ratio is known in either unit, it may be converted to the other unit as follows:

    1. *Weight ratio (PBW) to volumetric ratio (PBV)*

Known
    Specific gravity (sp gr) of materials A and B
    Weight ratio of A:B
Required to know
    Volumetric ratio of A:B
Then

$$\frac{\text{Parts B}}{\text{sp gr}} \times \frac{\text{sp gr} \times 100}{\text{parts A}} = \text{parts B per 100 parts A}$$

For example
    sp gr of A = 1.14
    sp gr of B = 1.04
    Weight ratio A:B = 10:1

$$\text{Volumetric ratio} = \frac{1}{1.04} \times \frac{1.14 \times 100}{10} = 10.95$$

Volumetric ratio A:B = 100:10.95 or 10:1.095

2. *Volumetric ratio (PBV) to weight ratio (PBW)*

Known
   Specific gravity (sp gr) of materials A and B
   Volumetric ratio of A:B
Required to know
   Weight ratio of A:B
Then

$$\frac{\text{Parts B}}{\text{Parts A}} \times \frac{\text{sp gr B}}{\text{sp gr A}} \times 100 = \text{parts B per 100 parts A}$$

For example
   sp gr of A = 1.08
   sp gr of B = 1.23
   Volumetric ratio A:B = 114:100

$$\text{Weight ratio} = \frac{100}{114} \times \frac{1.23}{1.08} \times 100 = 100$$

Weight ratio A:B = 100:100 or 1:1

Generally, metering equipment is sized by diameter and stroke length which, when calculated as area times stroke, gives known volumetric outputs. The PBV ratio is more useful if it is known.

Practical envelopes for high-viscosity metering equipment are as follows. However, manufacture of specialized units is feasible if the expense and the necessity of rigid maintenance procedures can be realized.

Viscosities (major or minor side): 40,000—4,000,000 cP
Ratios: 1:1—60:1
Pressures: Up to 2500 psi (17,500 kPa)

Some of the special features that may be added to widen the field of use of piston-type metering units are

1. High-pressure elastomeric seals or lapped metal-to-metal seals.
2. Ceramic coating for wear resistance with abrasive-filled materials.
3. Heating or cooling jackets for metering cylinders for viscosity stabilization.
4. Special electrical control packages. These packages may function to allow: total automatic sequencing of the machine, repetitive dispense sequences, counting and totaling shots or cycles, timing of dispense cycles, timing and sequencing of purge, flush and air-dry cycles, etc.

VI. PLURAL-COMPONENT MIXING

After the materials have been moved from their original containers
and metered in the correct proportion, it is necessary to mix them.
Several types of mixers are available, both static and dynamic,
which will handle high-viscosity materials. Because different materials have different mixing compatibilities, many devices have been
developed. For example, it is known that some polyester systems
will eventually cure even with very inefficient mixing. Other materials, such as urethane elastomers, require very accurate and
complete mixing to achieve their desired cure properties. Even today mixing is more of an art than a science.

A. Static Mixers

Static mixers are devices, usually tubular, that have within them
fixed, geometrically shaped elements which act as flow-splitting and
shear-energy--creating devices for the fluids which travel through
them (Fig. 6). Medium pricing and no moving parts are definitely
positive reasons for using static mixers. They work well if the ra-

Fig. 6.   Static mixers.   (Courtesy of Chemineer-Kenics, Dayton,
Ohio.)

tio and viscosity ranges of the two materials are not too wide. High-viscosity silicones, urethanes, and epoxies have all been successfully mixed through static mixing devices. Against the use of a static mixer is the fact that they are hard to efficiently flush or purge if they contain large volumes of material. Thus, considerable material may be wasted during purging or flushing whether mixed material, base material, or solvent is used. Removable-element mixers are available which obviously allow much more efficient cleaning.

B. Dynamic Mixers

In dynamic mixers, elements are rotated within a mixing chamber by means of rotary air, electric, or hydraulic motors (Fig. 7). Dynamic mixers, designed to create very high shear, can handle any combination of fluid ratio and viscosities with or without fillers at pressures up to 2500 psi (17,500 kPa). Their one big disadvantage is that they are dynamic! This means wear problems are inevitable and shaft seals are necessary. Wear can be very rapid in high-volume mixing, and replacement of shaft seals becomes a routine maintenance function.

Cleaning of dynamic mixers usually requires either base purging or dismantling. With high-viscosity materials especially, solvent flushing can be very ineffective. Unless the material is extremely soluble in the solvent, all that will happen is that the solvent will break through in a small channel and flow through the mixer without turbulence of the mixed material. (This is even more of a problem in static mixers.) Cleaning assistance may be given the solvent by introducing pressurized air in conjunction with the solvent to create turbulence within the mix chamber. This air blast may also be used to evaporate any remaining solvent after flushing has been completed.

VII. ACCESSORIES AND CONTROLS

For many high-viscosity fluids heating may be used to reduce the viscosity and, hence, reduce the "strain" on the system due to forces required to push the material through meters, hoses, valves, mixers, etc. On the other hand, cooling may be required to counteract shear-energy heating of fluids in dynamic mix chambers or to extend the pot life of mixed material.

Various manual or automatic dispense valves are available ranging from the manual pistol-grip flow gun to the multiple-orifice, automatic heads with antidrip or snuff-back action valving (Fig. 8).

Header systems consisting of rigid piping may be designed to carry materials long distances from the supply pump to the applica-

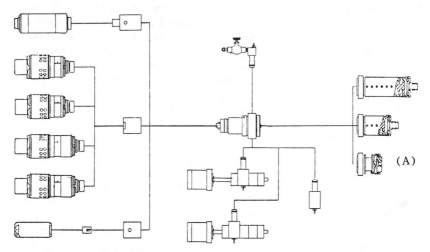

AIR MOTOR DRIVE & ADAPTOR + VALVING + BEARING/SEAL ASS'Y' + MIXING CHAMBER

Fig. 7. Dynamic mixers. (A) Schematic layout; (B) miscellaneous assemblies.

tion point. Distances of 200 ft (65 m) are not uncommon, with drops to points of application positioned as required by the produc- tion-line setup (Fig. 9).

Various types and sizes of material hoses are available to give flexibility in locating the dispensing head.

Control systems are available using pneumatic, hydraulic, or electric components, individually or in conjunction with each other. Systems range from simple, manual start-stop to totally automated with variable control of dispense frequency, dispense timing, flush or purge timing, etc. Dispensing-unit controls may easily be inte- grated into existing automation-line controls.

VIII. ELEVATORS

The elevator, a seemingly crude piece of equipment, is an important element in enabling the pump to extract material from a pail or drum. The primary function of an elevator is to assist in priming the pump by pressurizing the material into the pump throat. (The secondary function is to lift the pump into and out of the container.) The means by which this is done is as follows (Fig. 10): Application of air or hydraulic pressure to two pistons transfers the force to the follower plate which, in turn, pushes on the material surface within the pail or drum; the pressure thus created forces material into the pump throat.

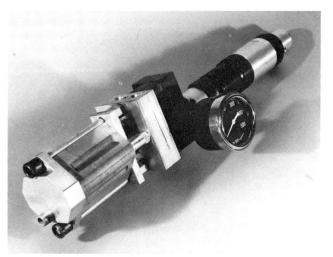

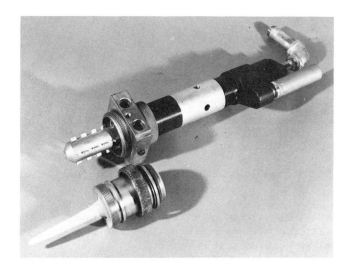

(B)

Fig. 7 (Continued).

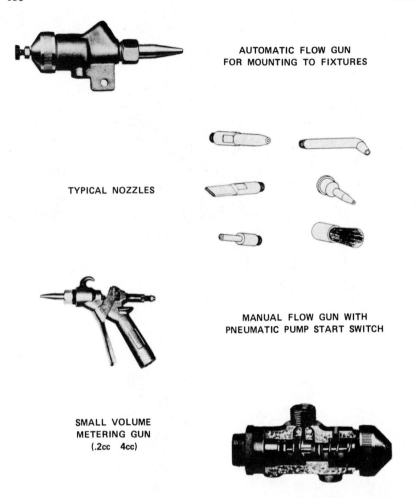

AUTOMATIC FLOW GUN
FOR MOUNTING TO FIXTURES

TYPICAL NOZZLES

MANUAL FLOW GUN WITH
PNEUMATIC PUMP START SWITCH

SMALL VOLUME
METERING GUN
(.2cc   4cc)

Fig. 8.   Manual and automatic dispensing valves.

In fact, although the elevator is a most important element in pump priming and material feed, the amount of assistance it provides is marginal as we shall see.

The ratio of the sum of the areas of the two elevator pistons to the area of the follower plate creates a power ratio which we would like to be as high as possible.   Generally, we have two 3-in.—diameter (7.6-cm) pistons working against an 11.25-in.—diameter (78.6 cm) follower, for 5-gal (19-liter) pail units, and two 3-in. (7.6-cm) or 5.5-in. (14-cm) diameter pistons working against a 22.5-in —diameter (57 cm) follower, for 55-gal (208 liter) drum units.

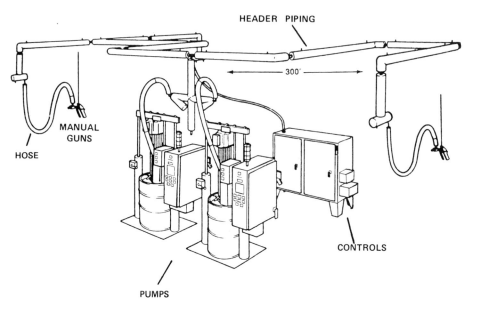

MODULAR HEADER SYSTEM

Fig. 9. Modular header system.

The total elevator down force (E_{df}) available can be calculated as follows:

$$E_{df} = 2[A_e - A_r] \tag{1}$$

where

A_e = piston area
A_r = rod area
($A_e - A_r$) = 5.8 in.^{2} for a 3-in. elevator
($A_e - A_r$) = 22.49 in.^{2} for a 5.5-in. elevator

The follower-plate area for a 5-gal pail is 99.4 in.; and the follower-plate area for a 55-gal drum is 397.4 in.^{2}. Using these figures we can calculate several power ratios.

1. *Power ratio for 5-gal pail, 3-in. elevator unit*

   (2 × 5.8):99.4 = 0.116:1

2. *Power ratio for 55-gal drum, 3-in. elevator unit*

   (2 × 5.8):397.4 = 0.029:1

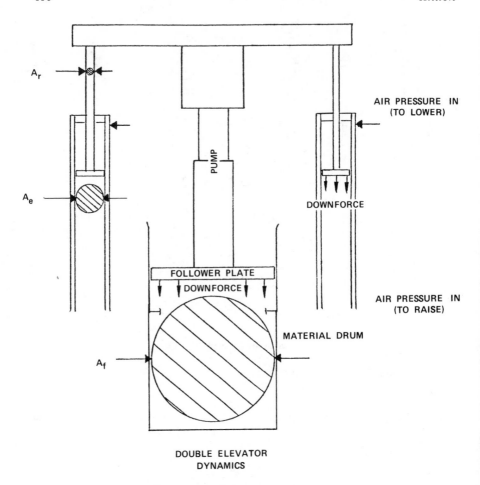

Fig. 10.   Cross section of a double elevator.

3.   *Power ratio for 55-gal drum, 5.5-in. elevator unit*

$(2 \times 22.49):397.4 = 0.113:1$

So in the first case, every 100 psi (689 kPa) of air pressure applied to the elevator creates 11.6 psi (80 kPa) material pressure with which to feed the pump; in the second case, every 100 psi (689 kPa) creates 2.9 psi (20 kPa); and in the third case every 100 psi (689 kPa) creates 11.3 psi (80 kPa).   As can be seen, these values are not very high, especially when we expect this force to be used to prime some very high viscosity materials.   The total force available to prime a pump will include the vacuum draw that the pump is able to create, as well as the elevator downforce. Obviously, the vacuum pressure cannot be more than atmospheric pressure. It is usually only on the order of 10–12 psi (70–84 kPa). Another restriction on the amount of pressure that can be exerted on a material is the strength of the container. Manufacturers of steel pails and drums do not recommend pressures above 14–20 psi (98–140 kPa) for fear of explosion. There is equipment available with hydraulically operated elevator rams, but it is not recommended for use without approval of the material supplier.

IX. GENERAL EQUIPMENT SELECTION GUIDE

A. Pumps

Requirement	Use
High-viscosity ($\geqslant$ 30,000 cP) materials	Double-acting chop and check
High output volume	Low power-ratio pump— less than 16:1[a]
High material output pressure	High power-ratio pump— greater than 16:1

[a]Pump power ratio is the ratio of the swept area of the air motor piston (A_p) to the swept area of the pump displacement rod (A_r). It will usually be expressed as 17:1, 21:1, 59:1, etc.

Increasing A_p and holding A_r constant will give

1. Increased power ratio
2. Increased pressure at the material outlet
3. Change in air consumption
4. No change in pump volumetric output per cycle.

B. Controls

Requirement	Use
Precise, repetitive volumetric output	Air or electric stroke counters; incremental stroke counter for very accurate output-volume control

Requirement	Use
Elimination of start-stop material pulsing	Accumulators; double regulation kit; side port checks; automatic, air-motor pressure dump valving
Heated material (viscosity reduction)	Heated pump jackets, hoses, tanks, etc.
Low-level indication for pail or drum units	Air or electrical (audio or visual) low-level indicator and pump shutoff valve
Monitoring of pressures; high or low limit shut down	Material pressure relief valve; air-motor air-pressure regulator; power ratio that will give less than maximum material pressure at maximum air pressure; pressure-switch sensing and feedback.

C. Hoses and Piping

Always use the largest hose or pipe I.D. possible in conjunction with a working pressure rating at least equal to the pump ratio times the maximum air pressure the pump air motor may ever see.

Length of hoses should be minimized to reduce material pressure drop through the system and consequent reduction in pumping efficiency.

In manual gun applications, if flexibility is required, use a large hose from the pump outlet to the farthest downstream point possible and then attach a short length of a smaller diameter, flexible hose.

Hose construction will depend on strength required, cost, flexibility, and material compatibility. One of the common constructions in high-viscosity pumping is the single, stainless-steel overbraid (or multiple braids) around a Teflon core. Hose fittings should also be selected to minimize pressure drop.

Shutoff or diverter valves (ball or spool types) used in fluid application systems are often very restrictive especially if designed for pressures above 3000 psi (21,000 kPa). Try to use full-ported ball valves, or go one or two sizes larger than required if using standard pipe-size valves.

X. MATERIAL CHARACTERISTICS VERSUS EQUIPMENT REQUIREMENTS

A. Viscosity

High viscosity often requires low-speed handling with complex high-pressure equipment. Mixing difficulties may arise when an extreme differential exists in the viscosities of the components. Pressure drops become critical. Pump priming becomes difficult.

B. Stickiness

Stickiness, often found in base liquid polymers, has an effect similar to high viscosity. It makes materials difficult to move, usually hinders cleaning, and makes clean-material cutoff in dispensing operations somewhat difficult.

C. Thixotropy

A thixotropic material has the ability to retain its shape in a static condition. However, the viscosity drops considerably when the material is moved or agitated. Many filler materials impart this characteristic to a fluid. Usually, such mixtures are easy to pump. On occasion, pump pressure or shear action of a mixer will temporarily destroy the thixotropic characteristic of a material.

D. Presence of Fillers

Fillers are usually added to increase viscosity, reduce material costs, or impart special physical properties to a product. Many fillers are abrasive in character and cause wear of piston cups, packing seals, and precision parts. The finer the filler particles, the less their effect on equipment will be. Some fillers settle so radically that they tend to pack in dead areas of valves, fittings, and seats, thereby making the use of dispensing equipment impractical. Compressible fillers, such as microballoons, make accurate metering extremely difficult.

E. Corrosiveness

The presence of acids, amines, and other corrosive elements in material systems requires the use of stainless-steel parts. These are costly and are difficult to machine. Action of solvents in material systems presents problems related to seals. There is no perfect elastomer in the sense of resistance to solvents. Even highly corrosion-resistant Teflon is not always a suitable seal because of its cold-flow tendencies and poor resistance to abrasion. When using elastomeric seals, care must be taken to identify the elements of a component and their effects on the elastomer. For example, it is impossible to operate a machine in a ketone peroxide and to flush the same machine with a chlorinated hydrocarbon. Although there are suitable elastomeric compounds for each of these situations, no one elastomeric compound will withstand both materials.

F. Operating Temperatures

Many components of a material system may require temperature control. Cooling of components may be necessary in order to extend the work life of mixed components. Heating of individual components may be required to reduce viscosity, to improve control of the curing process, or to improve the flow characteristics of the mixed product.

G. Component Ratios

Wide-ratio materials contain between 1 and 10 parts of catalyst to 100 parts of base material. Close-ratio materials contain between 10 and 100 parts of catalyst to 100 parts of base. In general, the wider the ratio of catalyst to base material, the more critical the accuracy of proportioning becomes. Many ratios in the 1:20 range require proportioning as accurate as ± 1%. This wide ratio requires closer tolerances, necessitating more costly and complex equipment. It also necessitates the use of more highly skilled operators and more accurate inspection of finished products. Some close-ratio materials, with ratios approaching 1:1, can tolerate proportioning errors as high as 10% without any serious effect on their cured-material properties.

H. Material Stability

Lack of stability of one component in a material system can create serious equipment problems. For example, evaporation of solvent from a component will increase its viscosity and change the mixing proportion or ratio. Materials like isocyanates, which react with moisture, crystallize to create hard deposits within the material and on exposed machine surfaces unless special provisions are made to counteract this effect. Certain urethane prepolymers will cure spontaneously given enough time or higher-than-average ambient temperatures.

I. Pot Life or Work Life

The pot life, or work life, of a material is the time available from the moment the components are mixed until the mixture has cured to the point that it cannot be applied. It can range from a few seconds to many hours. In some instances pot life, or work life, may be indefinite at room temperatures. In general, long-pot-life materials do not present equipment problems. It is desirable to use material with as long a pot life as possible—one consistent with production requirements such as time required to fill a mold, number of molds available, and storage space available for curing units. Short pot life, in the area of a few seconds, requires mixing equipment incorporating small-volume chambers, rapid-ejection means, and an adequate means for immediate flushing. Certain materials have a tendency to cure progressively in thin sections on the surfaces of a mixer. The use of these materials necessitates frequent flushing with solvent or air or both.

J. Hazardous Specifications

Many components are highly toxic, thus creating personnel problems. Of particular concern in this area are the curing agents, hardeners, or catalysts frequently used in multiple-component systems. Some materials produce toxic vapors, while many are highly irritating to the skin of sensitive people. Other problems arise in connection with fire hazards or with the possibility of explosions. Whenever problems of this kind are suspected, consult with material suppliers and equipment manufacturers.

XI. GLOSSARY

The following terms are widely used in fluid handling:

Abrasive materials: Fluids that contain, as a homogeneous mixture, small particles of abrasive solids, for example, silica, pecan shell, sand, clay, etc.

Accumulator: Device for storing potential fluid energy. Used to make up material flow when fluctuation in flow is created at the material supply pump.

Adhesive: A material, initially fluid, used to bond one solid to another.

Air consumption: Volume of air used by an air-powered device. Usually expressed in SCFM (standard cubic feet per minute).

Air motor: Rotary or reciprocating device for moving mechanical components using compressed air as its power source. For pumps, dynamic mixers, etc.

Auto pump crossover: A pneumatic or electrical control coupling two pumps together such that when pump A depletes its material source, it will signal pump B to start dispensing.

Balanced pump: One that displaces the same volume on the upstroke as on the downstroke.

Cavitation: The effect created when a reciprocating pump cycles faster than the speed at which the material can be fed into it.

A vacuum is drawn internally and results in intermittent material output and/or popping and spitting at the point of material application.

Centipoise (cP):   A unit of viscosity measurement.   It is one one-hundredth of a poise (P).

Check: The valve within a pump that allows positive-displacement pumping to take place on alternate strokes, in opposing directions.

Chop-and-check pump: A pump that uses plate-type check valves in lieu of balls and seats.   Usually used on high-viscosity materials.

Compatibility: The lack of a reaction between two liquids; that is, no chemical or physical reactions take place when two compatible liquids come into contact with each other.

Component: A chemical element of a plural system, for example, resin and catalyst.   When two components are mixed, chemical reactions usually take place.

Consistency: Comparative term for viscosity descriptions, for example, thick paste, creamy, watery.   Applies to high, medium, and low viscosity.

Corrosive: Pertaining to a material or fluid that reacts chemically with another to destroy or change the physical characteristics of the second material.

Cycle: One complete upstroke plus one complete downstroke of a reciprocating pump.

Dispense: To make material flow.

Displacement rod: The piston rod of a pump which, when reciprocated back and forth in a chamber, forces out the same volume of material as the volume of the rod.

Double-acting pumps: Pump that displaces material on both the upstroke and the downstroke.

Double elevator: The device which physically supports the pump and also assists pump priming by exerting down pressure on the material via the follower plate.

Elevator: See double elevator.

Filler: Particles used to increase or decrease the density of a fluid. Range from heavy clays to glass microballons.

Flow rate: The rate at which material is dispensed from a pump. May be expressed in volumetric or mass units per unit time.

Follower plate: Circular plate used to exert pressure on nonlevel-seeking materials contained in pails or drums in order to prime the supply pump.

Friction: Resistance to flow or other motion.   Creates pressure drops in systems which result in volumetric outputs which are less than expected.

F.R.L.:   Name given to combination of filter, regulator, and lubricator required to control pneumatically actuated devices, for example, air motors and elevators.

Header: Rigid tubular device used to transport fluids for long distances. Usually consists of steel pipe supported in an overhead position.

Heated: Having a device to increase the temperature of any component in a dispensing system.

High pressure: In the fluid-application field, pressure above 1000 psi (7000 kPa).

High viscosity: Viscosity above 60,000 cP.

Hose: Flexible tube capable of withstanding fluid flow at a wide range of pressure. Used to connect two points in fluid flow system.

Level-seeking: Pertains to the ability of a fluid to seek a level within a container.

Low viscosity: Viscosity below 10,000 cP.

Lubro: See F.R.L.

Moisture sensitive: Pertaining to a fluid whose properties change when in contact with water or water vapor (air).

Newtonian: Pertaining to a fluid whose viscosity does not change with changing applied shear rates.

Output: Flow rate of material in lb/min, gal/hr, cm^2/min, or other units of volume or mass per time.

Pot life: The allowable time for which a mixed material may be stored or worked before curing reactions begin to take place.

Pressure drop: Loss of pressure, in a fluid flow path, between pressure source and any downstream location, due to extensive energy forces required to move the fluid along this path.

Proportion: Ratio of one component to another in a plural-component system.

Ratio (material): See proportion.

Rheology: The study of viscosity and flow relationships as applied to fluids.

SCFM: Standard cubic feet per minute.

Specific gravity: The ratio of mass per unit volume of a material to that of water.

Stability: The ability of polymers, in liquid form, to remain in a state that is workable by the dispensing equipment.

Thixotropic: Pertaining to a material whose viscosity is shear-rate dependent.

Viscosity: A measure of the internal friction of a fluid. A material requiring a shear stress of 1 $dyne/cm^2$ to produce a rate of shear of 1 sec^{-1} has a viscosity of 1 P or 100 cp. Water at room temperature has a viscosity of one cP.

20
Hot-Melt Application Equipment

Werner Bohm *Nordson Corporation, Atlanta, Georgia*

I. INTRODUCTION

The decision to use hot-melt adhesives in any given industrial manufacturing process is usually based on more than one reason. The predominant one is the rapid setting of these adhesives once they have been applied. Because of this feature, hot melts allow operation speeds unsurpassed by any other group of adhesives.

Another prime reason for their success is that there is standardized, modular design, application equipment available for these adhesives. Providing maximum versatility, this equipment requires only a fraction of the front-end investment one could expect from a custom-designed, one-at-a-time solution.

In 1981 alone, more than 10,000 hot-melt applicators were put into use. The applications range from automatic case sealing to manu-

facturing car batteries, from pallet stabilization to insulating glass,
from hi-fi loudspeaker boxes to baby diapers. The end products vary
tremendously. And, even though there are hundreds of different hot-
melt formulations, a surprisingly broad number of jobs can be done
with just a few modular application-equipment concepts.

Modern hot-melt equipment fulfills three functions:

1. Transfer of heat into the adhesive to liquefy it
2. Movement of the liquid adhesive to its application point
3. Control of amount, pattern configuration, and application
   temperature

Fig. 1. Self-contained handgun. (Courtesy of Emhart Chemical
Group, Bostik Division, Middleton, Massachusetts.)

II. MELTING DEVICES

Melting devices perform the function of transferring heat into the adhesive to transform it from solid to liquid state. Their design is determined by two prime considerations:

Physical propertes of the adhesives—such as configuration in the solid state; and viscosity, cohesion, thermal stability, and oxygen stability in the molten state

Required heat transfer capacity or melting capacity necessary to meet the output requirements of an individual application.

A. Self-Contained Handguns

Self-contained handguns incorporate all three functions of a hot-melt applicator in one unit (Fig. 1). These hot-melt applicators rank at the low end of the scale in terms of melting capacity. However, at least the larger ones have justified their existence in industrial bonding operations.

The adhesive, usually in the cylindrical configuration of a slug, stick, or rope, is rammed into a melter, electrically heated, and forced through the extrusion nozzle after melting.

The simplest and smallest do-it-yourself guns ask for the use of the operator's thumb to provide ram pressure on the hot-melt stick; others provide simple ratchet-type mechanisms. Industrial self-contained handguns usually provide air-pressured rams or are of an "all-electric" design with small electrical drives.

Self-contained handguns rank at the very low end of the scale in terms of output capacity. They have distinct advantages, but also certainly have some very obvious disadvantages, as the following summary chart shows:

Pro	Con
Low capital investment	Usually high adhesive costs due to necessary slug pre-
Portability	parations
Ability to handle thermal- and oxygen-sensitive materials	Low output
	Use limited to hard and brittle adhesives
	Inability to handle pressure-sensitive adhesives
	Application limited to manual or semiautomatic applications methods

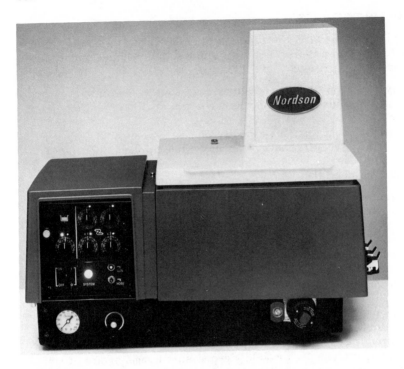

Fig. 2. Hot-melt applicator with melting tank.

B. Tank Melters

Tank melters are the most commonly used melting device in hot-melt
equipment today (Fig. 2). They are most versatile in accepting different hot-melt configurations. Whether the adhesive is granulated or
is in the form of slabs or blocks, the tank will accommodate it as long
as the total tank size and the opening permit.

Tanks are equally versatile in handling hot melts with different
physical properties of cohesion and adhesion.

Tank melters usually are made from either stainless steel or aluminum alloys. Electricity is the power source most frequently used
with heating elements which are either cast-in, slip-in cartridges or,
in a few instances, strap-on band heaters.

Tank holding capacities range from as low as 8 lb to the more
than 100 lb found in large premelting pots. The most common mistake
made in selection of applicators is to specify large holding capacity to
avoid the need for refilling during a shift. This decision not only is
costly as an initial investment but also might create additional burdens
for maintenance and adhesive costs. Selection of the proper tank size
is a simple task using the following criteria:

Any operation requires the application of a certain amount of molten adhesive at a certain application temperature. A consistent application temperature within a narrow range is a major requirement for obtaining consistent bond quality.

Too small a capacity creates serious potential problems, the most common one being fluctuations in application temperature due to insufficient heat-up time after refilling.

Too large a tank capacity, however, can create its own set of problems. Hot melts often degrade at elevated temperatures through the effects of heat or oxygen. Although protected by certain additives, like antioxidants, hot melts should not be kept in the molten state for an extended period of time. Most formulations do not show any significant degradation after several hours; however, others can break down so rapidly that they should not be used in tank melters at all.

Significant degradation not only will change the bonding characteristics of a hot-melt adhesive but also can cause charred particles to build up gradually on tank walls. This build-up can cause an insufficient application temperature due to the insulating effect of the charred adhesive.

Proper tank selection not only will help to avoid these shortcomings but also will provide a convenient refill cycle which fits smoothly into the production schedule (Fig. 3).

Some tank melters offer an additional holding capacity provided by a hopper mounted over the tank. The hopper stores a large amount of solid adhesive and, when triggered by a level control, delivers it on demand to the tank.

As a rule of thumb for most applications, a surplus of melting capacity is better and easier to handle than a deficiency.

Evaluation of the pros and cons of tank melters shows this:

Pro	Con
Very versatile in accepting different hot-melt configurations	Of limited use for oxygen-sensitive hot melts
Easy to clean and maintain	Limited in heat transfer when handling high-performance
Easy access for refilling	hot melts

C. Grid Melters

Grid melters derive their name from the design of the heating unit, a pattern of cone-shaped melting surfaces with openings in between to permit the flow of molten adhesive.

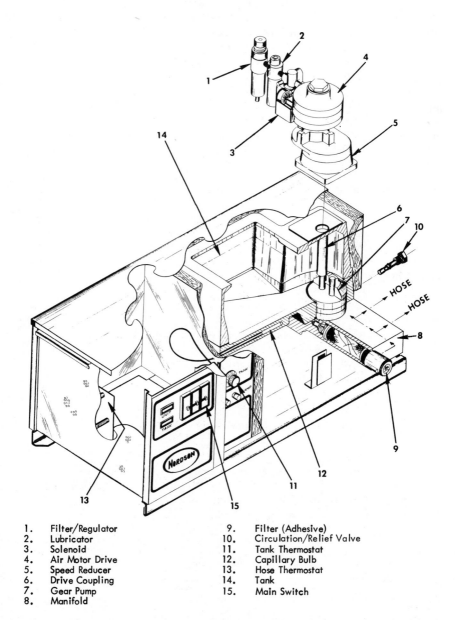

1.	Filter/Regulator	9.	Filter (Adhesive)
2.	Lubricator	10.	Circulation/Relief Valve
3.	Solenoid	11.	Tank Thermostat
4.	Air Motor Drive	12.	Capillary Bulb
5.	Speed Reducer	13.	Hose Thermostat
6.	Drive Coupling	14.	Tank
7.	Gear Pump	15.	Main Switch
8.	Manifold		

Fig. 3.   Cutaway of tank-type, hot-melt unit; HM X1 system.

Grid melters are usually designed to keep solid, unmolten material suspended in a hopper which gravity feeds the melting grid. As the material melts, it runs through the openings between the cones to a small reservoir next to the pump inlet.

Two major hopper configurations are available. One design, the "cool" hopper, keeps the material relatively unaffected by heat before it reaches the grid. The other, the "warm" hopper, utilizes the heat radiation from the grid for a stage-type melting process.

The main feature of a melting grid is that it can transfer heat more effectively to hot melts than a melt tank does. This is because the cones provide a larger surface area for the transfer of heat than a tank of comparable outer dimension does. However, this advantage is effective only if the hot-melt configuration provides a close contact with the cone surface as shaped small granules or solid blocks do.

The cool hopper is the first choice for handling sensitive materials — such as oxygen-sensitive, polyamide-based products or thermal-sensitive ones, such as some polyester-based materials. The cool hopper grid melts material rapidly, but melting occurs only on the grid. This design shortens the molten time of the material prior to its application and, therefore, works toward preventing degradation. Heat radiation from the grid to the hopper cannot be prevented totally. Therefore, hot melts with relatively high softening and melting points work ideally in the cool hopper.

Materials with lower melting points can create "bridging" in a cool hopper. This means that heat radiation softens the material enough to let individual pieces (chunks, granules, slats) stick together and to the hopper wall, but not enough to let the material slide down to the melting grid. Thus, the operation is interrupted due to lack of molten material (Fig. 4).

Formulations with medium to low melting points are better handled in a warm hopper where the heat radiation from the grid is used to melt the hot melt in stages as it moves closer to the melting grid (Fig. 5).

Grid melting devices offer some significant pros, but they are not a cure-all.

Pro	Con
Provide effective heat transfer when matched with proper hot-melt configuration.	Limited versatility in accepting hot-melt configurations.
Cool hopper versions effectively protect sensitive materials	More sophisticated in maintenance and repair than tank devices.
Warm hopper versions provide stage-melting capabilities.	

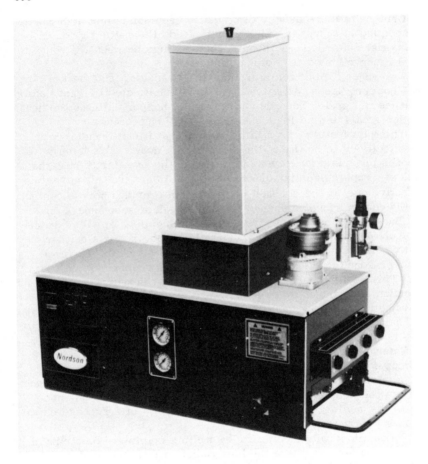

Fig. 4. Hot-melt applicator with grid melter and cool hopper.

D. Bulkmelters

Conceptually, there are two approaches for the occasions when large amounts of hot melt need to be applied.

1. The premelt tank

Heated by electricity, directly, by oil, or by steamheat, premelt tanks are basically larger versions of the tank melting devices discussed earlier. Some are equipped with pumping devices to feed the molten material to the area of application, usually via a smaller reservoir where it is picked up by or fed to the final application device (wheel,

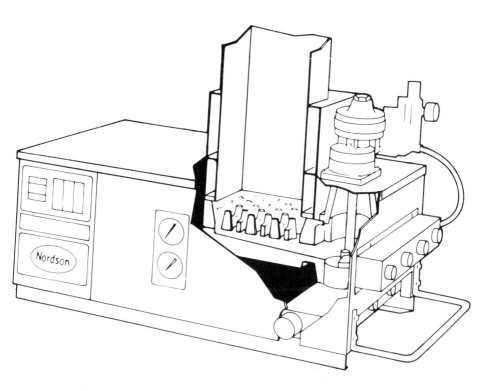

Fig. 5. Hot-melt applicator with grid-melter model.

roll, or gun). Other designs simply gravity feed the hot melt into the application reservoir. Some units use grid configurations in the tank bottom to increase the area of heat transfer.

The principle of operation of these tanks, however, is quite similar to that of small units. They melt larger quantities of hot melt (from 100 lb/hr to beyond 1000 lb/hr) and keep molten materials at a controlled temperature until they are needed for application.

Premelt tanks are versatile in handling a great variety of hot-melt configurations. However, they have two severe limitations:

1. By design, they keep hot melts in a molten state over extended periods of time. This limits their use to oxygen-stable and low-charring formulations. Even then, melting capacity and actual usage have to be well matched to avoid extensive charring and degradation.
2. Premelt tanks handle low viscosities best. Viscosities in the medium and high ranges can be very difficult to feed suffi-

ciently without well-designed siphoning and pumping devices.

A summary of the pros and cons of premelt tanks follows.

Pro	Con
Relatively low initial in- vestment	Keep hot melt in molten state over long periods of time
Versatility in handling various hot-melt con- figurations	Generally cannot be used to apply hot melt directly
Unsophisticated mainte- nance and repair requirements	Handle a limited range of vis- cosities
	Operate with only moderate energy efficiency
	Need material transferred from shipping container.

2. The drum and pail unloaders

Modern drum and pail unloaders melt and feed hot melt on demand directly out of the shipping container, a drum or pail.

The equipment usually features electrically heated grids or fin
sections mounted to a round platen. They are equipped with seals
which prevent leakage between the platen and the container wall. The
platen diameter must closely match the I.D. of the drum or pail.

Mounted to a crossbar element and supported by vertical pneumatic or hydraulic elevating posts, the heated sections of the platen
melt the hot melt gradually, while the siphoning effect of a pump, with
its inlet mounted in the platen, and the downward elevator pressure
move the molten material progressively out of the container (Fig. 6).

Drum and pail unloaders are designed to develop sufficient hydraulic pressures to be used as premelters that feed application devices directly as well as premelters that feed reservoirs of other hot-
melt applicators (Fig. 7).

Installations of two or more machines provide the capability of a
continuous output despite the necessary changeover of drums. Drum
and pail unloaders can handle practically any hot-melt adhesive or
sealant, from very low to very high viscosity. They melt instantly, on
demand, and therefore don't keep material in a molten state at elevated
temperatures. These units melt material directly from the shipping
container and thus ease the handling of material with high surface
tack (i.e., pressure sensitives).

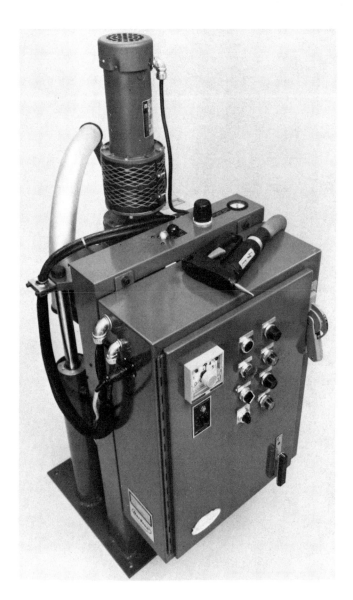

Fig. 6. Pail unloader with hose and handgun attached.

Fig. 7. Drum unloader with handgun attached.

The hot melt needs to be tapped, however, from a high-quality drum. The platen seals usually are flexible enough to overcome small dents (i.e., less than ½ in.), but larger dents can stop the platen or cause molten material to leak through the seals.

Drum and pail unloaders are the most versatile of all melting devices able to handle heavy-duty jobs, such as sealant applications or very precise low-viscosity pressure-sensitive adhesives. These units can melt and apply nearly all types of materials for almost every type of application.

The pros and cons of drum and pail unloaders are shown in the following table.

Pro	Con
Handle almost any viscosity and polymer base	Require relatively high initial investment
Melt on demand; cause practically no degradation of hot melt	Require one size high-quality container
Provide great versatility for possible system layouts to meet the needs of a wide range of applications	Require medium to high degree of technical sophistication in operation
Are relatively easy to maintain	

III. FEEDING MECHANISMS

The molten hot melt needs to be fed to the point from which it will be applied to the substrate. The simplest method is to use gravity.

For most applications however, this simple method is insufficient. In most modern hot-melt applicators the combined melting and feeding unit is located in a remote, easily accessible position. The molten material is fed by hydraulic pressure through heated, flexible hoses to the point of application. This design allows the distribution of material in any direction, provided the feeding system develops sufficient hydraulic pressure.

The most common feeding devices in modern hot-melt applicators are piston pumps and gear pumps.

A. Piston Pumps

Piston pumps, usually air-driven, are the backbone of thousands of hot-melt applicators in operation today. They have become refined over the years and are today very simple, extremely reliable, and easy to maintain. Several vertically mounted models have no wear-prone parts, such as packing glands, resulting in increased service life.

Piston pumps are offered in two basic designs — single-acting and double-acting pumps.

The single-acting pump siphons molten material during the up-stroke (the piston moving away from the inlet) and feeds it during the downstroke. This pump design can only be used for intermittent applications.

The double-acting piston pump siphons and feeds simultaneously and can be used in intermittent as well as continuous applications. In a continuous application, reduction in output will become visible due to a temporary loss of full hydraulic pressure while the pump switches stroke directions. Depending on the individual application pattern, this "wink" in the output may not be acceptable.

The good and bad features of piston pumps are

Pro	Con
Technically unsophisticated	Do not provide pulsation-free output
Easy to maintain	
Very reliable; can be designed practically without wear-prone parts	Cannot precisely key output to varying speeds of the substrate
Relatively insensitive to filler abrasion	Usually do not have the capability of handling high viscosities
Able to vary speed automatically on demand	

B. Gear Pumps

Gear pumps consist of two rotating elements. There are several basic concepts used for pumping and feeding hot melts, the most common being spur and gearotor pumps.

The gear pump operates continuously in three stages (Fig. 8):

1. Suction action to take material into the openings between the teeth of the two elements
2. Compression action by matching the teeth to the recesses
3. Feeding action by separating the teeth from the recesses

Unlike piston pumps, which create a pressure fluctuation when changing strokes, gear pumps maintain a constant pressure due to their continuous rotation.

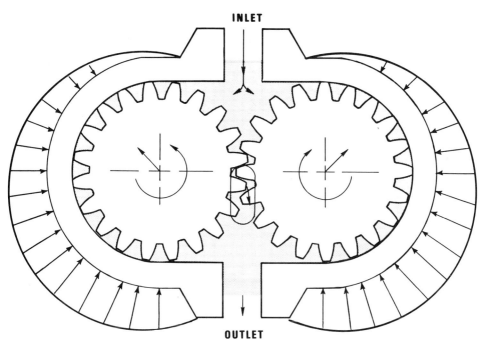

Fig. 8. Gear pump.

When powered by DC electrical motors or driven by a direct power-er takeoff in the parent machine, gear pumps provide the opportunity to key their speed exactly to the speed of the parent machine. Since the number of pump revolutions determines the amount of material the pump can feed, the output of a gear pump precisely matches the varying speed of the parent machine.

Gear pumps can handle a wide range of viscosities—anything from thin waxes to heavy sealants.

The pumping process within a gear pump is an extensive mechanical action. Abrasive fillers, such as those used in a few hot-melt formulations, can cause premature wear of the rotating parts. Some hot-melt formulations can break down due to mechanical impact. However, these potential limitations are seldom encountered in actual use. Gear pumps have become workhorses in hot-melt applications, as have the piston pumps. Each complements the other.

The pros and cons of gear pumps are

Pro	Con
Provide constant pressure, no fluctuation	Use wear-prone pump elements subject to deterioration by filled materials
Handle wide range of viscosities	Require medium-high maintenance efforts due to relative sophistication
Match output precisely to varying speed of parent machine	Need provisions to respond to varying output demand
Usually provide several drive options	Can break down adhesive formulation by mechanical action

Piston and gear pumps have been on the market for years and are very reliable systems. The individual needs of the application should determine which system to use. An application system featuring a gear pump usually requires slightly more sophisticated operation, maintenance, and repair.

Choosing between individual product lines becomes difficult partly because there are usually trade-offs. A low-pressure system allows operation with a low danger of nozzle clogging because large orifices can be used. A high-pressure system, however, allows a more accurate deposit, and proper filtration can provide trouble-free operation.

Hot-melt applicators are production assets. They are very often an integral part of sophisticated production lines. Downtime can become a major problem; therefore, it is worthwhile to assess the service and maintenance needs of a system.

Seals and packings in pumps are wear-prone parts. The pump with the lesser number of seals or packings usually requires less maintenance and provides longer service than its counterpart with a larger number of wear-prone parts does. If a system breaks down, serviceability of the system and the availability of both replacement parts and factory service personnel can make a difference of hundreds of thousands of dollars.

IV. APPLICATION DEVICES

To apply the molten hot melt to the substrate, various methods and equipment have been developed which can be structured into three major groups:

1. Bead-extrusion guns and nozzles (heads, valves, dispensers)
2. Web-extrusion guns and nozzles (heads, valves, dispensers)
3. Wheels and rolls

A. Bead-Extrusion Guns and Nozzles

Bead-extrusion guns and nozzles are the most common devices used for hot-melt bonding applications. Conceptually, they act as valves of a hydraulic pressure line which is formed between the pump and the gun.

If a ball-and-seat or needle-and-seat assembly is used, they can be actuated manually or automatically.

Automatic guns are usually designed to be actuated either elec-tropneumatically, via an electric solenoid and an air piston, or elec-

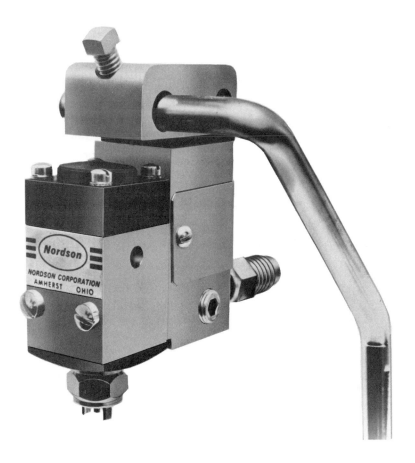

Fig. 9. Automatic, single-orifice gun for bead extrusion with nozzle attached.

tromagnetically, via a coil incorporated into the gun. In both cases, electric signals—created either mechanically by switches in the parent machine or optically by light-detection devices—either directly or in combination with electric or electronic timing devices actuate (open and close) the gun (Fig. 9).

The hot melt is extruded in the shape of a round string which will usually form a half-round bead once applied to the substrate. This configuration helps to prevent a rapid heat transfer from the hot melt to the substrate and the environment, to keep sufficient "opentime," and to achieve proper wetting of the second substrate. Upon compression of the substrates, the half-round beads will be pressed to flat patterns (Fig. 10).

Automatic guns are available with either single or multiple orifices. Multiple-orifice guns often are of modular design, and the number and arrangement of orifices can be customized.

The amount of the hot-melt deposit will be determined by pump speed, hydraulic pressure, duration of open cycle, and travel speed of the substrate. All these variables can easily be controlled and adjusted.

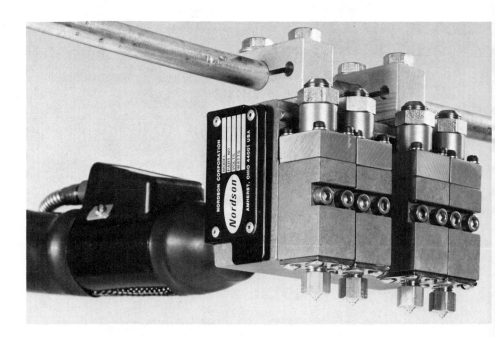

Fig. 10. Automatic gun with multiple orifices or bead extrusion with nozzles attached.

Fig. 11. Automatic bead-extrusion gun and right-angle nozzle for case sealing.

The finetuning of a hot-melt extrusion is provided by the nozzles attached to the gun opening(s). The orifice diameter and length of the nozzle help to achieve optimum control of the hot-melt deposit.

Multiple-orifice nozzles allow the application of more than one bead from a single-orifice gun. Right-angled nozzles allow the application of hot melt in a direction different from that of the gun orifice points, a feature which can eliminate positioning problems due to lack of space in many installations (Fig. 11).

Modern, automatic hot-melt guns have enormous cycling capabilities which, depending on design, range between several hundred and several thousand cycles (openings and closings) per minute. Response times are measured in milliseconds. Provided they are used within their design specifications, they are capable of handling many diverse applications (Fig. 12).

B. Web-Extrusion Guns or Slot Nozzles

Web-extrusion guns, or slot nozzles, allow the deposit of flat hot-melt patterns, either continuously or intermittently. Manufacturers offer various widths from less than 1 in. to 40 in. or more (Fig. 13).

The molten hot melt is extruded through a slot which opens to the substrate. Slot dimension and finetuning of the distance between slot and substrate allow control of the thickness of the application. Different deposit patterns are made possible by the use of a shim plate with the desired application pattern in the slot opening (Figs. 14 and 15).

Web-extrusion guns, also referred to as slot nozzles or slot guns, are mainly used to apply hot melts for coatings or for temporary or indirect bonds, such as those in pressure-sensitive coated objects, or to apply heat-sealable patterns. They can also be used for web lamination of various materials.

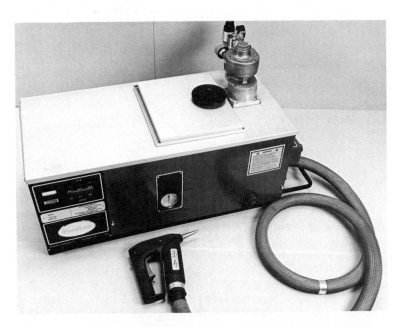

Fig. 12. Hot-melt applicator with tank melter and air-driven gear pump with hose, handgun, and extrusion nozzle attached.

Fig. 13. Slot nozzle applying double pattern.

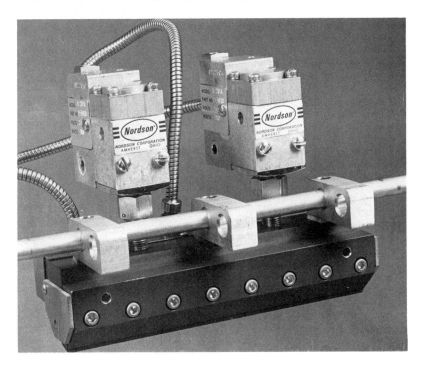

Fig. 14. Slot nozzle for 16-in.–wide pattern.

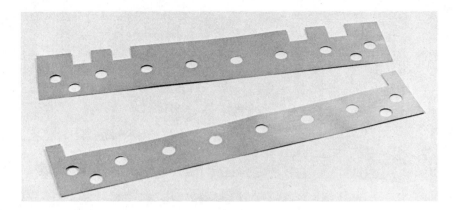

Fig. 15. Shim plate configurations.

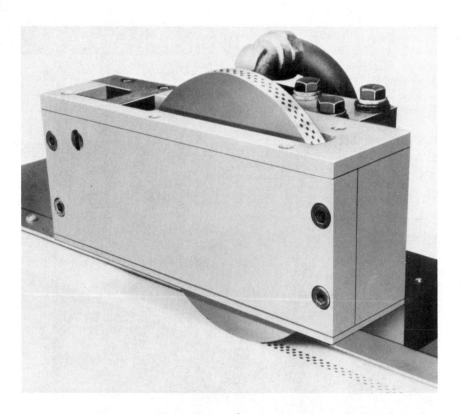

Fig. 16. Hot-melt wheel applicator for downward application.

C. Wheels and Rolls

Wheels and rolls can apply flat hot-melt patterns or, if their surfaces are gravured in the desired arrangement, structured patterns. The devices turn in a reservoir which contains molten material and which is fed by a melting device. Hot melt is picked up from the reservoir; a doctor blade scrapes off the surplus; and, in turning the wheels and rolls apply the material to the substrate. If a structured pattern needs to be applied, small round or rectangular recesses are drilled into the wheel or roll surface, and the blade scrapes the surface completely free of hot melt. Hot melt remains in the recesses and will be applied from them to the substrate (Fig. 16).

Roll applicators are usually an integral part of roll coaters. Supplemented by unwinding and rewinding units, they are the heart of lines used for paper coating and paper converting and laminating. Used in wide-web applications of pressure-sensitive hot melts, they serve in the carpet industry as well as in tape and label manufacturing (Fig. 17).

Wheel applicators serve for small-web applications and for application of precise lines and patterns. Driven by a power takeoff from

Fig. 17. Bolton Emerson 60-in. park-coating line running heat-activated, pressure-sensitive, and other hot-melt coatings. The line features (from right to left) unwind, coating module, chill-pull module (with operator at controls), and rewind. (Courtesy of Bolton Emerson, Lawrence, Massachusetts.)

the parent machine, wheel applicators have maximum repeatability and precision at varying speeds. However, wheels have very distinctive disadvantages:

Integration and synchronization with parent machines are much more sophisticated than they are with slot nozzles.

They are limited to relatively low substrate speeds, particularly gravured wheels. (Hot melt will spin off at high speeds.)

Wheels, especially in downward applications, are prone to leakage.

Web-extrusion guns in both intermittent and continuous applications now provide many of the features which had previously been in the domain of wheel applicators. Web-extrusion guns are far more versatile and are much easier to integrate into a parent machine than wheel applicators are.

V. HOSES

Most commercially available hot-melt applicators are modular systems. The melting and feeding unit can usually be placed in an easily accessible position, remote from the dispensing point where the aforementioned application devices—gun/nozzle combinations or wheel/reservoir combinations—are used. The missing link between these elements is the pipelines, which allow the movement of molten material from the melting station to the point of application.

Flexible, electrically heated hoses provide the vital link between the melting and application points. Almost all commercially available

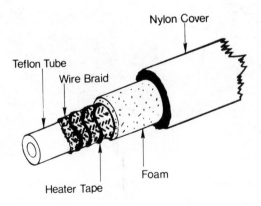

Fig. 18.   Cutaway diagram showing construction of a typical heated hose.

hot-melt applicators today feature these hoses. They are offered in standard lengths with increments from several feet up to 16 ft and more.

Hot-melt hoses are built on several layers of different materials: A Teflon tube on the inside is supported by a steel-mesh tube that provides pressure resistance. Heater tape is spiral wound around both and then covered with a thick layer of foam to provide insulation (Fig. 18).

Depending on the individual layout, hot-melt applicators provide one or several possible hose connections; thus, a single applicator can serve several application points. The flexible design allows bending of the hoses in any direction, thereby increasing installation versatility.

The temperature-control system of hoses is designed with consideration for heat radiation from the hoses into the environment. Therefore, the flexible hot-melt hoses should never be led through additional piping or have additional covering material added.

Hot-melt hoses, unlike the melting device and the hot-melt passages in the application devices, are not capable of adding temperature to the hot melt in a dynamic situation. Therefore, stage heating, increasing the temperature of the hot melt gradually between the melting and the application devices cannot be achieved. Hot-melt hoses, however, are perfectly capable of maintaining heat in a dynamic situation. Hot melt fed into the hose at application temperature will arrive at the gun at application temperature.

Hot-melt hoses will transfer heat into the hot melt under static conditions. Therefore, a complete, heated, thermostatically controlled circuit must be provided from the melting point to the application station. Since heat is the element which makes hot-melt application possible, a simple on-off power switch can put an entire hot-melt system in or out of operation.

VI. CONCLUSION

Hot-melt adhesives have experienced a very dynamic entry into the bonding, fastening, and sealing operations of the manufacturing industry. They have revolutionized numerous operations with their feature of rapid setting without emissions of solvents or water. While hot-melt application equipment was originally designed and marketed by companies primarily in the adhesive business, manufacturing and marketing of hot-melt application equipment soon became a separate business. Since those early days, equipment has undergone development dynamically similar, but not always parallel to the hot melts. Sometimes the equipment had to catch up with the opportunities created by the developmental successes of chemistry. Often, however, the equipment has been ahead of the adhesives—plowing new ground, fostering new formulations, and thus, pushing hot-melt adhesives to new frontiers. Hot-melt application equipment today is an above-average growth industry with widespread distribution and service networks and with multimillion-dollar budgets for research and development. Hot-melt equipment manufacturers are continuing to look toward future horizons and are preparing to add new chapters to the history book of hot melts.

ACKNOWLEDGMENTS

All figures, except Figs. 1 and 17, are courtesy of Nordson Corporation, Amherst, Ohio.

21
Adhesives for Specific Substrates

Gerald L. Schneberger *GMI Engineering and Management Institute,
Flint, Michigan*

I. INTRODUCTION

The major emphasis of this book is on the properties, selection, and
use of various kinds of adhesives. It is appropriate to include some
material which is specifically related to the surfaces which are commonly bonded. There is no universal adhesive simply because there is no
universal substrate. A trememdous number of metals, alloys, ceramics, plastics, and composites are widely used. In addition many materials, which nominally have the same composition, arrive at the point
of assembly with different thicknesses, thermal histories, microstruc-
tures, and chemicals deposited on their surfaces—with or without the
user's knowledge.

The service environment is at least as important in the selection
of an adhesive as the nature of the surfaces. The characteristics required of an adhesive which must withstand a low static load under
ambient temperature and low-humidity conditions are vastly different
from those required of a product which must endure large cyclic
stresses in a moist, acid or alkaline environment. It is instructive to
begin a consideration of adhesives for specific surfaces by grouping

commonly bonded materials on the basis of surface similarities. Section II discusses metal, ceramic, plastic, and elastomeric surfaces.

II. GENERAL SURFACE CHARACTERISTICS

A. Metals

Virtually all commonly bonded metal surfaces exist as hydrated oxides [1]. This situation, which is illustrated in Fig. 1, means that adhesives used for metal bonding are in reality used for metal-oxide bonding and that they must be compatible with the firmly bound layer of water attached to surface metal-oxide crystals. Even materials such as stainless steel and nickel or chromium are, in fact, coated with transparent metal oxides which tenaciously bind at least one layer of water.

METAL SURFACES

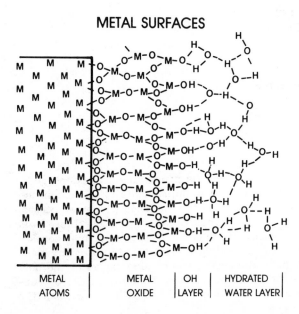

| METAL ATOMS | METAL OXIDE | OH LAYER | HYDRATED WATER LAYER |

NOTE: 1. The oxide layer is typically 40-80 Å thick.
        2. The hydrated layer is tightly bound.
        3. A pure metal surface is rarely available for engineering use.

Fig. 1. Metal surfaces are actually hydrated metal oxides.

The nature of the metal which lies beneath the hydrated-oxide layer does influence the properties of the surface. Thus, certain metals will possess surfaces which interact more effectively with one type of an adhesive than another. This is the reason why adhesives formulators need to know as much as possible about the surfaces to be assembled.

B. Ceramics

Ceramic materials are hard and brittle, and have high softening temperatures. Many have smooth, glasslike surfaces. Ceramics contain both ionic and covalent bonds, primarily involving various combinations of oxygen, silicon, sodium, aluminum, calcium, barium, boron, and lead. They may be crystalline or amorphous. Many commercially important ceramics have glazed (glasslike) surfaces. These surfaces are formed by high-temperature firing of either the ceramic material itself or of a liquid glaze applied to the surface.

The polar nature of the bonds between atoms in a ceramic means that there will be an adsorbed layer of water and hydroxide ions. This layer is often tightly held by the ceramic surface; traces of water are present even above 100°C (212°F).

Adhesives used with ceramics, like those used with metals, must be compatible with the surface moisture layer. Since ceramics generally have high surface energies, it is usually not difficult to bond them under ambient conditions.

C. Plastics

Plastics differ from metals and ceramics in a number of important ways. As a general rule they are much softer, more soluble, much less heat resistant, less inclined to react with atmosphere (corrode), and they exhibit a wider range of polarity. Some of these differences contribute to easier and more effective bonding; others serve to complicate the process.

Being softer and less solvent-resistant implies that plastic materials may be subject to solvent attack, which can vary from slight swelling to complete solution. Table 1 lists common plastic materials and solvents which they are unable to resist. It should be pointed out that the attack of adhesive solvents on plastics is not a problem for thermosetting materials. It is only thermoplastics, that is, those plastics which can be heat softened, that are subject to solvent attack. A more complete discussion of the solvent resistance of thermoplastic will be found in Chap. 3.

The reduced heat stability of polymers compared to metals and ceramics is the result of the significantly weaker forces of attraction which exist between and within polymer molecules relative to those of metallic and ceramic structures. The practical consequence of this

Table 1 Softening Temperatures and Solvent Resistance of Common
Plastics

Material	Softening temperature [°C(°F)]	Common attacking solvents
Polyethylene	82−121 (180−250)	None
Polypropylene	88−160 (190−320)	None
Vinyls	66−93 (150−200)	Ketones, toluol, xyol
Fluoroplastics	149−260 (300−500)	None
Acrylics	60−93 (140−200)	Ketones, esters
Epoxies	93−260 (200−500)	None
Polyesters	149−232 (300−450)	Ketones, esters
Polystyrene	60−93 (140−200)	Hydrocarbons, ketones
Polyurethanes	88−121 (190−250)	Resists most
Elastomers	121−204 (250−400)	Hydrocarbons
Thermosets, unfilled	71−260 (160−500)	None
Polycarbonate	121−135 (250−275)	Chlorinated hydrocarbons

Source: Data for softening temperature from Ref. 2.

difference is, primarily, the inability to adhesively assemble polymers
using processes that require temperatures much above 150−175°C
(302−347°F). Table 1 lists the heat distortion temperatures for important plastic materials. Bonding operations near or above these
temperatures may be impractical because of substrate distortion.

Unlike many common metals, plastics generally do not react with
the atmosphere, that is, they do not spontaneously oxidize to form
corrosion products. In most cases water, either vapor or liquid, has
little effect on the properties of a plastic surface. Nylons and polyesters, which sometimes violate this principal, are discussed in greater detail in Sec. III.C. 1 and 2.

Polarity is important in plastics and elastomers. Polarity indicates the degree to which a material (surface) possesses regions of
negative and positive charge. A nonpolar surface is essentially neutral and has no localized charged areas. Highly polar surfaces, on
the other hand, have regions in which certain atoms are more highly
charged (either positively or negatively) than the surrounding atoms.
Polarity is important because the presence or absence of electric
charges has a pronounced effect on the ability of one surface to attract another (adhere) and on the tendency of moisture or other
chemicals to react with a plastic. Table 2 groups a number of common
plastics on the basis of their polarity. Many of the materials which
we commonly associate with high strength are highly polar. The
presence of positive and negative charges within the molecules of

Table 2 Plastics Classified on the Basis of Their Polarity

Degree of polarity	Plastics
Low polarity	Polytetrafluoroethylene
	Silicones
	Polyethylene
	Polypropylene
	Some elastomers
	Polyvinyl chloride (PVC)
	Acetals
Medium polarity	EPDM
	Some elastomers
	Phenolics
	Acrylonitrile butadiene styrene (ABS)
High polarity	Acrylics
	Epoxies
	Urethanes
	Nylons
	Cellulose esters
	Polyesters
	Vinyls

these materials permits them to attract adjacent molecules very effectively and to resist tensile and bending stresses. The result is a material which we normally consider "strong." Nonpolar plastics, on the
other hand, do not have these internal attractive forces. As a result,
they are unable to resist applied stresses and, therefore, possess
much lower strength.

D. Elastomers

Elastomers may be viewed as a special type of plastic. They contain
long-chain polymer molecules but possess considerably lower strengths
than plastics. In addition, they are highly stretchable-compressible;
that is, when a force is applied to an elastic material, it stretches or
compresses and then recovers its original dimensions upon removal of
the force.

These differences between elastomers and plastics are essentially
due to the fact that the molecules of elastic materials are highly flexible. They can bend and stretch when a force is applied, but they do
not flow from one location to another. Thus, their stretching, or deformation, is not permanent. The molecules will "snap back" to their
original locations when the stress is removed.

Generally speaking, elastomers cannot be highly polar. If they were, their attractions for one another would prevent easy and recoverable deformation. The fact that most elastomers are of relatively low polarity and are easily deformed has two significant consequences for adhesives. The first is that some adhesives will not easily wet an elastic material. The second is that adhesives must either be able to prevent substrate distortion or be elastic themselves.

III. SPECIFIC SURFACE CHARACTERISTICS

A. Metals

1. Iron and steel

Because of their widespread use in industry, iron and steel are frequently bonded. Like the surfaces of most metals, their surfaces actually exist as a complex mixture of hydrated oxides and absorbed water [1]. Unfortunately, iron oxides are often not the best surface for adhesives because the oxides may continue to react with the atmosphere after an adhesive has been applied, thus forming weak crystal layers.

Bonding operations frequently require the mechanical or chemical removal of loose oxide layers from iron and steel surfaces before adhesives are applied. In order to guard against slow reaction of the surface with environmental moisture after the bond has formed, iron and steel surfaces are often phosphated prior to bonding. This process converts the relatively reactive iron atoms to a much more passive, "chemically stable" form which is coated with zinc or iron phosphate crystals. Such coatings are applied in an effort to convert a reactive and largely unknown surface to a relatively inert one whose structure and properties are reasonably well understood.

In order to be successful, any surface cleaning or conversion process must be well controlled. Attention must be paid to temperatures, times, and concentrations if uniformly acceptable results are to be obtained. If these process parameters are under good control, adhesive bonding of iron and steel surfaces is entirely practical.

2. Aluminum

Aluminum is an even more reactive material than iron. Aluminum oxide forms instantaneously when a freshly machined aluminum surface is exposed to the atmosphere. Unlike iron oxide, however, aluminum oxide tends to be very tightly held to the substrate aluminum, to be cohesively strong, and to be electrically nonconductive. These desirable characteristics of aluminum oxide as an adhesive substrate can be enhanced if the oxide layer is formed under carefully controlled conditions as is the case with anodizing. In this process aluminum is oxidized (corroded) under carefully controlled conditions of temperature and voltage. The resulting oxide is highly inert and will provide an excellent base for subsequent bonding operations if the anodized layer is not water-sealed first. If it is sealed, adhesion will be a much greater problem, as evidenced by poor adhesion of many paint coatings even with subsequent chromic acid cleaning dips.

Aluminum is also prepared for bonding with chromate chemical conversion coatings.

3. Zinc

Zinc is becoming an increasingly common substrate for bonding as a result of its widespread use in galvanized materials for corrosion resistance. Like aluminum, zinc is highly reactive and oxidizes spontaneously to form a stable oxide film. Freshly galvanized surfaces are often phosphated or chromated to convert the active zinc atoms to a more inert form prior to bonding. Otherwise, zinc atoms may react in an undesirable fashion with the adhesive.

4. Stainless steel

The stable oxide film which exists on stainless steel appears to tightly bind ions from prior manufacturing steps. Rinsing seems to be especially critical in ensuring the successful bonding of stainless-steel surfaces, but even bond quality on well-rinsed surfaces may not be comparable with that on other, more easily bonded materials.

B. Ceramics

Ceramics generally are not difficult to bond. They have high surface energies, so they are easily wet by polar and nonpolar organic materials, including adhesives. They are also wet by water without much difficulty.

Ceramic surfaces, like metal surfaces, have an adsorbed layer of water molecules which may cause difficulty in bonding. If this water is partially removed by heating prior to bond formation, atmospheric moisture may diffuse into the bondline and displace adhesive molecules from the surface. To overcome this problem, coupling agents have been developed to enhance the attraction between the adhesive and the surface. Coupling agents, which are typically silanols, have a high affinity for the ceramic and provide an outward surface which is more compatible with organic adhesives than with water. Coupling agents may also reduce the thermal-expansion stresses between ceramics and adhesives. They are routinely used with glass-reinforced plastics to prevent separation of the glass fiber from the polymeric matrix.

C. Plastics

1. Nylons

There are two important things to keep in mind when bonding nylons. The first is that nylon picks up water from the atmosphere to the extent of 0.5‒2% by weight, or more, at high humidity. This water absorption results in a slight swelling of the part which may induce stresses at the bondline. Nylon will give up its adsorbed water to the atmosphere and shrink slightly when placed in a dry environment. Therefore, adhesives which can accommodate slight changes in nylon-part dimensions must be chosen.

The second significant fact about nylon is that it will react with moisture in the presence of either acids or bases. The reaction is essentially a depolymerization (correctly termed hydrolysis). As a result, any acidic or alkaline bonding procedure should be carefully checked to be certain that no adverse effects are experienced.

2. Polyesters

A great many polyester materials are adhesively assembled. They include alkyds, sheet molding compound, and fiberglass-reinforced plastics. Generally, they are easily bonded if mold release agents have been removed and acid or alkaline chemicals (which may attack the polyester bonds) are used with care.

3. Polyolefins

Polyolefins, such as polyethylene and polypropylene, are frequently difficult to bond because they are inherently low-surface-energy materials which are not easily wet by adhesives. It may be necessary to modify the surfaces of these materials by flaming, acid etching, or ultraviolet treatment in order to make them bondable.

4. Fluorocarbons

Fluorocarbons present the same difficulties as polyolefins, only more so. They have very low surface energies and are extremely difficult to bond. Etching with sodium metal dispersed in hydrocarbons is sometimes used to permit surface adhesion. This process is somewhat sophisticated and requires careful attention for routine success.

5. Vinyls

The major problem associated with bonding vinyls relates primarily to flexible grades of this material. Flexible vinyls contain high levels of plasticizer. If the plasticizing material has a tendency to migrate to the surface of the vinyl, it may lift the adhesive and cause bond failure. Some adhesives are formulated specifically for use on flexible vinyls. These products should be considered when such materials are to be bonded.

6. Urethanes

Urethanes are generally quite easily bonded. Amine-cured urethanes, however, are somewhat less stable in acid and alkaline environments than polyol-cured varieties. Thus, acid and alkaline process chemicals should be used with discretion.

7. Elastomers

Stretchy, easily deformed surfaces are characteristic of elastomers. Elastomers range from nonpolar materials, like natural rubber, to the more polar nitrile rubber.

Expansion due to thermal or mechanical stresses is of major concern when bonding elastomers. Adhesives with elastic behavior matched as closely as possible to the surface being bonded should be used. Primers are sometimes used to provide a layer of material in the glueline with expansion behavior that lies between that of the elastomer and the adhesive.

IV. ADHESIVE SELECTION TABLES

Table 3 is an updated version of Weggemans' adhesive application charts [3]. It summarizes the experience of many users with various types of adhesives. It should not be used as the only adhesive selection criterion, however. Adhesives for specific applications should be chosen only after adequate testing.

Table 4 gives general information about the adhesive types referenced in Table 3. Table 5 further describes commercially available epoxy formulations.

Table 3 Adhesives Commonly Used for Joining Various Materials

Material	Adhesive	Table 4 reference
Acrylonitrile	Polyester	a
butadiene styrene	Epoxy	e
	Alpha-cyanoacrylate	c
	Nitrile-phenolic	b
	Acrylics	w
Aluminum and	Epoxy	e
its alloys	Epoxy-phenolic	d
	Nylon-epoxies	f

Table 3 (Continued)

Material	Adhesive	Table 4 reference
	Polyurethane rubber	g
	Polyesters	a
	Alpha-cyanoacrylate	c
	Polyamides	h
	Polyvinyl-phenolic	i
	Neoprene-phenolic	b
	Acrylics	w
Brick	Epoxy	e
	Epoxy-phenolic	d
	Polyesters	a
Ceramic	Epoxy	e
	Cellulose esters	j
	Vinyl chloride−vinyl acetate	k
	Polyvinyl butyral	l
	Acrylics	w
Celcon (see polyformaldehyde)		
Chromium	Epoxy	e
Concrete	Polyesters	a
	Epoxy	e
Copper and its alloys	Polyesters	a
	Epoxy	e
	Alpha-cyanoacrylate	c
	Polyamide	h
	Polyvinyl-phenolic	i
	Polyhydroxyether	m
	Acrylics	w
Delrin (see polyformaldehyde)		
Fluorocarbon (surface-treated)	Epoxy	e
	Nitrile-phenolic	b
	Silicone	t
	Acrylics	w
Glass	Epoxy	e
	Epoxy-phenolic	d
	Alpha-cyanoacrylate	c
	Cellulose esters	j
	Vinyl chloride−vinyl acetate	k
	Polyvinyl butyral	l
	Acrylics	w

Table 3 (Continued)

Material	Adhesive	Table 4 reference
Lead	Epoxy	e
	Vinyl chloride—vinyl acetate	k
	Polyesters	a
	Acrylics	w
Leather	Vinyl chloride—vinyl acetate	k
	Polyvinyl butyral	l
	Polyhydroxyether	m
	Polyvinyl acetate	n
	Flexible adhesives	g
Magnesium	Polyesters	a
	Epoxy	e
	Polyamide	h
	Polyvinyl-phenolic	i
	Neoprene-phenolic	b
	Nylon epoxy	f
Mylar (see polyethlene-terephthalate)		
Nickel	Epoxy	e
	Neoprene	g
	Polyhydroxyether	m
Paper	Animal glue	o
	Starch glue	p
	Urea-, melamine-, resorcinol-, and phenol-formaldehyde	q
	Epoxy	a
	Cellulose esters	j
Paper	Vinyl chloride—vinyl acetate	k
	Polyvinyl butyral	l
	Polyvinyl acetate	n
	Polyamide	h
	Flexible adhesives	g
Phenolic and melamine	Epoxy	e
	Alpha-cyanoacrylate	c
	Flexible adhesives	g
	Acrylics	w
Polyamide	Epoxy	e
	Flexible adhesives	g
	Phenol- and resorcinol-formaldehyde	q
	Polyester	a
	Acrylics	w

Table 3 (Continued)

Material	Adhesive	Table 4 reference
Polycarbonate	Polyesters	a
	Epoxy	e
	Alpha-cyanoacrylate	c
	Polyurethane rubber	g
	Acrylics	w
Polyester, glass-reinforced	Polyester	a
	Epoxy	e
	Polyacrylates	r
	Nitrile-phenolic	b
	Acrylics	w
	Urethanes	v
Polyethylene (surface-treated)	Polyester, isocyanate-modified	a
	Butadiene-acrylonitrile	g
	Nitrile-phenolic	b
	Acrylics	w
Polyethylene-terephthlate	Urethanes	v
	Polyesters	a
Polyformaldehyde	Polyester, isocyanate-modified	a
	Butadiene-acrylonitrile	g
	Nitrile-phenolic	b
	Acrylics	w
Polymethylmeth-acrylate	Epoxy	e
	Alpha-cyanoacrylate	c
	Polyester	a
	Nitrile-phenolic	b
Polyphenylene oxide	Epoxy	e
	Acrylics	w
Polypropylene (surface-treated)	Polyester, isocyanate-modified	a
	Nitrile-phenolic	b
	Butadiene-acrylonitrile	g
	Acrylics	w
Polystyrene	Vinyl chloride—vinyl acetate	k
	Polyesters	a
Polyvinyl chloride, flexible	Butadiene-acrylonitrile	g
	Polyurethane rubber	g
Polyvinyl chloride, rigid	Polyesters	a
	Epoxy	e
	Polyurethane	g

Table 3 (Continued)

Material	Adhesive	Table 4 reference
Rubber, butadiene-styrene	Epoxy	e
	Butadiene-acrylonitrile	g
	Urethane rubber	g
Rubber, natural	Epoxy	e
	Flexible adhesives	h
Rubber, neoprene	Epoxy	e
	Flexible adhesives	h
Rubber, silicone	Silicone	t
Rubber, urethane	Flexible adhesives	h
	Silicone	t
	Alpha-cyanoacrylate	c
Silver	Epoxy	e
	Neoprene	g
	Polyhydroxyether	m
	Acrylics	w
Steel	Epoxy	e
	Polyesters	a
	Polyvinyl butyral	l
Steel	Alpha-cyanoacrylate	c
	Polyamides	h
	Polyvinyl-phenolic	i
	Nitrile-phenolic	b
	Neoprene-phenolic	b
	Nylon-epoxy	d
	Acrylics	w
Stone	See brick	
Tin	Epoxy	e
Wood	Animal glue	o
	Polyvinyl acetate	n
	Ethylene−vinyl acetate	u
	Urea-, melamine-, resorcinol-, and phenol-formaldehyde	q
Zinc	Polysulfides	s
	Acrylics	w
	Epoxy	e
	Polyesters	a

Source: Data compiled from Ref. 3 and various, other sources. Consult adhesive suppliers for additional information.

Table 4 Adhesive Commentary

Adhesive type	Table 3 reference	Comments	Typical cure conditions
Polyesters and their variations	a	These adhesives are used primarily for repairing fiberglass-reinforced polyester resins, ABS, and concrete. Generally unsaturated esters are polymerized with a catalyst, such as methyl ethyl ketone (MEK) peroxide, and an accelerator, such as cobalt naphthenate. A coreactant-solvent, such as styrene, may be present. Bonds are strong. The adhesives are sometimes combined with polyisocyanates to control shrinkage stresses and reduce brittleness. Unreacted monomer, if present, keeps viscosity low for application, provides good wetting, and enhances crosslinking. Occasionally they are used on metals.	Minutes to hours at room temperature
Nitrile phenolic and neoprene phenolic	b	These adhesives are a blend of flexible nitrile or neoprene rubber with phenolic novolac resin. They combine the impact resistance of the rubber with the strength of the crosslinked phenolic. They are inexpensive, and produce strong, durable bonds which resist water, salt spray, and other corrosive media well. They are the workhorses of the adhesive-tape industry although they do require high pressure and relatively long, high-temperature cures. They are used for metals and some plastics including ABS,	Up to 12 hr at 250–300°F (120–150°C)

polyethylene, and polypropylene. Airframe components and automotive brakes are typical examples of their use.

Alpha-cyanoacrylate	c	These low-viscosity liquids polymerize or "cure" rapidly in the presence of moisture or many metal oxides. Thus, they can bond most surfaces. The bonds are fairly strong but somewhat brittle. These adhesives are used widely for the assembly of jewelry and electronic components.	0.5–5 min at room temperature
Epoxy phenolic	d	This adhesive is a combination of epoxy resin with a resol phenolic. It is noted for strength retention at 300–500 °F (150–250°C), strong bonds, and good moisture resistance. Normally, the adhesive is stored refrigerated. It is used for some metals, glass, and phenolic resins.	1 hr at 350°F (175°C)
Epoxy: amine-, amide, and anhydride-cured	e	As a class, epoxies are noted for high tensile and low peel strengths. They are crosslinked and in general have good high-temperature strength, resistance to moisture, and little tendency to react with acids, bases, salts, or solvents. There are important exceptions to these generalizations, however, which are often the result of the curing agent used. Primary amines give faster-setting adhesives which are less flexible and less moisture resistant than is the case when polyamide curing agents are used. Anhydride-cured epoxies generally have good high-temperature strength but are subject to hydrolysis, especially in the	See Table 5

Table 4 (Continued)

Adhesive type	Table 3 reference	Comments	Typical cure conditions
Epoxy: amine-, amide, and anhydride-cured		presence of acids or bases. Other important features of epoxies are their low shrinkage upon cure, their compatibility with a variety of fillers, their long life when properly applied, and their easy modification with other resins. Crosslink density is easily varied with epoxies; thus, some control over brittleness, vapor permeation, and heat deflection is possible. These resins are widely used to bond metal, ceramics, and rigid plastics (not polyolefins).	
Nylon epoxy	f	Tensile shear strengths above 6000 psi (41.4 MPa) and peel strengths above 100 lb/in. (18 kg/cm) are possible when epoxy resins are modified with special low-melting-point nylons. These gains, however, are accompanied by loss of strength upon exposure to moist air, a tendency to creep under load, and poor low-temperature impact behavior. A phenolic primer may increase bond life and moisture resistance. These adhesives are used primarily for aluminum, magnesium, and steel.	1 hr at 300–350°F (150–175°C)
Flexible adhesives: natural rubber, butadiene-	g	These adhesives are flexible. Thus, their load-bearing ability is limited. They have excellent impact and moisture resistance. They are easily	Pressure-sensitive tape for solvent cements, low-temperature bake

tackified and are used as pressure-sensitive tapes or as contact cements. Urethane and silicone adhesives are lightly crosslinked, as they have reasonable hot strength. They are also compatible with many surfaces but are somewhat costly and must be protected against moisture before use. They have good low-temperature tensile shear and impact strength. The urethanes are two-part products which require mixing before use. Silicones cure in the presence of atmospheric moisture. (See entries r and t.)

for urethane; ambient cure for silicones

acrylonitrile, neoprene, polyurethane, polyacrylates, and silicones

h

These adhesives, which are chemically similar to nylon resins, have good strength at ambient temperatures and are fairly tough. They are available in a variety of molecular weights, softening ranges, and melt viscosities. Often applied as hot melts, they adhere well to a variety of surfaces. The higher-molecular-weight varieties often have the best tensile properties. Lower-molecular-weight polyamides may be applied in solution.

Hot melt, cure by cooling

Polyamides

i

These resins, which combine a resol phenolic resin with polyvinyl formal or polyvinyl butyral, were the first important, synthetic structural adhesives. A considerable range of compositions is available with hot strength and tensile properties increasing at the expense of impact and peel strength as the phenolic content rises. The durability of vinyl-phenolics is generally excellent.

1 hr at 300°F (150°C)

Polyvinyl-phenolic

Table 4 (Continued)

Adhesive type	Table 3 reference	Comments	Typical cure conditions
Polyvinyl-phenolic		They are often selected for low-cost applications where heat and pressure curing can be used.	
Cellulose esters	j	Cellulose ester adhesives are usually high-viscosity, inexpensive, rigid materials. They do not have high strength and are sensitive to heat and many solvents. They are normally used for holding small parts or repairing wood, cardboard, or plastic items. Model airplane cement is a common example.	Air dry
Vinyl chloride–vinyl acetate	k	This adhesive is a combination of two resins which are sometimes used alone. They may be used as hot melts or as solution adhesives. Since thin films of vinyl chloride–vinyl acetate are somewhat flexible, they are often used for bonding metal foil, paper, and leather. A range of compositions is available with a corresponding variety of properties.	Hot melt, cure by cooling; solvent loss
Polyvinyl butyral	l	This adhesive is a tough, transparent resin which is used as a hot melt or heat-cured solution adhesive. It adheres well to glass, wood, metal, and textiles. It is flexible and can be modified with other resins or additives to give a range of properties. It is not generally used as a structural adhesive, although structural phenolics	Hot melt, cure by cooling; heat under pressure

Polyhydroxyether	m	sometimes incorporate polyvinyl butyral to give better impact resistance. These resins are based on hydroxylated poly-ethylene-oxide polymers. Generally used as hot melts, they have only moderate strength but are flexible and have fairly good adhesion.	Hot melt, cure by cooling
Polyvinyl acetate	n	This adhesive is generally supplied as a water emulsion (white glue) or used as a hot melt. It dries quickly and forms a strong bond. It is flexible and has low resistance to heat and moisture. Porous substrates are required when the resin is used as an emulsion.	Hot melt cure by cooling; emulsion, air dry
Animal glue	o	Chemically, animal glues are proteins; they are polar, water-soluble polymers with high affinity for paper, wood, and leather surfaces. They easily form strong bonds but have poor resistance to moisture. They are being replaced in many areas by synthetic-resin adhesives, but their low cost is often an important advantage. They are usually applied as highly viscous liquids.	Air dry under pressure
Starch glue	p	These products, based on corn starch, have high affinity for paper but are used for little else. They are moisture sensitive and are applied as water dispersions.	Dry at room temperature

Table 4 (Continued)

Adhesive type	Table 3 reference	Comments	Typical cure conditions
Urea-formaldehyde, melamine-formaldehyde, resorcinol-formaldehyde, and phenol-formaldehyde	q	These thermosetting resins are widely used for wood bonding. Urea-formaldehyde is inexpensive but has low moisture resistance. It can be cured at room temperature if a catalyst is used. Melamine-formaldehyde resins have better moisture resistance but must be heat cured. Phenol-formaldehyde adhesives form strong, waterproof, wood-to-wood bonds. The resorcinol-formaldehyde resin cures at room temperature while phenol-formaldehyde requires heating. These resins are often combined, resulting in an adhesive with intermediate processing or performance characteristics.	Dry at room temperature
Polyacrylate esters	r	These resins are *n*-alkyl esters of acrylic acid. They have good flexibility and are frequently used for high-quality, pressure-sensitive tapes and foams. They are not suitable for structural applications because of their poor heat resistance and their cold-flow behavior. They are frequently used on flexible substrates.	Pressure sensitive
Polysulfides	s	These resins have good moisture resistance and can range from thermoplastic to thermosetting depending on the degree of crosslinking which is developed during cure. They are two- or three-	Low pressure, moderate temperature

		part systems; the third part being a catalyst. Ventilation is generally required during use. They make excellent adhesive sealants for wood, metal, concrete, and glass. Polysulfide resins may be combined with epoxies to flexibilize the latter.	
Silicones (See also flexible adhesives)	t	These expensive adhesives have high peel strength and excellent property retention at high and low temperatures. They resist all except the most corrosive environments and will adhere to nearly everything. They are usually formulated to react with atmospheric moisture and form lightly crosslinked films.	Low pressure, room temperature
Ethylene–vinyl acetate	u	This copolymer is widely used as a hot-melt adhesive because it is inexpensive, adheres to most surfaces, and is available in a range of melting points. It is widely used for bookbinding and packaging.	Hot melt, cure by cooling
Urethanes, rigid	v	Rigid urethanes are highly crosslinked. While somewhat expensive, they adhere well to most materials, especially plastics, and have good impact strength. Structural urethanes are two-part systems and have good low-temperature strength retention.	Low pressure, up to 300°F (149°C)

Table 4 (Continued)

Adhesive type	Table 3 reference	Comments	Typical cure conditions
Acrylics	w	Acrylics are versatile structural adhesives which are becoming increasingly popular. They cure at room temperature and are applied as a conventional two-part system or by coating one substrate with the resin and the other with the catalyst. Impact resistance is controllable since acrylics may vary from rigid to flexible. Floor ventilation is often recommended during use because they may release heavier-than-air monomer vapors which are odorous.	Room temperature or up to 130°F (54°C), 10–20 min with only fixturing pressure

Table 5 Epoxy Adhesive Curing Conditions

Adhesive type	Time[a]	Temperature[a]	Pot life[a]
Two-part, room-temperature—cured			
Slow	8—24 hr	Room	2—4 hr
Fast	2—4 hr	Room	0.5 hr
Very fast	2—5 min	Room	1—2 min
Two-part, heat-cured			
Slow	3—5 hr	145°C (300°F)	Days
Fast	0.5—2 hr	124°C (250°F)	Hours
One-part, heat-cured			
Slow	0.5—1 hr	175°C (350°F)	NA[b]
Fast	5 min	150°C (300°F)	NA

[a]Time and temperature values given are approximate. They may vary considerably depending on the curing agent and the accelerator, if any.
[b]NA, not applicable.

REFERENCES

1. J. C. Bolger and A. S. Michaels, Molecular structure and electrostatic interactions at polymer-solid interfaces, in *Interface Conversion for Polymer Coatings* (P. Weiss and D. Cheever, eds.), Elsevier, New York, 1969.
2. *Modern Plastics Encyclopedia*, McGraw-Hill, New York, 1968.
3. D. M. Weggemans, *Adhesives Age 16*:31 (1973).

PERFORMANCE, DURABILITY,
AND TESTING

22

Testing Adhesives

Harold Koski and Gerald L. Schneberger *GMI Engineering and Management Institute, Flint, Michigan*

I. WHY TEST ADHESIVES

Adhesives are tested by manufacturing people to

1. Assist in selecting an adhesive for a particular use
2. Monitor the quality of an incoming product
3. Confirm the effectiveness of the bonding process

The adhesive selection process usually concentrates on tests of joint strength. The goal is to determine which of several adhesives is the most suitable. The quality of incoming adhesives is primarily assessed by checking adhesive properties such as viscosity, percent solids, infrared spectrum, and bond strength. Destructive tests of joint strength are usually used to verify that the bonding process, for example, cleaning, mixing, curing, etc., has been correctly performed.

Literally hundreds of tests of adhesives and bonds have been developed during the past 50 years. Some are widely used. Others have been forgotten or find very limited use. Section II describes some widely used tests which have applicability in manufacturing.

II. TYPES OF TESTS

A. General Requirements

Most adhesive tests of interest to manufacturers can be considered tests of the adhesive material or tests of adhesive bonds. The vast majority of these tests are based upon procedures approved by the American Society of Testing and Materials (ASTM). An entire volume of their test specifications is devoted to adhesive testing [1]. Whether ASTM tests or tests of your own design are used, it is imperative that the test conditions be rigorously specified in order to make comparative judgments based upon test results. Surface cleaning and other preparation, joint geometries, method and extent of mixing, method of application, fixtures utilized, and cure conditions are among the obvious variables that can affect joint strength. Repetitive tests must be carried out under identical conditions if useful results are to be obtained. See Chap. 24 for a discussion of the use of statistics in testing bonds.

B. Material Tests

The most commonly used tests for properties of adhesive materials measure viscosity, shelf life, pot life, tack, cure rate, and percent solids.

1. Viscosity

Viscosity is defined as the resistance of a liquid material to flow. Adhesive viscosity is an indication of how easily pumped or spread the product will be. Viscosity may also reflect errors in compounding or excessive age of the material. Adhesives may be Newtonian, thixotropic, or nonflowable liquids. Viscosity measurements for free-flowing Newtonian or near-Newtonian adhesives are usually based on one of the following methods, which are described in detail in ASTM Standard D-1084 [1].

a. Volume flow cups. Volume flow cups measure the time required for a given volume of adhesive to flow through an orifice in the bottom of a metal cup containing an accurately known volume of material. A typical viscosity cup is shown in Fig. 1. Cup size and orifice diameter are chosen so that the flow-out time is not more than approximately 60–90 sec. Results are reported as the time in seconds required for the adhesive to flow from a specified cup. Conversion values for various volume flow cups are given in Table 1.

b. Rotating spindle method. The rotating spindle method, which is used for products which range in viscosity from 50 to 200,000 cP, measures the rotational resistance encountered by a metal spindle of a particular size spinning at a predetermined rate [1]. The test is ac-

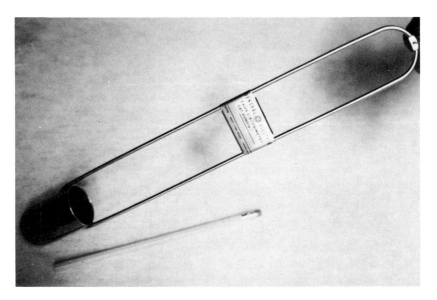

Fig. 1. A typical viscosity cup (Zahn No. 3).

curate and fairly rapid. It can be adapted to production-floor conditions although it is most conveniently performed in a laboratory. A
typical, rotating-spindle viscometer is shown in Fig. 2.

The viscosity of thixotropic adhesives is usually determined by
ASTM Standard D-2556 [1]. Thixotropic materials exhibit a viscosity
which is shear-rate dependent. The viscosity is determined at several
different shear rates, usually with an electrically driven spindle,
disk, T-bar, or coaxial cylinder rotated in the adhesive. A plot of
apparent viscosity versus rotational speed is prepared, and from this
plot the apparent viscosity associated with the particular rotation
speed and spindle shape is obtained.

c. *Extrusion*. The viscosity of nonflowable products is often determined on the production floor by an extrusion test. A test of this
type typically utilizes a Semco 440 nozzle attached to a standard extrusion cartridge filled with the adhesive to be tested. Air at a pressure of 90–95 psi (630–665 kPa) is used to extrude a bead of the product. Viscosity is reported as the extrusion rate in grams per minute based on the average of three 10-sec extrusion trials [2].

2. Shelf life

The shelf life of an adhesive is usually determined by noting changes
in viscosity or in bond strength which occur after periods of storage.
The usual methods of determining viscosity or bond strength are used.

Table 1 Conversion Times for Common Volume Flow Viscosity Cups

Fisher #1 (sec)	Fisher #2 (sec)	Ford cup #3 (sec)	Ford cup #4 (sec)	Zahn #1 (sec)	Zahn #2 (sec)	Zahn #3 (sec)	Zahn #4 (sec)	Zahn #5 (sec)
20								
25								
30	15	12						
35	17	15						
39	18	19						
50	21	25	5	30	16			
	24	29	8	34	17			
	29	33	10	37	18			
	33	36	12	41	19			
	39	41	14	44	20	10		
	44	45	18	52	22	12	10	
	50	50	22	60	24	14	11	
	62	58	25	68	27	16	13	
		66	28		30	18	14	
			31		34	20	16	
			32		37	23	17	10
			34		41	25	18	11
			41		49	27	20	12
			45		58	30	21	13
			50		66	32	22	14
			54		74	34	24	15
			58		82			
			62					
			65					
			68					
			70					
			74					

Fig. 2. A typical, rotating-spindle viscometer (Brookfield model).

Time and temperature of storage, as well as the size and nature of storage containers should be agreed upon in advance by the adhesive user and supplier. There is no universally accepted change in viscosity or average bond strength which denotes a nonusable product. Any decision relating to this question should be based upon agreement reached by the supplier and user of the adhesive at the time the original adhesive selection is made.

3. Pot life

The pot life of an adhesive is that period of time over which it is actually usable. Pot life may vary from minutes to hours and is clearly
an important parameter. Methods of determining pot life, which, like
the shelf-life tests, are based upon changes in either viscosity or
bond strength, are described by ASTM Method D-1138 [1]. Any accurate method of determining viscosity or of testing for bond strength
may be used. Decisions concerning the rejection of a product because
of changes in viscosity or bond strength should be based upon agreement reached between the adhesive supplier and user at the time the
adhesive is originally selected. To simply say that a given percent
change in viscosity makes the product unusable is not sensible. Considerable prior testing is usually required to make such a determination.

4. Tack

Tack is that characteristic of an adhesive which results in the easy
sticking of an adhesive-coated surface to another surface upon contact. High-tack adhesives are those which are "sticky." A number of
tack-measuring devices have been used. Many of them measure the
force required to separate two surfaces, one of which has been coated
with the adhesive, after they have been in contact over a specified
area at a specified pressure for a specified time. A typical tackmeter
is shown in Fig. 3.
     Another type of tack test (developed by the Douglas Aircraft
Company) measures the distance that a steel ball will roll over an adhesive-coated surface after the ball has traveled down an inclined
plane [3]. The higher the tack, the shorter the roll will be.

5. Cure rate

It is often desirable to know how the strength of a bond will vary with
the rate of cure. A recommended practice for determining the bond
strength at intervals of two-fifths, three-fifths, four-fifths, five-
fifths, and six-fifths of the prescribed cure time is described by
ASTM D-1144 [1]. The actual bond strength may be determined with
a standard ASTM tensile test (D-987) [1]. Results are commonly reported as a graph of average breaking load versus cure time.

6. Percent solids

The solids content of an adhesive should be checked to ensure that
formulation or dilution errors have not been made. Solids are usually
determined by weighing a sample of the adhesive and then heating or
curing it until a constant weight is obtained. Percent solids is equal
to the sample weight before heating or curing multiplied by 100 and
divided by the sample weight after heating or curing.

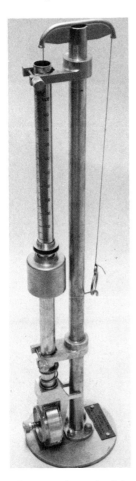

Fig. 3. A typical tackmeter.

C. Joint Tests

A number of tests of finished bonds have proved to be of value in checking the assembly process. If bond strengths suddenly decline using good ahesive, it normally indicates that something has gone awry with the surface preparation or the assembly process. Not all of the following tests need to be utilized in every situation. Lap shear and impact tests are probably the most commonly used, but the others might be chosen if they more accurately reflect the most critical stresses to which a particular product might be exposed in service.

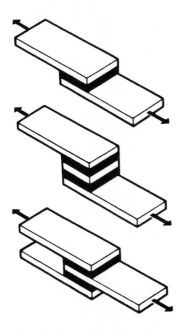

Fig. 4. The lap shear tests as described in ASTM D-1002. The bonds are 1 in. wide and have a ½-in. overlap. Adherend thickness is normally 0.064 in. Stressing the joint in tension introduces peel or cleavage stresses at the joint. (Courtesy of The Sira Institute, Ltd., Kent, England.)

1. Lap shear

Lap shear testing is very widely used to determine adhesive bond strength. The test [1], which utilizes a specimen such as that shown in Fig. 4, is easy to carry out and is, therefore, widely used. Lap shear generates peel or cleavage sources at the ends of the bond and is, thus, a stringent measure of bond strength. Pertinent details of the lap shear test are given by ASTM D-1002 [1]. Test results are reported in failing load per unit area.

2. Tensile tests

A number of tests, of which the ASTM D-897 standard is the most common, have been devised for determining the tensile strength of an adhesive bond. In all cases the specimen is loaded in nominally pure tension. Test results will be significantly lower if cleavage forces are allowed to develop in the loaded samples. Figure 5 indicates typical geometries for tensile testing. Cross-lap tensile tests are sometimes used with glass or plastic substrates. A device for carrying out such

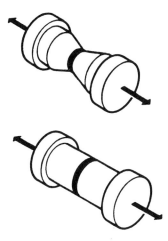

Fig. 5. Tensile testing as described in ASTM D-897. (Courtesy of
The Sira Institute, Ltd., Kent, England.)

a cross-peel test is illustrated in Fig. 6. Test results are reported in
failing load as psi.

3. Peel

Peel tests measure the force required to peel a flexible substrate away
from a rigid or flexible material. The general geometries utilized are
shown in Fig. 7. The geometries commonly used for the tests are illustrated in ASTM methods D-903, D-1876, and D-773 [1]. The
climbing drum peel test, which is widely used in the aircraft industry
to determine bond strengths involving honeycomb sandwich structures, is described in ASTM D-1781 (Fig. 8). In all cases, it is important to specify the peel rate. Peel strengths are reported in load
per unit bond width.

4. Cleavage

Adhesive bonds are usually tested for resistance to cleavage failure
with a specimen similar to that shown in Fig. 9. The test is described
in detail in ASTM D-1062 [1]. The bonding surfaces are 1 × 1 in.,
and the test requires that both adherends be rigid. The specimen is
usually loaded at 500–800 lb/min. The results are reported in breaking load per unit area.

5. Impact

Impact testing [1] measures the force required to break an adhesive
bond when one adherend is rigidly held and the other is impacted,

Fig. 6. A cross-peel tester. The bonded area is typically 1 × 1 in.

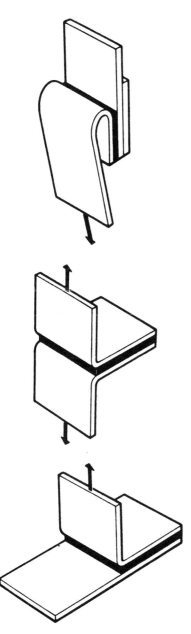

Fig. 7. Some common tests for adhesive-bond peel strengths. (Courtesy of The Sira Institute, Ltd., Kent, England.)

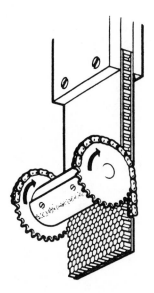

Fig. 8. The climbing drum peel test, ASTM D-1781. (Courtesy of
The Sira Institute, Ltd., Kent, England.)

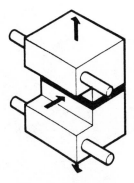

Fig. 9. Bonding resistance to cleavage stresses is often measured
with ASTM D-1062. (Courtesy The Sira Institute, Ltd., Kent, England.)

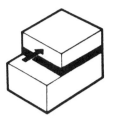

Fig. 10. The impact resistance of adhesive bonds is often measured with ASTM D-950. (Courtesy of The Sira Institute, Ltd., Kent, England.)

typically by a pendulum as described in Fig. 10. It is often difficult to achieve reproducible results with impact testing, and, as a result, the test is not widely used in production situations [2].

6. Fatigue

Fatigue testing of adhesive bonds is generally carried out using standard fatigue-test machines and lap shear bonds as described in Sec. II.C.1. The specimen is subjected to cyclic axial or bending stresses. It is important to specify the frequency, amplitude, temperature, and type of stress as well as its magnitude. Cycles to failure and the corresponding loads are plotted on coordinates of stress versus the logarithm of the cycle. The point at which the smooth curve connecting the points of minimum stress crosses the 10 million-cycle line is known as the fatigue strength.

7. Strength retention

It is often desirable to know the rate at which an adhesive bond will lose strength in service. These data are extremely difficult to obtain with much accuracy. Accelerated aging tests are difficult to extrapolate to the service conditions. Nonetheless, strength retention studies are frequently used as a basis for adhesive selection since they will often identify the most durable adhesive from a list of candidate materials.

Strength-retention values may be based upon the effect of applied stresses under ambient conditions or after exposure to a controlled environment.

Stressed aging tests are important because they more accurately simulate service conditions encountered by most bonded products than simple, on-the-shelf, static aging tests do. Bonds may be subjected to stress in a number of ways, the simplest of which is probably simply hanging the specimen from a hook and attaching an object of known mass so that a constant stress is experienced by the bond.

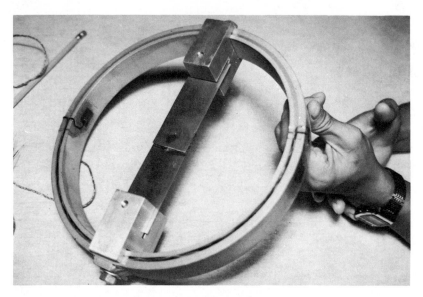

Fig. 11. The Alcoa Stress Test Fixture is a simple-to-use device for
applying controlled stresses to a lap shear or tensile test specimen.

The Alcoa Stress Test Fixture (Fig. 11) is a much more sophisticated,
yet simple-to-use, device for applying stresses of various magnitudes
to lap-shear—bonded samples. Specimens are removed at various intervals and tested in the normal way. A plot of strength versus time
can give informative data about the loss of bond strength under
stressed conditions.

    Environmental exposure is commonly used to simulate accelerated
aging of the bond. Exposure typically is to elevated temperature,
water, salt spray, or various chemical solutions which will simulate
the service conditions of the part. A number of standard chemicals in
which bonded specimens are soaked for seven days at room temperature are described in ASTM D-896 [1]. The concentration of the
chemical is important and should be specified. After suitable exposure, bonds may be tested in whatever manner seems appropriate for
the use at hand.

D. Nondestructive Testing

The ability to determine bond strength and/or durability nondestruc-
tively is a goal of obvious value to anyone using adhesives. Unfortunately, the reduction-to-practice testing (NDT) has been neither
widespread nor foolproof. The aircraft industry has been a leader
in the development of use of NDT methods although at least one American automobile manufacturer is now using the techniques to check

fiberglass-reinforced, plastic-part bonds in a production situation [4]. Nondestructive tests of adhesive bonds may be classified as sonic (top), ultrasonic, radiographic, eddy current, thermal, acoustic emission, and acoustic holographic. In each case, the technique depends on the difference in reaction to a physical stimulus between well bonded and poorly bonded areas. Usually, only unbonded areas can be detected although some techniques will identify areas of porosity in the adhesive film. There apparently is no good technique available to detect areas of low adhesion.

1. Sonic tests

Sonic tests use an aluminum rod or a coin to tap the surface of bonded metal sheets. Disbonds of about 1.5 in. in diameter can be detected by experienced personnel.

2. Ultrasonic tests

Ultrasonic techniques, which are widely used in the aircraft industry, depend on the detection of changes in frequency and amplitude which occur when an ultrasonic transducer is liquid-coupled to a bonded metal. Some of these testers can detect bondline porosity as well as areas of disbond. Schliekelmann [5,6] has established a correlation between ultrasonic measurements and lap shear bond strengths.

3. Radiography

Radiographic tests require that an x-ray photograph of the bond be prepared and evaluated. Voids and porosity in the bondline can be detected, especially if x-ray—opaque adhesives are used.

4. Eddy current test

The eddy current technique uses an oscillating current in a probe coil to induce eddy currents in bonded metals. These induced currents result in a mechanical vibration in the metal which changes in acoustic response when an unbonded area is encountered. Voids must normally be approximately 0.5 in. in diameter to be detected.

5. Thermal Inspection

Heat-sensitive detectors are used to locate differences in surface temperature which occur over unbonded areas when the entire surface is heated. Ideally, the difference in thermal conductivity between the bonded metal and the adhesive should be as great as possible so that disbonded regions will appear as hot spots. Infrared detectors, liquid crystals, heat-sensitive papers, and heat-sensitive dyes have all been used for this purpose [7]. Liquid crystal techniques are quite sensitive, and thermal-sensitive papers can be used to make a permanent record.

6. Acoustic emission techniques

In some instances detectable acoustic emission results when a bond is mechanically or thermally stimulated. The emisssions are usually detected with a piezoelectric sensor (resonant frequency around 175 kHz) and an amplifier.

7. Acoustic holography

The acoustic-holography approach to NDT uses pulsed echo ultrasound and focused transducers to produce a hologram of reflections from within the bond. The technique appears to be promising, but the equipment is expensive and the process somewhat time consuming.

REFERENCES

1. *1982 Annual Book of ASTM Standards, Part 22*, ASTM, Easton, Pennsylvania, 1982.
2. C. V. Cagle, in *Handbook of Adhesive Bonding*, McGraw-Hill, New York, 1973.
3. W. Parst, in *Handbook of Adhesives*, 2d ed. (I. Skeist, ed.), Reinhold, New York, 1977.
4. F. J. Meyer and G. B. Chapman, *Adhesives Age April*, 1980.
5. R. J. Schliekelmann, *Nondestructive Testing April and June*, 1972.
6. R. J. Schliekelmann, *Nondestructive Testing April*, 1975.
7. J. D. Minford, Testing of adhesive joints, a lecture presented at General Electric, Schnectady, September 1978.

23
Permanence of Adhesive–Bonded Aluminum Joints

J. Dean Minford   *Alcoa Laboratories, Alcoa Center, Pennsylvania*

I. INTRODUCTION

Although adhesive bonding of aluminum has been employed in military and civilian aircraft construction since the 1950s, the study of the factors affecting the durability of bonded metal structures continues to be actively pursued. Because a bonded joint involves the complex interaction of many factors, it has been an almost impossible task to accurately predict the final service life of bonded assemblies. In spite of this difficulty, there are many justifications for the continued use of bonding in virtually all phases of general manufacturing.

Substantial savings in weight reduction or manufacturing costs are often cited as general inducements for bonding. Or, bonding may be chosen because the particular materials of construction are not amenable to other forms of joining. This particularly is the case when it is desirable to join metals to nonmetals in composite structures. Unfortunately, even when bonding seems the most practical choice for joining, it may not be acceptable because of suspect joint durability. It needs to be shown that in many of these situations a previous, unacceptable performance was based largely on improper selection of adhesive, improper surface treatment of adherends, inappropriate curing conditions, or lack of understanding of the mechanical and chemical interactions of the joined materials in the interfacial area. Or, it may be that a joint configuration which was inadequate to handle the loads under service conditions or permitted interfacial damage to accumulate because of wrong joint design was used. Finally, there may have been a lack of understanding of how the environmental factors act to debond or degrade the interfacial area so that catastrophic joint failure occurred, causing the total bonding experience to be suspect.

It is the purpose of this work to subject the various parameters of adhesive bonding to scrutiny so that better judgments about the durability of bonded joints can be made by those seeking to manufacture bonded structures. This chapter is not intended as an exhaustive treatment of the subject area, but it should serve as an appropriate introduction to the field, and the references, as a springboard to more detailed study.

II. BONDLINE CHARACTERISTICS

A. Modulus of Elasticity

The high moduli of various materials, coupled with differences in their coefficients of thermal expansion, can be important factors in determining bond durability. When large differences exist between the coefficients of expansion of the adherends (Table 1), it is usually beneficial to use an adhesive of low modulus, such as a nitrile-rubber—phenolic adhesive, as a stress-relief interlayer.

B. Interfacial Imperfections

DeBruyne [1] and Bascom [2] have furnished examples in the literature where an imperfection, such as a trapped air bubble, was the site for high, localized stresses. These stresses can add to the effect

Table 1 Adherend-Adhesive Differences for Some Representative
Materials

Material	Modulus	Linear coefficient of expansion (in./in.°C)
Stainless steel	29,000,000	17×10^{-6}
Aluminum	10,500,000	$22-28 \times 10^{-6}$
Alumina	50,000,000	8×10^{-6}
Epoxy (unfilled)	500,000	$50-60 \times 10^{-6}$
Epoxy (filled)	2,000,000	$25-40 \times 10^{-6}$
Nitrile-rubber-phenolic	4,000	$50 + \times 10^{-6}$

Source: Courtesy of Alcoa, Alcoa Center, PA.

of any desorbing agent from the service environment, such as water,
in significantly reducing overall bond performance. Loew et al. [3]
have investigated the feasibility of detecting and classifying bond defects in the categories of unbond, voids, or porosity conditions. Ultrasound techniques were shown to be both a practical and a feasible
way to detect and classify defects even in multiply layered adhesively
bonded structures.

Some degree of interfacial imperfection can arise as a result of
the normal curing of a structural joint. After it has wet the surface
of the adherend, the adhesive develops recognizable strength by the
process of solvent evaporation, cooling, or chemical polymerization.
Plueddemann [4] has listed the stresses due to adhesive shrinkage
along with differences in coefficients of expansion as among the principal factors responsible for bond failure.

A different situation can develop if the bond interface is altered
due to swelling of the adhesive. For thermoplastic adhesives, the absorption of organic solvent can produce a volume change of sufficient
magnitude to create significant stresses in the interfacial area. Metal
adherends offer the worst situation since they neither absorb solvent
nor change in volume to help compensate for the volume change in the
adhesive. The failure of adhesive bondlines is inevitable and, usually,
rapid where swelling of the adhesive by water absorption is accompanied by water desorption of the adsorbed adhesive at the adhesive-
metal interface.

C. Heat Curing

The permanence of bonded joints can be positively affected in several ways by the heat curing of the adhesive. At elevated temperature the lowered viscosity of the adhesive can improve adhesion through better wettability of the surface (Fig. 1). Even with some soiling present, better bond durability may be achieved because the lower-viscosity adhesive can absorb, dissolve, disperse, or desorb the soil. Finally, the higher temperature could increase the degree of chemical interaction between the adhesive and a primer or the adherend surface.

In general, the higher the temperature of the cure, the greater the durability of the joint is. Among the present commercially available adhesives, those that cure at 177°C (350°F) are the most durable; those that cure at 121°C (250°F) show a lesser degree of durability. If an adhesive with a recommended cure temperature of 121°C (250°F) is cured at a higher temperature, an increase in durability will usually be demonstrated.

Whether this increase is due to an increase in modulus or a change in the number and/or type of attachments to the substrate is not clear. The fact that within a given family of adhesives the bond durability increases as the creep resistance increases seems to favor the argument for modulus increase [5].

D. Pressure

The positive effect of pressure in promoting bond permanence can occur in several ways. Applied with heating, pressure may promote better surface wetting and spreading of the adhesive. Pressure could force extra adhesive into the rough profile of an irregular surface or into the actual pores of a substrate such as wood or ceramic materials. The good durability of joints made with adhesive polymers which can form volatile products under elevated-temperature curing conditions, such as the vinyl-phenolic adhesives, is highly dependent on maintaining high pressure on the bondline. Failure to apply adequate and

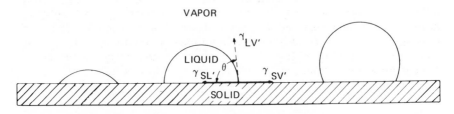

Fig. 1. Wetting and contact angle. (From Ref. 8.)

uniform pressure to a bondline may result in both nonuniformity of the bondline cross section and less than optimal adhesive flow.

E. Mechanical Energy

The mechanical energy expended in rubbing an adhesive into a metal surface has been correlated with measured increases in bond-strength values by Cassidy and Yager [6]. It was presumed that the mechanism might be displacement of a contaminant film on the surface. If this displacement is the true explanation for the effect, then it must be considered a minor factor in promoting the good bond durability of uncontaminated, acid-etched aluminum joints.

F. Cohesive Versus Adhesive Failure Observations

It is generally agreed that the bonded metal joints with the highest durability potential initially fail cohesively. The degree of durability is related to the retention of the cohesive failure property after long periods of exposure to weathering service conditions. All observations made at Alcoa in evaluating the bond permanence of aluminum structural joints have shown that the initial presence of a cohesive bond failure prior to exposure is never a reliable indication of good bond durability. Similarly, the initial level of bond strength with apparent cohesive bond failure without exposure is not predictive of any relative durability value [7−10]. An example of each of these situations is cited by DeLollis [11,12] in a two-year, water-sensitivity study of the relative bond permanence of bonded aluminum joints fabricated with nitrile-phenolic or nylon-epoxy adhesives. Whether exposed to continuous high humidity or water immersion, nitrile-phenolic joints showed a high degree of durability, exhibiting cohesive failure under all exposure conditions, while the nylon-epoxy joints with initial cohesive bond failure showed adhesive failure within two months. Initial lap shear strength of the nylon-epoxy joints was approximately three times higher than that of the nitrile-phenolic joints. Thus, regardless of how the bond fails initially or how high the initial bond strength may be, the rate of desorption of the adhesive by the action of water at the interface has a most significant influence on bond permanence. The most conclusive studies relating to the theory of bond desorption by water have been shown in investigations of adhesion and bond failure between adhesive resins and glass [13−16].

Prominent adhesive scientists such as Bikerman [17, 18] and Sharpe and Schonhorn [19] have stated that there is only the degree of cohesive failure to consider in failed joints since the attractive forces involved at the interface are invariably greater than the cohesive strength of the adhesive and, sometimes, greater than the cohesive strength of the adherend. Bikerman has pointed out that bond rupture so rarely proceeds exactly between adhesive and adherend

that these events (i.e., failure in adhesion) need not be considered
in any theory of adhesive joints. It is true that apparent adhesive
failures are common, but they take place so near the interface that the
adhesive remaining on the adherend after the bond rupture is not observable. Good [20] also has reported that his analysis of adhesive
joint failure has led him to conclude that interfacial separation seems
highly improbable, particularly when true chemical wetting of a surface has taken place.

Acceptance of these ideas complicates acceptance of the theory of
bond failure by desorption by water, described by a variety of investigators and mentioned previously. The 1965 study by Sterman and
Toogood [21] which showed the increase in bond permanence of glass
joints in high-temperature-water exposures through the use of silane
adhesion promoters, has furnished further proof that the desorption
of the adhesive by water and failure in adhesion at the adhesive-adherend interface are viable mechanisms. How fast the water can permeate the adhesive to the interface as compared to how fast it can
diffuse along the adhesive-adherend interface has been investigated
by Laird [22]. He has shown that diffusion of water along the interface can be as much as 450 times faster than permeation through the
adhesive.

Adhesive product data sheets and the literature often cite the
durability of bonded joints in various organic solvents. Because adhesives are often polar in nature, they can be sensitive to some of the
stronger, polar organic solvents. Any absorption of the solvent
which results in swelling of the adhesive can cause stress to develop
at the interface with metal or ceramic adherends since neither metal
nor ceramic will swell. The adhesive-adherend interface will, therefore, become a plane of significant stress concentration.

Another interfacial factor can be the presence and relative concentration of nonadsorbable or nondesorbable, contaminating films at
the interface. The most common contaminants are grease films, industrial oils, forming lubricants, and certain plasticizer chemicals. In
assembling metal structures by adhesive bonding, certain metal-form-
ing operations often precede the assembly of the parts. Thus surface
contamination by forming lubricants can result. The best bond durability can be expected only if a separate operation to remove the
lubricant has taken place prior to adhesive application. Also, metal
sheet or extrusions in storage will often have corrosion-resistant films
on their surfaces.

Industries such as automotive manufacturing choose adhesives
which have the highest tolerance for these surface contaminants.
There has been only limited success to date in achieving high-durabil-
ity bonds in the presence of these contaminants as compared to the
bonds formed between clean and etched adherend surfaces. The vinyl
plastisol adhesives remain the best commercially available adhesives

for achieving modest strength and reasonable bond durability at lowest cost in such situations. Adding vinyl plastisol to higher-strength
epoxy resins has been a further step toward achieving higher-bond-
strength joints with better tolerance for surface oil contaminants than
joints formed with epoxies alone have. More recently, high-performance structural acrylic adhesives have been introduced for bonding
to unprepared metals [23]. The significant lowering of bond durability by the presence of these oils indicates that they should be removed prior to bonding and that special surface preparations for promoting better bond durability should be encouraged.

Admittedly, it is difficult to determine after bond failure—even
using visual or microscopic examination—whether an apparent adhesive failure occurred at the original interface due to improper wetting
or at some new interface, leaving behind a thin layer of adhesive. In
comparing surface features after bond failure with the original adherend surface, the maximum resolution of about 1×10^{-8} m (100 Å) for
scanning electron microscopes may not always be sufficient to detect
a thin film of adhesive closely reproducing the original surface profile.
Optical and staining methods described by Brett [24] to determine the
presence of such films are only applicable to fairly thick films since the
optical technique uses interference phenomenon. Films a few ang-
stroms thick are still largely undetected.

In recent years the use of certain highly specialized analytical
tools, such as secondary ion mass spectroscopy (SIMS), ion scattering
spectroscopy (ISS), and Auger electron spectroscopy (AES), by investigators such as Baun, McDevitt, and Solomon [25—29] has vastly
improved the opportunity for deducing the surface compositional
changes. Even when the surface films are only on the order of atomic
dimensions or when the failure occurred near the original interface
and included parts of both the adhesive and adherend, these tools may
provide useful information on the locus of failure. Of these procedures, AES has been found to be the least useful primarily because of
heating effects of the electron beam which may cause decomposition of
adhesive surfaces. By itself, SIMS has been shown to be a powerful
tool for elemental surface characterization by Benninghoven [30] and
Schubert and Tracy [31]; however, uncertain or rapidly changing
secondary ion yields due to changes in chemical bonding make quantitative analysis virtually impossible using SIMS alone [32,33]. However, SIMS is most helpful when combined with other techniques,
such as ISS and AES, which use a beam of ions of correct energy for
combined use with SIMS. The majority of apparent adhesive failures
which have been examined using these tools show that cohesive failure
in both adhesive and adherend probably occurred. An elemental analysis method must be used to confirm whether the surface contains
both adhesive and adherend. The test for the amount of branched
polyethylene adhesive on a copper oxide surface as performed by Blair
et al. [34] does not, in itself, prove that the mode of failure was only

cohesive in the polymer. When some failures classed as purely adhesive have been examined by ISS and SIMS, it has been found that the adherend surface had obviously not been wet by the adhesive. No trace of adhesive on the adherend could be found. Further examination in this kind of situation has usually disclosed that a thin layer of contaminating elements had caused the adherend surface to be most accurately described as "dirty." Certainly this kind of bond failure at the interface should not be considered adhesive since bonding between the materials had never really occurred.

III. CHARACTERISTICS OF THE ALUMINUM SURFACE

What is usually referred to as a metal surface is actually a metal-oxide surface with various, spontaneously adsorbed overlayers of water, organics, and gases (Fig. 2); thus, the adhesive material in the joint is not really involved in any metallic bonding. The covering layer eliminates the attractive forces and interfacial effects ordinarily associated with the underlying metal—for the purpose of this discussion, aluminum—atoms. Only by knowing more about the surface chemistry of aluminum oxide can we deal with the interaction between the adhesive polymer and the bondable interfaces of aluminum adherends.

Since all adhesive polymers show some degree of mobility in the uncured state, the adhesive will flow over the interfacial area in a

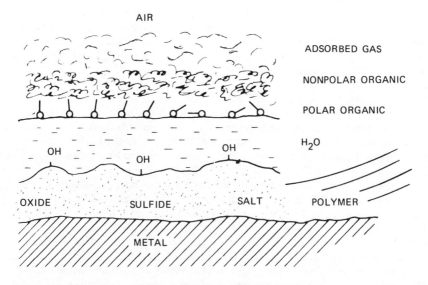

Fig. 2. Hierarchy of spontaneously adsorbed layers on a metal surface. (From Ref. 8.)

Table 2 Effect of Better Surface Wetting on Bond Performance

Surface treatment	Adhesive	Type cure	Range of bond survival times in seacoast exposure (days)
Vapor degreased	Two-part epoxy	R.T.	70−80
Vapor degreased	One-part epoxy	Heat	71−270
Alodine 1200	Two-part epoxy	R.T.	1158−1440
Alodine 1200	One-part epoxy	Heat	> 2920
Chromic-sulfuric	Two-part epoxy	R.T.	270−760
Chromic-sulfuric	One-part epoxy	Heat	760−1440

Source: Courtesy of Alcoa, Alcoa Center, PA.

kinetic process, eventually filling all void areas in the substrate surface. If the viscosity of the adhesive is very high, it is possible for it to trap air and create voids. If the viscosity is low, such entrapments are less likely. This is one explanation for the superior durability of many heat-cured epoxy joints as compared to room-temperature—cured epoxy joints in highly corrosive environments. The degree of improved chemical wetting of the aluminum surface—due to the lower-viscosity, melt condition of a one-part epoxy resin—during the high-temperature curing step is shown by the data in Table 2 which compares the durability of joints produced by room-temperature— and elevated-temperature—curing epoxies on variously treated aluminum surfaces.

In general, the wettability of an adhesive, or its tendency to adsorb on an adherend surface, is assumed to be intimately related to the bond-joint durability.

Various interactions between surface oxides and environmental factors—such as water, temperature, and atmospheric chemicals—can create bonding complications. Although the oxide surface can apparently be dehydrated at elevated temperatures, under normal, ambient, bonding conditions the outermost surface oxygens will hydrate to form a layer of hydroxyl groups which, in turn, will adsorb and retain molecular layers of bound water. On aluminum, iron, and copper surfaces, for example, the presence of up to 20 molecular layers of water has been shown on so-called dry surfaces. The presence of these various polar constituents on the aluminum surface can be fortunate for promoting bonding to polar organic resins, such as epoxies. If this polar surface is not present, as can happen on a fractured metal surface in a moisture-free environment, then the metallic surface can-

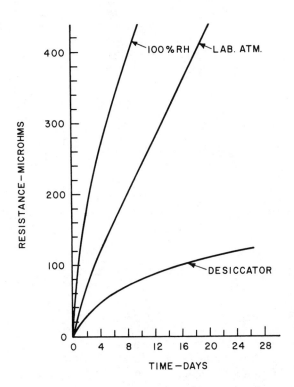

Fig. 3. Effect of storage environment on 1100 alloy treated in R5
Bright Dip. (Courtesy of Alcoa, Alcoa Center, Pennsylvania.)

not be wet by polar organic liquids. Admitting a small amount of oxygen to permit surface oxidation on an oxide-free surface permits the
polar liquid to spread out and wet the surface [35].

What happens to the native oxide on an aluminum surface when
the surface is mechanically abraded, or acid or alkaline etched? It is
possible to deoxidize an aluminum surface using any of these methods,
but new oxide regrows instantaneously so that an oxide-free surface
is never present in any practical bonding situation (Fig. 3).

The oxide layer on "as received," or vapor degreased, aluminum
has been identified as consisting mainly of the bayerite, or beta-aluminum trihydroxide form with an oxide thickness varying from 1
$\times 10^{-8}$, (100 Å) to 1.5×10^{-7} m (1500 Å). After alkaline cleaning at
76.7°C (170°F) or acid etching for spot welding, the surface oxide
layer remains primarily bayerite.

The surface produced by anodizing has a typical, thin, barrier
oxide layer at the base. The actual surface involved in bonding is a

much thicker oxide layer than the barrier layer and consists primarily of boehmite, or beta-aluminum oxide hydroxide as boehmite is more correctly described [36]. It is not unexpected that this form of bonding oxide bonds durably since a Forest Products Laboratory (FPL) etched aluminum surface, while showing a much thinner coating than the anodized oxide layer, has also been shown to be predominantly boehmite in character [37,38].

Minford [10] has offered evidence of the good durability of joints made with surfaces anodized by treatment with either chromic or phosphoric acid. Both stressed and nonstressed specimens were exposed for long time periods to water soaking, alternate saltwater immersion, or natural, atmospheric, weathering environments.

IV. EFFECT OF WATER

Probably the most important environmental factor in the general deterioration of bonded metal joints is the invasion of the boundary layer of the joint by water molecules. Two of the most effective ways to increase the stability of the boundary layer to humidity are the use of special, aluminum-surface pretreatments and the use of corrosion-resistant primers. However, it must be admitted that all these special surface-treatment and primer developments have been carried forward without an exact knowledge of the nature of the adhesive forces existing between the organic adhesives and the aluminum surface. While the bonds between structural adhesive and metal surface may be chemical in nature, they are not absolutely water-stable. Some of the earliest investigations on the unique effect of water on the durability of epoxy-bonded aluminum joints were by Falconer, MacDonald, and Walker [39] in 1964, with subsequent confirmation by many other investigators [11,40−42]. The 1967 experiment of Kerr, MacDonald, and Orman, comparing the relative effects of dry ethanol and water on the cured epoxy polymer (whether inside or outside the bond-joint area), furnished specific evidence for the preferential action of water at the aluminum-oxide−polymer interface. While alcohol was readily absorbed into the polymer matrix, reducing the tensile strength by 30%, there was no bond-strength deterioration in the corresponding aluminum-epoxy joint after soaking in alcohol for the same period. By contrast, water in the environment was barely absorbed by the cured epoxy, and the epoxy tensile strength was lowered by only 8%, yet the aluminum-epoxy joint strength was decreased by 50% after exposure to water (Fig. 4).

V. EFFECT OF SURFACE TREATMENT

The effect of surface pretreatment on the aluminum-oxide surface is both critical and intimately related to the bond permanence of alumin-

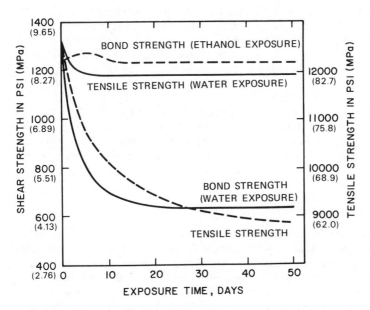

Fig. 4. Bond strength of aluminum-epoxy joints and tensile strength
of epoxy resin after exposure to water and ethanol at 90°C (194 °F).
(From Ref. 8.)

um joints [7-10]. It is difficult, however, to distinguish the benefits
of different surface treatments in the absence of some environmental
exposure since the joints will fail cohesively for many, different, sur-
face-treating conditions when tensile-shear tested without exposure.
A list of some of the testing environments used by Alcoa is shown below.

    1.   Static heat aging or cryogenic exposures
    2.   Water immersion exposures
    3.   Humidity exposure
    4.   Wet, freeze, thaw cycles
    5.   Continuous or intermittent salt spray
    6.   Natural atmospheric exposures
    7.   Simultaneous stress and environments.

Extended exposure to accelerated or natural weathering conditions
provides discriminating evaluation of most surface treatments. But
the order of effectiveness of different treatments in promoting maximum bond permanence can vary significantly based on which weathering condition is used for testing. For example, the acid-etching pretreatments rate high in promoting bond permanence in water soaking
testing. In corrosive weathering environments like salt water or sea-

Table 3 Relative Effectiveness of Different Surface Treatments on 6061-T6 Aluminum Bonded With a Two-Part, Room-Temperature—Cured Epoxy

	Exposure conditions							
	Industrial atmosphere lap shear strength [psi (MPa)]				Seacoast atmosphere lap shear strength [psi (MPa)]			
	Initial	2 yr	4 yr	8 yr	Initial	2 yr	4 yr	8 yr
Vapor degreasing	1,970 (13.5733)	1,530 (10.5417)	1,264 (8.7089)	907 (6.2492)	1,970 (13.5733)	0	0	0
Grit blasting and vapor degreasing	2,030 (13.9876)	2,060 (14.1934)	2,030 (13.9867)	1,930 (13.2977)	2,030 (13.9867)	1,800 (12.4020)	1,580 (10.8862)	1,350 (9.3015)
Chromate conversion coating	1,260 (8.6814)	1,230 (8.4747)	1,198 (8.2542)	1,130 (7.7857)	1,260 (8.6814)	835 (5.7532)	430 (2.9627)	0
Anodized[a]	2,530 (17.4317)	2,130 (14.6757)	1,868 (12.8705)	b	2,530 (17.4317)	1,660 (11.4374)	1,690 (11.6441)	b
Chromic-sulfuric acid	2,830 (19.4987)	2,900 (19.9810)	2,731 (18.8166)	2,370 (16.3293)	2,830 (19.4987)	0	0	0
Phosphoric acid-alcohol etch	2,660 (18.3274)	2,470 (17.0183)	2,450 (16.8805)	2,130 (14.6757)	2,660 (18.3274)	0	0	0
Acid paste	2,500 (17.2250)	2,300 (15.8470)	2,099 (14.4621)	2,030 (13.9867)	2,500 (17.2250)	0	0	0

[a] Anodized in 15% by weight sulfuric acid at 21.1°C (70°F) and 12 amps/ft² and sealed for 20 min in boiling, deionized water (0.85 mil coating).
[b] Joints still under test after 7 yr of exposure.
Source: Courtesy of Alcoa, Alcoa Center, PA.

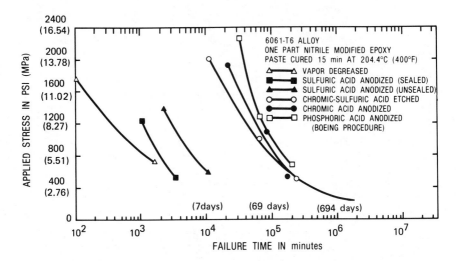

Fig. 5. Bond durability of vapor-degreased, chromic-sulfuric—etched, and various anodized surface joints stressed and exposed in 100% rh at 51.7°C (125°F). (Courtesy of Alcoa, Alcoa Center, Pennsylvania.)

coast atmospheres, on the other hand, the chromate conversion coating or anodizing pretreatments produce the more durable bonded joints. The high durability in the seacoast, shown in Table 3, for silica grit blasting is atypical for mechanical pretreatments.

Stressing plus environmental weathering can produce still a different rating among aluminum surface pretreatments (Fig. 5). At the highest stress levels the magnitude of the stress tends to be more significant than the degree of weathering, resulting in early bond failures even under relatively mild water-soaking conditions. At low stress levels, and particularly in high-humidity, high-temperature, or corrosive environments, the type of weathering condition can be the most significant factor in determining bond durability.

There has been increasing interest in recent years in nonaerospace, structural adhesive bonding for many manufacturing applications. For these situations there has been a need to develop bond-durability data for a wide variety of surface pretreatments with better cost effectiveness and less sophistication than FPL etching. Alcoa has long recognized that there could be more interest among the nonaerospace manufacturing producers in the simple, vapor-degreased surface condition than in the FPL etch. Both kinds of pretreated joints can constitute a standard for comparison with all other kinds of surface pretreatments. Bond durability studies at Alcoa have usually included comparative durability testing of a given alloy and adhesive combination with both vapor-degreased and FPL-type etched adherends. Most recently, interest has been increasing in evaluating bond permanence in the presence of deliberate contamination, such as the oils and lubricants used for forming parts, on the aluminum surface [43,44].

Several joining situations should also be mentioned which have potential applications both in aerospace and other manufacturing. They involve the combination of spot welding, riveting, or metal stitching with adhesive bonding. Once again, investigations have been conducted including a range from sophisticated, surface-cleaning pretreatments to those adding or leaving some surface contamination.

Over a period of years, a large variety of two-part, room-temperature−cured pastes, one-part, elevated-temperature−cured pastes, and heat-cured tape and film adhesives has been evaluated by Minford using three accelerated, laboratory, weathering environments and two natural, atmospheric weathering environments [8].

For applications outside the aerospace industry, it seemed more apropos to compare the standard chromic-sulfuric etching procedure to other surface treatments including vapor degreasing; grit blasting; amorphous-chromate conversion coating; anodizing; field-applied, acid-paste etch; and phosphoric-acid−alcohol etch. Tests were conducted for as long as 8 years under natural weathering conditions in an industrial setting and a seacoast environment (Table 3). Laboratory, accelerated-test conditions included continuous immersion in water or a wet-freeze-thaw cycle test for 2 years.

One of the other main approaches to improving the durability of aluminum bonded joints through alteration of the metal-adhesive interface has been the use of corrosion-inhibiting primers employing chromates. Sell [5] has pointed out the need for rather close control of the primer thickness as part of choosing the right combination of adhesive and mechanical characteristics necessary for applying corrosion-inhibiting primers. In general, the durability of primed joints increases with primer thickness up to the point where the primer begins to act as a separate layer. At that point the properties begin to level off since the primer is no longer as tough as the adhesive layer. Also, the primer thickness needs to be balanced with other physical properties, such as low-temperature peel, noting that the peel properties tend to increase as the primer thickness decreases.

VI. EFFECT OF ALLOY

Aluminum alloys have a wide variety of alloying compositions, strengths, and processing conditions that can affect bond permanence. First, there is a definite relationship between bond permanence and the general corrosion resistance of an aluminum alloy. This relationship can most readily be observed where bonded joints have

been exposed to a corrosive test environment like continuous or intermittent salt spray. For example, when one of the higher-strength
but less corrosion-resistant alloys, such as 2024 or 2036, is used to
make joints, their survival time at the seacoast may be only 25% as
long as that of 6061-T6 alloy joints.
     The effect of processing conditions on bond durability is related
to the variation in oxide buildup on the adherend surface, which can
result from special heat treatments. All the aircraft alloys fall into
this category. The thorough deoxidizing of the surface, which is
part of the standard process for adhesive bonding of these alloys,
automatically removes the oxide buildup. Where relatively high-
strength alloys may still be desired for general manufacturing purposes, this deoxidizing of the surface may not be affordable, and a
lower bond durability, based on comparative weathering testing, must
be accepted.

VII. EFFECT OF JOINT DESIGN

Significant differences exist in the ability of bonded joints to survive.
Much depends on the kind of stresses to which the joint is exposed
and the ability of the design of the joint to resist these stresses. In
general, we can conclude that bonded joints survive best when high
tensile or shear loads are present; in sharp contrast, they offer poor
resistance to high peel or cleavage loads (Fig. 6). The diagram of
the stresses imposed in a simple, lap-joint configuration shows clearly
that the combination of decreasing stiffness in the adherend and increasing brittleness in the adhesive is conducive to decreasing bond
durability when the bond is under stress.

VIII. EFFECT OF ADHESIVE SELECTION AND CURING

Choosing the right adhesive is vital to securing good bond durability.
Certain generic adhesive families are conducive to the creation of
higher bond durability than other families [5,8]. Also, there are
significant performance differences within a single family of similar
chemical description. These differences are readily distinguished by
long-term testing. Among a group of five one-part, heat-cured
epoxies evaluated by Alcoa in an aggressive, wet-freeze-thaw cycle
test, the joints in one product failed within 520 days; those in another
failed within 2 years; those in the three others all maintained bond
strengths in the vicinity of 13.78 MPa (2000 psi) after 2 years of aggressive, accelerated weathering. The most durable bond performances are generally given by the tape and film adhesives developed
for aerospace applications.

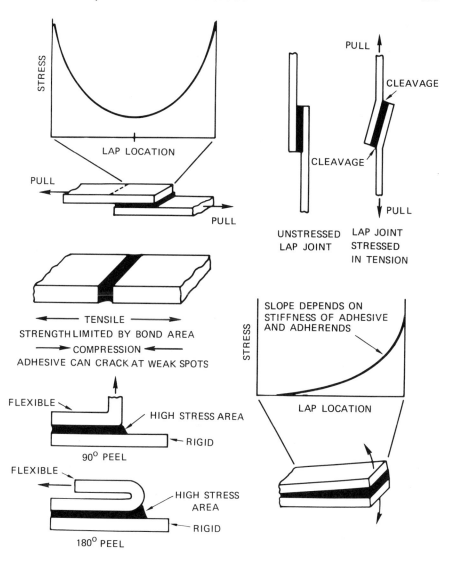

Fig. 6. Common stresses met in adhesive joints. (From Ref. 8.)

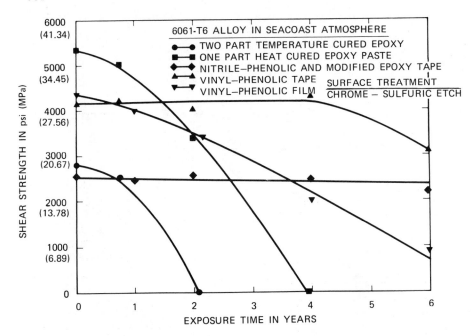

Fig. 7. Effect of adhesive on bond durability. (Courtesy of Alcoa, Alcoa Center, Pennsylvania.)

Significant variations between different chemical families have been shown by Minford in the results of exposure to the seacoast atmosphere (Fig. 7). Bond durabilities ranged from failure after 2 years for a two-part, room-temperature–cured epoxy to virtually 100% bond-strength retention after 6 years of exposure for a combination nitrile-phenolic and modified epoxy tape. Seventy-five percent bond-strength retention was obtained with a vinyl-phenolic tape.

IX. COMPARISON OF CLAD AND BARE ALLOYS

Clad aluminum alloys have been used in aircraft since World War II. It was not until the Vietnam War that a controversy arose regarding the durability of clad versus bare aircraft alloy joints. The location of bonded joints on exposed surface areas was responsible for this problem. In addition, the concentration of aircraft along the seacoast and on aircraft carriers during the Vietnam conflict accelerated the development of the problem. Cladding aluminum alloys was specifically developed to improve their corrosion resistance. It should not be

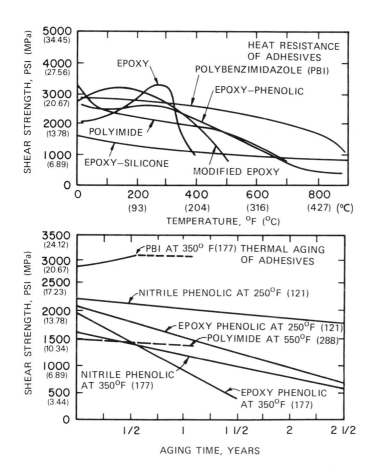

Fig. 8. Comparison of high-temperature structural adhesives shows that the polyaromatics (PBI and PI) have the highest shear strengths at temperatures above 800°F (427°C) (top) and best retention of strength on aging (bottom). (From Ref. 8.)

unexpected to find that Riel [45,46], Greer [47], and Rogers [48,49] have reported that clad alloys are more susceptible than bare alloys to bondline corrosion under aggressive, saltwater exposure conditions. It has been shown by McMillan [50] that bonds to clad 2024-T3 alloy adherends will perform more durably than bonds to clad 7075-T6 surfaces, indicating that a difference in performance between claddings also exists. McFarlen [51] has reported a comprehensive investigation of the performance of clad versus bare surfaces and found no

evidence for preferential attack of the clad interface. Different adhesives and primers can apparently produce different results.

X. EFFECT OF SERVICE TEMPERATURE

Because temperature is a common factor in all service environments,
its effect on bond permanence should be considered. Different adhesive families demonstrate wide variability in their responses to different temperatures, as shown in Figs. 8 and 9 from a Kausen review
article in 1964. At the time of World War II the service temperature
range of primary interest was from about −55°C (−67°F) to 93.3°C
(200°F). With the more recent development of aerospace applications,
the range of interest has expanded to include temperatures from
−252.8°C (−423°F) to 537.8°C (1000°F). Problems relating to bond
permanence with varying temperature are the result of stress concentrations and gradients developed within the joint. Some of the causes
of stress-concentration development are differences in the thermal
coefficients between adhesive and adherends; shrinkage of adhesive
during cure; trapped gases or volatiles evolved during bonding; differences in modulus of elasticity and shear strengths of adhesive and
adherends; differences in thermal conductivity of adhesive and adherends; and residual stresses in the adherends as a result of release
in bonding pressure.

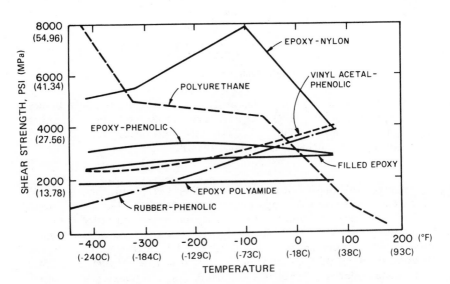

Fig. 9. Properties of cryogenic structural adhesive systems. (From
Ref. 8.)

The probable causes of and mechanisms for the deterioration of metal-adhesive bonds at temperatures up to 260°C (500°F) were investigated by Black and Blomquist [52] in 1958. These authors showed the significant effect of the conditions of the metal surface on the permanence of joints made with phenolic-epoxy adhesives. Operating at cryogenic temperatures, for example, can intensify stress concentrations generated as a result of the magnified differences in the physical and mechanical properties of adhesives and adherends at these very low temperatures. For example, a low-modulus adhesive may be able to stress relieve itself at room temperature by deformation, but at cryogenic temperatures the modulus of elasticity may increase to the degree that this is no longer possible.

The best bond permanence for temperatures up to 537.8°C (1000°F) has been provided by polybenzimidazone (PBI) and polyimide (PI) adhesives with the PI joints showing the most permanence after long-time exposure at temperatures from 204.4°C (400°F) to 315.6°C (600°F).

XI. EXPOSURE TO NATURAL WEATHERING ENVIRONMENTS

Beginning in 1955 and continuing through 1960, Eickner [53] began to report on the bond permanence of aluminum bonded joints exposed to exterior weathering conditions as compared to the laboratory type of accelerated weathering testing being conducted in various laboratories to qualify adhesives under Military Specification MIL-A-5909B or MIL-A-8431. Two sites chosen for their aggressive environment were a site 50 yd from the seashore in Miami, Florida, and another one ¼ mi from salt water in the Panama Canal Zone. General good performance of unstressed aluminum joints was found after several years of exposure if the aluminum surface had been prepared by etching in hot sulfuric-acid–dichromate solution.

A survey study of the effect of 1, 2, or 3 years of natural weathering on aluminum joints in South Florida was conducted by Hause, Pagel, and McKown [54] in 1965. Room-temperature–cured, two-part and heat-cured, one-part epoxy pastes were evaluated, as were heat-cured nitrile-phenolics, nylon-epoxies, modified epoxies, and vinyl-phenolics in film form. All joints were fabricated using acid-etched surfaces. Testing for overlap shear, honeycomb peel, and honeycomb beam flexure was conducted. The conclusion was that these bond-strength properties were generally unaffected after 3 years of exposure.

Many of the longest-time evaluations of aluminum-bond permanence conducted under natural, atmospheric weathering conditions have been reported by Minford beginning in 1972 [7–10,55].

XII. APPLICATION OF FRACTURE MECHANICS TO STUDY OF ADHESIVE JOINT PERMANENCE

Fracture mechanics as a science has made it possible to define the toughness of materials in the presence of flaws. While originally applied to the study of the toughness of solid materials, it has been adapted to study adhesively bonded sections which can fail by the propagation of a brittle crack through the bondline. It is probable that the adhesive joints in real structures are always blemished by flaws of a variety of types and sizes; for example, bubbles, dust particles, or unbonded areas. Furthermore, it is the progressive separation radiating from these flaws that basically controls the strength of the structure. Thus, it is somewhat unrealistic for investigators to evaluate structural-bond permanence by using as near-perfect joints as they know how to make to measure the stress needed to cause the joints to fail.

The above background was applied to adhesive joints by Ripling, Mostovoy, and Patrick in 1963 using a tapered, double-cantilever, beam-type specimen (Fig. 10). They concluded that the overriding factor in establishing the toughness of any bonded joint is the speed with which the crack moves along the bondline. Low toughness will be characterized by abrupt jumps of the crack while a slow-moving crack predicts a high degree of bond permanence, that is, toughness [56].

The general theory states that bonds, like homogeneous solids, fail because they contain flaws that cause load variations of the stress field to induce crack growth. Once this preexistence of cracklike flaws has been accepted, the problem of bond failure reduces to determining the condition or conditions under which a flaw will enlarge. If the flaw is at the loading edge of the bond, then the stress will be intensified at these locations. In studies, a crack is deliberately ini-

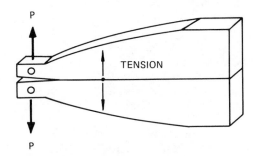

Fig. 10. Bonded, Tapered Double-Cantilever Beam (TDCB). Used for quantitative evaluations of fracture toughness and environmental durability effects on bonded systems. (From Ref. 8.)

Fig. 11. Thin adherend uniform double-cantilever beam (DCB) ("wedge test"). Not for quantitative evaluations because of possible plastic deformation of adherends. (From Ref. 8.)

tiated and allowed to arrest. The whole test device is placed in some environment, and the rate of crack growth with time is reported. The most widely used test specimen for conducting general crack-propagation evaluations is shown in Fig. 11. The ASTM has written a recommended procedure for conducting this test but warns that it is capable only of distinguishing between different surface pretreatments on the adherends; differences in joint durabilities cannot be estimated to predict service-life potential. The work of Bethune [57] and Marceau, Moji, and McMillan [58] should be consulted for more details about the development of the wedge-type joint for evaluating bond durabilities.

XIII. EFFECT OF SIMULTANEOUS STRESS AND ENVIRONMENTAL CONDITIONS

It is difficult to explain why so few bond-permanence-testing investigations involving sustained stress were described in the literature prior to Sharpe's technical disclosures in 1965 and 1966 [59,60]. The effect of simultaneous stressing and weathering on bond permanence

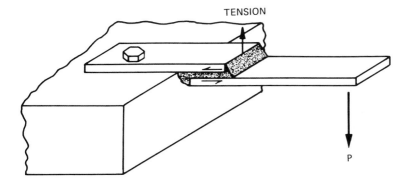

Fig. 12. Cleavage lap shear time to failure recorded. (From Ref. 8.)

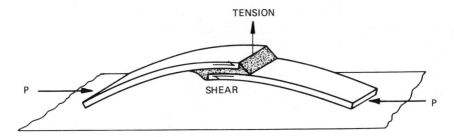

Fig. 13.   Standard lap shear in binding.   Specimen, 2.54-cm (1-in.)
wide, or entire assembly is bent and time to failure recorded.   (From
Ref. 8.)

was quite dramatic and needed to be studied in more detail.   It has
certainly always been recognized that periodic or sustained stress was
present in every bonded joint in an assembly.

Another early investigator in simultaneous stress and weathering
was Carter, who first placed stressed aluminum joints in a natural
weathering environment in Florida in June 1964 [61].   Because the
results of these test were not reported in the literature until 1967,
their significance from a historical viewpoint has been somewhat unrecognized.   It is clear from Carter's 1967 paper that he was aware of
the dramatic bond failures that could occur in a natural weathering
environment with epoxy-bonded aluminum joints if the stress on the
bond was high.

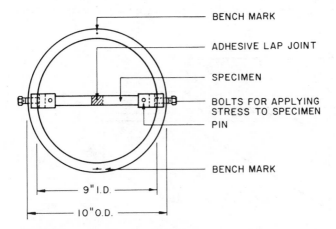

Fig. 14.   Alcoa stressing fixture details.   (Courtesy of Alcoa, Alcoa
Center, Pennsylvania.)

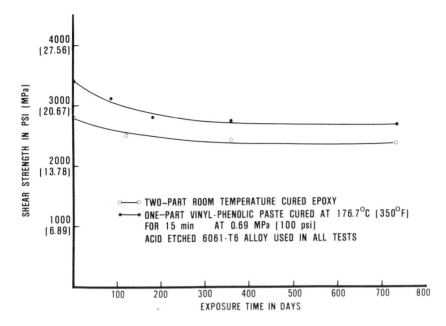

Fig. 15. Shear strength in psi (MPa). Unstressed specimens in 100% relative humidity at 51.7°C (125°F). (Courtesy of Alcoa, Alcoa Center, Pennsylvania.)

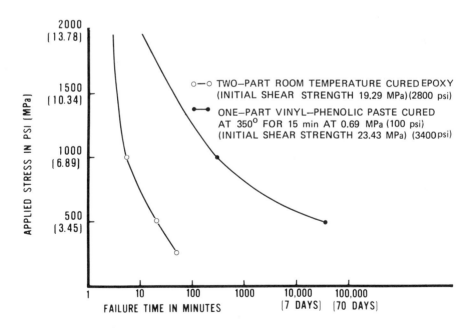

Fig. 16. Applied stress in psi (MPa). Stressed specimens exposed in 100% relative humidity at 51.7°C (125°F) acid etched 6061-T6 aluminum. (Courtesy of Alcoa, Alcoa Center, Pennsylvania.)

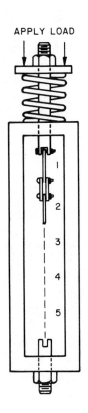

APPLY LOAD

Fig. 17. 3M's spring-loaded fixture. (From Ref. 8.)

In the earliest testing under stress, weights were hung from the joints or the specimens were bent between fixed support points (Figs. 12 and 13). Later, spring-loaded jigs were used [59,60], or a fixture like the Alcoa Stressing Ring (Fig. 14) was used. Data of the type shown in Figs. 15 and 16 representing the unstressed versus the stressed exposure situation were obtained by Minford using the Alcoa fixture in 1972 [7]. Many investigators, like those at Minnesota Mining and Manufacturing (3M) Company, have used a spring-loaded fixture (Fig. 17) to develop stress and environmental data.

All the above comments have referred to the use of a steady stressing situation which can produce durability definitions for different combinations of adherend, adhesive, and surface treatment. However, there are many structural situations where a steady load application does not represent the service condition. Certainly this is the case with aircraft and most vehicle structures in which loads are applied and relaxed at intervals. There has been, therefore, a trend in

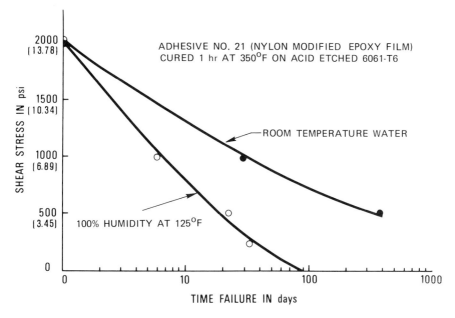

Fig. 18. Stressed specimens. (Courtesy of Alcoa, Alcoa Center, Pennsylvania.)

the durability testing of aerospace bonds toward testing procedures in which the load is repetitiously applied and relaxed. This cyclic stressing, in turn, should be applied under environmental conditions consistent with weathering factors in the service environment. Some of the most comprehensive studies in this regard, undertaken by Frazier and Lajoie at Bell Helicopter, have been summarized in an Air Force Materials Laboratory Technical Report [62].

The bond permanence is obviously longer at the same stressing level when the specimen is simply soaked in room-temperature water than when it is exposed to condensing humidity at 51.7°C (125°F) (Fig. 18). In one case the bond survival time at 3.45 MPa (500 psi) (8% of initial strength) was approximately 300 days, which was 15 times as long a survival time as at the same stress in 100% rh, at 51.7°C (125°F). Other Alcoa tests show that the bond permanence of joints under stress in a natural, industrial, atmospheric weathering condition is significantly longer than it is at the same stress level in a water-soaking environment. Mention should also be made of the work of McAbee and Levi [62]; Levi et al. [64], and Wegman et al. [65] in developing a technique for assessing the durability of structural adhesives involving the soaking of bonded joints in hot water under stressing conditions.

XIV. EFFECT OF SURFACE CONTAMINATION

Historically, it has been mandatory to clean metal surfaces chemically
if optimum bond durability is desired. More recently, it has been
recognized that the buildup of special oxide surfaces for bonding
after chemical deoxidizing of the surface, as in anodizing, generates
the highest order of joint permanence [10,66—68]. In contrast is the
situation in the automotive industry where it is common to attempt
bonding over lubricant- or oil-contaminated metal surfaces. Minford
and Vader [69] reported some of the earliest durability testing using
deliberately oiled adherends in 1974. The adhesive manufacturers
have attempted to modify their product to permit higher toleration of
oil contamination, but the durability of oil-contaminated joints obviously suffers in comparison to the durability of joints formed by aerospace bonding procedures. What the best compromise may be between
lower manufacturing costs and affordable bond durability in metal-part
forming remains to be determined based on exposure and service-type
testing. Considerable data are being developed at Alcoa to establish
which adhesive families offer the most realistic compromise of acceptable bond durability and low cost in the presence of practical levels
of lubricant. There seems to be a basis for believing that heat-cured
adhesives can generate more durable joints with oiled adherends than
room-temperature—cured adhesives and that the presence of oil on the
surface in the vicinity of the bondline can have a positive effect on
bond durability where corrosive saltwater service conditions exist
[70].

XV. DURABILITY OF COMBINATION-TYPE JOINTS

Considerable attention has been directed in recent years to combining
the properties of adhesive bonding with spot welding, riveting, or
metal clinching to form joints referred to as weldbonds, rivetbonds,
or clinchbonds. In each instance, the adhesive serves to increase the
interfacial area of the joint, resulting in significant improvements in
joint strength and fatigue resistance. In 1969, a Lockheed-Georgia
publication [71] discussed their earliest conclusions about the possible
advantages of weldbonds, focusing on comparative, sonic-fatigue endurance data. In 1971 another Lockheed-Georgia report described
the evaluation of a variety of surface pretreatments, primers, adhesives, and processing variables [72].

Considerable interest in weldbond joints has spread among truck
and trailer manufacturers, leading to the weldbond and adhesive joint
durability comparisons by Hall [73] in 1974. Minford et al. extended
the investigations to include aluminum, automotive-body-sheet alloys
in 1975 [74]. They showed a distinctly positive effect on both initial
strength and durability in weathering exposures for weldbond joints

Table 4 Effect of Weldbond on Vinyl Plastisol Heat-Cured Joint Durability in High-Humidity Exposures[a]

Type joint	Average initial bond strength (lb)		Overall strength retention of metal joints (%)			
			2036-T4 alloy		X5058-H111 alloy	
	2036-T4	X5085-H111	85% rh at 24°C (75°F)	100% rh at 51.7°C (125°F)	85% rh at 24°C (75°F)	100% rh at 51.7°C (125°F)
Adhesive	910 (409.5)	860 (387.0)	76	70	78	70
Weldbond	1450 (652.5)	1510 (679.5)	99	97	89	93

[a]Exposure period was 3 months.

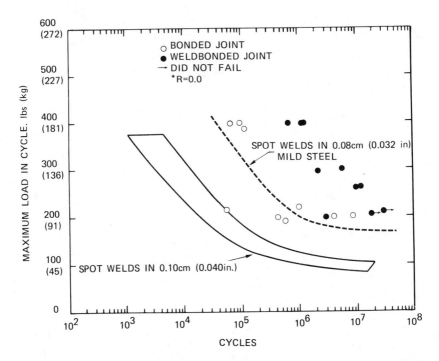

Fig. 19. Fatigue strength of lap joints in 0.10-cm (0.040-in.) 2036-T4 sheet room-temperature—curing polysulfide-modified epoxy. (From Ref. 8.)

using modest-strength adhesives like the vinyl plastisols and polysulfide-modified epoxy adhesives (Table 4). Similar effects have been shown for rivetbonds and clinchbonds [75]. Two significant property improvements can be found for weldbond joints as compared to spot welded or adhesively bonded joints. The improved fatigue strength is shown in Fig. 19. More details of this type of data can be cited [74,76]. Finally, bond permanence data are now being generated for dissimilar material joints [77,78].

XVI. SUMMARY

The predictable bond strength and potential service life of adhesively bonded joints, based on adhesion factors alone, should be much higher and longer, respectively, than the measurements obtained in the real world. This is better understood when it is recognized that the surface of the materials to which the organic polymer adhesives must

adhere is entirely different from the idealized surface used for theoretical calculations of the interacting forces between individual or collected pools of atoms, dipoles, and ions. Consequently, the best
bonding practice has been to generate the cleanest and most highly
reactive surface on aluminum that can be practically established and
accept the bond permanence that results from that practice. More
recently, however, the importance on bond permanence of deoxidizing
an aluminum surface and, subsequently, building up a new surface of
specific chemical composition by a process such as phosphoric acid
anodizing has been recognized. Also, testing methods, such as the
wedge-type crack propagation test, have been developed to distinguish the relative bond-durability potentials of different surface-
treating conditions. The importance of including simultaneous stressing and environmental exposure in assaying the potential durability
of bonded aluminum joints has been widely demonstrated, and a
variety of stressing fixtures have been mentioned in the literature.
It still remains to compare the results obtained from many of the deliberately accelerated durability tests with the existing results from
long-term durability tests in order to establish the confidence level of
these accelerated techniques.

REFERENCES

1. N. A. DeBruyne, The extent of contact between glue and adherend, in *Bull. 168* Aero Res., Cambridge, England, December 1956.
2. W. C. Bascom, The origin and removal of microvoids in fila-
   ment wound composites, in *NRL Rep. 6268*, 24 May 1965.
3. M. H. Loew, J. M. Fitzgerald, R. K. Elsley, and G. A.
   Alers, Exploratory development of adhesive bond flaw detection, in *AFML-TR-78-206*, December 1978.
4. E. P. Plueddemann, Adhesion through silane coupling agents,
   at *25th SPI Reinforced Plastics Composites Division Conf.*,
   Washington, D.C., February 1970.
5. W. D. Sell, Some analytical techniques for durability testing of
   structural adhesives, at *19th Natl. SAMPE Symp.*, April 1974.
6. P. E. Cassidy and B. J. Yager, A review of coupling agents as
   adhesion promoters, *NASA Contract NASA-24073*, September,
   1969.
7. J. D. Minford, *Metals Eng. Quart. 12(4)*:48 (1972).
8. J. D. Minford, Durability of adhesive bonded aluminum joints,
   in *Treatise of Adhesion and Adhesives*, Vol. 3 (R. Patrick,
   ed.), Dekker, New York, 1973.
9. J. D. Minford, *Adhesives Age 17*:24 (1974).
10. J. D. Minford, Comparison of aluminum adhesive joint durability
    as influenced by etching and anodizing pretreatments of bonded

surfaces, in *J. Appl. Polym. Symp. 32*:91 (1977).

11. N. J. DeLollis, *Appl. Polymer Sci. June* (1967).

12. N. J. DeLollis, *Adhesives Age 11*:21 *and 12*:25 (1968) and (1969).

13. J. A. Laird and F. W. Nelson, *SPE Trans.* 4:120 (1964).

14. P. W. Erickson, A. Volpe, and E. R. Cooper, *Mod. Plast. 41* 141 (1964).

15. J. Outwater and D. Kellogg, A simple experiment to show the origin of water debonding of resin against glass, in *Rep. for Contract V-3219(01) (X)*, 15 September 1961.

16. W. A. Zisman, Surface chemistry of glass-fiber reinforced plastics, in *NRL Rep. 6083, Symp. on Glass-Resin Interface of SPI Exhibit and Conf.*, Chicago, 6 February 1964.

17. J. J. Bikerman, in *The Science of Adhesive Joints*, Academic, New York, 1961.

18. J. J. Bikerman, in *Recent Advances in Adhesion* (Lieng-Huang Lee, ed.), Gordon, New York, 1973.

19. L. H. Sharpe and H. Schonhorn, *Adv. Chem. Ser. 43*:189 (1964).

20. R. J. Good, in *Recent Advances in Adhesion* (Lieng-Huang Lee, ed.), Gordon, New York, 1973, p. 357.

21. S. Sterman, and J. B. Toogood, *Adhesives Age 8*:34 (1965).

22. J. A. Laird, Glass surface chemistry for glass reinforced plastics, *Navy Contract W-0679-C (FBM)*, June 1963.

23. D. J. Zalucha, *Adhesives Age 22*:21 (1979).

24. C. L. Brett, *J. Appl. Sci. 18*:315 (1974).

25. W. L. Baun, N. T. McDevitt, and J. S. Solomon, Chemistry of metal and alloy adherends by secondary ion mass spectroscopy, ion scattering spectroscopy, and auger electron spectroscopy, in *ASTM STP 596*, March 1975.

26. N. T. McDevitt and W. L. Baun, Some observations of the relation between chemical surface treatments and the growth of anodic barrier layer films, in *AFML-TR-76-74*, June 1976.

27. N. T. McDevitt, W. L. Baun, G. Fugate, and J. S. Solomon, Surface studies of anodic aluminum oxide layers formed in phosphoric acid solutions, in *AFML-TR-77-55*, May 1977.

28. N. T. McDevitt, W. L. Braun, G. Fugate, and J. S. Solomon, Accelerated corrosion of adhesively bonded 7075 aluminum using wedge crack specimens, in *AFML-TR-77-184*, October 1977.

29. W. L. Baun, N. T. McDevitt, and J. S. Solomon, Pitting corrosion and surface chemical properties of a thin oxide layer on anodized aluminum, in *AFML-TR-78-128*, September 1978.

30. A. Benninghoven, *Surface Sci. 28*:541 (1971).

31. R. Schubert and J. C. Tracy, *Rev. of Sci. Instr. 44*:487 (1973).

32. R. Schubert, *J. Vacuum Sci. Tech.* *12*(1):505 (1975).
33. R. Werner, *Dev. Appl. Spectroscopy* *7A*:297 (1969).
34. H. E. Blair, S. Matsuoka, R. G. Vadimsky, and T. T. Wange, *J. Adhesion* *3*:89 (1971).
35. J. C. Bolger, The chemical composition of metal and oxide surfaces and how these interact with polymeric materials, in *30th Ann. Tech. Conf. SPE*, Chicago, 1972.
36. K. Wefers and G. M. Bell, Oxides and hydroxides of aluminum, in *Tech. Paper 19*, Alcoa Res. Lab., Alcoa Center, 1972.
37. B. B. Bowen, *SAMPE Materials Rev.* *7*:374 (1975).
38. T. P. Remmel, Characteristics of surfaces prior to adhesive bonding, in *AFML-TR-76-118*, July 1976.
39. D. J. Falconer, N. C. MacDonald, and P. Walker, *Chem. Ind.* *27*: 1230 (1964).
40. C. Kerr, N. C. MacDonald, and S. Orman, *J. Appl. Chem.* *17*: 62 (1967).
41. J. A. Scott, Durability of bonded joints, in *ASTM STP 401*, 1966.
42. S. Orman and C. Kerr, The effect of water on aluminum epoxide bonds, in *Aspects of Adhesion*, No. 6 (D. Alner, ed.), Univ. London, 1971, p. 64.
43. J. D. Minford, Comparative effect of surface contamination on the strengths and performance of aluminum spot-welded or adhesive-bonded joints, in *ASM/ADDRG Conf.—Tech. Impact of Surfaces: Relationship to Forming, Welding and Painting*, Dearborn, April 1981.
44. J. D. Minford, SAE Paper 810816, Dearborn, 8–12 June 1981.
45. R. J. Riel, Corrosion of adhesive bonded aluminum alloy panels, in *Rohr Rep. 24-2047*, August 1968.
46. R. J. Riel, *SAMPE J.* *7*(5) (1971).
47. R. H. Greer, Corrosion resistant adhesive bonding, in *Nat. SAMPE Tech. Conf. Ser.*, Vol. 2, 1970, p. 561.
48. N. L. Rogers, Corrosion of adhesive bonded clad aluminum, at *SAE Nat. Business Aircraft Meeting 720,344*, Wichita, 1972.
49. N. L. Rogers, A comparative test for bondline corrosion: clad versus bare aluminum alloys, in *Proc. 5th Nat. Tech. Conf.*, SAMPE, Kiemesha Lake, New York, 1973.
50. J. C. McMillan, Surface preparation—the key ot bondment durability, *AGARD Lecture Series 102*, October 1979.
51. W. T. McFarlen, Environmental testing of adhesive bonded metallic structures, in *Composite Materials in Engineering Design* (B. Noton, ed.), ASM, 1973, p. 678.
52. J. M. Black and R. B. Blomquist, *Ind. Eng. Chem.* *50*:918 (1958).

53. H. W. Eickner, Weathering of adhesive bonded lap joints of
     clad aluminum alloys, in *WADC Tech. Rep. 54-447*, Part 1,
     Forest Products Lab, Madison, Wisconsin, 1955.
54. C. I. Hause, W. C. Pagel, and A. G. McKown, The effects of
     South Florida weather aging on structural adhesive joints, in
     3M Tech. Rep., 3M Company, St. Paul, Minnesota, October 1965.
55. J. D. Minford, *Int. Adhesion Adhesives January* (1982).
56. E. J. Ripling, S. Mostovoy, and R. L. Patrick, Application of
     fracture mechanics to adhesive joints, in *Adhesion*, ASTM,
     1964.
57. A. W. Bethune, *SAMPE J. 11(3)*:4 (1975).
58. J. A. Marceau, Y. Moji, and J. C. McMillan, A wedge test for
     evaluating adhesive bonded surface durability, *Adhesives
     Age 20*:28 (1977).
59. L. H. Sharpe, Some aspects of the permanence of adhesive
     joints, at *Symp. Structural Adhesives Bonding*, Stevens Inst.
     Tech., Hoboken, 16 September 1965.
60. L. H. Sharpe, *Appl. Polymer Symp. 3*:353 (1966).
61. G. F. Carter, *Adhesives Age 10*:32 (1967).
62. T. B. Frazier and A. D. Lajoie, Durability of adhesive bonded
     joints, in *AFML-TR-74-26*, Bell Helicopter (March 1974).
63. E. McAbee and D. W. Levi, Use of a reaction rate method to
     predict failure times of adhesive bonds at constant stress, in
     Tech. Rep. 4105 (December 1970) Picatinny Arsenal.
64. D. W. Levi, R. F. Wegman, M. C. Ross, and E. A. Garnis,
     SAMPE Quart. 7(3):1 (1976).
65. R. F. Wegman, M. J. Bodnar, and M. C. Ross, *SAMPE J.
     14(1)*:20 (1978).
66. J. C. McMillan, J. T. Quinlan, and R.A. Daves, *SAMPE
     Quart. 7(3)*:13 (1976).
67. J. A. Marceau and Y. Moji, Development of environmental
     stable aluminum adhesive bonds, in *Doc. D6-41145*, Boeing
     Company, Seattle, Washington.
68. Common bonding requirements for structural adhesives, in
     Doc. BAC 5514, Boeing Company, Seattle, Washington.
69. J. D. Minford and E. M. Vader, SAE Paper 740,078, SAE,
     Detroit, February 1974.
70. J. D. Minford, Unpublished data, Alcoa Lab., Alcoa Center,
     Pennsylvania.
71. *Iron Age April*:104 (1969).
72. Weldbond flight component design/manufacturing program in
     First Quarterly Contract F33615-72-C-1716, Lockheed-Georgia,
     October 1971.
73. R. C. Hall, *SAMPE J. 10*:10 (1974).
74. J. D. Minford, F. R. Hoch, and E. M. Vader, SAE Paper
     750,462, SAE, Detroit, February 1975.

75. J. D. Minford, Unpublished data, Alcoa Lab., Alcoa Center, Pennsylvania.
76. G. E. Nordmark, SAE Paper 780,397, at SAE Exposition, February-March 1978.
77. J. D. Minford, Adhesives joining aluminum to engineering plastics, I. polyester fiberglass composite, *Physicochem. Aspects of Polymer Surfaces*, Vol. 2 (K. L. Mittal, ed.), Plenum, New York, 1983.
78. J. D. Minford, Adhesive joining aluminum to engineering plastics, II. engineering grade styrene and cross-linked styrene, *Physicochem. Aspects of Polymer Surfaces*, Vol. 2 (K. L. Mittal, ed.), Plenum, New York, 1983.

24
Mathematical Modeling and Statistical Inference

Dale Meinhold *GMI Engineering and Management Institute, Flint, Michigan*

I. INTRODUCTION

The engineer is frequently involved in activities involving testing and measuring. The test results are recorded, and the analyst faces the task of extracting information from these records. There is an almost endless variety of statistical procedures one can apply to aid in this extraction process. The selection of the appropriate procedures depends on what kind of questions the analyst has. For example, estimating the average bonding strength of a new adhesive requires a technique different from that required to test the validity of a manufacturer's claim concerning bonding strength. Many commonly asked questions deal with the problem of sample size. How many replicates must be run in order to obtain an average value which is "correct" to within ±10%? How narrow must the range of sample values be before adding another replicate becomes unimportant?

Fortunately, the most frequently asked questions can be answered on the basis of a few rather straightforward statistical concepts and procedures. Experience indicates that even though these basic procedures are widely used, considerable confusion exists as to the interpretation of the information provided by these tests. I have also noted that a certain amount of vagueness exists concerning the precise meaning of some of the terms used in statistical jargon.

One of the objectives of this chapter is to present and illustrate a few common statistical procedures. At the same time I wish to provide the reader with a review of some of the fundamental concepts in statistical inference and the correct meaning of some commonly used terms. I intend to focus less on the mathematical justification of the procedures than on the correct interpretation of the test results.

II. SAMPLE VERSUS POPULATION

Samuel Johnson once said, "You don't have to eat the whole ox to know that the meat is tough." The essence of sampling is to study the part in order to gain information about the whole. Statisticians use the term *population* to refer to the whole and *sample*, or random sample, to refer to the part.

For example, suppose a chemical corporation has developed a new adhesive suitable for pressure-sensitive tape. In particular, the corporation might want to estimate the tack strength of the tape. To make such an estimate, the laboratory would devise a procedure to measure the tack strength of a piece of tape and then repeat the procedure many times. The set of measurements thus obtained constitutes a *sample* from the *population* of all tack strengths for this particular adhesive tape measured under the same conditions.

Proper sampling techniques are essential. Well-conceived and carefully carried out sampling procedures should lead to a set of measurements which can be thought of as a miniature replica of the population itself. Common sense dictates that all sample measurements be obtained under identical conditions. The surface should be homogeneous and uniformly cleaned following each measurement; exactly the same pressure should be applied; and a consistent method to pull the tape should be devised. Failure to follow such guidelines can cause sufficient errors in measurement to lead to erroneous conclusions. This is known as experimental error.

In order to generate some data to use for purposes of illustration, a mechanical spring-loaded device called a tackmeter was used to separate $\frac{1}{2}$-in. masking tape from a glass surface. Following a few dozen practice runs to establish proper technique, the following set of 80 measurements of the force in grams required to separate the tape from the glass was recorded as shown in Table 1.

The sample itself is of interest only in that it can provide information about the population distribution. To begin with, we would like to know the average value of all the items in the population. This quantity is called the *population mean* μ. Our best estimate of μ is the average, or mean, of all the items in the sample. For a sample of size n we define the *sample mean* $\bar{x}$ as follows:

$$\bar{x} = \frac{1}{n}(x_1 + x_2 + \cdots + x_n) = \frac{1}{n}\sum_{i=1}^{n} x_i$$

Table 1 Sample of 80 Tack Strengths

470	480	495	505	505	505
510	515	515	520	520	525
525	530	530	530	535	535
540	540	545	545	545	545
550	550	550	550	550	555
555	555	555	560	560	560
565	565	565	565	565	570
570	570	570	570	570	575
575	575	575	580	580	580
585	585	585	585	590	590
590	595	595	600	600	605
605	605	610	615	615	620
620	620	620	625	630	630
630	650				

Specifically, for our sample of size 80 we calculate (see Table 2)

$$\bar{x} = \frac{1}{80}(470 + 480 + 495 + 505 + \cdots + 650) = 565.56$$

A second desirable piece of information is a measure of the variability among items in the population. The most commonly used measure is the average of the squared deviations of items from the population mean. This quantity is called the *population variance* σ^2. Its square root is called the *population standard deviation* σ. Our best estimate of this quantity is obtained from sample data and is called the *sample variance* s^2. Its square root is called the *sample standard deviation* s. In general, we define

$$s^2 = \frac{1}{n - 1}[(x_1 - \bar{x})^2 + (x_2 - \bar{x})^2 + \cdots + (x_n - \bar{x})^2]$$

$$= \frac{1}{n - 1}\sum_{i = 1}^{n}(x_i - \bar{x})^2$$

The details of the calculations for calculating s^2 for our sample are indicated in Table 2.

For some types of analysis the most important question of all concerns the range of values in the population. We can easily see that our sample ranges from 470 to 650, but surely there are population values outside this interval. Specifically, we might ask, "What proportion of the population has values in excess of 650? less than 500? between 545 and 585?" Or, "The weakest 1% of the tape has tack strength less than what number?" To answer these questions we must

Table 2   Calculations for Sample Mean and Sample Variance

1	470	470 - 565.56 = - 95.56	9131.72
2	480	480 - 565.56 = - 85.56	7320.51
3	495	495 - 565.56 = - 70.56	4978.71
•	•	• • •	•
•	•	• • •	•
•	•	• • •	•
79	630	630 - 565.56 = 64.44	4151.51
80	650	650 - 565.56 = 84.44	7130.11
Totals	45,245	0	113,500.34

$$\overline{x} = (1/n) \sum x = (1/80) \, (45,245) = 565.56$$

$$s^2 = [1/(n-1)] \sum (x - \overline{x})^2 = (1/79) \, (113,500.34) = 14326.71$$

$$s = \sqrt{1436.71} = 37.904$$

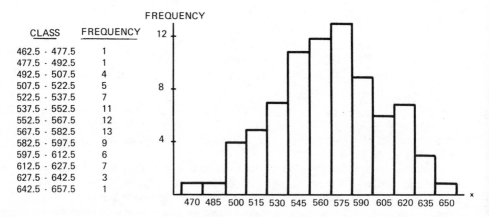

CLASS	FREQUENCY
462.5 - 477.5	1
477.5 - 492.5	1
492.5 - 507.5	4
507.5 - 522.5	5
522.5 - 537.5	7
537.5 - 552.5	11
552.5 - 567.5	12
567.5 - 582.5	13
582.5 - 597.5	9
597.5 - 612.5	6
612.5 - 627.5	7
627.5 - 642.5	3
642.5 - 657.5	1

Fig. 1.   Distribution of sample of 80 tack strengths.

look beyond the simple estimates of the population mean μ and the
population variance σ^2. We need some indication of what the population looks like. Toward this end, we organize the sample of 80 tack
strengths into a *frequency distribution* illustrated in Fig. 1. A pic-
torial representation of the frequency distribution is called a *histogram*.

If the techniques used to gather data are relatively free from
experimental error, we can be assured that the sample is reasonably
representative of the population from which it was drawn. In fact, if
many samples, each of size 80, were selected from the same population, we would expect their histograms all to exhibit the same general
shape. The smooth curve superimposed on our histogram represents
this "general shape," and we refer to the curve as the *population
distribution* (Fig. 2).

Over the years numerous curves have been developed to serve
as models for population distributions. These models are constructed
in such a way that the area under the curve can be equated with probability. For instance, the probability that a randomly selected population item will have a value between x_1 and x_2 is simply equal to the
area under the curve between x_1 and x_2. Of course, the total area
under such a curve is 1. It is for this reason that these models are
called *probability distributions*. The actual equation of the curve defines what is referred to as a *probability density function*.

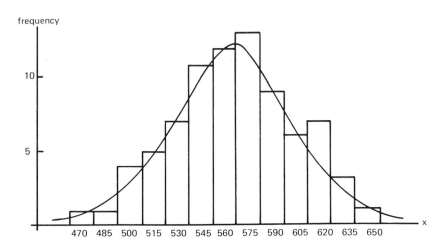

Fig. 2. Hypothesized distribution for population of tack strengths.

III. POPULATION DISTRIBUTIONS

Before continuing, it may be necessary to illustrate the meaning of
two of the mathematical terms which we will be using. We define a
family of curves as a set of curves, all of which are defined by the
same basic equation. For instance, the single equation y = mx + b
defines a whole family of straight lines. In this case the slope m and
the y-intercept b are referred to as *parameters*. Different choices for
m and b produce different lines, but all are members of the same
family.

The models which have been developed for population distributions can be categorized into families. Each family has been designed
to fit some particular type of data. For example, reliability engineers
often set up tests whereby a number of bonded assemblies are used for
a given period of time or until a certain proportion of the bonds fail.
This "time to failure" type of data is known to be modeled quite satisfactorily by the Weibull family of probability distributions. Its pro-
bability density functions is $y = (\alpha\beta)t^{\beta-1} \exp(-\alpha t^{\beta})$ where β is the
shape parameter and α is a scale parameter.

A common type of distribution is one in which one "tail" is
stretched out further than the other. For example, consider the distribution of annual incomes in America (Fig. 3). We would describe
this distribution as being *skewed* to the right. The *Pearson* family of
distributions is often used to model populations such as this, which
lack symmetry.

Probably the most widely used of all models is the family of *normal* distributions. Measurements whose variability can be attributed
to a large number of chance factors can be modeled quite accurately
with a normal distribution. These measurements include such things
as lengths of "4-in." bolts, weights of "25-lb" bags of dog food, the
number of amps required to blow a "25-amp" fuse, and so on. It
would be unreasonable to expect that our population of tack strengths
might not be modeled successfully by a normal distribution. The differ-

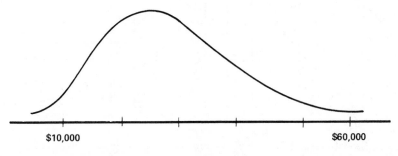

$10,000 $60,000

Fig. 3. Distribution of annual incomes.

ences in the individual sample measurements might be due to a variety of chance factors. As careful as the analyst might have been, there were probably slight differences in the times between fixing and pulling the tape, small variations in the pressure used to fix the tape to the surface, and some unavoidable inconsistencies in reading the tackmeter. The surface itself possibly contained small irregularities. Even changes in temperature and humidity could conceivably have contributed to the variability in measurements. The random errors arising from these and many other sources combine to produce the bell-shaped curve which is characteristic of the normal distribution. The effects of many of the random errors tend to cancel each other, explaining why the bulk of the measurements are clistered about the mean. Less often, the errors may accumulate in one direction or the other producing measurements comprising the tails of the distribution.

In addition to serving as a model for population distributions, the normal model plays a key role in statistical inference, so it is worthwhile to study it in some detail. The probability density function for the family of normal curves is defined as

$$y = \frac{1}{\sigma\sqrt{2\pi}} e^{-(1/2)\left(\frac{x-\mu}{\sigma}\right)^2} \quad -\infty < x < +\infty$$

The two parameters μ and σ in this theoretical distribution correspond exactly to the mean μ and the standard deviation σ of the physical population it is intended to model. Methods of calculus used on the above density function show

1. The curve achieves its maximum value of $1/(\sigma\sqrt{2\pi})$ at the point where $x = \mu$. Note that this maximum height depends only on the value for σ.

2. The curve is symmetric about the line $x = \mu$.

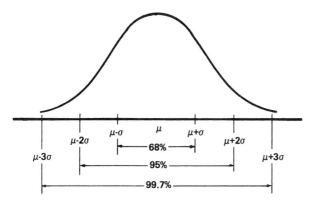

Fig. 4. Areas for the normal distribution.

3. Inflection points occur at $x = \mu \pm \sigma$, that is, the curvature changes at the points located one standard deviation from the mean.
4. The area under the curve between points 1, 2, and 3 σ's from the center is indicated in Fig. 4.

To illustrate the above notions and to outline the methods used to compute various probabilities, let us suppose we have a population of peel strengths x, normally distributed with mean $\mu = 50$ g and standard deviation $\sigma = 4$ g. (This population, of course, is purely hypothetical. In "real world" situations we can only estimate these values from sample data.) The appropriate normal model would be obtained by substituting $\mu = 50$ and $\sigma = 4$ into the general probability density function [Eq. (3)], yielding the normal model shown in Fig. 5. The equation of the curve is $y = (1/4\sqrt{2\pi}) \exp \{-\frac{1}{2}[(x-50)/4]^2\}$.

IV. PREDICTIONS

Suppose we wish to compute the probability that an item chosen at random from the population would have a value less than, say, 45.2. This would be equivalent to computing the proportion of the population having values less than 45.2. Consistent with the fact that probability equals area, we seek the area under the curve to the left of $x = 45.2$. The procedure for finding this area involves *standardizing* the variable x. Standardization of a variable is accomplished by subtracting the mean μ and dividing by the standard deviation σ. The resulting variable z is referred to as the *standard normal* variable. For our example,

$$z = \frac{x - \mu}{\sigma} = \frac{x - 50}{4}$$

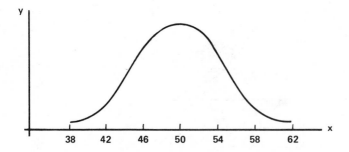

Fig. 5. Normal distribution of peel strengths.

It is important to note that this transformation produces a variable whose units are in terms of standard deviations. Integration techniques from the calculus show that areas under the x curve correspond exactly to areas under the z curve, which can be found by referring to a table such as Table 3.

To illustrate the procedure, let us compute the probability that a randomly selected value from our population of peel strengths will be less than 45.2.

$$z = \frac{x - 50}{4} = \frac{45.2 - 50}{4} = -1.2$$

It is helpful and quite meaningful to note that this transformation, in effect, tells us that x = 45.2 is located 1.2 standard deviations to the left of the center value of μ = 50. Referring to Table 3 we note that the area to the left of z = − 1.2 is 0.1151. Using probability statements we can write

$$P(x < 45.2) = P(z < -1.2) = 0.1151$$

Thus we can say that 11.51% of our population of peel strengths has a value of less than 45.2.

Another problem might be to locate the particular peel strength that is exceeded by only 5% of the population. Statisticians often refer to this point as the 95th percentile. From our z table we note that the area to the left of z = 1.645 is 0.95, that is, 95% of the area under the normal curve is to the left of the point 1.645 standard deviations to the right of the mean. We convert this information to peel strengths (see Fig. 6) as follows:

$$z = \frac{x - 50}{4} = 1.645$$

$$x = 4(1.645) + 50$$

$$= 56.58$$

Thus, we can say that only 5% of all peel strengths should exceed 56.58 g.

Let us return to our problem of selecting and utilizing an appropriate mathematical model for our distribution of tack strengths for adhesive tape. The choice of one of the family of normal distributions to model our population is quite reasonable in view of the nature of the data as well as the shape of our histogram for our sample of size 80. The particular model that best fits our population is, of course,

Table 3 Area Under the Standard Normal Curve

z	0.00	0.01	0.02	0.03	0.04	0.05	0.06	0.07	0.08	0.09
-3.4	0.0003	0.0003	0.0003	0.0003	0.0003	0.0003	0.0003	0.0003	0.0003	0.0002
-3.3	0.0005	0.0005	0.0005	0.0004	0.0004	0.0004	0.0004	0.0004	0.0003	0.0003
-3.2	0.0007	0.0007	0.0006	0.0006	0.0006	0.0006	0.0006	0.0005	0.0005	0.0005
-3.1	0.0010	0.0009	0.0009	0.0009	0.0008	0.0008	0.0008	0.0008	0.0007	0.0007
-3.0	0.0013	0.0013	0.0013	0.0012	0.0012	0.0011	0.0011	0.0011	0.0010	0.0010
-2.9	0.0019	0.0018	0.0017	0.0017	0.0016	0.0015	0.0015	0.0015	0.0014	0.0014
-2.8	0.0026	0.0025	0.0024	0.0023	0.0023	0.0022	0.0021	0.0021	0.0020	0.0019
-2.7	0.0035	0.0034	0.0033	0.0032	0.0031	0.0030	0.0029	0.0028	0.0027	0.0026
-2.6	0.0047	0.0045	0.0044	0.0043	0.0041	0.0040	0.0039	0.0038	0.0037	0.0036
-2.5	0.0062	0.0060	0.0059	0.0057	0.0055	0.0054	0.0052	0.0051	0.0049	0.0048
-2.4	0.0082	0.0080	0.0078	0.0075	0.0073	0.0071	0.0069	0.0068	0.0066	0.0064
-2.3	0.0107	0.0104	0.0102	0.0099	0.0096	0.0094	0.0091	0.0089	0.0087	0.0084
-2.2	0.0139	0.0136	0.0132	0.0129	0.0125	0.0122	0.0119	0.0116	0.0113	0.0110
-2.1	0.0179	0.0174	0.0170	0.0166	0.0162	0.0158	0.0154	0.0150	0.0146	0.0143
-2.0	0.0228	0.0222	0.0217	0.0212	0.0207	0.0202	0.0197	0.0192	0.0188	0.0183
-1.9	0.0287	0.0281	0.0274	0.0268	0.0262	0.0256	0.0250	0.0244	0.0239	0.0233
-1.8	0.0359	0.0352	0.0344	0.0336	0.0329	0.0322	0.0314	0.0307	0.0301	0.0294
-1.7	0.0446	0.0436	0.0427	0.0418	0.0409	0.0401	0.0392	0.0384	0.0375	0.0367
-1.6	0.0548	0.0537	0.0526	0.0516	0.0505	0.0495	0.0485	0.0475	0.0465	0.0455
-1.5	0.0668	0.0655	0.0643	0.0630	0.0618	0.0606	0.0594	0.0582	0.0571	0.0559
-1.4	0.0808	0.0793	0.0778	0.0764	0.0749	0.0735	0.0722	0.0708	0.0694	0.0681
-1.3	0.0968	0.0951	0.0934	0.0918	0.0901	0.0885	0.0869	0.0853	0.0838	0.0823
-1.2	0.1151	0.1131	0.1112	0.1093	0.1075	0.1056	0.1038	0.1020	0.1003	0.0985
-1.1	0.1357	0.1335	0.1314	0.1292	0.1271	0.1251	0.1230	0.1210	0.1190	0.1170
-1.0	0.1587	0.1562	0.1539	0.1515	0.1492	0.1469	0.1446	0.1423	0.1401	0.1379

z	.00	.01	.02	.03	.04	.05	.06	.07	.08	.09
−0.9	0.1841	0.1814	0.1788	0.1762	0.1736	0.1711	0.1685	0.1660	0.1635	0.1611
−0.8	0.2119	0.2090	0.2061	0.2033	0.2005	0.1977	0.1949	0.1922	0.1894	0.1867
−0.7	0.2420	0.2389	0.2358	0.2327	0.2296	0.2266	0.2236	0.2206	0.2177	0.2148
−0.6	0.2743	0.2709	0.2676	0.2643	0.2611	0.2578	0.2546	0.2514	0.2483	0.2451
−0.5	0.3085	0.3050	0.3015	0.2981	0.2946	0.2912	0.2877	0.2843	0.2810	0.2776
−0.4	0.3446	0.3409	0.3372	0.3336	0.3300	0.3264	0.3228	0.3192	0.3156	0.3121
−0.3	0.3821	0.3783	0.3745	0.3707	0.3669	0.3632	0.3594	0.3557	0.3520	0.3483
−0.2	0.4207	0.4168	0.4129	0.4090	0.4052	0.4013	0.3974	0.3936	0.3897	0.3859
−0.1	0.4602	0.4562	0.4522	0.4483	0.4443	0.4404	0.4364	0.4325	0.4286	0.4247
−0.0	0.5000	0.4960	0.4920	0.4880	0.4840	0.4801	0.4761	0.4721	0.4681	0.4641
0.0	0.5000	0.5040	0.5080	0.5120	0.5160	0.5199	0.5239	0.5279	0.5319	0.5359
0.1	0.5398	0.5438	0.5478	0.5517	0.5557	0.5596	0.5636	0.5675	0.5714	0.5753
0.2	0.5793	0.5832	0.5871	0.5910	0.5948	0.5987	0.6026	0.6064	0.6103	0.6141
0.3	0.6179	0.6217	0.6255	0.6293	0.6331	0.6368	0.6406	0.6443	0.6480	0.6517
0.4	0.6554	0.6591	0.6628	0.6664	0.6700	0.6736	0.6772	0.6808	0.6844	0.6879
0.5	0.6915	0.6950	0.6985	0.7019	0.7054	0.7088	0.7123	0.7157	0.7190	0.7224
0.6	0.7257	0.7291	0.7324	0.7357	0.7389	0.7422	0.7454	0.7486	0.7517	0.7549
0.7	0.7580	0.7611	0.7642	0.7673	0.7704	0.7734	0.7764	0.7794	0.7823	0.7852
0.8	0.7881	0.7910	0.7939	0.7967	0.7995	0.8023	0.8051	0.8078	0.8106	0.8133
0.9	0.8159	0.8186	0.8212	0.8238	0.8264	0.8289	0.8315	0.8340	0.8365	0.8389
1.0	0.8413	0.8438	0.8461	0.8485	0.8508	0.8531	0.8554	0.8577	0.8599	0.8621
1.1	0.8643	0.8665	0.8686	0.8708	0.8729	0.8749	0.8770	0.8790	0.8810	0.8830
1.2	0.8849	0.8869	0.8888	0.8907	0.8925	0.8944	0.8962	0.8980	0.8997	0.9015
1.3	0.9032	0.9049	0.9066	0.9082	0.9099	0.9115	0.9131	0.9147	0.9162	0.9177
1.4	0.9192	0.9207	0.9222	0.9236	0.9251	0.9265	0.9278	0.9292	0.9306	0.9319
1.5	0.9332	0.9345	0.9357	0.9370	0.9382	0.9394	0.9406	0.9418	0.9429	0.9441
1.6	0.9452	0.9463	0.9474	0.9484	0.9495	0.9505	0.9515	0.9525	0.9535	0.9545
1.7	0.9554	0.9564	0.9573	0.9582	0.9591	0.9599	0.9608	0.9616	0.9625	0.9633
1.8	0.9641	0.9649	0.9656	0.9664	0.9671	0.9678	0.9686	0.9693	0.9699	0.9706
1.9	0.9713	0.9719	0.9726	0.9732	0.9738	0.9744	0.9750	0.9756	0.9761	0.9767

Table 3 (Continued)

z	0.00	0.01	0.02	0.03	0.04	0.05	0.06	0.07	0.08	0.09
2.0	0.9772	0.9778	0.9783	0.9788	0.9793	0.9798	0.9803	0.0208	0.9812	0.9817
2.1	0.9821	0.9826	0.9830	0.9834	0.9838	0.9842	0.9846	0.9850	0.9854	0.9857
2.2	0.9861	0.9864	0.9868	0.9871	0.9875	0.9878	0.9881	0.9884	0.9887	0.9890
2.3	0.9893	0.9896	0.9898	0.9901	0.9904	0.9906	0.9909	0.9911	0.9913	0.9916
2.4	0.9918	0.9920	0.9922	0.9925	0.9927	0.9929	0.9931	0.9932	0.9934	0.9936
2.5	0.9938	0.9940	0.9941	0.9943	0.9945	0.9946	0.9948	0.9949	0.9951	0.9952
2.6	0.9953	0.9955	0.9956	0.9957	0.9959	0.9960	0.9961	0.9962	0.9963	0.9964
2.7	0.9965	0.9966	0.9967	0.9968	0.9969	0.9970	0.9971	0.9972	0.9973	0.9974
2.8	0.9974	0.9975	0.9976	0.9977	0.9977	0.9978	0.9979	0.9979	0.9980	0.9981
2.9	0.9981	0.9982	0.9982	0.9983	0.9984	0.9984	0.9985	0.9985	0.9986	0.9986
3.0	0.9987	0.9987	0.9987	0.9988	0.9988	0.9989	0.9989	0.9989	0.9990	0.9990
3.1	0.9990	0.9991	0.9991	0.9991	0.9992	0.9992	0.9992	0.9992	0.9993	0.9993
3.2	0.9993	0.9993	0.9994	0.9994	0.9994	0.9994	0.9994	0.9995	0.9995	0.9995
3.3	0.9995	0.9995	0.9995	0.9996	0.9996	0.9996	0.9996	0.9996	0.9996	0.9997
3.4	0.9997	0.9997	0.9997	0.9997	0.9997	0.9997	0.9997	0.9997	0.9997	0.9998

Source: Ref. 1.

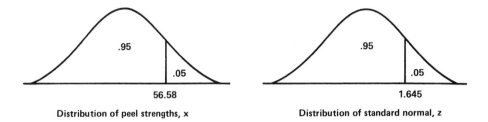

Distribution of peel strengths, x   Distribution of standard normal, z

Fig. 6. Distribution of peel strengths x and distribution of standard normal z.

the model whose parameters μ and σ agree exactly with the mean μ and standard deviation σ of our population of tack strengths. But these values are unknown! The best we can do is to estimate these values from our sample data. The question of how well the resulting model will fit our population is paralleled by the questions: How well does the sample mean $\bar{x}$ estimate the population mean μ, the center of the distribution? And how well does the sample standard deviation s estimate the population standard deviation σ, which determines the spead of the distribution? We will have more to say concerning the precision of these estimates later when we discuss estimates of *confidence intervals* for these population parameters.

Recall that for our sample of size n = 80 tack strengths, we calculated the sample mean $\bar{x}$ = 565.56 and the sample standard deviation s = 37.904. For now, let us assume that these numbers are fairly accurate estimates of μ and σ, respectively. (In view of our relatively large sample size, n = 80, this is not an unreasonable assumption.)

Note that exactly 3 of the 80 measurements in our sample, 3.75% of our *sample* had a value less than 500. Suppose we ask what proportion of our *population* would have a value less than 500. To answer this question, we transform x = 500 into units of standard deviation as follows:

$$z = \frac{500 - \mu}{\sigma} \backsim \frac{500 - \bar{x}}{s} = \frac{500 - 565.56}{37.904} = -1.73$$

So in our population model, x = 500 is a point 1.73 standard deviations to the left of center. Referring to our z Table 3 we see that the area to the left of z = − 1.73 is 0.0418. Hence, we can say that 4.18% of our *population* has a tack strength of 500 g or less.

Similarly, we could find the value exceeded by only 5% of the population by calculating

$$z = 1.645 = \frac{x - 565.56}{37.904}$$

$$x = (37.904)(1.645) + 565.56 = 628$$

Thus, only 5% of all tack strengths should exceed 628. (Coincidentally, exactly 5%, 4 out of 80, of our sample values exceeded 628.)

The above procedures must not be used indiscriminately! They are appropriate only when a population can reasonably be assumed to be normally distributed. Procedures for handling non-normal populations are beyond the scope of this paper.

V. ESTIMATION OF POPULATION MEAN

An issue which is very important, not only as it pertains to the above discussion, but also in its own right is the precision with which the sample mean $\bar{x}$ estimates the population mean μ. In our example, we used $\bar{x} = 565.56$ as our estimate of μ. It is almost certain, however, that this value is not exactly correct. In fact, if we were to select many, many samples, each of size 80, we could compute many, many sample means; each of these sample means could provide a different estimate of μ. Statisticians have discovered an amazing pattern for these estimates which, at least for large sample size, is almost independent of the nature of the population distribution. Their conclusions are stated in the following theorem:

> Central Limit Theorem. If many samples, each of size n, are randomly selected from a population with mean μ and variance σ^2, the distribution of these sample means can be modeled quite precisely by a normal distribution with mean μ and variance σ^2/n provided that n is sufficiently large.

The answer to the question, "How large is 'sufficiently large'?" depends on the shape of the distribution of the parent population, that is, the population being sampled. If the parent population is itself normally distributed (or nearly so), then the theorem is valid for any sample size whatsoever. If the parent population is symmetric about its mean, then samples of 5 or more will suffice. The more the parent population departs from symmetry, the larger the sample size must be. It is generally conceded that samples of size 30 or more will suffice even for skewed population distributions.

To illustrate the implications of the Central Limit Theorem, let us suppose a sample of size n (n sufficiently large to validate the Central Limit Theorem) is to be selected from some population with mean μ and variance σ^2. The knowledge that $\bar{x}$ can be modeled with a normal distribution with mean μ and variance σ^2/n enables us to make definite

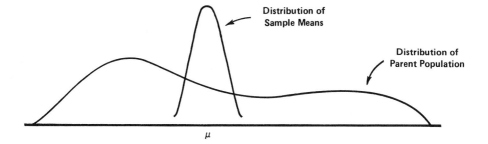

Fig. 7. Distribution of sample means and the population distribution.

statements as to the likelihood that $\bar{x}$ will "miss" the true population mean by some specified amount. For example, suppose that a population (not necessarily normal) has unknown mean μ but that somehow we do know the population variance σ^2 is 81. We intend to select a random sample of size 64 from the population and use the sample mean $\bar{x}$ to estimate the population mean μ. The Central Limit Theorem assures us that if many samples of size 64 were selected, their *means* would be approximately normally distributed about the population mean μ with standard deviation $\sqrt{\sigma^2/n} = \sqrt{81/64} = 1.125$. This is displayed graphically in Fig. 7.

VI. CONFIDENCE INTERVALS

We could, for instance, calculate the probability that a particular sample mean would be within (say) ± 2 of the population mean μ.

$$P(\mu - 2 < \bar{x} < \mu + 2)$$

$$= P\left(\frac{(\mu - 2) - \mu}{1.125} < \frac{\bar{x} - \mu}{1.125} < \frac{(\mu + 2) - \mu}{1.125}\right)$$

$$= P(-1.78 < z < 1.78) \quad \text{where } z = \frac{\bar{x} - \mu}{1.125}$$

$$= 1 - P(z > 1.78) - P(z < -1.78) = 1 - 0.0375 - 0.0375 = 0.9250$$

Thus, with probability 0.9250, a particular sample mean would "miss" the true population mean by 2 or less. In other words, 92.5% of *all possible* sample means would miss μ by 2 or less. If we select a particular sample of size 64 from this population we can be 92.5% *confident* that $\bar{x}$ as an estimate of μ will be in error by less than 2.

This notion can be generalized to what is known as a confidence interval estimate of a population mean. If we use E to denote the error of estimation and $z_{\alpha/2}$ to denote the particular z value with area $\alpha/2$ to its right, we can write

$$\bar{x} - E < \mu < \bar{x} + E \quad \text{with } (1 - \alpha)\ 100\% \text{ confidence}$$

where $E = (z_\alpha/2)\,\sigma/\sqrt{n}$.

In our current example, suppose a particular sample of size n = 64 from a population with standard deviation $\sigma = \sqrt{81} = 9$ yielded a sample mean $\bar{x} = 53$. We could construct a 90% confidence interval estimate for μ as follows:

$$(1 - \alpha)(100\%) = 90\% \quad 1 - \alpha = .90 \quad \alpha/2 = .05$$

$$z_{\alpha/2} = z_{.05} = 1.645$$

$$E = (z_{\alpha/2})\,\frac{\sigma}{\sqrt{n}} = (1.645)\,\frac{9}{\sqrt{64}} = 1.85$$

Thus $(53 - 1.85) < \mu < (53 + 1.85)$ with 90% confidence

$$51.15 < \mu < 54.85 \quad \text{with 90\% confidence}$$

In words, the mean of the population from which our samples was drawn is between 51.15 and 54.85. We can make this assertion with 90% confidence due to the fact that if many such samples were selected from this population, and 90% confidence intervals were constructed from each, 90% of the intervals would be expected to contain the true mean. Of course, this also means that 10% of the assertions would be false.

You probably noted in the previous example that it was rather unrealistic that the population variance $\sigma^2 = 81$ was assumed to be known. In real world problems we are forced to substitute the sample variance s^2 for σ^2 in our calculations.

The probability statements used in the derivation of the confidence interval formula involved the transformation of $(\bar{x} - \mu)/(\sigma/\sqrt{n})$ into the standard normal variable z. If we do not know σ, we are forced to approximate the transformation with $(\bar{x} - \mu)/(s/\sqrt{n})$. Around the turn of the century a statistician named W. S. Gossett investigated the behavior of $(\bar{x} - \mu)/(s/\sqrt{n})$ under the assumption that $\bar{x}$ was normally distributed. (The Central Limit Theorem makes this assumption a reasonable one.) He devised a family of probability curves known collectively as the t distribution to model the behavior of $t = (\bar{x} - \mu)/(s/\sqrt{n})$. This model is symmetric about its center t = 0, and its extent is determined by the number of degrees of freedom associated with the sample. For our purposes the number of degrees

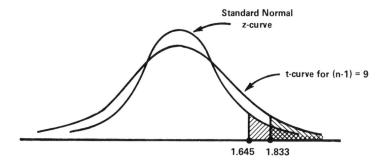

Fig. 8. Comparison of the t distribution to the standard normal.

of freedom is always $n - 1$ where n is the number of items in our sample. The individual members of the t family of curves become more and more similar to the standard normal z curve as the sample size increases. When s is computed from smaller sample sizes, the corresponding t curve is somewhat "fatter in the tails" than the standard normal curve. This, in effect, serves to compensate (in computation of probabilities) for the additional uncertainty introduced by using s, an estimate of σ, rather than σ itself. For instance the 95th percentile of the z distribution is $z_{.05} = 1.645$. Based on a sample of size $n = 10$ (see Table 4), the corresponding point of the t distribution is $t_{.05, n-1} = t_{.05, 9} = 1.833$. This is illustrated graphically in Fig. 8.

The general form of the confidence interval is as previously stated except that the error term is written in terms of the sample standard deviation s instead of σ and the t variable instead of z. We write

$$\bar{x} - E < \mu < \bar{x} + E \quad \text{with } (1 - \alpha) \ 100\% \text{ confidence}$$

but where $E - (t_{\alpha/2, n-1}) s / \sqrt{n}$

To illustrate this, let us return to our original sample of 80 tack strengths with sample mean $\bar{x} = 565.56$ and sample standard deviation $s = 37.904$. Let us construct a 90% confidence interval for μ.

$$(1 - \alpha) \ 100\% = 90\% \quad \alpha/2 = .05$$

$$t_{\alpha/2, n-1} = t_{.05, 79} \stackrel{\sim}{=} 1.667 \quad \text{(using linear interpolation)}$$

$$E = t_{\alpha/2, n-1} \frac{s}{\sqrt{n}} = (1.667) \left(\frac{37.904}{\sqrt{80}} \right) = 7.06$$

Table 4 Area Under the t Curve.

ν	0.10	0.05	0.025	0.01	0.005
1	3.078	6.314	12.706	31.821	63.657
2	1.886	2.920	4.303	6.965	9.925
3	1.638	2.353	3.182	4.541	5.841
4	1.533	2.132	2.776	3.747	4.604
5	1.476	2.015	2.571	3.365	4.032
6	1.440	1.943	2.447	3.143	3.707
7	1.415	1.895	2.365	2.998	3.499
8	1.397	1.860	2.306	2.896	3.355
9	2.383	1.833	2.262	2.821	3.250
10	1.372	1.812	2.228	2.764	3.169
11	1.363	1.796	2.201	2.718	3.106
12	1.356	1.782	2.179	2.681	3.055
13	1.350	1.771	2.160	2.650	3.012
14	1.345	1.761	2.145	2.624	2.977
15	1.341	1.753	2.131	2.602	2.947
16	1.337	1.746	2.120	2.583	2.921
17	1.333	1.740	2.110	2.567	2.898
18	1.330	1.734	2.101	2.552	2.878
19	1.328	1.729	2.093	2.539	2.861
20	1.325	1.725	2.086	2.528	2.845
21	1.323	1.721	2.080	2.518	2.831
22	1.321	1.717	2.074	2.508	2.819
23	1.319	1.714	2.069	2.500	2.807
24	2.318	1.711	2.064	2.492	2.797
25	1.316	1.708	2.060	2.485	2.787
26	1.315	1.706	2.056	2.479	2.779
27	1.314	1.703	2.052	2.473	2.771
28	1.313	1.701	2.048	2.467	2.763
29	1.311	1.699	2.045	2.462	2.756
inf.	1.282	1.645	1.960	2.326	2.576

Source: Ref. 1.

$\bar{x} + E = 565.56 + 7.06 = 572.62$

$\bar{x} - E = 565.56 - 7.06 = 558.50$

Hence

$558.50 < \mu < 572.62$ with 90% confidence

In words, we are 90% confident that the average tack strength of the population lies somewhere in the interval from 558.50 to 572.62. There is a corresponding 10% chance that the population average lies outside this interval. It should be clear that the level of confidence, 90% in this case, should always be included in the statement of the interval estimate.

In some cases, the researcher may wish to construct an interval estimate with a higher degree of confidence. An increase in confidence can be achieved at the expense of precision, that is, interval width. To illustrate this type of trade-off, suppose we construct a 98% confidence interval estimate for μ. The only change from the previous calculations would be to substitute $t_{.01,79} \simeq 2.380$ for $t_{.05,79} = 1.667$. This substitution would result in an increased error term E $= 10.09$. Hence we could say $555.47 < \mu < 575.65$ with 98% confidence.

Another factor in the determination of the width of the confidence interval is sample size n. To illustrate this, suppose that the mean tack strength of the sample $\bar{x} = 565.56$ and the standard deviation of the sample s$= 37.904$ had been calculated from a sample of only n $= 10$ measurements instead of the actual 80. The fact that s as an estimate of σ is considerably less reliable based on only 10 observations is compensated for by an increase in the size of t. Previously, we used $t_{.05,79} = 1.667$, but for sample size 10 we would use $t_{.05,9}$ $= 1.833$. An even more dramatic increase occurs in the computation $s/\sqrt{n}$. In this case $s/\sqrt{n} = 37.904/\sqrt{10} = 11.99$ rather than $37.904/\sqrt{80}$ $= 4.24$. Thus, the appropriate error term is E $= (1.833)(11.99)$ $= 21.98$. This is roughly three times as large as the E $= 7.06$ computed on the basis of sample size n $= 80$.

In practice, however, the analyst may not be able to take as large a sample as is desirable. Cost and time considerations often impose severe restrictions on sample size. It is essential to point out that in practice, when confidence intervals are constructed from small samples, the analyst and subsequent users of the information may well be disappointed to find that the error associated with the estimate of μ is considerably larger than they would like. Nevertheless, the width of the confidence interval accurately reflects the amount of information available in the sample. We must live with the fact that small samples simply do not provide much information about the population.

We are now able to address the specific question of, How large a sample should one take to obtain a reasonable estimate for μ? Careful analysis of the construction of the error term E = $(t_{\alpha/2,n-1})(s/\sqrt{n})$ reveals that the size of the error term is dependent, not only on sample size n alone, but also on the confidence coefficient $1 - \alpha$ and on a measure of the variability of the data s. To shed a bit more light on the situation we solve the above equation for n yielding

$$n = \left(\frac{t_{\alpha/2,n-1}s}{E} \right)^2$$

To determine sample size from the equation, the first step is to decide on what level of confidence you wish to place on your final estimate. In effect, you must choose a value for $1 - \alpha$. It is common to choose a value between 0.90 and 0.99. Next, you will have to make an estimate of s. Quite often you will have had some experience with similar data to form the basis of an estimate for s. Finally, choose a value for E that you consider "reasonable." Unfortunately, the correct t value depends not only on the α you have chosen, but also on the sample size; hence, a certain amount of trial and error may be necessary to select an n and corresponding t so that the equation is reasonably satisfied.

A somewhat simpler procedure might be to obtain some minimal size sample and go ahead and construct a confidence interval estimate for μ. If your confidence interval is too wide, simply add more replicates to your sample and then construct your new interval estimate based on the larger sample. Continue in this way until you are satisfied that the estimate is sufficiently precise. As a general rule, it is necessary to quadruple sample size in order to halve the width of the confidence interval.

At the risk of overkill concerning a point that probably is already crystal clear, I would like to reemphasize that the sample mean $\bar{x}$ and the population mean μ are not the same thing. All too often, researchers make statements like, "The average bonding strength has been determined to be 25 lb." The implication is, of course, that μ is 25 lb when, in fact, 25 lb is merely a sample average. As we have seen, the size of the error in purporting $\bar{x}$ to be μ depends heavily on the size of the sample as well as the degree to which individual measurements vary. The fallacy of this kind of statement becomes obvious in the light of the additional information contained in confidence interval estimates for μ which reflect neither more nor less information than is provided by the sample data.

Recall our earlier discussion when we attempted to find the value of tack strength exceeded by only 5% of the population of tack strengths. Our calculations ignored the problem of correctly centering our theoretical distribution of tack strengths by assuming the center was at $\bar{x}$ = 565.56 rather than at μ as it should be. Now, we know (with 90% confidence) that $\bar{x}$ might have missed the true mean by as much as E = 7.06 (as calculated on p. 622). An additional source of error is the use of the sample standard deviation s in place of the population standard deviation σ in our calculations. This error can be partially compensated for by using t in place of z. Nevertheless, the error in estimating the 95th percentile will probably be even larger than the error of using $\bar{x}$ to estimate μ.

Most computer facilities have statistical packages available that are designed to perform most of the commonly used analyses. Thus, virtually all of the drudgery of hand calculations is eliminated. The user merely reads in the sample data and asks for a specific analysis to be performed. This very simplicity, however, can lead to gross misuse of the statistical tests. Especially in cases where the sample sizes involved are small, the user is well advised to carefully read the user's manual or relevant material in a good statistics text. For instance, a computer might print out a confidence interval for the population mean μ based on the sample data read in. The interval, however, is valid only to the extent to which $\bar{x}$ is normally distributed as per the Central Limit Theorem. If the sample size is small and the population is asymmetrical, the interval should be properly interpreted as only a rough approximation—probably narrower than a valid confidence interval.

Situations may arise in which the analyst wishes to compare one population to another. Suppose, for example, that some minor change, which may or may not affect bonding strength, is being considered in the production process for a particular type of adhesive. Let us use μ_1 for the average strength of the old process and μ_2 for the average strength of the new process. We are not really interested in separate estimates for μ_1 and μ_2. We really want an estimate for the difference in bonding strengths, $\mu_1 - \mu_2$. Our best point estimate will, of course, be $\bar{x}_1 - \bar{x}_2$. Statisticians have investigated the difference in sample means when the samples are each selected from a different population. If both $\bar{x}_1$ and $\bar{x}_2$ have normal distributions (as per the Central Limit Theorem), then $\bar{x}_1 - \bar{x}_2$ will also be normally distributed. If this is so, the general form of the confidence interval is

$$(\bar{x}_1 - \bar{x}_2) - E < \mu_1 - \mu_2 < (\bar{x}_1 - \bar{x}_2) + E \quad \text{with } (1 - \alpha)\ 100\% \text{ confidence}$$

The formula used to calculate the error term differs depending on whether or not it is reasonable to assume the population variances are equal.

Case 1

If we can reasonably assume $\sigma_1^2 = \sigma_2^2$, the sample variances s_1^2 and s_2^2 are pooled in order to estimate the common variance:

$$E = t_{\alpha/2, \nu} \sqrt{\frac{(n_1 - 1)s_1^2 + (n_2 - 1)s_2^2}{n_1 + n_2 - 2} \left(\frac{1}{n_1} + \frac{1}{n_1}\right)}$$

where the number of degrees of freedom used in the t distribution is $\nu = n_1 + n_2 - 2$.

Case 2

If it is not reasonable to assume $\sigma_1^2 = \sigma_2^2$:

$$E = t_{\alpha/2, \nu} \sqrt{\frac{s_1^2}{n_1} + \frac{s_2^2}{n_2}}$$

where the number of degrees of freedom is approximated by the following formula for ν:

$$\nu = \frac{(s_1^2/n_1 + s_2^2/n_2)^2}{[(s_1^2/n_1)^2/(n_1 - 1)] + [(s_2^2/n_2)^2/(n_2 - 1)]}$$

Let us illustrate with an example. Suppose we sample each process and generate the following data:

New process	Old process
$n_1 = 12$	$n_2 = 16$
$\bar{x}_1 = 84$ g	$\bar{x}_2 = 72$ g
$s_1^2 - 41$	$s_1^2 = 45$

Since the sample sizes are relatively small, we need some assurance that the distributions of the parent populations are at least fairly symmetric in order to validate the Central Limit Theorem. The fact that the sample variances do not differ too much indicates that the assumption that $\sigma_1^2 = \sigma_2^2$ is reasonable. (We will shed a bit more light on this later.)

Hence, the equation at the top of this page is appropriate:

$$\nu = n_1 + n_2 - 2 = 12 + 16 - 2 = 26$$

$$t_{\alpha/2, \nu} = t_{.025, 26} = 2.056 \quad \text{for a 95\% confidence level}$$

$$E = 2.056 \sqrt{\frac{(11)(41) + (15)(45)}{12 + 16 - 2} \left(\frac{1}{12} + \frac{1}{16}\right)} = 5.17$$

So, with $\bar{x}_1 - \bar{x}_2 = 84 - 72 = 12$, we write

$$6.83 < \mu_1 - \mu_2 < 17.17 \quad \text{with 95\% confidence}$$

We are 95% confident that μ_1, the mean bonding strength of process 1, exceeds μ_2, the mean bonding strength of process 2, by at least 6.83 g but not by more than 17.17 g.

VII. STATISTICAL TESTS OF HYPOTHESIS

To aid the analyst in the decision-making process, statisticians have developed procedures known as statistical tests of hypothesis. Essentially, the analyst assumes something concerning a population to be true. This assumption is referred to as a null hypothesis H_0, which is then rejected or not, depending on the degree to which the sample evidence *contradicts* H_0. The rejection of H_0 leads to the acceptance of an alternative hypothesis H_1. To illustrate the rationale underlying the test procedures, suppose you are involved in a coin tossing game in which you lose whenever a head shows. You might decide to toss the coin 400 times in order to test the assumption that the coin is fair. If we use p to denote P(head) we can state

$$H_0: \quad p = 0.5 \quad \text{Coin is fair.}$$

$$H_1: \quad p > 0.5 \quad \text{Coin is biased favoring heads.}$$

We will reject the assumption that the coin is fair, that is, reject H_0, if the number of heads in 400 tosses is *significantly* more than the 200 we would expect if, indeed, H_0 were true. Suppose we observe 205 heads in the 400 coin tosses. According to the laws of probability theory, the probability of obtaining 205 or more heads in 400 tosses of a *fair* coin is approximately 0.326. In other words, if 400 fair coins were tossed repeatedly, we would expect that 205 or more heads would occur 32.6% of the time. Such an occurrence seems consistent with H_0 (the coin is fair), so we do not reject the assumption.

On the other hand, suppose we had observed 225 heads. It can be shown that 225 or more heads in 400 tosses of a *fair* coin would occur only 0.7% of the time—a very rare event indeed. In this case, we would reject H_0 and conclude that the coin is biased in favor of heads. Using statistical terminology, we refer to these probabilities as the *level of significance* of the sample results. We say that 205 heads is significantly more than the 200 expected at a 32.6% level of significance. The occurrence of 225 heads, at a significance level of 0.7%, is considerably more significant. (Although it seems contradictory, we say a result is *highly* significant when the *level of significance* is small.) The decision to reject H_0 with the occurrence of 225 or more heads (0.7% significance level) and the decision not to reject H_0 with the occurrence of 205 or fewer heads (36.2% significance level) suggest that the significance level of the sample result is a reasonable criterion to use in the decision-making process.

The established procedure is to choose *a priori* a critical level of significance to represent the dividing line between rejecting and not rejecting H_0. This critical value is called the *significance level of the test* and is usually denoted by α. Stating it in the form of a decision rule, we write

> *Decision Rule.* Reject H_0 if the significance level of the sample result is less than α; otherwise, do not reject H_0.

Suppose for our coin tossing example, we choose $\alpha = 5\%$. (Values between 1 and 10% are common.) The particular sample result that would be significant at the 5% level is 217 heads. So, if we use x to represent the number of heads in 400 tosses of a fair coin, we can rephrase our decision rule in terms of the *test statistic* x.

> *Decision Rule.* Reject H_0 if $x \geqslant 217$; otherwise, do not reject H_0.

A more practical application of this statistical procedure is illustrated in the following hypothetical situation. A furniture manufacturer uses Stik-Tite adhesive to secure the arms of an upholstered chair. A bottleneck in the assembly process occurs at this point due to the fact that 3 hours are required for Stik-Tite to set up properly. An adhesive salesman claims that his brand, Fast-Lok, has the same bonding strength but will set up in less time. To test the claim, the furniture manufacturer uses Fast-Lok on a sample of 25 chairs, carefully noting the setup time for each chair. The 25 times average $\bar{x}$ = 2.8 hr with a sample standard deviation of s = 0.4 hr. Is this sufficient evidence to *reject* an assumption that the setup time for Fast-Lok is the same 3 hr as that to Stik-Tite? That is, does the sample evidence *support* the salesman's claim of shorter setup time?

The appropriate statistical hypotheses would be stated in terms of μ, the average setup time for the *population* of Fast-Lok setup times.

H_0: $\mu = 3$ hr   Fast-Lok same as Stik-Tite

H_0: $\mu < 3$ hr   Fast-Lok is quicker than Stik-Tite

Suppose we take the significance level of the test to be $\alpha = 5\%$. We can state

Decision Rule. Reject H_0 if the significance level of $\bar{x} = 2.8$ is less than 5%.

To determine the significance level of $\bar{x} = 2.8$ we must compute the probability that a sample mean $\bar{x}$ would be 2.8 hr or less given that H_0 is true and $\mu = 3$ hr. Once again relying on the Central Limit Theorem we can state that $\bar{x}$ has an approximately normal distribution with mean $\mu_{\bar{x}} = \mu = 3$ and

standard deviation $\sigma_{\bar{x}} = \sigma/\sqrt{n} \simeq s/\sqrt{n} = 0.4/\sqrt{25} = 0.08$.

We compute $P(\bar{x} \leqslant 2.8) = P[t \leqslant (2.8 - 3)/0.08] = P(t \leqslant -2.5)$.

Due to the abbreviated nature of available t tables we are not able to locate this value precisely. However, with $n - 1 = 24$ degrees of freedom we *can* locate $P(t < -2.492) = $ (using symmetry) $P(t > 2.492) = 0.01$

Consequently $P(\bar{x} \leqslant 2.8) \simeq P(t < -2.492) = 0.1$. Thus, we can say that $\bar{x} = 2.8$ is significantly less than 3 at approximately a 1% level of significance. Since this value is less than the $\alpha = 5\%$ significance level of the test, we can reject H_0 and conclude that Fast-Lok adhesive sets up in less than 3 hr.

Due to the difficulty of obtaining values for the exact probability distribution of t, an alternate but equivalent form of the decision rule is commonly used. The practice is to compare the t computed from the sample mean $\bar{x}$ to the value of t significant at the $\alpha = 5\%$ level. In our case, compute

$t = (\bar{x} - \mu)/(s/\sqrt{n}) = (2.8 - 3)/0.08 = -2.50$.

From the t table $-t_{.05,24} = -1.711$.

We would write

Decision Rule. Reject H_0 if $t < -t_{.05,24} = -1.711$.

Decision: Reject H_0 since $t = -2.50 < -1.711$.

The broad concept of hypothesis testing is a very powerful statistical tool that can be applied to a wide variety of problems. The previous example involving testing an assumption about the value of a population mean is only one of many existing applications of the general concept. Among the many tests that have been developed are two which are relevant to what we have covered so far. Under the general heading of goodness-of-fit tests is one which tests the validity of the assumption that a particular sample was drawn from a normal population. Another test relating to Eqs. (8) and (9) allows us to test whether or not it is reasonable to assume $\sigma_1^2 = \sigma_2^2$. The details of the various tests can be found in any good textbook on statistics. In most cases the analyst will rely on the computer and a "canned" statistical package to perform the actual calculations. Nevertheless, the analyst should be aware of the conditions under which the particular test is valid. The computer output should be taken literally only to the extent that the required conditions are met.

Regardless of how much the computational aspects of the various tests differ, the rationale and interpretation of the results remain constant. In every case the test is carried out under the *assumption* that H_0 is true. This assumption is rejected in favor of the alternate hypothesis H_1 only when sample results strongly *contradict* H_0, that is, strongly *support* H_1. This strong evidence manifests itself in the form of a near-zero level of significance of the sample results. (Recall, the significance level of the sample is actually the probability that if indeed H_0 were true, the sample result could be as contradictory to H_0 as was actually observed.)

Consequently, when we reject H_0, we can *accept* H_1 based on some solid evidence that H_0 is false. On the other hand, if we fail to reject H_0 when the sample results only mildly contradict it, as evidenced by a lower significance level of the sample result, we should *not* conclude that H_0 is true. The proper conclusion is that the sample evidence is simply not sufficient to warrant the rejection of H_0. A suitable analogy might be H_0: defendant is innocent versus H_1: defendant is guilty. The jury's verdict is either "guilty" (accept H_1) or "not guilty" (do not reject H_0). The "not guilty" verdict does not imply innocence; it only implies insufficient evidence to "prove" guilt.

Thus, the alternate hypothesis H_1 must always be the one that the analyst hopes to be able to *support* with solid sample evidence.

VIII. SAMPLES CONTAINING EXTREME VALUES

Suppose an engineer tests a new type of adhesive and records the following 10 test measurements in order of increasing magnitude:

418   429   439   442   448   455   460   469   480   542

The largest value, 542, seems to differ markedly from the test of the sample items. Extreme values such as these are generally referred to as outliers. The presence of one or more outliers in a sample can be quite disconcerting to the analyst. Should the outlier be included in the computation of an estimate of μ, or should the outlier be rejected from further consideration? Common sense dictates that the experimenter should carefully review the testing procedures in an effort to explain the existence of the outlier in terms of experimental error. It is possible that the extreme test value of 542 was caused by a more

generous coating of adhesive than the others; or, possibly, the conditions under which the curing took place were somehow different.
The outlier may even have resulted from the failure of the measuring
equipment. If the experimenter has any good reason to believe that
the outlier was caused by experimental error, it should be discarded
from any further consideration.

For the case where no physical reason can be found for the existence of the outlier, the decision to retain or reject the outlier can be
made within the framework of statistical hypothesis testing.

The statistical hypotheses relating to the outlier problem are as
follows:

H_0: All sample values come from the same *normal population*.
The seemingly large deviation of the outlier from the
rest of the sample is explainable in terms of ordinary variability in the population.

H_1: The deviation of the outlier from the bulk of the sample
items is too large to be explained by ordinary variation.
There is some (even though unknown) physical reason to
explain the existence of the outlier.

If, on the basis of statistical evidence, one decides to reject H_0, the
outlier should be rejected. The population mean should then be estimated using only the remaining sample values. Otherwise, all sample
values—including the outlier—should be used in any further calculations.

Over the years statisticians have developed various criteria for
the rejection of H_0. The statistic that I recommend is the ratio of the
distance between the outlier and the sample mean to the sample standard deviation. Let the n sample values be arranged in increasing
order, $x_1 \leqslant x_2 \leqslant \cdots \leqslant x_n$. If x_n is the suspected outlier, form the
ratio T_n (see Table 5):

$$T_n = \frac{x_n - \bar{x}}{s}$$

where

$$\bar{x} = (1/n) \sum x$$

$$s^2 = \sum [(x - \bar{x})^2/(n - 1)]$$

$$s = \sqrt{s^2}$$

Similarly, if x_1 rather than x_n is the suspected outlier, use
$T_1 = (\bar{x} - x_1)/s$. Table 5 shows the critical values of T at the 5%,
2.5%, and 1% significance levels. The table values are interpreted as
follows.

Table 5 Table of Critical Values for T (One-sided Test) When Standard Deviation Is Calculated from the Same Sample

Number of observations[a] n	5% significance level	2.5% significance level	1% significance level
3	1.15	1.15	1.15
4	1.46	1.48	1.49
5	1.67	1.71	1.75
6	1.82	1.89	1.94
7	1.94	2.02	2.10
8	2.03	2.13	2.22
9	2.11	2.21	2.32
10	2.18	2.29	2.41
11	2.23	2.36	2.48
12	2.29	2.41	2.55
13	2.33	2.46	2.61
14	2.37	2.51	2.66
15	2.41	2.55	2.71
16	2.44	2.59	2.75
17	2.47	2.62	2.79
18	2.50	2.65	2.82
19	2.53	2.68	2.85
20	2.56	2.71	2.88
21	2.58	2.73	2.91
22	2.60	2.76	2.94
23	2.62	2.78	2.96
24	2.64	2.80	2.99
25	2.66	2.82	3.01
30	2.75	2.91	
35	2.82	2.98	
40	2.87	3.04	
45	2.92	3.09	
50	2.96	3.13	
60	3.03	3.20	
70	3.09	3.26	
80	3.14	3.31	
90	3.18	3.35	
100	3.21	3.38	

$$T_n = \frac{x_n - \bar{x}}{s} \quad s = \left\{ \frac{\sum (x_i - \bar{x})}{n-1} \right\}^{1/2} = \left\{ \frac{n \sum x_i^2 - (\sum x_i)^2}{n(n-1)} \right\}^{1/2}$$

$$T_2 = \frac{\bar{x} - x_1}{s} \quad x_1 \leqslant x_2 \leqslant \cdots \leqslant x_n$$

[a]For n > 25, the values of T are approximated. All values have been adjusted for division by n − 1 instead of n in calculating s.

Source: Ref. 1.

If a sample of size (say) 6 is drawn randomly from a *normal population*, that is, assuming H_0 is true, then the T_n (or T_1) computed from the sample will exceed 1.82 only 5% of the time, exceed 1.89 only 2.5% of the time, and exceed 1.94 only 1% of the time.

To illustrate the procedure to be followed, consider the 10 test values for the adhesive mentioned earlier:

418   429   439   442   448   455   460   469   480   542

We would state

H_0: All 10 items were drawn from the same normal population

H_1: Only the first 9 items were drawn from this population. The value 542 is from a different population altogether.

Suppose we decide to carry out the test at a 1% level of significance. The critical value in the table for n = 10 at a 1% significance level is T = 2.41. Thus we can say that if indeed H_0 is true, there is only a 1% chance that a randomly selected sample would yield a T_n of 2.41 or more. We would write

Decision Rule. Reject H_0 (at the 1% significance level) if $T_n > 2.41$.

You should be able to verify from the sample data that $\bar{x} = 458.2$ and s = 34.69 and, thus, to calculate $T_n = (542 - 458.2)/34.69 = 2.416$. Since our computed T_n exceeds the critical value, we can reasonably reject H_0 at the 1% level. We conclude that 542 is indeed an outlier and that some physical reason exists for the 542 to be so far removed from the rest of the sample values. You should keep in mind that we are rejecting H_0 fully aware that there is a 1% chance that T_n *could* be that large even with H_0 true. Armed with the knowledge of this unavoidable risk and its consequences, the experimenter must decide at what significance level he is willing to reject H_0 and, consequently, reject the outlier.

It may be that a sample contains more than one extreme value. Consider the following sample of 10 items:

2.02   2.11   3.04   3.23   3.59   3.73   3.94   4.05   4.11   4.13

It appears that 2.02 and 2.11 are both possible outliers. The recommended procedure in this case is to work from the inside out, testing 2.11 as an outlier in the sample of 9 values which remain while ignoring the 2.02. Your should be able to verify that $T_1 = (\bar{x} - x_1)/s = (3.548 - 2.11)/0.6635 - 2.167$. From Table 5 for n = 9, we note that a T_1 as large as 2.167 would occur by chance (from a normal population) with a probability between 2.5% and 5%. We would take this as evidence that 2.11 is probably not from the same population as the

other eight sample values. Since the 2.02 is even more extreme than
2.11, the 2.02 is clealy an outlier and we have no need to test further.
We simply state that both of these extreme values came from a population which is different from the remaining eight values.

It should be emphasized that working from the inside out is absolutely essential. Suppose we simply test the 2.02 first as a single
outlier in the entire sample or size 10. We would compute $T_1 = (3.395
- 2.02)/0.7904 = 1.74$. Since this is considerably less than $T = 2.18$
from our table with $n = 10$ at the 5% level, we would be led to accept
H_0 and to conclude that 2.02 came from the same population as the
rest of the sample.

The sequential inside-out procedure illustrated above is perfectly acceptable in the case where both outliers are subsequently rejected. However, when two observations are closer to each other than to
the bulk of the observations, they produce a masking effect in which
case the sequential procedure may fail to identify even a single outlier. Procedures are available for the simultaneous testing of the two
largest or two smallest observations [1]. This article also details more
sensitive tests in the special case where there is some prior knowledge
of the variation of the population being samples. The most general
case of several possible outliers occurring at both the upper and lower
ends of the sample is discussed in some detail in the literature [2].

Most of the procedures we have discussed require certain assumptions. For example, we needed the assumption of a normal population in order to guarantee the validity of a confidence interval estimate for μ in the case where sample size was small. Fortunately, most
of the procedures are still reasonably reliable for slight departures
from the necessary assumptions. A number of test procedures have
been developed that assume no knowledge whatsoever about the distribution of the underlying population. These tests are called *non-
parametric* or *distribution-free* tests. Among the advantages of such
tests is the ease with which the necessary calculations can be performed. Also, the results are guaranteed to be valid since there are
no assumptions that could be violated. There are also numerous disadvantages to these tests. Primarily, they do not utilize all of the
information provided by the sample; consequently, these test are less
efficient than the corresponding parametric procedures. Generally,
these tests require a larger size sample if any reasonable conclusion
is to be reached. The interested reader can obtain further information from most standard statistics texts.

REFERENCES

1. F. E. Grubbs, *Technometrics II* (1969).
2. G. L. Tietjen and R. H. Moore, *Technometrics 14* (1972).
3. R. E. Walpole, in *Introduction to Statistics*, Macmillan, New York, 1968.

25

Adhesive Specifications and Quality Control

Garry O. DeFrayne *Chrysler Corporation, Highland Park, Michigan*

I. SPECIFICATIONS

Specifications, the cornerstone of any quality control program, are sometimes referred to as the standards of quality. Specifications can be defined as the implement for establishing, in measurable terms, a property or group of properties which determine the quality of the item in question. All of the above seems very straightforward and simple until you have to prepare a specification. Before presenting a format for the preparation of a specification, it is necessary to consider the following:

Objective
Dos and don'ts of specification writing
Language of specification
Justification of requirements

A. Objective

The basic objective of a specification is to define what is needed for all parties involved. These parties include, but are not limited to, purchasing agents, suppliers, quality-control people, product development departments, and manufacturing personnel.

B. Dos and Don'ts

The following are short lists of dos and don'ts for specification writing:

1. Do

   Use simple words.
   Use the exact meaning of technical terms.
   Use short sentences.
   Use the same word throughout, never synonyms.
   Use the same mathematical system throughout.

2. Don't

   Use words having a double meaning.
   Make the specification too voluminous.
   Put in requirements that cannot be justified.
   Use trade names.
   Make limits too tight or too loose.
   Define requirements loosely.

In summary, the language of a specification should be clear, definite, exact, brief, technical, and to the point. In writing a specification, don't do anything that will cause confusion and defeat its purpose.

C. Justification of Requirements

Requirements are the backbone of the specification. Requirements should be set forth definitely, completely, and clearly so as to leave no area of misunderstanding between the seller and the buyer. Every requirement should have a logical justification, or it should not be included in the specification. This simple statement, if it were followed judiciously, would eliminate a great many decisions regarding the specification. Unfortunately, in many cases the specification writer puts in so-called standards tests that are not applicable to the material or includes items only because they have always been included in the past.

D. Types of Specifications

There are many ways of cataloging specifications, such as

Source: federal, military, corporation, and technical area
Product: Adhesives, paints, fabrics, caulks, and elastomers
Performance: Equipment—such as welders, spray guns, compu-
ters, etc.

This list represents just a few of the criteria; it could be expanded to include many more types.

E. Specification Sources

The following list presents a few sources of specifications for adhesives and related equipment:

1. Military

   Naval Publication and Forms Center
   5801 Tabor Avenue
   Philadelphia, PA   19120

2. Federal standards

   Specification Sales
   3FRSBS Bldg. 197
   Washington Navy Yard
   General Services Administration
   Washington, DC   20407

3. Society of Automotive Engineers (SAE)

   SAE, Inc.
   400 Commonwealth Drive
   Warren, PA   14096

4. American Society for Testing and Materials (ASTM)

   ASTM
   1916 Race Street
   Philadelphia, PA   19103

F. Language of a Specification

The prime consideration of a specification is its technical content,
which should be presented in language that is free of vague and am-
biguous terms and that uses the simplest words and phrases that will
convey the intended meaning. Incorporation of necessary information
should be complete. Consistency in terminology and origination of
material, as well as short and concise sentences, all assist in the
preparation of a clear and precise specification.

   Punctuation should assist in the development of the specification;
however, it should be kept to a minimum. To avoid ambiguity abbre-

viations should be avoided wherever possible. For example, "Eng."
could be interpreted as either engineer or engine. Proprietary names
should be avoided along with trade names and copyrighted names.
Use the emphatic form of verbs; for example, write, "Apply a 10-mm
film of adhesive," rather than, "A 10 mm-film of adhesive should be
applied." Decimals should be used instead of fractions. A figure,
such as a graph or a picture, can be used if it is properly identified
and related to the text of the specification. All figures should be tit-
led and numbered consecutively. Tables should be used when they
can present pertinent data more clearly than the text can.

G. Specification Format

A typical specification has the following format:
    Title
    Scope
    General requirements
    Performance requirements
    Test methods
    Controls
    Reference documents
    Approved source list

Appendix 1 illustrates a typical automotive adhesive specification de-
fining a structural adhesive for bonding hood inner panels to hood
outer panels.

II. QUALITY CONTROL

Quality control is defined as an aggregate of activities designed to ensure adequate quality in manufactured products. As they relate to
adhesives, these activities can be any control procedures (in-process
controls, whether visual or physical) that are utilized to ensure the
reliability of the finished product. In this chapter we will address
ourselves to those controls implemented by the user of the adhesive
in the manufacture of a finished product.

A. Total Control Program

The total control program monitors the adhesive from the time it is
received at the user's facility to the time the finished product is ready
for shipment. Each of the activities within the manufacturing process
has its own control program. These activities are

    Receiving inspection
    Quality control
    Manufacturing control
    Final inspection

B. Receiving-Inspection Control Program

Receiving-inspection is the first step in the total control program. This program generally consists of a comparison of the purchase order with what is received, including such details as identification of material specification; part number; container type, size, and quantity; manufacturing data; and expiration (do-not-use-after) date. All of the above activities are normally visual and the results are recorded on appropriate receiving-inspection forms. The adhesive is then tagged, "Hold for quality control disposition." At this point, the quality control department's responsibility begins.

C. Quality Control Program

Before the adhesive can be released for use by production, quality control, utilizing the appropriate materials specification as a guide, must inspect the adhesive in order to ensure its performance in the manufacturing process. The inspection consists of an evaluation of physical and chemical properties such as

Color
Viscosity
Percent solids
Weight per gallon
Flash point
Pot life
Open time
Flow
Surface tack
Weldability

Note that all of the above tests are not applicable to all adhesives. They merely represent a general listing of the types of tests that could be performed.

Performance characteristics, such as the following, may also be evaluated:

Adhesion (peel, shear, tensile, cleavage)
Impact resistance
Resistance to environmental effects (heat, condensing humidity, salt spray, temperature cycles, Fade-Ometer, and Weather-Ometer)
Flexibility
Paintability (exposed joint)

These tests that define both the chemical and physical properties of the adhesive are relatively simple and can be performed in a comparatively short period of time. Those tests that define the performance characteristics of the adhesive are generally more difficult to carry out and usually require a greater period of time. These tests are defined in the applicable material specification and generally relate to the specific end product. They are defined by the user and may take the form of the user's own test methods or other methods, such as those of ASTM, SAE, Military Specifications (Mil. Specs.), and others.

It is also interesting to note that test procedures vary between companies, even in the same industry having the same end product. A typical example is the test methods used in the automotive industry to determine the shear strength of a structural metal-to-metal adhesive (Table 1). After testing by the quality control group has been completed, the adhesive is released to production for manufacturing the finished article.

D. Manufacturing Control Program

In order to ensure that the finished parts will be adequately bonded and perform satisfactorily, the following items must be considered as integral parts of the manufacturing control program:

> Fitting of component parts
> Surface preparation
> Application of adhesive
> Assembly of parts
> Curing of adhesive

1. Fitting of component parts

As might be expected, all parts are fitted together without benefit of adhesive prior to actually starting the production process in order to minimize production problems. In some cases, such as attaching a nameplate to a metal surface, the fit can be established by bonding the nameplate to a piece of glass and examining the bond through the glass. In almost all cases the suitability of the fit is evaluated by visual examination. Additionally, it is recommended that the extremes in dimensional tolerance of the parts be used for the initial fit. This procedure will simulate the worst possible condition. There are numerous ways to establish the fit, most of which simply involve the use of common sense. Regardless of the method used, it is important for the fit to be satisfactory, for if it is not, the finished parts will not perform as expected.

2. Surface preparation

Much has been written about the different methods of surface preparation and their advantages and disadvantages. The important thing to remember is that whatever surface preparation is specified, it is intended to give the best possible bond under actual manufacturing conditions. The following is a short list of various types of surface treatment for metallic surfaces:

Table 1 Shear-Adhesion Test Conditions for Determining the Shear Strength of Metal-to-Metal Structural Adhesives in the Automobile Industry[a]

	ASTM D-1002	Auto company A	Auto company B	Auto company C
Substrate	Metal	Metal	Metal	Metal
Surface preparation	Per mfgr's instruction	Ketone wipe and drawing compound	VM&P naphtha wipe and drawing compound	VM&P naphtha wipe and drawing compound
Adhesive film thickness	Per mfgr's instruction	0.010 in. (0.025 cm)	0.032 in. (0.08 cm)	0.020 in. (0.05 cm)
Bond area				
Width	1 in. (2.54 cm)	1 in. (2.54 cm)	1 in. (2.54 cm)	1 in.^{2} (6.25 cm^2)
Overlap	0.5 in. (1.27 cm)	0.5 in. (1.27 cm)	1 in. (2.54 cm)	
Metal thickness	0.064 in. (0.16 cm)	0.032 in. (0.08 cm)	variable	
Cure cycle				
Time	Per mfgr's instruction	12 min at 360°F (180°C)	20 min at 325°F (160°C) + 30 min at 250°F (120°C)	30 min at 300 °F (160°C)
Temperature				
Pressure		Clamping		
Test equipment	Shall conform to ASTM E-4	Instron	Tensile tester	Instron
Crosshead speed	0.05 in./min (0.13 mm/min)	0.1 in./min (0.25 cm/min)	2 in./min (5.1 cm/min)	2 in./min (5.1 cm/min)
Test method	Lap shear tension	Lap shear tension	Lap shear tension	Lap shear tension
Test method title	D-1002			
Conditioning period	Per mfgr's instruction	4 hr at R.T. prior to test	1 hr at R.T. prior to test	2 hr at R.T. prior to test
Number of specimens	5	3	3	3

[a]The metal surface is cleaned and then purposely contaminated with drawing compound.

Sandblasting	Treating with
Abrading	1,1,1-Trichloroethane
Machining	Perchloroethylene
Scouring with sandpaper	Methylene chloride
Surface degreasing	Acetone
Cleaning with alkaline solutions	V, M, & P naphtha

For most nonmetallic surfaces wiping the surface with a suitable solvent, abrading the surface, or both will provide an appropriate surface for bonding.

3. Application of adhesive

Adhesives are generally applied by one of the following methods:

Spray (airless or atomized air)
Roller coat
Transfer film
Brush
Manifold (individual drops of adhesive)

In most cases the adhesive is applied directly to the prepared surface. However, in some cases, such as structural bonding, a primer may first be applied to ensure the ultimate performance of the adhesive. By contrast, in some industries structural adhesives are applied directly over drawing compounds (oil). The primary items of concern at this point are proper adhesive quantity, application in the designated area, and using an adhesive which has an unexpired shelf life.

4. Assembly of parts

After having been coated with adhesive, the parts are assembled and held together by such means as clamps, tools, welding, or other fixtures. The primary concerns are to ensure that

The open time of the adhesive is not exceeded.
The parts are put together in the proper sequence.
The bonding is performed under specified environmental conditions.
The parts are held together until cured.

5. Curing of adhesive

Curing of the adhesive is accomplished with pressure and either the application of heat or the addition of a catalyst. All of these are variables which, if allowed to deviate from the prescribed condition, can result in poor bonding and, ultimately, poor performance of the finished parts. The items that must be controlled during this process are

Pressure
Maximum temperature
Time at temperature
Temperature buildup
Cool-down period

E. Final Inspection

Basically, final inspection determines the quality of the finished part.
Numerous tests are performed to evaluate the result of the total process, that is, adhesive material, surface preparation, adhesive application, fitting of the parts, and curing of the assemblies. Much has
been written describing various test methods, their advantages and
disadvantages, and what characteristics of the bonded assembly they
can evaluate [1,2].

It is interesting to note that while all industries use the same
basic test methods, each industry has its own special modifications.
Likewise, within an industry, companies vary in the manner in which
they run a certain test.

III. APPENDIX: A TYPICAL AUTOMOTIVE ADHESIVE SPECIFICATION

FIG 1: A Typical Automotive Adhesive Specification

T Y P I C A L   A U T O   C O M P A N Y

E N G I N E E R I N G   S P E C I F I C A T I O N

MATERIAL STANDARD

STRUCTURAL ADHESIVE - VINYL PLASTISOL - GALVANIZED BONDING

A. - General

This standard defines a modified vinyl plastisol type structural
adhesive, suitable for such applications as hem flange bonding of
lift gates. The adhesive shall exhibit good adhesion to galvanized
metal and to steel surfaces coated with approved metal forming
lubricants.

B. - General Requirements

1 - This adhesives shall be suitable for production use for at
least 2 months when stored at a temperature not to exceed
38°C (100°F). When tested at any time during this 2 month
interval, the adhesive shall meet the requirements herein
specified. If stored in excess of the above conditions, a
sample of the adhesive shall be submitted to the Stamping
Division, Materials Engineering, for approval prior to use.

2 - The adhesive shall be capable of being applied with conventional
production pumps and shall exhibit no breakdown due to the
pumping action or to passage through hoses, guns, and
ejectors.

3 - The adhesive shall not have an objectionable or noxious odor.
It shall be the suppliers' responsibility to determine
through the Industrial Hygiene and Safety Department, that the
materials contained in the adhesive and the contemplated use
thereof are deemed to be non-objectionable.

4 - The material shall satisfactorily pass the corrosion resistance
requirements, as established by the Metallurgical Engineering
Department.

C - Detailed Requirements (See Section D for Methods of Testing)

1 - Specific Gravity ---------------------------1.3 to 1.5

2 - Non-Volatile Content----------------------Minimum 97%

E N G I N E E R I N G   S P E C I F I C A T I O N

C - Detailed Requirements (continued)

   9 - a - (4)   6 Minutes at 176°C (350°F), 12
                 Minutes at 121°C (250°F), 20
                 Minutes at 163°C (325°F), and 20
                 Minutes at 121°C (250°F)-------2225 kPa (325 psi)

      b - Accelerated Aging

         All failures shall be predominantly cohesive and the shear
         values shall not be less than specified.

CURE	BOND THICKNESS	INITIAL	3 WEEKS AT 70°C(158°F)	5 ENVIRON-MENTAL CYCLES	3 WEEKS CONDENSING HUMIDTY	250 HOURS SALT SPRAY
20 Min. at 163°C (325°F) and 30 Min. at 121°C (250°F)	5.3 mm (0.21 in.)	1380 kPa (240 psi)	1650 kPa (240 psi)	1450 kPa (210 psi)	1300 kPa (190 psi)	1100 kPa (160 psi)
	0.76 mm (0.33 in.)	2200 kPa (320 psi)	2500 kPa (365 psi)	2680 kPa (390 psi)	2070 kPa (300 psi)	2200 kPa (320 psi)

         **Each cycle consists of 24 hours at 107°C
         (225°F), 24 hours condensing humidity, and
         24 hours at -29°C (-20°F).

   10 - Metal Draw Down ----------------------No visible depression or
                                               deformation of the metal
                                               panel.

D - Methods of Testing

      In all tests requiring faying surfaces to be coated with metal forming
      lubricants, use only those lubricant compounds (for galvanized metal/steel)
      which are approved for the application. For all step-wise cures involving
      more than 1 oven cure schedule, cool down the specimen for 1 hour at
      room temperature prior to subjecting to additional oven cure schedules.

                                               LABORATORY PROCEDURE

      1 - Specific Gravity   Typical Auto Co. Inc.   ABC-001

      2 - Non-Volatile Content   Typical Auto Co. Inc.   ABC-002

Fig. 1. (Continued)

ENGINEERING   SPECIFICATION

D - <u>Methods of Testing</u> (continued)

LABORATORY PROCEDURE

3 - Viscosity----------------------------Typical Auto Co. Inc. ABC-003
(3.18 + 0.002 mm [0.125 + 0.001 in]
diameter orifice and 276 kPa [40 psi]
pressure)

4 - Durometer - Shore 'A'-------------------Typical Auto Co. Inc. ABC-004
(Instantaneous readings on sample
cured 20 minutes at 121°C [250°F])

5 - Flow --------------------------------Typical Auto Co. Inc. ABC-005

6 - Flash Point---------------------------Typical Auto Co. Inc. ABC-006

7 - <u>Cold Impact</u>

a - For Steel to Steel------------------Typical Auto Co. Inc. ABC-007

b - For Steel to Galvanized Metal-------Typical Auto Co. Inc. ABC-007

8 - <u>Phosphate Coating System Resistance</u>

a - For Steel to Steel------------------Typical Auto Co. Inc. ABC-008

b. For Steel to Galvanized Metal-------Typical Auto Co. Inc. ABC-009

9 - <u>Adhesion - Shear</u>

For preparing specimens with a 0.76 mm (0.03 in) bond thickness,
substitute a suitable spacer bar which will provide the final bond
thickness of 0.76 mm (0.03 in) instead of the 6.35 mm (0.25 in)
bar specified in the laboratory procedure.

a - Steel to Steel----------------------Typical Auto Co. Inc ABC-10

b - Steel to Galvanized Metal-----------Typical Auto Co. Inc. ABC-10
except that the shear sample
will consist of galvanized
metal bonded to steel.

10 - <u>Metal Draw Down</u>

a - Fill 4 aluminum cups in such a manner that there is a slight
excess of adhesive.

b - Invert the cups and attach to phosphate coated steel panel
exerting sufficient pressure to achieve metal-to-metal contact.

c - Remove excessive adhesive which has been squeezed beyond the
confines of the cup.

Fig. 1. (Continued)

E N G I N E E R I N G   S P E C I F I C A T I O N

C - Detailed Requirements (continued)

    3 - Viscosity

       a - Initial----------------------------90 to 130 seconds

       b - Aged-------------------------------Maximum 250 seconds
          [72 hours at 43°C (110°F)]

    4 - Durometer - Shore 'A'-------------------Maximum 55

    5 - Flow

       a - 4 Hours at Room Temperature---------Maximum 12.7 mm (0.5 in)

       b - 20 Minutes at 191°C (375°F)---------Maximum 3.2 mm (0.125 in)
                                   additional flow

    6 - Flash Point --------------------------Minimum 224°C (435°F)

    7 - Cold Impact

       No adhesive failure or cracks when applied to either steel or
       galvanized panels.

       a - 20 Minutes at 141°C (325°F) and 121°C (250°F)

       b - 20 Minutes at 191°C (375°F) and 121°C (250°F)

    8 - Phosphate Coating System Resistance-----No displacement, wash-out, or
                                      damage by impingement.

    9 - Adhesion - Shear

       All failures shall be predominately cohesive and the shear adhesion
       values shall not be less than specified.

       a - Initial Cure [Bond Thickness 0.76 mm (0.03 in)]

          (1) 6 Minutes at 176°C (350°F)------1725 kPa (250 psi)

          (2) 6 Minutes at 176°C (350°F),
              12 Minutes at 121°C (250°F)-----1725 kPa (250 psi)

          (3) 6 Minutes at 176°C (350°F), 12
              Minutes at 121°C (250°F) and 20
              Minutes at 163°C (325°F)-------2225 kPa (325 psi)

Fig. 1. (Continued)

ENGINEERING   SPECIFICATION

D - METHODS OF TESTING (continued)

    10 - d - Place the assembly in an oven, so that the panel is
        horizontal and cups are on the underside. Bake 20 minutes
        at 163°C (325°F).

      e - Coat the side of the assembly not having the cups, with
        black exterior enamel. Bake for 30 minutes at 121°C
        (250°F).

      f - Allow to cool to room temperature and visually inspect for
        depressions or deformation of the painted panel.

E - CONTROL

    1 - Full approval under this standard demands compliance with all the
    requisites contained herein plus satisfactory performance in the
    designated plant.

    2 - MATERIALS COVERED BY THIS STANDARD SHALL BE PURCHASED ONLY FROM
    THE SOURCES WHICH APPEAR ON THE ENGINEERING APPROVED SOURCE LIST
    ATTACHED TO THIS STANDARD.

    3 - No change in construction, composition, or finishing of this
    material shall be made without written approval of the Engineering
    Office. Failure to comply shall be sufficient cause to warrant
    rejection of shipment and/or removal from the Engineering Approved
    Source List.

    4 - For additional technical information, address inquiries to the
    Engineering Office.

Fig. 1. (Continued)

REFERENCES

1. Annual Book of ASTM Standards, ASTM, Philadelphia, 1981.
2. Handbook of Adhesives, 2d ed. (I. Skeist, ed.), Reinhold,
    New York, 1977.

26

Adhesives and the Standardization Program Within the Department of Defense (DOD)

Richard Chait and Edward T. Clegg *Army Materials and Mechanics Research Center, Watertown, Massachusetts*

I. INTRODUCTION

The Defense Standardization and Specification Program (DSSP) plays a significant role in the Department of Defense (DOD) acquisition policy [1]. This policy has undergone continuous review in an attempt to get quality weapon systems consistent with military needs while keeping the cost of such systems within reason. Just as there have been innovations in the fabrication of materials such as adhesives, with advanced fastening techniques, there have been innovations in the standardization procedure in the DOD. This chapter will emphasize current DOD standardization procedures and policies. However, to discuss them properly, it is first necessary to provide a brief historic background.

II. IMPORTANT EVENTS LEADING TO A DOD
STANDARDIZATION PROGRAM

After World War I in 1919, the Departments of War and Navy established a joint aeronautical board to address such matters as operating
procedures, tactics, flight safety, and aircraft design. This board
can be considered a forerunner of a military specification and standardization program. Another important event leading to such a program took place in 1941 when the Aeronautical Standards Board began
publishing a joint army-navy series of specifications and standards,
some of which are still in existence. The Aeronautical Board was replaced by the Aeronautical Standards Group which, in turn, was discon-
tinued in 1969 when a study indicated that the Defense Standardization Program would provide the necessary coordinating function.

During World War II, the Joint Army-Navy Specification Board
was formed to provide cooperation between the services in both procurement and standardization. The board established procedures and
format controls over the joint army-navy series of specifications which
continued to be used until 1953. During this period, the standardization within and among the military services was undergoing various
organizational and procedural changes.

Hearings by the House Armed Services Committee were embodied
in Public Law 436 of the 82nd Congress, dated 1 July 1952. The law
established the Defense Supply Management Agency, which would report to the Secretary of Defense. The law also assigned responsibility
for the federal cataloging program to the Secretary of Defense. The
secretary was directed to develop a related standardization program
for supplies.

Public Law 436, the Cataloging and Standardization Act (codified
title 10 U.S.C. Chapter 145, Section 2451-6) was translated into DOD
Directive 4120.3 in February 1953. This Defense Standardization Program established the military series of specifications and standards to
replace the joint army-navy series and all other series in use by the
military departments. It was at this time that the use of military
specifications was made mandatory for procurement.

The basic DOD Directive 4120.3 was revised in 1954, in 1965, and
again in 1978. Each revision was made to effect improved standardization management within the military services. Among other factors,
the 1965 revision provided for coordination of the Defense Standardization Program with the General Services Administration and other
government agencies and bureaus.

By and large, the DOD standardization program, now called
DSSP, has been successful, and it has served as a model for many
foreign countries. Currently, there are over 46,000 documents in the
DSSP. These documents are listed in the DOD Index of Specifications
and Standards, commonly referred to as the DODISS. This document,
a must for those involved with DOD specifications and standards, can

be obtained from the Navy Publications and Forms Center, 5801 Tabor Avenue, Philadelphia, Pennsylvania 19120.

III. DEFINITIONS AND PROCEDURES IN THE DSSP

A block diagram (Fig. 1) depicts the delegation of authority for the DSSP. As mentioned previously, the office of the Secretary of Defense is responsible for the DSSP. Responsibility for DSSP policy guidance and administration rests with the office of the Under Secretary of Defense for Research and Engineering (OUSDR&E). The office of the Deputy Under Secretary of Defense for Acquisition Policy assigns portions of the DSSP to each military department. The delegation of authority to act in the resolution of problems within the area of assignment is accomplished through the Defense Materiel Specifications and Standards Office (DMSSO). The military department of the defense agency organizational unit charged with overall management

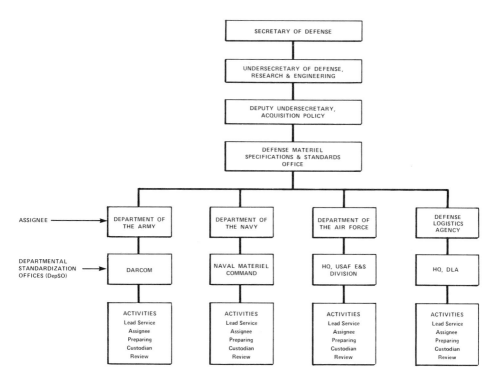

Fig. 1. Delegation of authority for the DSSP. Standardization and specification program.

and administration of standardization responsibilities within its department or agency is the Departmental Standardization Office
(DepSO). These DepSOs, which are located in the Washington, D. C.
area, are identified as follows: for the army, the Army Development
and Readiness Command; for the navy, the Naval Material Command;
for the Air Force, the Air Force Systems Command; and for the Defense Supply Logistics Agency, Headquarters—Defense Logistics
Agency. Responsibility for Federal Supply Classes (FSCs) and/or
AREAs is delegated by the DepSOs to agencies within their departments which are called assignee or lead-service activities. (These
terms will be defined later.) The agency having assignee responsibility for the adhesive Federal Supply Class (FSC 8040) is the Naval
Air Engineering Center at Lakehurst, New Jersey.

The mission of the DOD with respect to standardization is to develop, establish, and maintain a comprehensive and integrated system
of technical documentation in support of design, development, engineering, procurement, manufacturing, maintenance, and supply management. Specifications, standards, handbooks, engineering draw-
ings, and related documents constitute the basic types of documentation that are developed under the DSSP. Specifications are prepared
in accordance with MIL-STD-961, "Outline of Forms and Instructions
for the Preparation of Specifications and Associated Documents"; standards and handbooks are prepared in accordance with MIL-STD-961,
"Outline of Forms and Instructions for the Preparation of Military
Standards and Military Handbooks."

Preparing activities and/or military coordinating activities are
responsible for coordinating and otherwise managing standardization
projects. In addition to the responsibility for document preparation,
preparing activities also have the continuing responsibility for maintaining standardization documents under their jurisdiction current
with the state of the art and the requirements of the military departments. It is mandatory that standardization documents be reviewed
every 5 years. Some of these documents can be validated because
they reflect accurate and current requirements. Justification for major revisions must be evaluated by the preparing activity, taking into
consideration such factors as cost, degree of improvement in technical currency, and possible consolidation with other documents.

When revisions are necessary, the preparing activity prepares a
draft of the proposed revision and circulates it to review activities
and custodians. Definitions of preparing activities, review activities,
etc., can be found in Appendix 1.

The time allotted for coordination is 60—90 days unless the preparing activity determines that a different schedule is necessary.
When this allotted time has expired and all essential comments have
been resolved, the preparing activity publishes the revised document
if it is in the military series. Federal documents (such as MMM-A-180)
are submitted to the General Services Administration for publication.
New standardization documents are treated the same as revisions.

MIL-A-46050C
18 December 1979
SUPERSEDING
MIL-A-46050B
11 December 1970 (see 6.5)

MILITARY SPECIFICATION
ADHESIVES, CYANOACRYLATE, RAPID ROOM-TEMPERATURE CURING, SOLVENTLESS

This specification is approved for use by all Departments
and Agencies of the Department of Defense.

1.   SCOPE AND CLASSIFICATION

1.1   Scope. This specification covers solventless, room-temperature curing
cyanoacrylate adhesives for use, with or without an activator, when speed
of curing is a primary consideration. It also covers the activator which
may be used to provide still faster curing and to enable the adhesives to
bond to otherwise inhibiting surfaces. These adhesives are used for nonstructural applications requiring one-component bonding of small, well-
mated surfaces where heat and/or pressure cannot be applied (see 6.1).

1.2   Classification.

1.2.1 The adhesives shall be of the following types and classes, as specified
(see 6.2 and 6.4).

          Type I   - Methyl-2-cyanoacrylate
          Type II   - Ethyl-2-cyanoacrylate
          Type III   - Isobutyl-2-cyanoacrylate
          Type IV   - n-Butyl-2-cyanoacrylate

          Class 1 - Low viscosity
          Class 2 - Medium viscosity
          Class 3 - Medium-high viscosity
          Class 4 - High viscosity

1.2.2 The surface activators shall be of the following types, as specified
(see 6.2):

          Type IA   - for use with methyl-2-cyanoacrylate
          Type IIA   - for use with ethyl-2-cyanoacrylate
          Type IIIA   - for use with isobutyl-2-cyanoacrylate
          Type IVA   - for use with n-Butyl-2-cyanoacrylate

Beneficial comments (recommendations, additions, deletions) and any pertinent
data which may be of use in improving this document should be addressed to:
Director, US Army Materials and Mechanics Research Center, ATTN· DRXMR-LS,
Watertown, MA 02172 by using the self-addressed Standarization Document
Improvement Proposal (DD Form 1426) appearing at the end of this document or by
letter.

FSC 8040

Fig. 2. First page of a typical DOD specification.

It should be noted that each military specification in the DODISS
is uniform in its format and contains different sections devoted to the
scope of the specification, what other standardization documents are
applicable, important requirements, quality-assurance provisions to
ensure requirements are met, delivery aspects, and miscellaneous
notes.

The first sheet of a DOD specification is shown in Fig. 2. (The
document chosen is MIL-A-46050C, "Adhesives, Cyanoacrylate, Rapid
Room-Temperature Curing, Solventless.")

In the specification number, MIL-A-46050C, the "A" is the first
letter in the first word of the title, and the "C" after the number indicates the third revision. The preparing activity is identified by a
symbol which appears on the last page of a specification. In the bottom right corner of the first page is the FSC identification of adhesives, FSC 8040. All documents that are part of the DSSP are identified with either an FSC or an AREA. An FSC is a standardization
category encompassing commodities, such as adhesives, and an AREA
is a standardization category encompassing a subject not applicable to
a single FSC, such as Nondestructive Testing and Inspection. As-
signments to AREAs are made for broad engineering disciplines and
practices. There is a total of 641 FSCs and 28 AREAs in the DSSP.
Each of the more than 46,000 documents in the DODISS belongs to
either an FSC or an AREA. The Army Materials and Mechanics Research Center (AMMRC) has preparing-activity responsibility for a
number of documents in the FSCs and AREAs shown in Table 1.

Table 1 FSCs or AREAs for which the AMMRC has Standardization
Responsibility

FSC no. or AREA	Title of FSC or AREA
1305	Ammunition through 30 mm
1395	Miscellaneous ammunition
3439	Misc. welding, soldering, and brazing
4710	Pipe and tube
5330	Packing and gasket materials
5345	Disks and stones, abrasives
5350	Abrasive materials
6850	Msc. chemical specialties

Table 1 (Continued)

FSC no. or AREA	Title of FSC or AREA
8010	Paints, dopes, and varnishes
8030[a]	Preservative and sealing compounds
8040	Adhesives
8470	Armor, personal
9320	Rubber fabricated materials
9330	Plastic fabricated materials
9340	Glass fabricated materials
9350	Refractories and fire surfacing materials
9390	Misc. fabricated nonmetallic materials
95GP	Metal bars, sheets and shapes
9505	Wire, nonelectric, iron, and steel
9510	Bars and rods, iron, and steel
9515	Plate, sheet and strip, iron, and steel
9520	Structural shapes, iron, and steel
9525	Wire, nonelectric, nonferrous
9530	Bars and rods, nonferrous
9535	Plate, sheet and strip, nonferrous
9540	Structural shapes, nonferrous
9545	Plate, sheet and strip, foil, and wire precious metals
9630	Additive metal materials and master alloys
9640[a]	Iron and steel, primary and semi-finished products
9650	Nonferrous refinery and intermediate forms
9660	Precious metal, primary forms
THJM[a]	Thermal joining of metals
MECA[a]	Metal castings
MISC	Miscellaneous
MFFP[a]	Metal finishes and finishing processes and procedure
FORG	Metal forging
NDTI[a]	Nondestructive testing and inspection

[a]Assignee or lead service responsibility for DOD.

IV. CURRENT DOD POLICIES AND DIRECTIONS

In 1977, the Defense Science Board examined DOD standardization activities and issued a report [2]. Some of the major findings of this
study were as follows:

Specifications and standards are essential to technical procurement.

The present body of military specifications and standards is adequate to the DOD needs.

Specifications and standards contain corporate history of lessons
learned.

Major payoff for the improvement in specifications and standards
will come initially from their method of application followed by longer-
range improvement in control.

The major thrusts were to improve the climate of application, upgrade
the existing body of documents, and provide and maintain high-level
management attention.

Also in 1977, the National Materials Advisory Board (NMAB) examined that part of the DSSP which deals with materials and materials
processes. The result was NMAB Report No. 330 [3] which recommended the following:

Exploit cost-effectiveness potential of standardization.

Increase DOD emphasis on specifications and standards.

Use specifications and standards as a means to cope with materials shortages.

Take advantage of voluntary consensus standards system.

The voluntary consensus standards system was given further im-
petus by a recent publication of the Office of Management and Bud-
get (OMB) entitled, "OMB Circular No. A-119" [3a]. This circular,
which pertains to federal participation in the development and use of
voluntary standards, states that it is the general policy of the federal
government to rely on voluntary standards, both domestic and international, with respect to federal procurement whenever feasible and
consistent with existing laws and regulations.

Both the Defense Science Board study [2], which examined the
entire DSSP, and the NMAB report [3] which examined only those
documents pertaining to materials and materials processes, have led
to important new initiatives in DOD standardization. These initiatives
will now be discussed.

First, DOD Instruction 4120.20, "Development and Use of Nongovernment Specifications and Standards" [3b], emphasizes the adoption and use of nongovernment documents, such as those developed by
the private sector standards writing bodies, that is, American Society
for Testing and Materials (ASTM), Society for Automotive Engineers/
Aerospace Materials Specifications (SAE/AMS), and American Society

for Nondestructive Testing (ASNT). In particular, this document places greater emphasis on the following:

Adoption of nongovernment specifications and standards.

Use of nongovernment specifications and standards in the design and development of materials.

Development and adoption of nongovernment documents is preferred to the development of a new military document.

Participation in nongovernment bodies, activities engaged in the promulgation of nongovernment documents.

The Defense Standardization Manual indicates that the primary criteria for determining whether or not nongovernment standardization documents will be adopted in lieu of military documents are as follows:

They should fully satisfy the military with respect to technical sufficiency and metric practices.

There should be assurance that there will be sufficient copies available to satisfy DOD needs.

The nongovernment standards should be acceptable to interested government and nongovernment elements.

Adoption of private sector documents by DOD as a function of time is shown in Fig. 3. Note that it shows a significant increase from

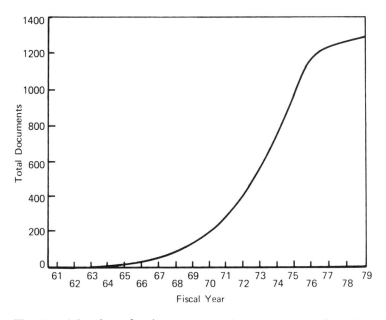

Fig. 3. Adoption of private-sector documents as a function of time.

the original handful of industry standards in 1965 to the current
total of approximately 1288 industry standards listed in the DODISS.
In reviewing the curve, it can be seen that a substantial increase in
the number of standards accepted began in 1971. This increase was
due, in part, to the acceptance of a large number of documents ur-
gently needed for space-program activities.

It cannot be expected that the original rate of adoption will be
maintained since there is only a finite number of documents in these
areas that are still candidates for DOD acceptance. A leveling of the
curve is readily observable. It should be pointed out, however, that
this leveling is not an indication of a decrease in activity in the gov-
ernment/industry program; rather, it merely points out that the saturation point for acceptance of present industry and professional-so-
ciety documents is being approached. As new nongovernment documents of concern to DOD are prepared, they, too, will be reviewed
for DOD adoption.

It is still a tremedous task for the various government agencies
to maintain the standards currently listed in the DODISS. Maintenance of these DOD-adopted, nongovernment documents requires par-
ticipating in the work of voluntary-consensus writing bodies to discuss and evaluate proposed revisions to these standards so that DOD
interests are protected, coordinating proposed revisions with interested government agencies, and preparing acceptance notices for these
revisions.

In April 1977, a second initiative, also in the form of an official
DOD policy [3c] was advanced. This policy required there to be a
specific and continuing effort to reduce cost. Therefore, DOD has
identified—through the Defense Science Board report—categories that
are candidates for misapplication and misinterpretation. These cost
driver areas are

General design requirement specifications
Environmental requirements and test methods
Reliability and maintainability
Quality control
Human factors and safety
Documentation
Configuration control
Integrated logistic support
Packing, packaging, preservation, and transport

Lastly, there is a strong move to procure more and more items on
an "off-the-shelf" basis, that is, without full dependence on specification. This direction stems from a 1976 Office of Federal Procurement
Policy statement that the government will purchase commercial off-the-
shelf products when such products will adequately serve the government's requirements, provided such products have an established
commercial-market acceptability [3d]. Items such as gasoline have

commercial-market acceptability. Items such as gasoline have been the subject of pilot programs. Evaluation of these programs is now underway.

V. ADHESIVES-RELATED DOD ACTIVITIES

The interest in adhesives stems from the many advantages that accrue from bonding and fastening materials and parts with adhesives. Various adhesives have replaced screws, rivets, and other types of mechanical fasteners in practically all areas of manufacturing. Today, few industries are untouched by the trend toward adhesive bonding in product assembly.

The major function of adhesives is, of course, to fasten parts together. They do this by transmitting stresses from one member of a joint to another, in a manner that distributes the stresses much more uniformly than conventional, mechanical fasteners can. Consequently, adhesives often permit the fabrication of structures that are mechanically equivalent to, or stronger than, conventional assemblies but that cost and weigh less. For example, thin adherends can be adhesively bonded to thick sections so that the full strength of the member is utilized.

It is important to note that DOD and other federal agencies are supporting active programs involving adhesives. Many problems have had to be solved in the NASA space-shuttle effort because it deals with the first space vehicle ever flown with a reusable thermal-protection system. Tiles that make up the thermal-protection system are bonded to the orbiter with adhesives. The air force has funded the Primary Adhesively Bonded Structure Technology (PABST) Program. The objective of this program is to develop and validate an aircraft adhesive system and processing techniques that can be relied upon to achieve the necessary structure. The army is considering anaerobic adhesives for use in the mechanical assembly of tanks, and the navy is doing work on the fracture mechanics of adhesive joints.

It is obvious that much in the area of adhesives technology, from a DOD standpoint, is still emerging. However, the effort for utilizing standardization to bring this technology into focus has been initiated. A list of federal and military specifications devoted to adhesives as well as those private-sector documents adopted for use in the DSSP is provided in Appendix 2. An adhesive specification, like all material specifications, is a document which specifies values for all the important properties of an adhesive and gives limits of variability and methods for determining these values. Specifications are looked upon differently by producers and consumers. The producer is concerned with the ability to meet the specification and to supply the adhesive competitively. The consumer relies on the specification to assist in

procuring a satisfactory adhesive at a reasonable cost. In general,
a procurement specification can be prepared based on composition or
performance requirements or both.

Standardization involving adhesive technology has not yet been
extended to non-product-oriented disciplines, such as nondestructive
testing and inspection (NDTI). There are many nondestructive inspection (NDI) techniques for which standards applicable to metallic
materials have been written. Examples of these inspection techniques
are MIL-STD 1265(MR), "Radiographic Inspection, Soundness Requirements for Steel Castings," and MIL-STD 2154 "Inspection Ultrasonic,
Wrought Metals, Process for." Similar examples for composite materials, and in particular adhesives, are not to be found. Current NDI
practices for these materials stem from NDI techniques that are used
for metallic materials. They rely primarily on ultrasonic or tap tests
to detect voids or disbonds. They may give a gross indication of material or materiel quality, but additional quantification is needed.
These and other nondestructive investigations may be modified to
yield acceptable and meaningful test methods for nondestructive testing of adhesives. Thus, the problem may not be the need for new
methods or techniques, but the need for more research and improvement on present methods. Nondestructive testing of adhesive bonds
is realistic, and in the near future it will be necessary in both aerospace and automobile industries. In the automotive industry, for example, the demand for weight-saving measures requires the use of
bonded plastic parts. As stated in a recent article, unless quality
assurance of bonded plastic parts is made feasible and put on stream,
efforts to improve fuel economy by replacing metal with plastic will
be delayed [4]. Efforts utilizing ultrasonics are under way to meet
this challenge.

Although the ultimate strength of bonds is not presently measured by nondestructive techniques, there are some useful properties
of adhesives—such as the dynamic elastic, viscous, or complex moduli, as well as the dielectric constants—which have been measured non-
destructively. Dietz has determined, by means of ultrasonics, the
moduli of polyvinyl butyralphenolic adhesive bonds in metal rod speci-
ments after various periods of heating to degrade the adhesive layer
[5,6]. The dynamic moduli and dielectric constants may be used to
differentiate between bonds of high and low tensile strength; thus,
they were used to determine the deteriorating effect of prolonged
heating. D'Agostino has explored the photoelastic properties of adhesives [7]. Norris has investigated dielectric constants by measuring
the change in electric capacitance of the adhesive layer as stress is
applied to the test specimen [8].

VI. CONCLUDING REMARKS

The DOD may not face the high-volume production that is present in the automotive and aerospace industries, but it does face the severest of service requirements. Described previously were DOD efforts to evaluate adhesive-forming techniques in meeting these requirements.

It is important to note that in designing for adequate adhesive quality assurance, realistic accept-reject criteria must be established. What distribution of defects (voids, disbonds, etc.) can be tolerated for adequate service life? What level of material anomaly can the NDI procedures detect? How must the design be changed to accommodate these NDI techniques? Like the automotive and aerospace industries, the DOD must take a systematic approach since utilization of adhesives involves the consideration of many variables [4]. Until more information is available a realistic set of specifications and standards bringing the use of adhesives to the level of the use of metallic materials may be premature. However, the framework within the preview of the DSSP for formulating, reviewing, circulating, and publishing these documents once they are ready is in place.

VII. APPENDIX 1: DEFINITIONS

Assignee: The military department or agency delegated responsibility for standardization in an FSC.

Assignee activity: The activity to which the responsibility for standardization of an FSC has been delegated by the cognizant assignee (DepSO).

Coordination: Participation by activities and representative segments of industry, having an interest in a standardization project, to obtain agreement.

Custodian: The activity responsible for coordination of standardization projects within its own department or agency.

Essential comment: A comment covering requirements or provisions which must be adopted or reconciled if the document is to be usable by the commenting activity.

Lead service: The military department or agency delegated responsibility for the development, preparation, and implementation of the standardization program in an assigned area.

Preparing activity: The military activity responsible for preparation and maintenance of standardization documents.

Review activity: An activity having an essential technical interest in the standardization document.

VIII. APPENDIX 2: DOD STANDARDIZATION DOCUMENTS LISTED IN FSC 8040

A. Government Specifications

1. Federal specifications

Document no.	Title	Prep. act.
MMM-A-100D	Adhesive, Animal-Glue	%a
MMM-A-105	Adhesive & Sealing Compounds Cellulose Nitrate Base, Solvent Type	/AS
MMM-A-110B	Adhesive, Asphalt, Cut-Back Type (For Asphalt & Vinyl Asbestos Tiles)	*MR
MMM-A-115C	Adhesive, Asphalt, Water Emulsion Type (For Asphalt & Vinyl Asbestos Tiles)	*YD
MMM-A-121	Adhesive, Bonding Vulcanized Synthetic Rubber to Steel	/SH
MMM-A-122C	Adhesive, Butadiene Acrylonitrile Base, Medium Solids, General Purpose	*MR
MMM-A-125C	Adhesive, Casein-Type, Water & Mold Resistant	/MR
MMM-A-130B	Adhesive, Contact	*ME
MMM-A-131A	Adhesive, Glass to Metal (For Bonding of Optical Elements)	*AR
MMM-A-132	Adhesive, Heat Resistant, Airframe Structural, Metal to Metal	/AS
MMM-A-134	Adhesive, Epoxy Resin, Metal to Metal Structural Bonding	/AS
MMM-A-137D	Adhesive, Linoleum	*MR
MMM-A-138A	Adhesive, Metal to Wood, Structural	/AS
MMM-A-00150B	Adhesive for Accoustical Materials	+ +
MMM-A-179B	Adhesive, Paper Label, Water-Resistant Water Emulsion Type	/MR
MMM-A-180B	Adhesive, Polyvinyl Acetate Resin Emulsion (Alkali Dispersable)	AS
MMM-A-181C	Adhesive, Phenol, Resorcinol, or Melamine Base	/MR

Document no.	Title	Prep. act.
MMM-A-182A	Adhesive, Rubber	/EA
MMM-A-185B	Adhesive, Rubber (For Paper Bonding)	*AS
MMM-A-187B	Adhesive, Epoxy Resin Base Low & Intermediate Strength, General Purpose	*AS
MMM-A-188C	Adhesive, Urea-Resin Type (Liquid & Powder)	/AS
MMM-A-189B	Adhesive, Synthetic Rubber, Thermoplastic, General Purpose	/MR
MMM-A-193C	Adhesive, Vinyl Acetate Resin Emulsion	/MR
MMM-A-250C	Adhesive, Water-Resistant (For Closure of Fiberboard Boxes)	/MR
MMM-A-260B	Adhesive, Water-Resistant (For Sealing Waterproofed Paper)	/MR
MMM-A-1058	Adhesive, Rubber Base (In Pressurized Dispensers)	*AS
MMM-A-1617	Adhesive, Rubber Base, General Purpose	/AS
MMM-A-1754	Adhesive & Sealing Compound, Epoxy, Metal Filled	*AS
MMM-A-1931	Adhesive, Epoxy, Silver Filled, Conductive	*AS
MMM-A-001993	Adhesive, Epoxy, Flexible, Filled (For Binding, Sealing, and Grouting)	/AS
MMM-A-002015	Adhesive, Semi-Solid, Stick Form, With Dispensers	+ +
MMM-A-2048	Adhesive, Fire-Resistant, Thermal Insulation	+ +

[a]See Sec. VII.C for explanation of symbols.

2. Military specifications

Document no.	Title	Prep. act.
MIL-G-413B	Glue, Marine, & Aviation Marine (Waterproof)	SH[a]
MIL-C-2399B	Cement, Liquid Tent Patching	GL

Document no.	Title	Prep. act.
MIL-A-3167A	Adhesive (For Plastic Inhibitors)	OS
MIL-A-3316B	Adhesive, Fire Resistant, Thermal Insulation	SH
MIL-A-3562B	Adhesive, Sealing (For Filters)	EA
MIL-A-3920C	Adhesive, Optical, Thermosetting	AR
MIL-A-5540B	Adhesive Polychloroprene	AS
MIL-A-8576B	Adhesive, Acrylic Base, For Acrylic Plastic	AS
MIL-A-9067C	Adhesive Bonding, Process & Inspection Requirements For	AS
MIL-A-9117D	Adhesive Sealing, For Aromatic Fuel Cells & General Repair	99
MIL-A-13374D	Adhesive, Dextrin, For Use in Ammunition Containers	MR
MIL-C-14064B	Cement, Grinding Disk	MR
MIL-A-17682D	Adhesive, Starch	MC
MIL-A-21026E	Adhesive, Resilient Deck Covering	SH
MIL-A-21366A	Adhesive, For Bonding Plastic Table Top Materials to Aluminum	SH
MIL-A-22010A	Adhesive, Solvent Type, Polyvinylchloride	SH
MIL-A-22397	Adhesive, Phenol & Resorcinol Resin Base (For Marine Service Use)	SH
MIL-A-22433A	Adhesive, Polyester, Thixotropic	OS
MIL-A-22895	Adhesive, Metal Identification Plate	SH
MIL-C-23092B	Cement, Natural Rubber, Magnetic Mine-sweeping Cable Repair	SH
MIL-A-24179A	Adhesive, Flexible Unicellular-Plastic Thermal Insulation	SH
MIL-A-24456	Adhesive for Plastic Vibration-Damping Tile	SH
MIL-A-25463A	Adhesive, Film Form, Metallic Structural Sandwich Construction	AS
MIL-A-43316A	Adhesive, Patching for Chloroprene Coated or Chlorosulphonated Polyethylene Coated Fabrics	GL
MIL-A-45059B	Adhesive for Bonding Clipboard to Terne-plate, Tinplate, and Zincplate	MR

Document no.	Title	Prep. act.
MIL-G-46030D	Glue, Animal (Protective Colloid)	MR
MIL-A-46050C	Adhesive, Special, Rapid Room-Temperature Curing, Solventless	MR
MIL-A-46091A	Adhesive, Brake Lining to Metal	MR
MIL-A-46106A	Adhesive, Sealants, Silicone, RTV, General Purpose	MR
MIL-A-46146	Adhesives-Sealants, Silicone, RTV, Non-Corrosive (For Use with Sensitive Metals & Equipment)	MR
MIL-A-46864	Adhesive, Epoxy, Modified, Flexible Two Component	MI
MIL-A-46869	Adhesive, Silicone Resin Base, Room-Temperature Curing	MI
MIL-A-47040	Adhesive-Sealant, Silicone, RTV, High Temperature	MI
MIL-A-47073	Adhesive, Plastic Resin	MI
MIL-A-47074	Adhesive System, Epoxy, For Dissimilar Metal Bonding	MI
MIL-A-47089	Adhesive, Metal Filled, Conductive, Electrical & Thermal	MI
MIL-P-47125	Primer for Silicone Rubber Insulating Material	MI
MIL-A-47126	Adhesive (Viscous) Epoxy Resin, Metal to Metal Bonding and Sealing	MI
MIL-P-47170	Primer, Silicone Rubber Sealant	MI
MIL-C-47171	Curing Agent, Mixed Amine	MI
MIL-A-47172	Adhesive, Modified Epoxy	MI
MIL-P-47216	Primer, Polyurethane	MI
MIL-L-47274	Liquid Imine Curing Agent	MI
MIL-P-47275	Primer, Silicone	MI
MIL-P-47276	Primer, Bonding	MI
MIL-P-47277	Primer, Special Purpose	MI
MIL-P-47278	Primer, Silicone Rubber	MI
MIL-P-47279	Primer, Silicone Adhesive	MI

Document no.	Title	Prep. act.
MIL-A-47280	Adhesives, Epoxy	MI
MIL-A-47284	Adhesive, Epoxy Resin Base	MI
MIL-A-47295	Adhesive Composition, Ha-polymer	MI
MIL-A-48611	Adhesive System, Epoxy-Elastomeric, For Glass to Metal	MI
MIL-A-42194A	Adhesive, Epoxy (For Bonding Glass Reinforced Polyester)	MR
MIL-A-81236	Adhesive, Epoxy Resin with Polyamide Curing Agent	OS
MIL-C-81247	Curing Agent, Bicyclo Anhydride Type	OS
MIL-A-81253	Adhesive, Modified Epoxy Resin with Polyamide Curing Agent	OS
MIL-A-81270	Adhesive, Synthetic Rubber	OS
MIL-A-82484	Adhesive & Sealing Compounds, Cellulose Nitrate Base, Solvent Type (For Ordnance Use)	OS
MIL-A-82636	Adhesive, Butyl, Two Component	OS
MIL-A-83376A	Adhesive Bonded Metal Faced Sandwich Structures, Acceptance Criteria	11
MIL-A-83377B	Adhesive Bonding (Structural) for Aerospace & Other Systems, Requirements for	11
MIL-A-87134	Adhesive, Contact, For Custom Fit Helmet Liners	11
MIL-A-87135	Adhesives, Nonconductive, For Electronics Application	11

[a]See Sec. VII.C for explanation of symbols

3. Federal standards

Document no.	Title	Prep. act.
FED-STD-175A	Adhesives, Methods of Testing	/MR[a]

[a]See Sec. VII.C for explanation of symbols.

4. Military standards

None

5. Military handbooks

Document no.	Title	Prep. act.
MIL-HDBK-691A	Adhesives	/MR[a]
MIL-HDBK-725	Adhesives: A Guide to Their Properties & Uses as Described by Federal & Military Specifications	/MR

[a]See Sec. VII.C for explanation of symbols.

B. Industry Documents

1. American Society for Testing & Materials

Document no.	Title	Prep. act.
ASTM D816	Rubber Cements, Testing	MR[a]
ASTM D896	Resistance of Adhesive Bonds to Chemical Reagents, Test for	MR
ASTM D897	Tensile Properties of Adhesive Bonds, Test for	MR
ASTM D898	Applied Weight Per Unit Area of Dried Adhesive Solids, Test for	MR
ASTM D899	Applied Weight Per Unit Area of Liquid Adhesive, Test for	MR
ASTM D903	Peel or Stripping Strength of Adhesive Bonds, Test for	MR
ASTM D906	Strength Properties of Adhesives in Plywood Type Construction in Shear by Tensile Loading, Test for	MR
ASTM D950	Impact Strength of Adhesive Bonds, Test for	MR

Document no.	Title	Prep. act.
ASTM D1002	Strength Properties of Adhesives in Shear by Tension Loading (Metal-to-Metal) Test for	MR
ASTM D1062	Cleavage Strength of Metal-to-Metal Adhesive Bonds, Test for	MR
ASTM D1084	Viscosity of Adhesives, Tests for	MR
ASTM D1146	Blocking Point of Potentially Adhesive Layers, Test for	MR
ASTM D1151	Effect of Moisture & Temperature on Adhesive Bonds, Test for	MR
ASTM D1183	Resistance of Adhesives to Cyclic Laboratory Aging Conditions, Tests for	MR
ASTM D1184	Flexural Strength of Adhesive Bonded Laminated Assemblies, Test for	MR
ASTM D1488	Amylaceous Matter in Adhesives, Test for	MR
ASTM D1489	Nonvolatile Content for Aqueous Adhesives, Test for	MR
ASTM D1579	Filler Content of Phenol, Resorcinol, & Melamine Adhesives, Test for	MR
ASTM D1583	Hydrogen Ion Concentration of Dry Adhesive Films, Test for	MR
ASTM D1875	Density of Adhesives in Fluid Form, Test for	MR
ASTM D1876	Peel Resistance of Adhesives (T-Peel Test) Test for	MR

[a]See Sec. VII. C. for explanation of symbols.

2. Aerospace materials specifications

Document no.	Title	Prep. act.
AMS 3106	Primer, Adhesive, Corrosion Inhibiting	AS[a]
AMS 3686	Adhesive, Polyimide Resin, Film & Paste High Temperature Resistant	AS
AMS 3688	Adhesive, Foaming, Honeycomb Core Splice, Structural, Minus 67 to Plus 188 Deg F	AS
AMS 3689	Adhesive, Foaming, Honeycomb Core Splice, Structural, Minus 67 to Plus 350 Deg F	AS

[a]See Sec. VII.C for explanation of symbols.

C. List of Symbols

1. Activity symbols

AR: U.S. Army Armament Research and Development Command (Dover, New Jersey).

AS: Naval Air Systems Command.

EA: U.S. Army Armament Research and Development Command (Aberdeen Proving Ground, Maryland).

GL: U.S. Army Natick Research and Development Laboratory.

MC: U.S. Marine Corps.

ME: U.S. Army Mobility Equipment Research and Development Command.

MI: U.S. Army Missile Research and Development Command.

MR: U.S. Army Materials and Mechanics Research Center.

OS: Naval Sea Systems Command (Ordnance Systems).

SH: Naval Sea Systems Command (Ship Systems).

YD: Naval Facilities Engineering Command.

11: U.S. Air Force Aeronautical Systems Division, AFSC.

99: Air Force Logistic Command Cataloging and Standardization Office (Battle Creek).

2. Codes used in preparing-activity column

%: DOD waived coordination interest.

$\frac{+}{+}$: No official DOD interest registered.

*: DOD has coordination interest.

/: Prepared by a DOD activity.

ACKNOWLEDGMENTS

Appreciation is expressed to Dr. Stanley E. Wentworth of the Polymer Research Division, Army Materials & Mechanics Research Center, for review and comments and to Miss Karen L. Celia of the Engineering Standardization Division, AMMRC, for editing and typing.

REFERENCES

1. J. S. Gansler, *Army Res. News Mag.* *18*(3) (1977).

2. Report of the Task Force on Specifications and Standards, in *Defense Science Board ADA040455*, April 1977, Office of the Director of Defense Research and Engineering, Washington, D.C.

3. Materials and Process Specification and Standards, in *Natl. Materials Advisory Board Rep. No. 330*, Natl. Acad. Sci., Washington, D.C., 1977.

3a. OMB Circular No. A-119, OMB, Washington, D.C., Jan. 17, 1980.

3b. Instruction 4120.20, DOD, Washington, D.C., Dec. 29, 1976.
3c. Directive 4120.21, DOD, Washington, D.C., April 19, 1977.
3d. Office of Federal Procurement Policy Memorandum, Washington, D.C., May 24, 1976.
     23(4) (1980).
4.   F. J. Meyer and G. B. Chapman, *Adhesives Age* (1980).
5.   A. G. H. Dietz, H. N. Backstruck, and G. Epstein, in *ASTM Spec. Tech. Publ. No. 138*, ASTM, 1952, p. 40.
6.   A. G. H. Dietz, P. J. Closmann, G. M. Kavanagh, and J. N. Rossen, *ASTM Proc. 50*:1414 (1950).
7.   J. D'Agostino, G. C. Drucker, C. K. Liu, and C. Mylonas, *Proc. Soc. Exp. Stress Analysis 12*:115 (1955).
8.   C. B. Norris, W. L. Jones, and J. T. Drow, in *ASTM Bull. No. 218*, ASTM, 1956, p. 40.

Index